Laser Physics at the Limits

Springer-Verlag Berlin Heidelberg GmbH

Physics and Astronomy ONLINE LIBRARY

http://www.springer.de/phys/

Hartmut Figger Dieter Meschede
Claus Zimmermann (Eds.)

Laser Physics at the Limits

With 245 Figures and 9 Tables

Springer

Dr. Hartmut Figger
Max-Planck-Institut für Quantenoptik
Hans-Kopfermann-Str. 1
85748 Garching
Germany
E-mail: hartmut.figger@mpq.mpg.de

Professor Dr. Dieter Meschede
Universität Bonn
Institut für Angewandte Physik
Wegelerstr. 8
53115 Bonn
Germany
E-mail: meschede@iap.uni-bonn.de

Professor Dr. Claus Zimmermann
Universität Tübingen
Physikalisches Institut
Auf der Morgenstelle 14
72076 Tübingen
Germany
E-mail: clz@ipit.physik.uni-tuebingen.de

Library of Congress Cataloging-in-Publication Data applied for.

Die Deutsche Bibliothek - CIP-Einheitsaufnahme
Laser physics at the limits ; with 9 tables / Hartmut Figger ... (ed.). - Berlin ; Heidelberg ; New York ; Barcelona ; Hong Kong ; London ; Milan ; Paris ; Tokyo : Springer, 2002
(Physics and astronomy online library)

http://www.springer.de

DOI 10.1007/978-3-662-04897-9

Originally published by Springer-Verlag Berlin Heidelberg New York in 2002.
MyCopy version of the original edition 2002

Typesetting: LE-TEX Jelonek, Schmidt & Vöckler GbR, Leipzig
Cover design: *design & production* GmbH, Heidelberg using a picture from © Frank Ossenbrink

Printed on acid-free paper SPIN 10848141 56/3141/di 5 4 3 2 1 0
www.springer.com/mycopy

Dedicated to Theodor W. Hänsch
on the Occasion of his 60th Birthday

A Passion for Physics

On the occasion of Professor Theodor W. Hänsch's 60th birthday

"A Passion for Precision" was the title of an essay written by Daniel Kleppner 12 years ago for the Reference Frame section of Physics Today. It describes the intensely pleasurable sensation an experimentalist feels when measuring something useful to umpteen significant figures. One of the most prominent experimentalists in this context is Theodor Hänsch, who must have noticed this sensation many times in his life. Indeed, the quest for attaining higher and higher precision in fundamental-physics experiments is one of the central themes in the scientific work of Theodor Hänsch, a true pioneer in high-resolution laser spectroscopy, first at the University of Heidelberg in the late 1960s, then at Stanford University, and since 1986 at the Max-Planck-Institute for Quantum Optics in Garching and the Ludwig-Maximilians-University in Munich.

For example, he and his collaborators cultivated laser-saturated absorption spectroscopy and two-photon Doppler-free absorption spectroscopy as a brilliant means to suppress the first-order Doppler effect. Already in the early 1970s, this made possible the measurement of the ultra-violet resonance line of hydrogen with an unprecedented precision of a few parts in 10^8. Some of his measurements from this time are now included in major atomic physics textbooks. To reduce the second-order Doppler effect, he, together with colleague and Nobel Prize winner Arthur Schawlow, had the ingenious idea to cool a gas of atoms by means of light pressure, nearly a decade before its first experimental realisation in the mid-1980s. Laser-cooled atoms are now an essential part of the world's best atomic fountain clocks, operating in the microwave domain with a remarkable accuracy of a few parts in 10^{15}. To transfer such a precision into the optical domain without the need for a sophisticated frequency chain, Hänsch and his co-workers recently invented the frequency-comb generator. Providing millions of closely spaced and stable reference frequencies ranging from the infrared regime across the optical spectrum, this revolutionary device allowed him to count the frequency of the fundamental 1S–2S two-photon transition in atomic hydrogen with an astonishing precision of a few parts in 10^{14} – a more than 10-fold improvement over previous measurements, which electrified the community. Largely due to Hänsch's longlasting dedication to high-precision experiments, the Rydberg constant is presently the best-known fundamental constant in physics, and optical atomic clocks with a potential for vast improvements in accuracy seem to be realisable in the not-too-distant future.

In addition to precise laser spectroscopy and fundamental physics, Theodor Hänsch has also pioneered the research field of laser cooling and trapping, another exciting area of present-day physics already mentioned above. Here, his widespread activities range from optical lattices to atom interferometers, atom lasers and atom chips. His achievements in this research field are too numerous to describe all of them in detail, so that only a few can be highlighted. For example, he and his co-workers were among the first to trap atoms in the motional ground state of a periodic lattice made of interfering light beams. In atom interferometry, the wave nature of atoms is used to explore fundamental quantum physical phenomena and measure atomic properties like the mass or the transition frequency with high precision. The atom laser realised in his laboratory is a quasicontinuous and monochromatic source of atoms which are ejected from a Bose–Einstein condensate as a highly collimated beam. Last, but not least, the atom chip is the result of the group's more-than-a-decade long research program devoted to the magnetic trapping of atoms. It employs an array of micron-sized current-carrying wires on the surface of a chip to store and guide atoms in a strongly confining magnetic field. Arguably the most spectacular result here is the recent achievement of Bose–Einstein condensation, which opens up the possibility to realise, for example, a miniature atom interferometer.

Most of the spectacular experiments mentioned above would have been impossible without a true "passion for precision". But the broad spectrum of research activities would also have been impossible without a deep interest in physics in general. Indeed, the laser wizard Theodor Hänsch has an everlasting passion for physics. For more than 30 years now, he has constantly inspired quantum optics and laser spectroscopy researchers with exciting ideas, novel techniques and fabulous experiments, sometimes originating in a small semiprivate laboratory at the University of Munich. His way of uncovering the secrets of physics reminds me of the way my children are discovering the world: "simply" by asking questions nobody has expected. Many of his achievements were believed to be impossible at the time they were proposed. It is only due to his persistence and steadiness that theoretical visions could become experimental reality.

It comes without saying that the Max-Planck Society is happy to have a leading scientist like Theodor Hänsch among its scientific members. He and his group make excellent use of the wonderful framework provided by the Society, demonstrating in a spectacular way that basic research supported over a long period of time is indeed a good investment for the future. Together with my co-directors and all co-workers of the Max-Planck Institute for Quantum Optics, it is a pleasure for me to wish him health and happiness, and many more years of his remarkably great scientific creativity. Happy birthday!

Garching, October 2001 *Gerhard Rempe*

Preface

Laser physics has always been driven by a genuinely scientific quest to extend existing limits, limits of physical knowledge and limits of physical methods. Enhancing precision, sensitivity, and resolution has been an essential topic of this evolution. Experiments and concepts of laser physics have also earned a central place in modern physics. More than 100 years after the invention of quantum physics, subtleties of light–matter interaction can be beautifully illustrated through the application of lasers, an invaluable highlight of our current presentation of modern microscopic physics.

The scientific advancement of physics is impossible without the ingenuity of experimenters and of theoreticians. Among those, Theodor Hänsch has occupied a commanding position for many decades now, contributing numerous original and fundamental contributions to the field of laser physics. This volume was thus initiated on the occasion of his 60th birthday. It demonstrates the influence of his scientific activities at present and during the recent past. Current developments in fields such as atomic clocks, precision measurements of fundamental constants, nonlinear optical effects, Bose–Einstein condensation, and atomic quantum engineering underline the virtue of concepts and results derived in his laboratories.

Authors were invited from among his present coworkers in Munich and Florence, prominent collaborators and close colleagues from his time at Stanford University before 1986, and colleagues working on closely related subjects. The authors were asked to write about their current research work and also, if possible, to include recent as yet unpublished work and ideas. They responded enthusiastically, and the result again proves the wide acceptance and recognition of Theodor Hänsch's work in atomic and optical physics and beyond.

We wish to thank all contributing authors for their fine articles, which had to be written in a rather short time, and for their warm response and cooperation. We realise that their contributions in this volume well reflect the present status in the fields mentioned above and the developments to be expected in the near future relating to "Laser Physics at the Limits".

All three editors met Theodor Hänsch while working with him at the Max-Planck-Institute für Quantenoptik in Munich, and the inspiration from this time will not end. We benefited not only from the scientifically fruitful spirit encouraged by Theodor in the group, but also from the free and relaxed atmosphere. One of the secrets of his success is relying on the ability, ambition and absolute will of all staff members and students to do only first-rate research.

We also gratefully acknowledge our constructive and friendly cooperation with the Springer-Verlag editors, Dr. Werner Skolaut, Ms. Gertrud Dimler and Dr. Hans J. Kölsch.

Munich,
Bonn,
Tübingen,
September 2001

Hartmut Figger
Dieter Meschede
Claus Zimmermann

Contents

Part III Precision Investigations of Fundamental Physical Problems

List of Contributors

Maria Allegrini
Dipartimento di Fisica
Università di Pisa
Via Buonarroti 2
56126 Pisa
Italy
allegrin@mail.df.unipi.it

Ennio Arimondo
Dipartimento di Fisica
Università di Pisa
Via Buonarroti 2
56126 Pisa
Italy
ennio.arimondo@df.unipi.it

Sergey N. Bagayev
Acad. Lavrent'ave. 13/3
Institute of Laser Physics
Russian Academy of Sciences
Siberian Branch
Novosibirsk 630090
Russia
bagayev@laser.nsc.ru

Christoph Balzer
Universität Hamburg
Institut für Laser-Physik
Jungiusstr. 9/9a
20355 Hamburg
Germany
balzerc@physnet.uni-hamburg.de

Davide Bassi
INFM and Dipartimento di Fisica
Università di Trento
38050 Trento-Povo
Italy
bassi@science.unitn.it

Marco Bellini
Istituto Nazionale
di Ottica Applicata
LENS and INFM
Largo E. Fermi 6
50125 Florence
Italy
bellini@ino.it

Andrea Bertoldi
INFM and Dipartimento di Fisica
Università di Trento
38050 Trento-Povo
Italy

François Biraben
Laboratoire Kastler-Brossel
Ecole Normale Supérieure et
Université Pierre et Marie Curie
4 Place Jussieu
75252 Paris Cedex 05
France
biraben@spectro.jussieu.fr

Rainer Blatt
Universität Innsbruck
Institut für Experimentalphysik
Technikerstrasse 25
6020 Innsbruck
Austria

Immanuel Bloch
Universität München
Sektion Physik
Schellingstr. 4/III
80799 München

and

Max-Planck-Institut
für Quantenoptik
Hans-Kopfermann-Str. 1
85748 Garching
Germany

Hans J. Briegel
Universität München
Sektion Physik
Theresienstrasse 37
80333 München
Germany

Sven Burger
European Laboratory
for Non-linear Spectroscopy (LENS)
University of Florence
Largo E. Fermi 2
50125 Florence
Italy

Pablo Cancio Pastor
University of Florence
Institute of Applied Optics
Largo E. Fermi 2
50125 Florence
Italy

Francesco Cataliotti
European Laboratory
for Non-linear Spectroscopy (LENS)
University of Florence
Largo E. Fermi 2
50125 Florence
Italy
fsc@lens.unifi.it

Steven Chu
Stanford University
Physics Department
Stanford, CA 94305
USA

Paolo De Natale
University of Florence
Institute of Applied Optics
Largo E. Fermi 2
50125 Florence
Italy

Wolfgang Demtröder
Universität Kaiserslautern
Fachbereich Physik
Erwin-Schrödinger-Str.
67663 Kaiserslautern
Germany
demtroed@physik.uni-kl.de

Alexander K. Dmitriyev
Acad. Lavrent'ave. 13/3
Institute of Laser Physics
Russian Academy of Sciences
Siberian Branch
Novosibirsk 630090
Russia

Kjeld S.E. Eikema
Laser Centre
Vrije Universiteit
Faculty of Physics
and Astronomy
De Boelelaan 1081
1081HV Amsterdam
The Netherlands
kjeld@nat.vu.nl

Jürgen Eschner
Universität Innsbruck
Institut für Experimentalphysik
Technikerstrasse 25
6020 Innsbruck
Austria

Tilman Esslinger
Universität München
Sektion Physik
Schellingstr. 4/III
80799 München
and
Max-Planck-Institut
für Quantenoptik
Hans-Kopfermann-Str. 1
85748 Garching
Germany

William M. Fairbank, Jr.
Department of Physics
Colorado State University
Fort Collins, CO 80523
USA

Allister I. Ferguson
University of Strathclyde
Department of Physics
and Applied Physics
Glasgow G4 0NG
UK
a.i.ferguson@strath.ac.uk

Hartmut Figger
Max-Planck-Institut
für Quantenoptik
Hans-Kopfermann-Str. 1
85748 Garching
Germany
hartmut.figger@mpq.mpg.de

Chiara Fort
European Laboratory
for Non-linear Spectroscopy (LENS)
University of Florence
Largo E. Fermi 2
50125 Florence
Italy

Sebastian Fray
Max-Planck-Institut
für Quantenoptik
Hans-Kopfermann-Str. 1
85748 Garching
Germany

Tim Freegarde
Dipartimento di Fisica
Università di Trento
Via Sommarive 14
38050 Povo (TN)
Italy
freegarde@science.unitn.it

Christoph Gahn
Max-Planck-Institut
für Quantenoptik
Hans-Kopfermann-Str. 1
85748 Garching
Germany

Giovanni Giusfredi
University of Florence
Institute of Applied Optics
Largo E. Fermi 2
50125 Florence
Italy

Victor Gomer
Universität Bonn
Institut für Angewandte Physik
Wegelerstr. 8
53115 Bonn
Germany
gomer@iap.uni-bonn.de

Günter R. Guthöhrlein
Universität der Bundeswehr
Fachbereich Elektrotechnik
Holstenhofweg 85
22043 Hamburg
Germany
guenter.guthoerlein@unibw-hamburg.de

John L. Hall
JILA, University of Colorado
and NIST
Boulder, CO 80309-0440
USA

Thilo Hannemann
Universität Hamburg
Institut für Laser-Physik
Jungiusstr. 9/9a
20355 Hamburg
Germany
hannemann@physnet.uni-hamburg.de

Stephen E. Harris
Stanford University
Edward L. Ginzton Laboratory
Stanford, CA 94305
USA
seharris@ee.stanford.edu

Wolfgang Hänsel
Max-Planck-Institut
für Quantenoptik
Hans-Kopfermann-Str. 1
85748 Garching
and
Universität München
Sektion Physik
Schellingstrasse 4
80799 München
Germany

Wolfgang M. Heckl
Universität München
Institut für Kristallographie
and
Center for NanoScience (CeNS)
and GeoBio-Center
Theresienstr. 41
80333 München
w.heckl@lrz.uni-muenchen.de

Andreas Hemmerich
Universität Hamburg
Institut für Laser-Physik
Jungiusstrasse 9
20355 Hamburg
Germany

Wim Hogervorst
Laser Centre
Vrije Universiteit
Faculty of Physics
and Astronomy
De Boelelaan 1081
1081HV Amsterdam
The Netherlands
wh@nat.vu.nl

Massimo Inguscio
European Laboratory
for Non-linear Spectroscopy (LENS)
University of Florence
Largo E. Fermi 2
50125 Florence
Italy
inguscio@lens.unifi.it

Lucile Julien
Laboratoire Kastler Brossel
Ecole Normale Supérieure et
Université Pierre et Marie Curie
4 Place Jussieu
75252 Paris Cedex 05
France

Reinald Kallenbach
International Space Science Institute
Hallerstrasse 6
3012 Bern
Switzerland

Savely G. Karshenboim
D.I. Mendeleev
Institute for Metrology
St. Petersburg
Russia;
and
Max-Planck-Institut
für Quantenoptik
Hans-Kopfermann-Str. 1
85748 Garching
Germany
sek@mpg.mpg.de

Matthias Keller
Max-Planck-Institut
für Quantenoptik
Hans-Kopfermann-Str. 1
85748 Garching
Germany
matthias.keller@mpq.mpg.de

Daniel Kleppner
Massachusetts Institute
of Technology
Department of Physics
77 Massachusetts Ave., Room 26-237
Cambridge, MA 02139-4301
USA
kleppner@mit.edu

Alexandre A. Kolomenskii
Department of Physics
Texas A&A University
College Station, TX 77843-4242
USA
a-kolomenskir@physics.tamu.edu

Christian Kurtsiefer
Universität München
Sektion Physik
Schellingstraße 4/III
80799 München
Germany
christian.kurtsiefer@
physik.uni-muenchen.de

Astrid Lambrecht
Laboratoire Kastler-Brossel
Université Pierre et Marie Curie
Case 74, Campus Jussieu
75252 Paris Cedex 05
France
lambrecht@spectro.jussieu.fr

Wulfhard Lange
Universität Münster
Institut für Angewandte Physik
Correnstr. 2–4
48149 Münster
Germany
w.lange@uni-muenster.de

Siu Au Lee
Department of Physics
Colorado State University
Fort Collins, CO 80523
USA
salee@lamar.colostate.edu

Dietrich Leibfried
Time and Frequency Division
National Institute
of Standards and Technology
325 Broadway
Boulder, CO 80305
USA

Gerd Leuchs
Universität Erlangen
Physikalisches Institut
Lehrstuhl für Optik
Staudstr. 7/B2
91058 Erlangen
Germany
leuchs@physik.uni-erlangen.de

Burghard Lipphardt
Physikalisch-Technische-
Bundesanstalt
Bundesallee 100
38166 Braunschweig
Germany
burghard.lipphardt@ptb.de

Shannon K. Mayer
Department of Physics
Bucknell University
Lewisburg, PA 17837
USA

David H. McIntyre
Oregon State University
Department of Physics
Weniger Hall 301
Corvallis, OR 97331-6507
USA
mcintyre@ucs.orst.edu

Michael Mei
Max-Planck-Institut
für Quantenoptik
Hans-Kopfermann-Str. 1
85748 Garching
Germany

Dieter Meschede
Universität Bonn
Institut für Angewandte Physik
Wegelerstr. 8
53115 Bonn
Germany
meschede@iap.uni-bonn.de

Francesco Minardi
University of Florence
Institute of Materials Science
Largo E. Fermi 2
50125 Florence
Italy

Nancy S. Minarik
513 Dartmouth Lane
Schaumburg, IL 60193
USA

Giovanna Morigi
Max-Planck-Institut
für Quantenoptik
Hans-Kopfermann-Str. 1
85748 Garching
Germany

Munir H. Nayfeh
University of Illinois
Department of Physics
Urbana-Champaign
1110 W. Green Street
Urbana, IL 61801-3080
USA
m-nayfeh@uiuc.edu

Werner Neuhauser
Universität Hamburg
Institut für Laser-Physik
Jungiusstr. 9/9a
20355 Hamburg
Germany
neuhauser@physnet.uni-hamburg.de

Sile Nic Chormaic
Dept. of Applied Physics
and Instrumentation
Cork Institute of Technology
Rossa Avenue
Bishopstown, Cork
Ireland

Markus Oberparleiter
Universität Innsbruck
Institut für Experimentalphysik
Technikerstr. 25
6020 Innsbruck
Austria
markus.oberparleiter@uibk.ac.at

Krzysztof Pachucki
Warsaw University
Faculty of Physics
Institute of Theoretical Physics
IFT UW, ul. Hoża 69
00-681 Warsaw
Poland
krzysztof.pachucki@fuw.edu.pl

Domagoj Pavicic
Max-Planck-Institut
für Quantenoptik
Hans-Kopfermann-Str. 1
85748 Garching
Germany

Francesco Pavone
European Laboratory
for Non-linear Spectroscopy (LENS)
University of Florence
Largo E. Fermi 2
50125 Florence
Italy

Thorsten Platz
AGFA AG
München
Germany
torsten.platz.tp@
germany.agfa.com

Georg Pretzler
Max-Planck-Institut
für Quantenoptik
Hans-Kopfermann-Str. 1
85748 Garching
Germany

Norman F. Ramsey
Lyman Laboratory of Physics
Harvard University
Cambridge, MA 02138
USA

Robert Raussendorf
Sektion Physik
Universität München
Theresienstrasse 37
80333 München
Germany

Jakob Reichel
Max-Planck-Institut
für Quantenoptik
Hans-Kopfermann-Str. 1
85748 Garching

and

Universität München
Sektion Physik
Schellingstrasse 4
80799 München
Germany
jakob.reichel@
physik.uni-muenchen.de

Dirk Reiß
Universität Hamburg
Institut für Laser-Physik
Jungiusstr. 9/9a
20355 Hamburg
Germany
reiss@physnet.uni-hamburg.de

Gerhard Rempe
Max-Planck-Institut
für Quantenoptik
Hans-Kopfermann-Str. 1
85748 Garching
Germany

Leonardo Ricci
INFM and Dipartimento di Fisica
Università di Trento
38050 Trento-Povo
Italy
ricci@science.unitn.it

Christian Roos
Laboratoire Kastler-Brossel
24, rue Lhomond
75231 Paris Cedex 05
France

Karsten Sändig
Max-Planck-Institut
für Quantenoptik
Hans-Kopfermann-Str. 1
85748 Garching
Germany

Axel Schenzle
Sektion Physik
Universität München
Theresienstrasse 37
80333 München
Germany

and

Max-Planck-Institut
für Quantenoptik
Hans-Kopfermann-Str. 1
85748 Garching
Germany

Ferdinand Schmidt-Kaler
Universität Innsbruck
Institut für Experimentalphysik
Technikerstrasse 25
6020 Innsbruck
Austria
ferdinand.schmidt-kaler@
uibk.ac.at

Heinz W. Schrötter
Universität München
Sektion Physik
Schellingstrasse 4
80799 München
Germany
heinz.schroetter@physik.uni-muenchen.de

Hans A. Schüssler
Texas A&A University
Department of Physics
College Station, TX 77843-4242
USA
schuessler@physics.tamu.edu

Mark H. Shroyer
Emory University
Department of Physics
Atlanta, GA 30322
USA

Alexei V. Sokolov
Stanford University
Edward L. Ginzton Laboratory
Stanford, CA 94305
USA

Harald R. Telle
Physikalisch-Technische-Bundesanstalt
Bundesallee 100
38166 Braunschweig
Germany
harald.telle@ptb.de

Peter E. Toschek
Universität Hamburg
Institut für Laser-Physik
Jungiusstr. 9/9a
20355 Hamburg
Germany
toschek@physnet.uni-hamburg.de

George D. Tsakiris
Max-Planck-Institut
für Quantenoptik
Hans-Kopfermann-Str. 1
85748 Garching
Germany

Wim Ubachs
Laser Centre
Vrije Universiteit
Faculty of Physics
and Astronomy
De Boelelaan 1081
1081HV Amsterdam
The Netherlands
wimu@nat.vu.nl

Thomas Udem
Max-Planck-Institut
für Quantenoptik
Hans-Kopfermann-Str. 1
85748 Garching
Germany
thomas.udem@mpq.mpg.de

Wim Vassen
Laser Centre
Vrije Universiteit
Faculty of Physics
and Astronomy
De Boelelaan 1081
1081HV Amsterdam
The Netherlands
wim@nat.vu.nl

Jürgen Volz
Ludwig-Maximilians-Universität
München
Sektion Physik
Schellingstraße 4/III
80799 München
Germany
juergen.volz@physik.uni-muenchen.d

Vladan Vuletić
Varian Physics Bldg
382 Via Pueblo Mall
Stanford, CA 94305-4060
USA
vladan.vuletic@stanford.edu

David R. Walker
Stanford University
Edward L. Ginzton Laboratory
Stanford, CA 94305
USA

Herbert Walther
Max-Planck-Institut
für Quantenoptik
Abt. Laserphysik
Am Coulombwall 1
85748 Garching
Germany
walther@mpq.mpg.de

Jochen Walz
CERN
1211 Geneva 23
Switzerland

Harald Weinfurter
Universität München
Sektion Physik
Schellingstraße 4/III
80799 München
Germany
harald.weinfurter@
physik.uni-muenchen.de

Antoine Weis
Université de Fribourg
Institute de Physique
Chemin du Musée 3
1700 Fribourg
Switzerland
antoine.weis@unifr.ch

Martin Weitz
Universität Tübingen
Physikalisches Insitut
Auf der Morgenstelle 14
72076 Tübingen
Germany

Carl E. Wieman
JILA/University of Colorado
Campus Box 440
Boulder, CO 80309-0440
USA
cwieman@jila.colorado.edu

Lorenz Willmann
Kernfysisch Versneller Instituut
9747 AA Groningen
The Netherlands
willmann@kvi.nl

Klaus J. Witte
Max-Planck-Institut
für Quantenoptik
Hans-Kopfermann-Str. 1
85748 Garching
Germany

Christof Wunderlich
Universität Hamburg
Institut für Laser-Physik
Jungiusstr. 9/9a
20355 Hamburg
Germany
wunderlich@physnet.uni-hamburg.de

Robert Wynands
Universität Bonn
Institut für Angewandte Physik
Wegelerstr. 8
53115 Bonn
Germany
wynands@iap.uni-bonn.de

Deniz D. Yavuz
Edward L. Ginzton Laboratory
Stanford University
Stanford, CA 94305
USA

Guang-Yu Yin
Edward L. Ginzton Laboratory
Stanford University
Stanford, CA 94305
USA

Claus Zimmermann
Universität Tübingen
Physikalisches Institut
Auf der Morgenstelle 14
72076 Tübingen
Germany
clz@pit.physik.uni-tuebingen.de

Part I

Atomic and Optical Clocks

Theodor Hänsch and Norman Ramsey

Application of Atomic Clocks

Norman F. Ramsey

Although radio astronomy originally had much worse angular resolution than optical astronomy, with the use of atomic clocks, arrays of distant radio telescopes can be synthesized together in Very Long Baseline Interferometry (VLBI) to give angular resolutions of 250 micro arc sec which is 250 times better than the best optical telescopes including the improved Hubble telescope. VLBI also provides the most accurate measurements of distances along the surface of the Earth and thereby aids in the study of the Earth's crustal dynamics. Atomic clocks are needed to measure the stability of pulsars, of the Earths rotation and of other periodic phenomena. The constancy of the fundamental constants can be tested by comparing different kinds of clocks at different times. The Global Positioning System (GPS) and Differential GPS (DGPS) use atomic clocks to locate positions to within a meter and these systems are extensively used in navigation, geographical exploration, geology, archeology, paleontology, environmental studies, etc. Navigation in outer space also utilizes accurate atomic clocks. The unit of length is now defined in terms of time and the representation of the volt is in terms of frequency. Theories of relativity are tested by measurements on clocks in high-altitude rockets, by the time delays of signals passing close to the Sun and by measurements of small changes in the orbital periods of binary pulsars. The rates of change of the periods of the binary pulsars agree with the Einstein form of general relativity to one half a percent, but do so only when one includes the radiation of gravity waves and strong field aspects of relativistic gravity.

1 Introduction

Theodor Hänsch by his fundamental precision measurements has contributed to physics in many ways, but I shall limit myself just to his contributions to improved atomic clocks. He and his collaborators developed laser-saturated absorption spectroscopy [1] and two-photon Doppler-free absorption spectroscopy [2] as methods for eliminating the first-order Doppler shift. In 1975 he and Schalow [3] first proposed laser cooling of atoms as a means of reducing the second-order Doppler shift. In addition to these revolutionary advances, he has recently invented the frequency comb [4] which makes possible the accurate comparison of low and high frequencies which in turn makes possible the use of optical frequency standards in most frequency ranges. In addition to the development of these fundamental methods, Hänsch and his associates have made such accurate measurements of the basic $1S$–$2S$ transition frequency in atomic hydrogen that this frequency could be a serious contender for the next definition of the second. Because of Theodor Hänsch's many contributions to atomic clocks, I have chosen as the title of this paper honoring his 60th birthday: "Application of Atomic Clocks".

The most accurate physical measurements are those of time and frequency made with atomic clocks and this precision is increasing with current developments. As a result atomic clocks are powerful tools for both fundamental research and practical applications in many fields.

2 Applications

2.1 Spectroscopy

One of the most direct and important applications of atomic clocks has been in spectroscopy, since the frequency standards used in spectroscopy are based on atomic clocks. Furthermore many of the technical advances for making better atomic clocks have been directly applied to the improvement of spectroscopic techniques and *vice versa.*

2.2 Very Long Baseline Interferometry (VLBI) in Radio Astronomy

In radio astronomy one looks with a parabolic reflector at the radio waves coming from a star, just as in optical astronomy one looks with a telescope at the star's light waves. Unfortunately, the wavelength of the microwave radiation is about a million times longer than the wavelength of light, so the resolution of a single radio telescope is about a million times worse than that of an optical telescope, depending as it does on the ratio of the wavelength to the telescope aperture. However, if there are two radio telescopes on opposite sides of the Earth looking at the same star and if the radio waves entering each are matched in time, it is equivalent to a single telescope whose aperture is the distance between the two telescopes, and the resolution of such a combination exceeds that of even the largest single optical telescope. However, each of the two telescopes must have highly stable atomic clocks, to beat their signals against, so that the reduced frequency signals can be transmitted to a central station without significant loss of phase. Usually hydrogen masers have provided such Very Long Baseline Interferometers (VLBI) with free running clocks of high stability and sufficient power. Arrays of up to 12 radio telescopes and atomic clocks have been matched together with their detected signals synthesized by high-speed computers into a two-dimensional sky map with an angular resolution of 200 μarc s which is 250 times better than the best optical telescopes, including the Hubble Space Telescope. Although the angular resolution gained in this way is limited by the size of the Earth, this limitation can be overcome by placing the telescope arrays in different widely spaced spacecraft, but such large baselines will require atomic clocks of even greater stability.

2.3 Pulsar Periods

Precision clocks are also needed to measure the periods of pulsars – stars that emit their radiation in short pulses – and the changes in their periods, which sometimes occur smoothly, sometimes abruptly. Of particular interest are millisecond pulsars, whose remarkably constant periods rival the stability of the best atomic clocks. In fact, one of these pulsars is so stable that it may eventually be suitable as a standard of time over long periods [5].

2.4 Variability of Earth's Rotation Rate and of Other Periodic Phenomena

Accurate measurements of time permit measurements of the variability of quantities that were once thought to be constant. As we have seen, the rotation period of the Earth, which once served as the basis for defining the unit of time, is now known to vary by a few parts in a hundred million from winter to summer and from year to year. Some of the variation is regular and some unpredictable. Atomic clocks are also used to test the stabilities of many other periodic phenomena such as oscillatory crystals and planetary orbital periods.

2.5 Tests of Constancy of the Fine Structure Constant, α

Different atomic clock rates have been accurately compared over long periods of time to see if there might be changes in their relative rates which could correspond to a change with time in the fundamental physical constants, but no such change has yet been discovered. The first such tests were carried out by Wineland, Kleppner and Ramsey, who compared the hydrogen and deuterium hyperfine separations at times several years apart. Since then a number of others have carried out more sensitive experiments of which the most recent and best is the French comparison of laser cooled Rb and Cs clocks which show to a 90% C.L. that

$$\frac{\mathrm{d}\alpha/\mathrm{d}t}{\alpha} < 8 \times 10^{-15}/\mathrm{yr}.$$

This is not yet the most sensitive limit since analysis of a prehistoric fission reaction in Africa gives

$$\frac{\mathrm{d}\alpha/\mathrm{d}t}{\alpha} < 10^{-15}/\mathrm{yr}$$

and analysis of the spectra of a quasar 4×10^9 light years away also gives

$$\frac{d\alpha/dt}{\alpha} < 10^{-15}/\text{yr}.$$

However, atomic clock tests should eventually become the most sensitive, since the accuracy of atomic clocks is improving and the sensitivity of the atomic clock tests improve with the square of the measuring time which so far is only one year for the most sensitive measurement.

2.6 Precision Navigation

Accurate clocks make possible an entirely new and more accurate navigational system, the global positioning system or GPS. Twenty-four satellites containing accurate atomic clocks transmit synchronized coded signals so that any observer receiving and analyzing the signals from four or more such satellites can determine his position to within 20 m with inexpensive receivers and within 8 m with better receivers. If the measurements are averaged at one location over long periods of time, as in continental drift measurements, the measurement can be accurate to 0.001 m. The measurement accuracy for shorter-term measurements can also be increased with differential GPS (DGPS) by having local receiving stations at known locations continually broadcasting corrections that are automatically used to correct GPS readings in the neighborhood. This now gives positions to within 3 m over wide areas and there is hope that this will eventually drop to 1 m. For surveying, the relative locations of two nearby GPS receivers can be measured to about 0.01 m. The FAA hopes soon to use DGPS as the primary aircraft navigation system. The FAA is also supporting work on a localized DGPS to give locations in the vicinity of an airport 0.1m and eventually for close approach and perhaps landing. In addition the US Coast Guard is installing DGPS station in many locations to aid coastal navigation.

GPS and DGPS are proving to be of great value in many sciences including geographical exploration, geology, archeology, paleontology, civil engineering, surveying and environmental studies.

2.7 Earth's Crustal Dynamics

The most accurate measurements of distances and changes of distances on the surface of the Earth are those between the radio telescopes of VLBI, so these give valuable information on the Earth's crustal dynamics. For example the distances from the Haystack radio telescope in Massachusetts to Owens Valley, California, was found to vary less than 0.03 m over an entire year whereas the distance to Palo Alto, on the other side of the San Andreas fault, grew by 0.3 m in one 11 week period. The GPS can also be used to study the Earth's crustal dynamics and has been used to detect contemporary continental drift. When the GPS can be kept at a single location for long times, much better accuracy can be obtained than when the observer is either moving or in single locations for only short periods of time. A valuable application of GPS to Earth's crustal dynamics occurred in the recent (1999) Turkish earthquake. A careful GPS survey of various sites was made shortly before the earthquake and another afterwards and from these measurements the Earth movements during the earthquake could be determined.

2.8 Navigation in Outer Space

Atomic clocks are essential to navigation in outer space. In the Voyager mission to Neptune, its location was determined by three radar telescopes each with two hydrogen masers to measure the time for the radar signals to go from each telescope to the satellite and return. Similar observations were made by VLBI telescopes, also based on atomic clocks.

2.9 Expressions of Other Physical Quantities in Terms of Time

Time can now be measured so accurately that wherever possible other fundamental measurements are reduced to time measurements. Thus the unit of length has recently been defined as the distance light travels in a specified time, and the practical realization for the unit of voltage is now expressed in terms of frequency.

2.10 Tests of the Special and General Theories of Relativity

Accurate clocks have provided important tests of both the special and general theories of relativity. The periodic rate of a hydrogen maser carried in a rocket to 10 000 km changed with speed and altitude by the amounts predicted by the special and general theories [6, 7]. In other experiments observers have measured the delays predicted by relativity for radio waves passing near the Sun. Future improvements in the stability of clocks should make possible even more rigorous tests of fundamental theories.

The most severe tests of general relativity have been those of Taylor and his associates [5, 8] on millisecond pulsars that are one member of a binary pair (two stars so close together that the period of their orbital motion about the common center of mass is typically 10 h or less). Although these binary stars are too distant to be resolved with either an optical or a radio telescope, their orbits can be measured with exquisite accuracy by the modulation of the pulsar rate through the Doppler shift as the pulsar in its orbit successively approaches and recedes from the Earth. They found that the orbital period of the binary is changing by just the amount expected from the loss of energy by the radiation of gravity waves predicted by the Einstein general theory of relativity – the first experimental evidence for the existence of gravity waves. The orbit is measured so well that it tests the strong field as well as the radiative aspects of relativity. The measurements confirm the Einstein form of the general theory of relativity and leave little room for alternative theories. The ratio of the observed rate of change of the binary period to that calculated from the Einstein theory of relativity including strong field aspects and gravity waves is 1.0032 ± 0.0050. In a sense these binary pulsar measurements permit a determination of the velocity of propagation of gravity waves since the excellent agreement between observations and calculations disappears when the velocity of propagation of gravity waves in the theory is allowed to differ from the velocity of light c by more than 1%.

3 Conclusions

This review covers only a portion of the important applications of atomic clocks and more applications continue to be invented. Some applications, such as VLBI from bases in outer space, will require clocks of even greater precision, but with the expected advances in clock technology there are strong reasons to believe these demands can be met. The next few decades should be an exciting period for atomic clocks and their applications.

References

1. T.W. Hänsch, M.D. Levenson, A.L. Schalow: Phys. Rev. Lett. **26**, 946 (1988)
2. T.W. Hänsch et al.: Opt. Commun. **11**, 50 (1974)
3. T.W. Hänsch, A.L. Schalow: Opt. Commun. **13**, 68 (1975)
4. J. Reichert, M. Niering, R. Holzwarth, M. Weitz, Th. Udem, T.W. Hänsch: Phys. Rev. Lett. **8**, 3232 (2000)
5. L.A. Rawley, J.H. Taylor, M.M. Davis, D.W. Allan: Science **238**, 761 (1987)
6. R.F.C. Vessot, M.W. Levine, E.M. Mattison et al.: Phys. Rev. Lett. **45**, 2081 (1980)
7. J. P. Turneaure, C. M. Will, B.F. Farrell, E.M. Mattison, R.F.C. Vessot: Phys. Rev. D **27**, 1705 (1983)
8. T. Damour, J.H. Taylor: Phys. Rev. D **45**, 1840 (1992)

Achievements in Optical Frequency Metrology

Thomas Udem and Allister I. Ferguson

We describe the recent developments in high-precision spectroscopy with mode-locked lasers that have eventually culminated in a simple and compact apparatus that makes it possible to directly measure optical frequencies of a few hundred THz. With this set-up even small-scale laser laboratories are enabled to perform measurements of the highest precision. It is possible to apply this principle in reverse as an all-optical clock which is projected to allow for a three orders of magnitude improvement in accuracy over existing cesium fountain clocks.

1 Introduction

Even though nobody really understands what time is, it is nevertheless the physical quantity that can be measured with, by far, the highest accuracy. Whenever the highest accuracy is needed, it is wise to try to convert to a time or frequency measurement if possible. Provided one counts the number of cycles correctly, a frequency measurement is nothing but the measurement of the time between two, not necessarily adjacent, zero crossings. Thus a frequency measurement possesses the same potential as a time measurement. Since the value of the speed of light in vacuum c_{o} was defined in 1983 it is in principle easy, by using $\lambda = 2\pi c_{\text{o}}/\omega$, to convert between wavelength λ and frequency ω without loss of accuracy.[1] The problem however was, when trying to perform a frequency measurement rather than a wavelength measurement, that optical frequencies are quite high. So a fast counter had to be designed. Recently an all-new concept based on an idea that Theodor Hänsch had back in the late 1970s became a reality. His approach has now resulted in a compact device that runs with sufficient reliability for it to be used as an all-optical clock providing a clockwork mechanism with a gear ratio operating from the visible to the rf region of the spectrum. It can be used as an optical clockwork system in a real optical clock that would tick for much longer than just a few seconds. These clocks use sharp optical, rather than radio frequency transitions, to slice time into thinner pieces and thus potentially improve substantially on the accuracy.

[1] This possibility was actually the reason for that definition.

These optical frequency counters are also valuable tools for basic research, as they allow for the most precise measurements of transition energies in atomic hydrogen performed today in Theodor Hänsch's laboratory in Garching [1]. These numbers are needed for the most accurate tests of quantum electrodynamics and to derive the Rydberg constant that is needed for a general adjustment of all natural constants. The Rydberg constant became the most precisely known natural constant because of these efforts and the absolute frequency measurements performed by Biraben and co-workers [2]. With new and more precise optical clocks it may also become possible to verify the predictions of general relativity. These clocks should be sensitive to altitude differences as small as 1 cm due to the gravitational red-shift. Also, one might find the natural constants slowly drifting in value as discussed by some theoreticians [3, 4].

2 The Traditional Way

When Evenson and co-workers designed the first optical counter back in 1973 [5], it was certainly one of the most difficult optical experiments at that time. Unfortunately, not much changed in that respect until the end of the 1990s. They used a harmonic frequency that consisted of oscillators in series with increasing frequency. It started with a radio frequency that is derived from an atomic clock and uses nonlinear devices successively to create harmonics. Because nonlinear frequency conversion is usually weak, a transfer oscillator was needed after each step. The chain of oscillators, as shown in Fig. 1, covers the whole electromagnetic spectrum from rf to uv. In each step either a phase-locked loop forces the oscillator to stay in phase with a harmonic of the preceding oscillator or, where this is not possible, a frequency counter measures the beating between them. Besides the complexity (each oscillator in the chain might be worth a PhD degree), its application was limited to a handful of optical frequencies at the most.

3 New Attempts

At the end of the 1980s Theodor Hänsch came up with a whole new idea of how to takle that problem [8]. It was based on the measurement of frequency differences between different harmonics of the same laser. In its simplest form it would just be the difference between an optical frequency ω and its own second harmonic 2ω. The idea was to divide that optical octave successively until it becomes accessible to rf frequency counters. The method devised by Theodor Hänsch to do the division was the so-called optical frequency interval divider (OFID). As shown in Fig. 2, such an OFID receives two input laser frequencies ω_1 and ω_2, and it forces a third laser to oscillate phase coherently at the precise midpoint ω_3. This is accomplished by first summing the frequencies of the two input lasers in a nonlinear crystal and then forcing

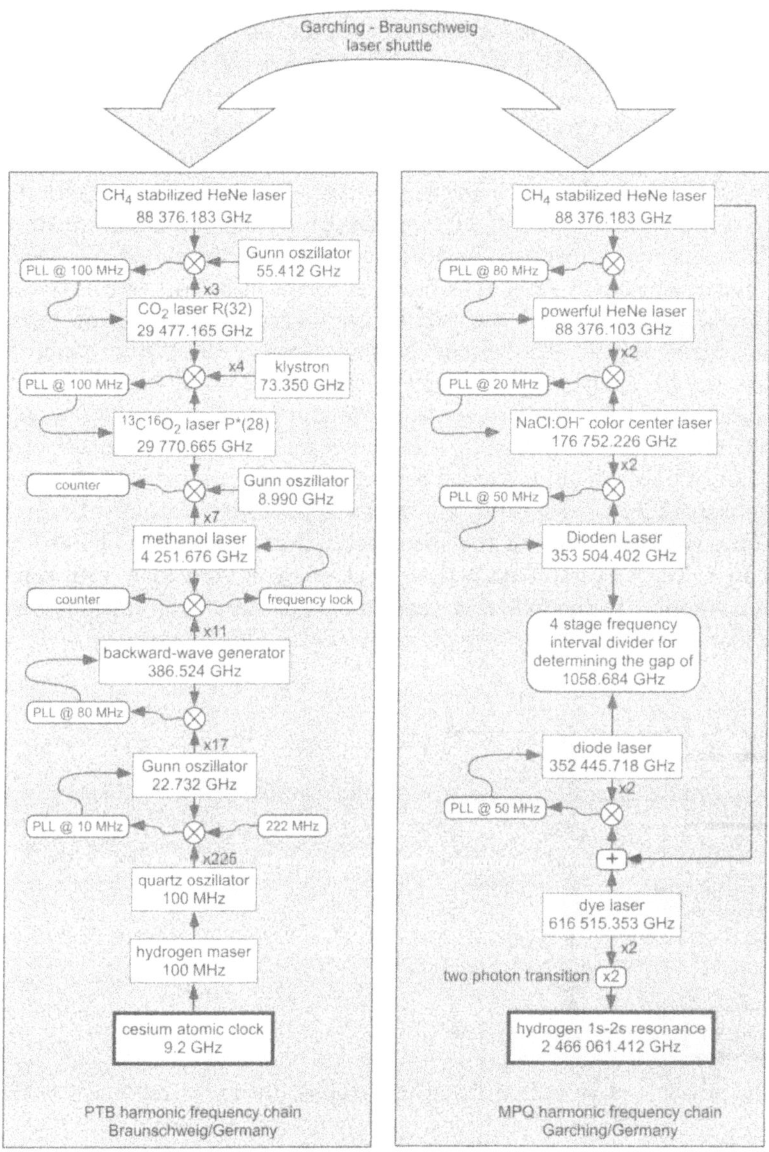

Fig. 1. A traditional harmonic frequency chain that was used in 1997 to determine the hydrogen 1S–2S interval [6, 7]. This set-up used to fill two large-scale optical labs, one at the *Physikalisch Technischen Bundesanstalt* (PTB) in Braunschweig/Germany (*left-hand side*) and the other at the *Max-Planck Institut für Quantenoptik* (MPQ) in Garching/Germany (*right-hand side*). The connecting piece was a methane-stabilized HeNe laser at 88.376 THz (3.39 μm) which was shuttled back and forth between Braunschweig and Garching. The frequency mixers (⊗) were Shottky or MIM diodes and photodiodes, depending on the wavelength region. The Shottky or MIM diodes also acted as the nonlinear frequency multiplier, whereas in the optical region nonlinear crystals had to used. Most of the oscillators shown were phase-locked to the preceding oscillator. A remaining gap in the chain of $\Delta f \approx 1$ THz, shown on the right-hand side, was closed with a 4-stage frequency-interval divider (not shown), consisting of six additional phase locked lasers

the second harmonic of the third laser to oscillate in phase with the sum frequency. This establishes that $(\omega_1+\omega_2)/2$ is equal to ω_3, or, in other words, the laser at ω_3 oscillates at the precise midpoint between ω_1 and ω_2. Phase control is accomplished by an electronic phase-locked loop. With a chain of n cascaded divider stages, a given frequency interval can be divided by 2^n. With a 12-stage OFID chain the gap between $\omega = 2\pi \times 300\,\mathrm{THz}$ and 2ω could be divided to a measurable 73 GHz. The advantage of using 12 OFIDs instead of, say, 12 frequency multipliers, as in the traditional harmonic frequency chain, is that basically the same technique, in terms of lasers, detectors and nonlinear mixers, could have been used in every stage, as this set-up never leaves the optical region. Another desirable feature of the OFID chain is that, unlike the harmonic chain, it permits some freedom of where to place the laser oscillators. After the first stage, made up of the frequencies (ω_1, ω_2) and ω_3, the fourth laser may form the second stage by locking it in between ω_1 and ω_3 or in between ω_2 and ω_3. Theodor Hänsch's original layout made use of the possibility of designing an "artificial hydrogen atom". Because of the simple gross structure of the spectrum it was possible to choose the intermediate OFID oscillators such that most of them coincided with some transition frequency. In the following years some groups [9–13], including his Garching team, has started to work on variations of OFID chains.

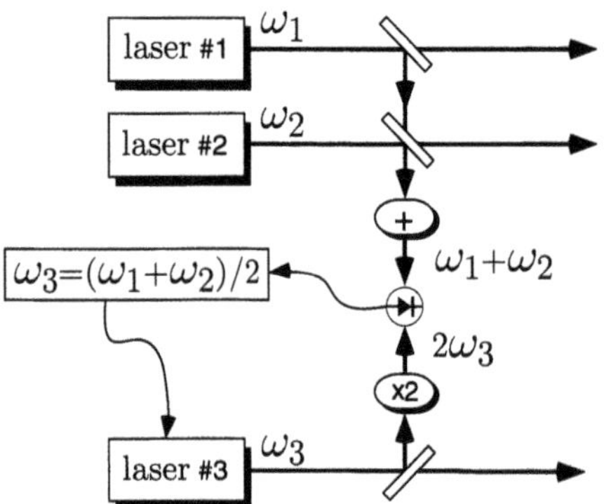

Fig. 2. The principle of an optical frequency-interval divider as explained in the text

4 The Comb

At the same time that new techniques were being developed to generate frequencies at precise intervals, significant progress was made to increase the maximum measurable frequency gap. Frequency combs that impose a large spectrum of optical sidebands on a continuous-wave laser, by efficient electro-optic modulation, were being produced by different groups. Such a comb of equally spaced modulation sidebands can be used as a ruler in frequency space to measure large optical frequency gaps. It turned out that the rate of progress on frequency combs was advancing faster than the rate at which OFID chains grew in length. With each factor of two in comb widths, one OFID could be dropped. The widest-frequency combs, of up to 30 THz, have been created by M. Kourogi and co-workers [14]. They used the pulsed output of a such a comb generator for self-phase modulation in an optical fiber after boosting the intensity in a fiber amplifier. Interestingly, a similar device that incorporated these components – optical amplification and self-phase – modulation already existed: the Kerr-lens mode-locked laser.

5 The Prehistory of the Laser Comb

The use of a train of short pulses from a mode-locked laser for optical frequency metrology can be traced back to the mid-1970s. At this time Theodor Hänsch along with Carl Wieman had just shown that it was possible to measure the ground-state Lamb shift in hydrogen by using the legendary Hänsch grating-tuned dye laser pumped by a nitrogen laser [15]. The frequency calibration in this case was provided by nature in that the energy of the 1S to 2S transition in hydrogen is almost exactly four times that of the Balmer-β line. This removed the need for a precise measurement of the Rydberg constant. This was the start of a journey lasting more than a quarter of a century for Theodor Hänsch to reach the ultimate precision on the 1S to 2S transition.

It was clear that the next level of precision in hydrogen spectroscopy would require the resolution normally associated with continuous-wave sources. How could this be achieved when there were no continuous-wave sources at the required 243 nm and when the 1S to 2S transition was so weak? Two approaches to overcoming this problem were pursued. In the first it was pointed out that, inspired by Norman Ramsey's ideas of obtaining high resolution by use of spatially separated fields and atomic beams, it would be possible to observe Ramsey fringes by time-separated fields using coherent pulses. The resolution would be given by the inverse of the time between pulses just as in the case of Ramsey fringes, where the resolution is given by the time it takes for the atoms to move from one field to the next. It was further realized that, since light was being used, it would be possible to have many coherent interactions simply by multiple bouncing of the beams through the atomic sample. Each successive bounce would improve the resolution. This led Theodor Hänsch, Rich Teets and Jim Eckstein to inject pulses from the nitrogen laser-pumped dye laser into an optical cavity, thereby giving many coherent interactions with the atoms [16]. If two pulse interactions are equivalent to a Young's slit interference experiment, giving a sinusoidal interference pattern, then the multiple-pulse coherent interaction would be equivalent to a diffraction grating and the resolution would be dictated by the number of pulses. Thus a pulsed laser could be used to achieve a resolution below the bandwidth of a single pulse.

This idea was further extended by going to the ultimate in a train of coherent optical pulses – the continuous-wave mode-locked laser. The use of a mode-locked laser to observe the 1S to 2S transition had been proposed by Chebotaev [17]. Although there were no mode-locked lasers operating at the required 243 nm, it was conceivable that a laser could be mode-locked at 486 nm and frequency doubled. The high peak power associated with the mode-locked pulses would enhance the frequency doubling. The prospect of obtaining a 1S to 2S spectrum with continuous-wave resolution was now very real. The use of a mode-locked laser to obtain narrow resonances was first demonstrated by in 1978 by Eckstein, Hänsch and one of the authors [18]. Although this was applied to the sodium atom – as all new spectroscopic techniques were at the time – it showed for the first time that linewidths of a few MHz could be observed using a train of pulses, each member of which had a bandwidth of many GHz. The laser that was used was a synchronously pumped dye laser. Theodor Hänsch and his co-workers can be seen looking at this laser in Fig. 3. One feature of the coherent multiple-pulse spectroscopy technique is that a simple spectrum could become complex, since all of the spectral lines appear within a frequency corresponding to the repetition rate of the laser. This is just like a Fabry–Pérot interferometer with overlapping orders. As with a Fabry–Pérot interferometer, these overlapping orders can be used to provide calibration. It was absolutely clear that to some extent this technique was self-calibrating – but just how equally spaced are the modes?

Steve Harris and Tony Siegman were among the pioneers of the early days of mode-locking of lasers and brought an electrical engineering approach to the issue. They were consulted on the properties of mode-locked lasers and in particular, just how equally spaced were the modes? Their view was that the mode-locked laser, could be considered as a series of single-frequency continuous-wave lasers equally spaced in frequency and locked in phase such that at some point in time all of the mode amplitudes added in phase to give a pulse of high peak intensity. The dark bits between the pulses were caused by destructive interference of the modes. In this model, dispersion and other mode-pulling and pushing effects were overcome by the mode-locking process, giving rise to a perfectly uniform comb of modes. Tony Siegman also pointed out another mode of laser operation called FM operation, in which the carrier frequency of the laser swept sinusoidally back and forth with a period determined by an intracavity phase modulator. In this case the equally spaced modes adjust in phase such as to give a constant amplitude. This is another mode that permits frequency calibration [19].

Although the major benefit of the use of mode-locked pulse trains was at first perceived as being due to the high peak power permitting efficient frequency doubling, it is what appeared at first to be a by-product that has been shown to have been the major step-forward. For a time Theodor Hänsch's

Fig. 3. (*From right to left*) Theodor Hänsch, Erhart Weber, Jim Eckstein, John Goldsmith and AIF looking into a home-constructed synchronously pumped dye laser. For a brief time it held the record for short-pulse generation. They are looking at a particle trapped in the laser beam, something that Askin and Chu were working on at Bell Labs – but that is a different and long story! At this time Weber and Goldsmith were making a precision measurement of the Rydberg constant on the Balmer$-\alpha$ line in hydrogen

group held the record for the shortest pulse from a synchronously pumped dye laser, but even then the bandwidth was a mere 500 GHz. The advent of the Kerr-lens mode-locked laser and photonic-crystal fibers have completely revolutionized the use of coherent pulse trains for frequency metrology and has brought the technique into the mainstream of metrology. This was made possible by the fact that these techniques are capable of generating bandwidths of more than one optical octave while maintaining a high repetition rate. As will be explained below, this technique is now used as a simple optical frequency counter.

6 The Return of the Pulsed Laser Comb

The output of a mode-locked laser in the time domain consists of a train of short pulses. The spectral width associated with pulses is roughly given by the inverse of the (Fourier-limited) pulse duration, so that 10 fs pulses should yield a spectral width of 100 THz. The useful width could be even larger, as the usefulness is not limited to the full width at half maximum. In the frequency domain, a pulse that circulates in a laser cavity can only be stable in shape if, and only if, all higher-order dispersion terms beyond the linear term are exactly canceled. The phase round-trip condition for the nth cavity mode with frequency ω_n is $2k(\omega_n)L = n2\pi$, where L is the cavity length, $k(\omega_n)c_o/n_i(\omega) = \omega_n$ is the wave number of that mode and $n_i(\omega)$ denotes the index of refraction. Expansion of $k(\omega_n)$ about some carrier frequency ω_c yields:

$$2L\left[k(\omega_c) + k'(\omega_c)(\omega_n - \omega_c) + \frac{k''(\omega_c)}{2}(\omega_n - \omega_c)^2 + \ldots\right] = 2\pi n\,. \tag{1}$$

When calculating the mode spacing $\omega_r \equiv \omega_{n+1} - \omega_n$ between the $(n+1)$th and the nth mode,

$$k'(\omega_c)\omega_r + \frac{k''(\omega_c)}{2}\left((\omega_{n+1} - \omega_c)^2 - (\omega_n - \omega_c)^2\right) + \ldots = 2\pi/2L\,, \tag{2}$$

it turns out that exactly the terms in the expansion that disturb the constant mode spacing, i.e. everything beyond linear dispersion (the group velocity dispersion $k''(\omega_c)$ (GVD) and higher-order contributions), are the ones that reshape the pulse envelope. Obviously, the longer the pulse stays together with a constant envelope, for whatever reason, the more precisely should the dispersion be linear and thus the mode spacing be constant, i.e. independent of n. Then

$$\omega_r = \frac{\partial \omega}{\partial k}\frac{1}{2L} = \frac{v_g}{2L} \tag{3}$$

is nothing but the inverse pulse round trip time with v_g being the group velocity. In fact, in a realistic mode-locked laser the same pulse can circulate for many hours (or even days) and can provide more than 10^{14} copies of itself at the output coupling mirror. So, these mode-locked lasers should be considered as ideal frequency generators. Although this was already pointed out by Theodor Hänsch and co-workers in the late 1970s [18], and they were able to perform the first measurement of a frequency difference in the optical region using this technology, they were only able to use picosecond pulse lengths and the application was limited to rather small frequency gaps.

7 Finally, the Breakthrough

Even though all the techniques would already have been available in the early 1980s, when colliding-pulse mode-locked lasers [20] reached into the fs regime, the real breakthrough in fs-frequency metrology had to wait until the end of the 1990s. At that time a group at the University of Bath, UK [21] and a group at Lucent Technologies [22] designed what the Bath people call a photonic-crystal fiber (PCF). This single-mode optical fiber is made from a single material. The low index of refraction in the fiber cladding is produced by air holes that run in parallel and close to the fiber core. No special equipment other than a regular fiber-drawing tower is necessary to produce these fibers. Because a fraction of the guided-mode field penetrates the air holes, the dispersion properties of such a fiber can be designed to some extend. It is possible to shift the zero of the GVD to the center emitting wavelength of the Ti:sapphire laser at 800 nm. Pulses from such a laser launched into this fiber stay focused both in space and time, giving rise to strong nonlinear interaction. Most interesting here is again self-phase modulation. Kourogi has already demonstrated that the intensity-dependent index of refraction, which is what causes self-phase modulation, can broaden the frequency comb [14]. Through this effect the optical path length of the fiber changes rapidly with the rate of change in intensity of the pulses, imposing additional sidebands on all the modes. These additional sidebands have the same separation and thus effectively add new modes onto the extended grid of laser frequencies, as Fig. 4 shows. Unlike within the laser oscillator, higher-order dispersion is not exactly

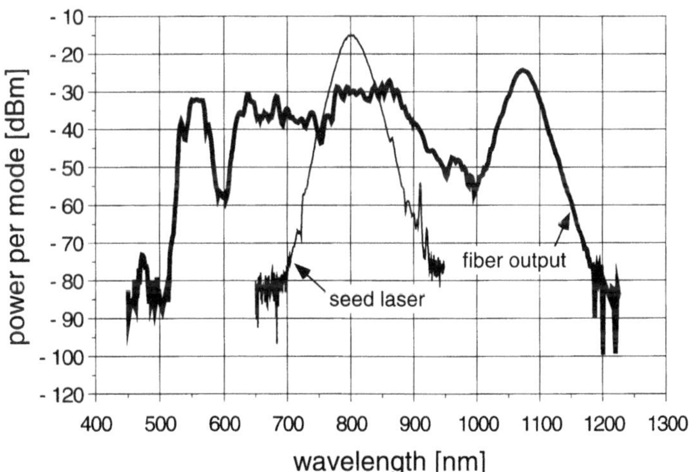

Fig. 4. Spectral broadening of a frequency comb ($\omega_r = 2\pi \times 750$ MHz) by coupling 180 mW time averaged power through a 30 cm-long PCF

zeroed inside the fiber, so that the pulses are expected to be significantly distorted when they leave the fiber. However, each pulse travels through the fiber only once. It is fair to assume that the fiber action, whatever that precisely is, is the same for each pulse and thus preserves the comb structure. From (2) it was concluded that the mode spacing is constant. However, because of the linear dispersion inside the laser cavity the comb is free to shift along the frequency axis, which may be expressed by the following formula:

$$\omega_n = n\omega_\mathrm{r} + \omega_\mathrm{o} \,, \tag{4}$$

where ω_o is an unknown radio frequency that may be confined to $0 \leq \omega_\mathrm{o} \leq \omega_\mathrm{r}$ with the right choice of mode numbering n.

8 The Plug & Play Optical Synthesizer!

The key to turning (4) into something extremely useful is the octave-spanning comb shown in Fig. 4. In general this equation maps two radio frequencies ω_r and ω_o onto a full octave of optical frequencies ω_n. Whereas the pulse repetition rate ω_r is readily accessible, there seems to be only one way to obtain the frequency offset ω_o with radio-frequency accuracy. Figure 5 shows how. To measure the offset frequency ω_o, a mode of frequency $n\omega_\mathrm{r} + \omega_\mathrm{o}$ on the

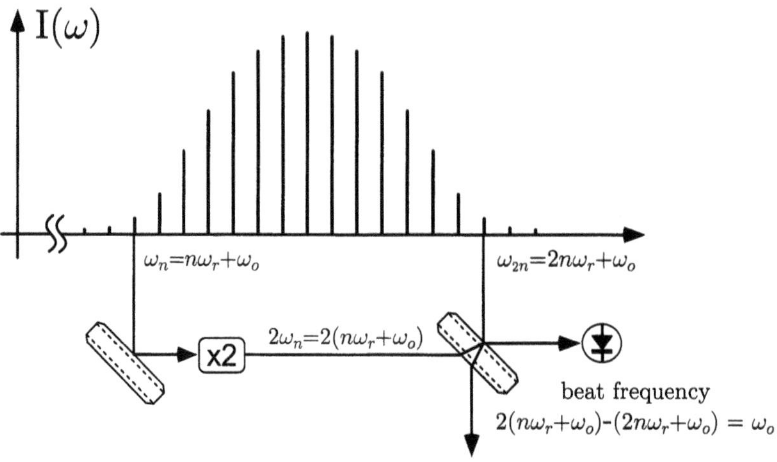

Fig. 5. Self-referencing the octave-wide frequency comb produces a beat note at ω_o. After phase locking of ω_r and ω_o to an atomic clock or a GPS receiver, a million precisely calibrated lasers are available

"red" side of the comb is frequency doubled to a frequency $2(n\omega_\mathrm{r} + \omega_\mathrm{o})$. If the comb contains more than an optical octave there will be a mode with the mode number $2n$ oscillating at frequency $2n\omega_\mathrm{r} + \omega_\mathrm{o}$. As sketched in Fig. 5, the frequency comb is made "self-calibrated" by observing the beat note between the frequency-doubled mode n and the mode $2n : 2(n\omega_\mathrm{r} + \omega_\mathrm{o}) - 2n\omega_\mathrm{r} + \omega_\mathrm{o} = \omega_\mathrm{o}$. The repetition rate and the offset frequency are both phase-locked to an atomic clock by controlling the laser cavity length and the power that is used to pump that laser. This method has allowed the construction of a very simple optical frequency synthesizer [23–26] that operates with only a single laser. It supplies us with a reference-frequency grid across much of the visible and infrared spectrum. As Fig. 6 shows, the set-up only covers minimal space on a table (roughly one square meter) with some potential for further miniaturization.

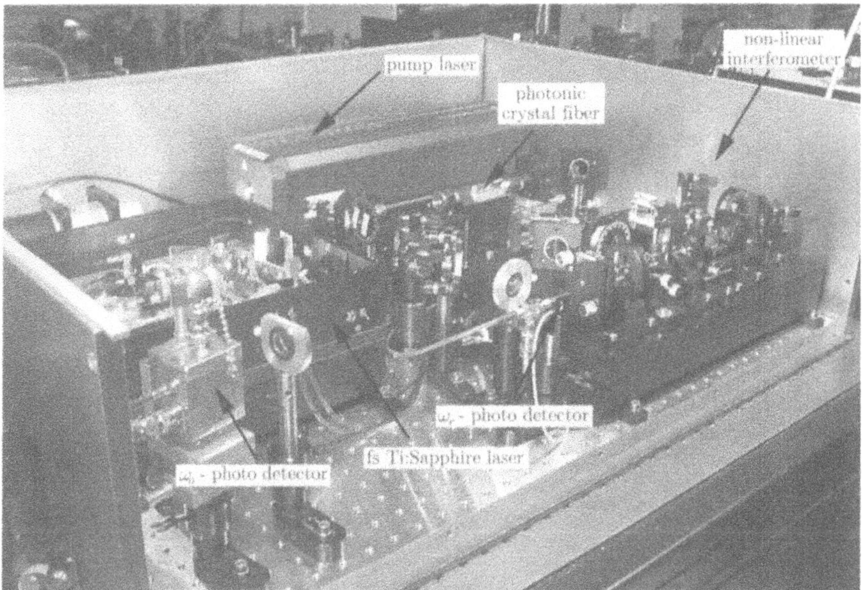

Fig. 6. The self-referenced optical synthesizer in Theodor Hänsch's laboratory. For higher efficiency not only one mode from the "red" side of the comb is frequency doubled but a few thousand. To bring all those beat notes (all at frequency ω_o) in phase an interferometer is used that matches the path lengths for the frequency-doubled and the "blue" pulse

Before the realization of that optical synthesizer in J. Hall's group at JILA Boulder/USA and in Hänsch's laboratory, a preliminary version of it was constructed also in Hänsch's laboratory. Lacking a PCF, this one relied on a 44 THz-frequency comb between $4\,f$, and $3.5\,f$, where f=88 THz was the output from an infrared HeNe laser ($\lambda = 3.39\,\mu$m). This optical synthesizer was used for an order of magnitude improvement on the hydrogen 1S–2S transition frequency [1] and for testing the new technology by comparing it with another synthesizer of the $f:2f$ type. Both synthesizers were referenced to the same 10 MHz quartz oscillator, which was multiplied to 353 THz, where a beat note was recorded. After averaging for about $4\frac{1}{2}$ hours the deviation was measured to be 71±179 mHz, which corresponds to a relative accuracy of 5.1×10^{-16} [26]. This rather encouraging result, and previous tests also performed in Theodor Hänsch's group, was the starting gun for the rush to that new technology. Meanwhile, all standard institutes of the main industrialized nations are either operating an optical synthesizer or working on setting one up.

9 The All-Optical Clock

They are all heading for the long awaited dream of using sharp optical transitions to set rapid time ticks in an optical clock. Ignoring systematic effects, which could be as low as 10^{-18} [27] for a well-chosen optical transition, the accuracy of a clock is roughly inversely proportional to the frequency of the "pendulum" that it uses. This has already been known for some time, but in order to build an optical clock that could run reliably an optical counter had to be designed. The traditional frequency chains never reached the stage where one should use the term reliable, while the femtosecond frequency synthesizers, due to their simplicity, can already run for hours and may eventually become turn-key systems. The system that comes very close to deserving the term "optical clock" uses a transition at 1064 THz in a trapped single mercury ion and is operated at NIST [28]. PTB/Germany is using a single trapped Yb^+ ion [29], and both NRC/Canada and NPL/UK are preparing fs-combs for their Sr^+ ions [30]. Another promising candidate seems to be a single trapped indium ion [27]. These clocks may prove to be very useful, not only for industrial applications such as satellite communication and network synchronization but also for playing an important role in basic research. The quest for natural constants that may drift slowly in time, as discussed by some theoreticians [3, 4], is one example. Also, these clocks may help to refine general relativity, which still poses one of the major problems in physics. This theory still remains untested beyond a level of 7×10^{-5} [31], which may be the reason why corrections due to a quantized theory have not been discovered so far.

10 Conclusion

For almost thirty years now Theodor Hänsch has led the way in introducing new methods for reaching the ultimate in precise measurements using lasers. His innovative and adventurous approach has inspired and engaged generations of students and colleagues with his excitement and passion for the subject. The all-optical clock has been a dream of metrologists for many years and he has played a crucial role in realizing that dream. As with all developments with which he has been involved, the latest developments will not be the end of the story but rather the beginning of the next chapter of new and exciting discoveries. Wherever these ventures might lead, we can be sure that Theodor Hänsch will be there, out in front, inspiring our efforts.

References

1. M. Niering, R. Holzwarth, J. Reichert, P. Pokasov, Th. Udem, M. Weitz, T.W. Hänsch, P. Lemonde, G. Santarelli, M. Abgrall, P. Laurent, C. Salomon, A. Clairon, Phys. Rev. Lett. **84**, 5496 (2000)
2. See Francois Biraben and Lucile Julien in this volume
3. S.G. Karshenboim, Can. J. Phys. **78**, 639 (2000); also in this volume
4. P.A.M. Dirac, Nature (London) **139**, 323 (1937)
5. The first harmonic frequency chain was started in 1967 and grew successively to higher frequencies. By 1970 it reached 30 THz (CO_2 laser) and by 1973 it ended at 88 THz (methane-stabilized HeNe): K.M. Evenson, J.S. Wells, F.R. Petersen, B.L. Danielson, G.W. Day, Appl. Phys. Lett. **22**, 192 (1973)
6. Th. Udem, A. Huber, B. Gross, J. Reichert, M. Prevedelli, M. Weitz, T.W. Hänsch, Phys. Rev. Lett. **3**, 66 (1988)
7. H. Schnatz, B. Lipphardt, J. Helmcke, F. Riehle, G. Zinner, Phys. Rev. Lett. **71**, 18 (1997)
8. T.W. Hänsch, High Resolution Spectroscopy of Hydrogen, in *The Hydrogen Atom*, ed. by G.F. Bassani, M. Inguscio, T.W. Hänsch (Springer, Heidelberg 1989) pp. 93–102
9. Th. Udem, J. Reichert, R. Holzwarth, M. Niering, M. Weitz, T.W. Hänsch, Measuring the Frequency of Light with Mode-Locked Lasers, in *Advanced Techniques for Frequency Measurement and Control*, ed. by A. Luiten (Springer, Heidelberg 2000) pp. 275–294
10. A.N. Luiten, R.P. Kovacich, J.J. McFerran, IEEE Trans. Instrum. Meas. **48**, 558 (1999)
11. T. Ikegami, S. Slyusarev, S. Ohshima, E. Sakuma, CW Optical Parametric Oscillator for Optical Frequency Measurement, in *Proc. 5th Sympos. on Frequency Standards and Metrology*, ed. J.C. Bergquist (World Scientific, Singapore 1996) pp. 333–338
12. K. Nakagawa, M. Kourogi, M. Ohtsu, Appl. Phys. B **57**, 425 (1993)
13. N.C. Wong, Opt. Lett. **17**, 1155 (1992)
14. K. Imai, M. Kourogi, M. Ohtsu, IEEE J. Quantum Electron. **34**, 54 (1998)
15. C.E. Wieman, T.W. Hänsch, Phys. Rev. Lett. **36**, 1170 (1976)
16. R. Teets, J.N. Eckstein, T.W. Hänsch, Phys. Rev. Lett. **38**, 760 (1977)
17. Y. Baklanov, V.P. Chebotayev, Appl. Phys. **12**, 97 (1977)
18. J.N. Eckstein, A.I. Ferguson, T.W. Hänsch, Phys. Rev. Lett. **40**, 847 (1978)
19. S.R. Bramwell, D.M. Kane, A.I. Ferguson, Opt. Commun. **61**, 87 (1987), Opt. Lett. **12**, 666 (1987)
20. R.L. Fork, B.I. Greene, C.V. Shank, Appl. Phys. Lett. **38**, 671 (1981)
21. W.J. Wadsworth, J.C. Knight, A. Ortigosa-Blanch, J. Arriaga, E. Silvestre, P.St.J. Russell: Electron. Lett. **36**, 53 2000
22. J.K. Ranka, R.S. Windeler, A.J. Stentz: Opt. Lett. **25**, 25 (2000)
23. J. Reichert, M. Niering, R. Holzwarth, M. Weitz, Th. Udem, T.W. Hänsch: Phys. Rev. Lett. **84**, 3232 (2000)
24. S.A. Diddams, D.J. Jones, J. Ye, S.T. Cundiff, J.L. Hall J.K. Ranka, R.S. Windeler, R. Holzwarth, Th. Udem, T.W. Hänsch: Phys. Rev. Lett. **84**, 5102 (2000)
25. D.J. Jones, S.A. Diddams, J.K. Ranka, A. Stentz, R.S. Windeler, J.L. Hall, S.T. Cundiff: Science **288**, 635 (2000)

26. R. Holzwarth, Th. Udem, T.W. Hänsch, J .C. Knight, W.J. Wadsworth, P.St.J. Russell: Phys. Rev. Lett. **85**, 2264 (2000)
27. Th. Becker, J. v. Zanthier, A.Yu. Nevsky, Ch. Schwedes, M.N. Skvortsov, H. Walther, E. Peik: Phys. Rev. A **63**, 051802(R) (2001)
28. Th. Udem, S.A. Diddams, K.R. Vogel, C.W. Oates, E.A. Curtis, W.D. Lee, W.M. Itano, R.E. Drullinger, J.C. Bergquist, L. Hollberg, Phys. Rev. Lett. **86**, 4996 (2001)
29. J. Stenger, Ch. Tamm, N. Haverkamp, S. Weyers, H.R. Telle: Opt. Lett. in press and physics/0103040
30. A. Madej (NRC), P. Gill (NPL), private communication
31. R.F.C. Vessot, M.W. Levine, E.M. Mattison, E.L. Blomberg, T.E. Hoffmann, G.U. Nystrom, B.F. Farrell, R. Decher, P.B. Eby, C.R. Baugher, J.W. Watts, D.L. Teuber, F.D. Wills: Phys. Rev. Lett. **45**, 2081 (1980)

Towards an Optical Hydrogen Clock

Lorenz Willmann and Daniel Kleppner

The special role of hydrogen in science has been recognized ever since Lavoisier identified the chemical properties of hydrogen in the eighteenth century. The names associated with the history of hydrogen are familiar: Balmer, whose empirical formula had startling predictive power; Bohr, whose early papers on hydrogen pointed the way to the creation of quantum mechanics; Dirac, whose relativistic quantum theory confounded the enigma of the fine structure of hydrogen; Rabi and Lamb, whose measurements of the hyperfine and fine structure of hydrogen pointed to the relativistic quantum electrodynamics of Schwinger, Feynman and Tomanga.

Hydrogen has also been a generating force for new experimental techniques: the hydrogen fine structure was discovered by Michelson in his quest for better metrological standards; the hydrogen hyperfine structure played a crucial role in the development of radio astronomy and became a primary tool for astrophysicists; and the hydrogen maser has played a continuing role in the development of atomic clocks and is an integral element in the global positioning system.

In the early 1970s spectroscopy was revolutionized by the creation of laser techniques. Theodor W. Hänsch helped to start this revolution and continues to be a leader. For over three decades the spectroscopy of hydrogen, particularly two-photon spectroscopy of the $1S$–$2S$ transition, has been a continuing theme in his research. During these decades the accuracy of that measurement increased by about one million. There is every reason to believe that the improvements will continue, inspired by Hänsch's accomplishments and enabled by the techniques of optical frequency metrology that he recently developed. We describe here some recent developments made possible by the creation of ultracold hydrogen that can lead to yet higher spectroscopic precision and by employing Hänsch's technique, possibly the development of a hydrogen optical atomic clock.

1 Research on Trapped Hydrogen

The history of cooling and trapping of hydrogen goes back some 25 years and is discussed in several articles [1]. The primary goal – the search for Bose–Einstein condensation (BEC) in hydrogen – was achieved in 1998 [2]. However, this quest has also laid the foundations for ultrahigh resolution spectroscopy of trapped hydrogen with applications to atomic theory – including both fundamental structure and atomic interactions – and to a possible optical frequency standard.

The work on the two-photon spectroscopy of the $1S$–$2S$ transition in trapped hydrogen commenced in 1989 and some years later the first signal was reported [3]. Since then the Doppler-free as well as the Doppler-sensitive signals have provided a wealth of information on the trapped gas [4] which culminated in a unique technique for detecting the Bose–Einstein phase transition [2]. In addition, the research demonstrated high densities of trapped metastable $2S$ atoms [5], which are well suited for spectroscopic studies.

The "eyes" of two-photon spectroscopy revealed the first information about a trapped cloud of cool hydrogen through the characteristic cusp-shape resonance line [3, 6]. The width is determined by the transit time of an atom through the laser beam – the lower the temperature, the narrower the linewidth. In case of a harmonic trap the atoms repeatedly cross the laser beam with the trap frequency. This results in a Ramsey-like structure, adding sidebands to the spectrum [6].

At very low temperatures the effect of interactions between the atoms becomes apparent in a shift of the transition frequency. This rather small shift, known as the cold collision frequency shift, is important in precision frequency measurements and it had been observed previously in the hydrogen maser [7] and atomic fountain clock [8, 9] experiments. From the point of view of precision measurements, the cold collision shift is a nuisance, since it limits the density that can be usefully employed. However, in BEC experiments, the shift is helpful, since it provides a diagnostic for the density.

2 High-Resolution Spectroscopy of Ultracold Hydrogen

Interest in the spectroscopy of ultracold hydrogen primarily lies in two directions. The first is in a precise determination of the $1S$–$2S$ transition frequency. The second consists of numerous studies made possible by exciting metastable ultracold $2S$ atoms to higher states.

The $1S$–$2S$ transition frequency is so well known from Hänsch's work [10] that further precision at this time will not yield a better value for the Lamb shift and nuclear shape corrections. Nevertheless, further precision is desirable both for advancing the frontier of optical metrology, and because the transition has potential applications for an optical frequency standard. The experimental resolution that has been achieved with a cooled atomic beam is far short of the ultimate resolution permitted by the natural linewidth. Thus, major improvements are possible.

The advantage of ultracold trapped hydrogen is that one may be able to achieve a coherence time comparable with the natural lifetime, 122 ms. The major source of broadening – Zeeman, AC Stark shifts and photoionisation – appear to be controllable on the level of the natural linewidth [11].

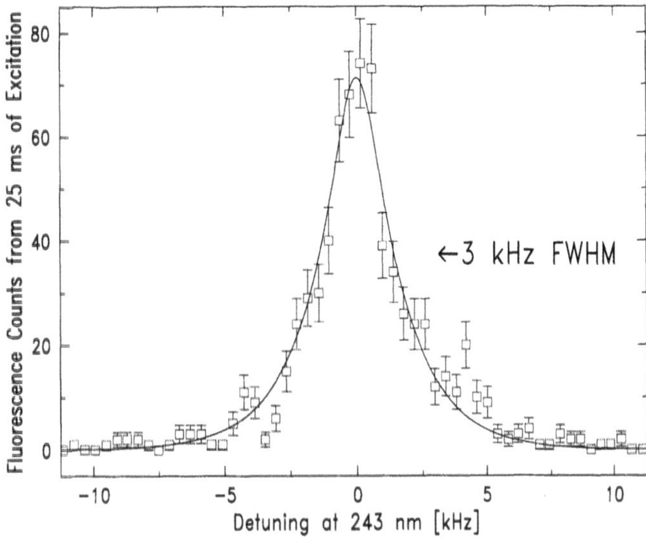

Fig. 1. $1S$–$2S$ two-photon spectrum of ultracold hydrogen, 25 ms laser excitation per point, 50% duty cycle. The resolution in limited by the stability of the laser. The *fitted line* is a convolution of the laser line width and the cusp line shape

Currently the resolution in our work is limited by the frequency jitter of the 243 nm laser source. Nevertheless, we have observed lines as narrow as 3 kHz (FWHM). Already, the signal rate is high enough to determine the center frequency to a fractional accuracy of about 5×10^{-14}, with 1 s integration time (Fig. 1).

Ultrahigh resolution spectroscopy requires ultrastable lasers. Fortunately, there has been major progress in this area. A laser locked to an external reference cavity [12] has yielded a resolution of a few parts in 10^{15} for an Hg^{+} ion in a trap [13].

2.1 Cold Collision Frequency Shift

Interactions between neighboring atoms shift and broaden the line. This can be described from a many-body picture as a result of the mean field energy shift, $\Delta E_{\mathrm{e}} = (4\pi\hbar^2 a_{\mathrm{e,g}}/m)\, n_{\mathrm{g}}$, where $a_{\mathrm{e,g}}$ is the s-wave scattering length of atoms in state e and g, m is the atomic mass, and n_{g} is the density of g-state atoms. From an atomic standpoint this results in a cold collision frequency shift which is the sum of the mean field shifts of the atomic states involved. For the hydrogen $1S$–$2S$ transition this is simple because the most of the atoms remain in the $1S$ state.

For hydrogen, the $1S$–$1S$ interaction is extremely small and the $1S$–$2S$ interaction is the dominant source of the observed frequency shift (Fig. 2). We have measured a shift of $3.8(8) \times 10^{-10}$ Hz/cm^{-3} [4]. This measurement

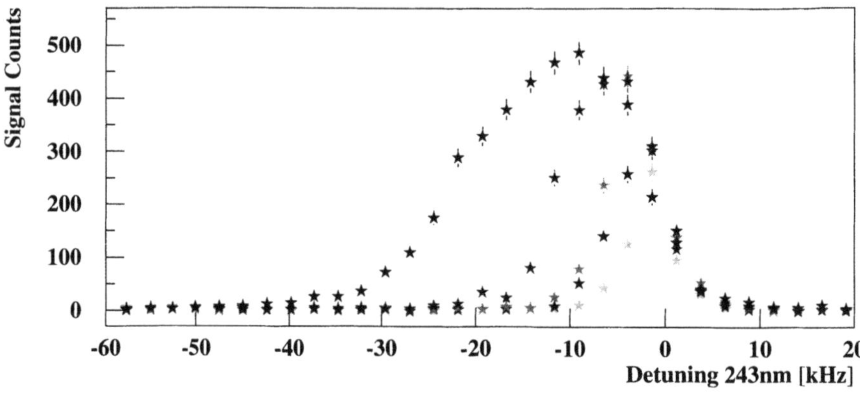

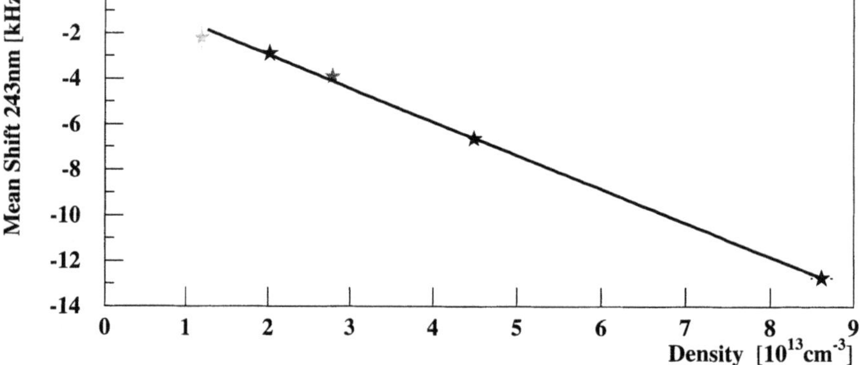

Fig. 2. Cold collision frequency shift observed in the spectra of a single 60 μK sample with initial maximum density of $8.8 \times 10^{13}\,\mathrm{cm}^3$. The farthest red-shifted spectrum corresponds to the largest density. The spectrum at the lowest density shows the characteristic double-exponential line shape

is in fair agreement with a theoretical calculation of $a_{1S\text{–}2S} = -2.3\,\mathrm{nm}$ [14] or a more recent one of $-3.0\,\mathrm{nm}$ [15]. We use this shift for measuring the density of the sample on our way to the Bose–Einstein phase transition.

2.2 $2S$–nS Transitions

The spectrum of hydrogen is composed of a gross structure, determined by the Rydberg constant, Lamb shift and other QED corrections and, finally, the nuclear shape contributions. More than two transitions must be measured to unscramble these contributions. The $1S$–$2S$ transition, which is most sensitive to the nuclear effect and was measured by the Munich group, provides one of these. However, the ultimate precision is limited by the second transition, currently one of the two-photon $2S$–nS/nD, $n = 8, 10, 12$, transitions that have been studied extensively by the Paris group [16]. These experiments employ a metastable hydrogen beam. The best accuracy achieved for one of these transitions is 8×10^{-12} or 5 kHz. Cold trapped hydrogen holds the potential for improving the accuracy of two-photon $2S$–nS transitions by an order of magnitude.

The precision of the metastable atomic beam experiments is fundamentally determined by the short interaction time of the metastable atoms. For a beam of length 0.6 m and a mean velocity of the atoms of 3000 m/s the interaction time is only about 200 μs. An efficient excitation rate requires laser intensities of typically 5 kW/cm^2. This intensity causes AC-Stark shifts of a few hundred kHz. The accuracy of the measurement is limited by the uncertainty in this shift.

Because ultracold trapped $2S$ atoms can interact with laser light for extended times, laser intensities as low as 100 W/cm^2 are sufficient to drive the two-photon transition. Thus, the primary systematic effect of the beam experiments is greatly reduced.

The cold collision frequency shifts of $2S$–nS transitions are a potential source of uncertainty. There are no theoretical predictions, and so they will have to be measured. However, if the scattering lengths are comparable to the $1S$–$2S$ scattering length, the cold collision shift will not be a limiting factor.

The starting point for the proposed experiment is a cloud of cold $2S$ atoms. The numbers for this look favorable, for more than 10^{10} $2S$ atoms/s have been produced in the experiments at MIT. A measurement of the metastable lifetime sets an upper limit on residual electric fields in the trap of 20 mV/cm, which gives negligible broadening.

One of the many possible configurations for the frequency measurement is depicted in Fig. 3. A frequency-doubled diode laser at 972 nm is locked to the dye laser at 486 nm, which is the primary laser for driving the $1S$–$2S$ transition. A frequency comb generated by a mode-locked laser is used to measure the frequency difference between the 972 nm diode laser and the 759 nm laser needed for the $2S$–$10S$ transition. Note that this experiment provides its own frequency standard, for the $1S$–$2S$ transition serves as the optical frequency reference.

This simple approach to the frequency measurement should allow us to observe other $2S$–nS transitions for $n = 3$ and higher. Measuring a series of these transitions should provide sensitive cross-checks.

In addition to precision frequency metrology, $2S$–$nS/P/D$ spectra can provide a wealth of information about scattering lengths for excited atoms. Metastable collision processes, and photoassociation processes, should also be observable. In summary, there are lots of scientific opportunities for trapped ultracold hydrogen.

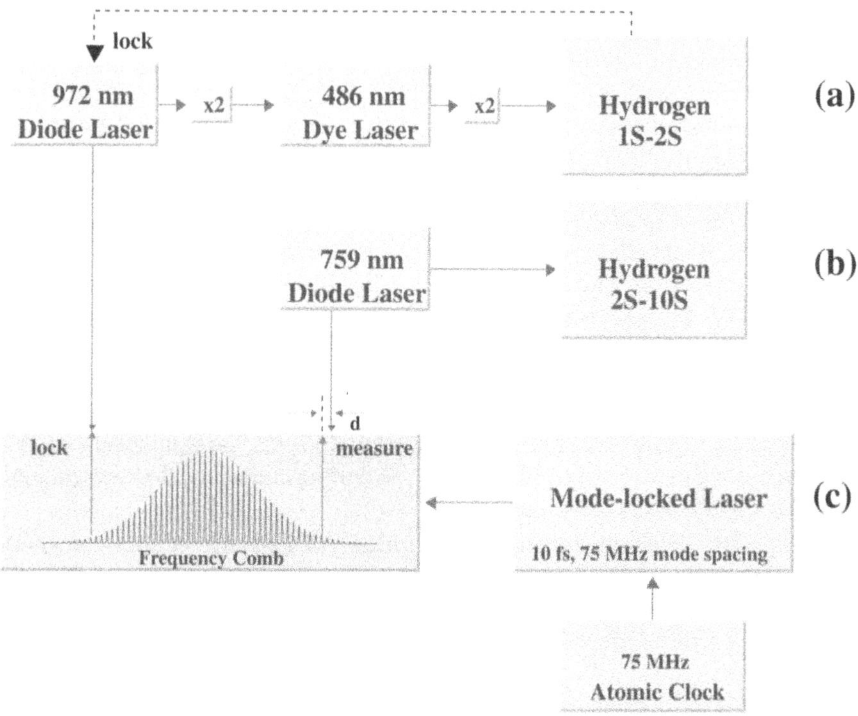

Fig. 3a–c. Possible setup for a frequency measurement of the $2S$–$10S$ transition relative to the $1S$–$2S$ transition utilizing the new developments with frequency comb generation by mode-locked lasers. The measurements are done at the same time in the same trap. The $1S$–$2S$ transition is used as the frequency reference (**a**). The $2S$–$10S$ transition is driven by a diode laser (**b**). The frequency difference between $\nu_{1S\text{-}2S}/8$ and $\nu_{2S\text{-}10S}/2$ is measured with the help of an optical comb. The scheme can be applied to other $2S$–nS transitions as well

3 An Optical Hydrogen Frequency Standard

The intrinsic properties of the $1S$–$2S$ transition makes it a suitable candidate for an optical frequency standard. The advantages are the natural linewidth, the insensitivity to magnetic fields and the Doppler-free two-photon spectroscopy. The results from the ultracold hydrogen work demonstrate the feasibility of an optical hydrogen clock.

The rapid development in the field of frequency metrology by the introduction of frequency combs generated by mode-locked lasers not only opens new possibilities for optical frequency measurements but makes it possible to generate a microwave frequency with the stability and precision of an optical frequency [17]. Recent experiments have shown that the spectrum of a mode locked laser can span more than an octave [18]. The second harmonic of the low-frequency side of the comb can be locked to the high-frequency side. Locking one mode of such a comb to a stable optical frequency not only determines the absolute frequency of all the modes but also transfers the stability of the optical transition to the repetition rate of the comb, which can be chosen around 1 GHz.

It is now recognized that the cold collision frequency shift is a crucial issue for every high-precision atomic frequency standard, microwave or optical. For hydrogen at a density of 10^9 cm^{-3} the shift of the $1S$–$2S$ transition is about 0.4 Hz [4], or a fractional shift of 1.7×10^{-16}. A rubidium fountain clock operated at a typical density of 3×10^6 cm^{-3} has the same fractional shift [19, 20].

3.1 The Ultimate Precision: a Two-Photon Laser Clock

The essence of an atomic clock or frequency standard is for close to ideal observation of an atomic resonance with intrinsically high resolution. To realize the full potential of optical clocks with accuracy in the range of 1 in 10^{18}, optical or ultraviolet laser sources with short-term stability in the mHz or sub-mHz range are required. This presents a formidable challenge. The challenge could be eliminated, however, if the optical standard is itself active rather than passive, that is, if the standard itself is a laser. The continuing usefulness of the hydrogen maser in the microwave regime proves the advantage of an active device.

In principle, a two-photon laser operating on the $2S$–$1S$ transition in hydrogen could serve as an optical frequency standard. Its natural linewidth is attractive for this purpose. The *bête noir* of two-photon lasers is spontaneous one-photon radiation to an intermediate state. However, for hydrogen there is only one such state, $2P_{1/2}$, and the transition frequency is so low that spontaneous emission is negligible.

The development of techniques for creating large populations of $2S$ atoms in ultracold hydrogen makes it worthwhile to examine the possibilities of realizing a two-photon laser. We present here an order of magnitude estimate for the operation of such a device.

We suppose that a complete population inversion of $2S$ atoms is maintained, with a density of n_{2S}. The power radiated is

$$P_{\mathrm{rad}} = \hbar\omega \Gamma_{\mathrm{rad}} n_{2S} V \,, \tag{1}$$

where Γ_{rad} is the radiation rate and V is the volume of the system. The radiation rate for the $1S$–$2S$ transition by two-photon absorption in a standing wave is [21]

$$\Gamma = \frac{85.7}{\Delta\omega} I^2 \mathrm{cm}^4\,\mathrm{W}^{-2}\,\mathrm{s}^{-2} \tag{2}$$

where $\Delta\omega$ is the homogeneous linewidth. Taking $\Delta\omega$ to be the natural decay rate, $\gamma_{2s} = 8.2\,\mathrm{s}^{-1}$, then the transition is saturated ($\Gamma_{\mathrm{rad}} = \gamma_{2s}$) for $I_{\mathrm{sat}} = 0.89\,\mathrm{W/cm}^2$.

We consider a resonator of area A, length L and finesse F. Then, the radiation intensity is $I_{\mathrm{rad}} = P_{\mathrm{rad}} F/A$. By taking $I_{\mathrm{rad}} = I_{\mathrm{sat}}$, and $\Gamma_{\mathrm{rad}} = \gamma_{2s}$, we obtain

$$n_{\mathrm{o}} = \frac{0.89\,\mathrm{W/cm}^2}{\hbar\,\omega\,\gamma_{2s}\,L\,F}\,. \tag{3}$$

Inserting numerical values, and taking the rather optimistic values $F = 100$ and $L = 100\,\mathrm{cm}$, we obtain $n_{\mathrm{o}} = 1.5 \times 10^{15}\,\mathrm{cm}^{-3}$. Densities greater than this have been achieved in Bose condensates in the $1S$ state, and one might think of adiabatically inverting the population to achieve laser action. However, such a density of $2S$ atoms would decay almost instantaneously due to quenching collisions. For instance, recent calculations suggest a quenching rate of about $10^{-9}\,\mathrm{cm}^3/\mathrm{s}$ [22]. Measurements under way in our laboratory are consistent with this [5]. At a density of $10^{15}\,\mathrm{cm}^{-3}$, the $2S$ lifetime would be catastrophically short, about one microsecond.

To achieve two-photon laser action, the product LF would need to be increased by a factor of approximately 10^6. Alas, the prospects for this are remote, and the two-photon hydrogen laser must remain no more than a dream for the foreseeable future. Nevertheless, although the future for a two-photon hydrogen laser looks dim, the prospects for a hydrogen optical clock appear to be bright.

Acknowledgements

This work has been supported by the Office of Naval Research, the National Science Foundation and the Center for Ultracold Atoms at MIT and Harvard.

References

1. For reviews of the early work on spin-polarized hydrogen see T.J. Greytak, D. Kleppner, in *New Trends in Atomic Physics*; ed. by G. Grynberg, R. Stora (North-Holland, Amsterdam 1984); I.F. Silvera, J.T.M. Walraven, in *Progress in Low Temperature Physics*, Vol. X, ed. by D.F. Brewer (North-Holland, Amsterdam 1986) Vol. X; J.T.M. Walraven, in *Quantum Dynamics of Simple Systems*, ed. by G.-L. Oppo, S.M. Barnett, E. Riis, M. Wilkinson (Institute of Physics Publishing, Bristol 1994)
2. D.G. Fried, T.C. Killian, L. Willmann, D. Landhuis, S.C. Moss, D. Kleppner, T.J. Greytak, Phys. Rev. Lett. **81**, 3811 (1998)
3. C.L. Cesar, D.G. Fried, T.C. Killian, A.D. Polcyn, J.C. Sandberg, I.A. Yu, T.J. Greytak, D. Kleppner, J.M. Doyle, Phys. Rev. Lett. **77**, 255 (1996)
4. T.C. Killian, D.G. Fried, L. Willmann, D. Landhuis, S.C. Moss, T.J. Greytak, D. Kleppner, Phys. Rev. **81**, 3807 (1998)

5. D. Landhuis, L. Matos, S.C. Moss, J. Steinberger, K. Vant, T. Greytak, D. Kleppner, L. Willmann, to be published
6. T.C. Killian, D.G. Fried, C.L. Cesar, A.D. Polcyn, T.J. Greytak, D. Kleppner, in *Atomic Physics 15*, ed. by H.B. van Linden van den Heuvell, J.T.M. Walraven, M.W. Reynolds (World Scientific, Singapore 1996) p. 158
7. B.J. Verhaar, J.M.V.A. Koelman, H.T.C. Stoof, O.J. Luiten, Phys. Rev. A **35**, 3825 (1987)
8. K. Gibble, S, Chu, Phys. Rev. Lett. **70**, 1771 (1993)
9. S.J.J.M.F. Kokkelmans, B.J. Verhaar, K. Gibble, D.J. Heinzen, Phys. Rev. A **56**, R4389 (1997)
10. For latest report on the Munich cold hydrogen beam experiment see, M. Niering, R. Holzwarth, J. Reichert, P. Pokasov, Th. Udem, M. Weitz, T.W. Hänsch, P. Lemonde, G. Santarelli, M. Abgrall, P. Laurent, C. Salomon, A. Clairon, Phys. Rev. Lett. **84**, 5496 (2000)
11. D. Kleppner, in Proc. Symposium *The Hydrogen Atom*, ed. by G.F. Bassani, M. Inguscio, T.W. Hänsch (Springer, Heidelberg 1989)
12. B.C. Young, F.C. Cruz, W.M. Itano, J.C. Bergquist, Phys. Rev. Lett. **82**, 3799 (1999)
13. R.J. Rafac, B.C. Young, J.A. Beall, W.M. Itano, D.J. Wineland, J.C. Bergquist, Phys. Rev. Lett. **85**, 2462 (2000)
14. M. Jamieson, A. Dalgarno, J.M. Doyle, Mol. Phys. **87**, 817 (1996)
15. T. Orlikowski, G. Staszewska, L. Wolniewicz, Mol. Phys. **96**, 1445 (1999)
16. C. Schwob, L. Jozefowski, B. de Beauvoir, L. Hilico, F. Biraben, O. Acef, A. Clairon, Phys. Rev. Lett. **82**, 4960 (1999), references therein
17. S.A. Diddams, T. Udem, J.C. Bergquist, E.A. Curtis, R.E. Drullinger, L. Hollberg, W.M. Itano, W.D. Lee, C.W. Oates, K.R. Vogel, D.J. Wineland, Science **293**, 825 (2001)
18. S.A. Diddams, D.J. Jones, J. Ye, S.T. Cundiff, J.L. Hall, J.K. Ranka, R.S. Windeler, R. Holzwarth, T. Udem, T.W. Hänsch, Phys. Rev. Lett. **84**, 5102 (2000); R. Holzwarth, T. Udem, T.W. Hänsch, J.C. Knight, W.J. Wadsworth, P.S.J. Russell, Phys. Rev. Lett. **85**, 2264 (2000)
19. C. Fertig, K. Gibble, Phys. Rev. Lett. **85**, 1622 (2000)
20. Y. Sortais, S. Bize, C. Nicolas, A. Clairon, C. Solomon, C. Wiliams, Phys. Rev. Lett. **85**, 3117 (2000)
21. F. Bassani, J.J. Forney, A. Quattopani, Phys. Rev. Lett. **39**, 1070 (1977)
22. Jonsell et al., private communication

Methane Frequency Standard for Precision Measurements

Sergey N. Bagayev and Alexander K. Dmitriyev

1 Introduction

The first experiments on the frequency stabilization of a He–Ne laser at 3.39 μm using nonlinear resonance at the $F_2^{(2)}P(7)\nu_3$ methane line showed that this system is promising [1, 2]. High frequency stability and reproducibility arouse interest in the use of this laser as an optical frequency standard in various precision measurement schemes.

Prof. Hänsch started the experiments on the frequency measurements of the 1S–2S transition of hydrogen atoms and more precise determinations of the Rydberg constant at Stanford University. In the mid-1980s he continued these investigations at the Max-Planck Institute for Quantum Optics. At that time, the relative error of measurements of the frequency of the 1S–2S transition of the hydrogen atom was 6×10^{-10} [3]. Prof. Hänsch evaluated the possibility of increasing the accuracy of the frequency measurement of the 1S–2S transition of the hydrogen atom up to $10^{-13}-10^{-14}$. At the time these predictions seemed too audacious. However, with time and after diligent work, his estimates were confirmed. In order to increase the accuracy of the experiments, it was necessary to modernize the entire setup and to increase the frequency stability and reproducibility of the He–Ne/CH_4 laser to an accuracy better than 10^{-14}. The laser served as a reference in the new measurement scheme. Scientists at the Institute of Laser Physics in Novosibirsk took upon themselves this task.

2 Principles of Operation and Parameters of the Frequency Standard

In the general case, in order to achieve high frequency stability and reproducibility of the laser standard, it is necessary to obtain a narrow optical resonance. For this, optical selection of cold particles in a low-pressure gas was used [4, 5]. For a small saturation parameter ($\kappa < 1$) of the absorbing medium, one can detect nonlinear resonance with the homogeneous half-width Γ determined only by collision broadening.

To obtain a small saturated parameter, the frequency modulation mode [6] was used. An additional contribution to the resonance amplitude is given by crossing resonances [7], which permits an increase in the signal/noise ratio.

To obtain simultaneously a narrow radiative line, high long-term stability, and frequency reproducibility of the laser frequency standard, we used the conventional scheme [8] consisting of three lasers (Fig. 1). All the lasers were mounted on a frame 1.5 m long. The high-voltage power supply of the lasers had a stability not worse than 10^{-3}.

A narrow lasing line and short-term frequency stability of the transportable optical standard were provided by reference laser 1, whose cavity was formed from a dense spherical mirror ($R \cong 70\,\mathrm{m}$) and a flat mirror (transmission coefficient $T = 10\%$). The mirrors were fixed on piezoceramics (PC). The methane pressure in the absorbing cell was about 2×10^{-3} Torr. Frequency stabilization was derived from the resonance with a half-width of about 50 kHz and with an amplitude and contrast of 10 μW and 5%, respectively. To detect the radiation of laser 1 and all the other lasers, we used an InSb photoresistor (D) cooled by liquid hydrogen. The automatic frequency-control system (AFC1) of the reference laser operated with the signal of the first harmonic at a frequency of 15 kHz. The noise in the servo-loop with a regulation band of 3 kHz at a modulation frequency of 15 kHz was mostly caused by the noise of the photodetector. The static gain coefficient reached a value of 10^9.

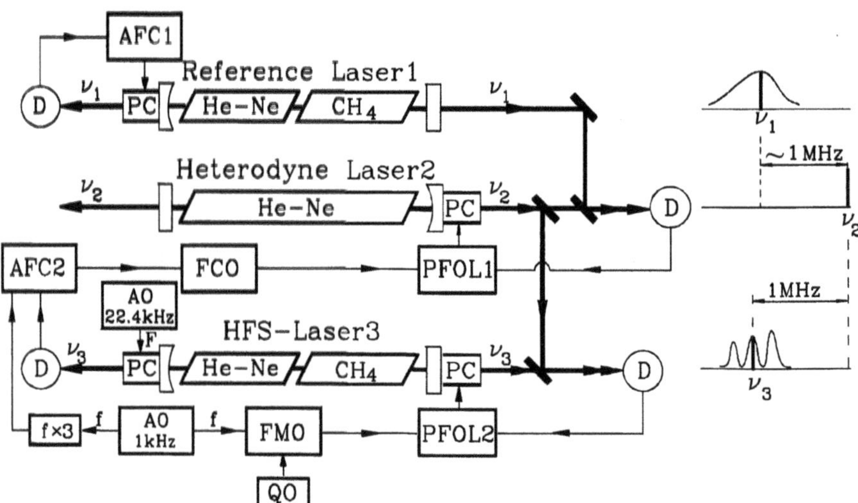

Fig. 1. Scheme of the laser frequency standard: reference laser 1 (frequency ν_1), heterodyne laser 2 (frequency ν_2), HFS-laser 3 (frequency ν_3), automatic frequency control systems (AFC1) and (AFC2), phase–frequency offset-lock systems (PFOL1) and (PFOL2), photodetectors (D), piezoceramics (PC), audio oscillators (AO 22.4 kHz) and (AO 1 kHz), quartz oscillator (QO), frequency-modulated oscillator (FMO), frequency-controlled oscillator (FCO)

The phase locking of the frequencies of heterodyne laser 2 and reference laser 1 was achieved using the phase–frequency offset-lock system (PFOL1) that makes it possible to transfer the frequency stability characteristics of laser 1 to laser 2. The regulation band in the PFOL1 loop was about 5 kHz. On the one hand, this allowed us to suppress completely acoustic, mechanical, and thermal perturbations. On the other hand, this made it possible "to cut off" the modulated part of the spectrum of laser 1 and to obtain monochromatic radiation of the heterodyne laser without side spectral components. The output power of the frequency standard was determined by heterodyne laser 2 and exceeded 1 mW.

Laser 3 provided the resolution of the hyperfine structure (HFS) of this line. The frequency of laser 3 was stabilized over the central 7–6 transition of the $F_2^{(2)}P(7)\nu_3$ methane line. The cavity of HFS-laser 3 was the same as that of laser 1, with only one difference: the curvature of the spherical mirror was 150 m and the light beam diameter in the absorption medium was 0.8 cm. The methane pressure in the absorbing cell of the HFS-laser was 2×10^{-4} Torr. The frequency of the HFS-laser was tuned to the zero signal of the third harmonic of the modulation frequency of 1 kHz. The frequency modulation of the HFS-laser was carried out by an audio oscillator (AO) through the frequency-modulated oscillator (FMO) of the PFOL2-system. The frequency of the FMO setting the difference frequency between the HFS and heterodyne lasers was always equal to 1 MHz owing to phase locking to the frequency of the quartz oscillator (QO). The error signal in the radiation power of laser 3 was detected using a phase detector of the AFC2 system. Then it arrived at the frequency-controlled oscillator (FCO), varying the lock frequency of the PFOL1-system. Finally, it shifted the frequency of the heterodyne laser so that the frequency of laser 3 corresponded to the maximum of the central 7–6 component (to the zero of the third harmonic).

To obtain the maximal signal/noise ratio in laser 3, the deviation of the frequency modulation at 1 kHz was optimized. In the case of the resolved HFS $F_2^{(2)}$ methane absorption line (it is a triplet with intervals of ~ 11 kHz), the optimal deviation differs from the value obtained for a single line. Our calculations and measurements have shown that the optimal deviation practically does not depend on the resonance half-width Γ, and is ~ 7.6 kHz if the hyperfine structure is resolved, but splitting is not observed due to the recoil effect [9].

A record of the HFS signal in the regime with additional frequency modulation at $F = 22.4$ kHz and a modulation index $\beta \cong 3.1$ is shown in Fig. 2. At this modulation index, the radiation spectrum has 7 major components. The three central components are shown in Fig. 2 only. They have equal amplitudes, and this was used to control β. The signal/noise ratio obtained is much greater than in the case without the above modulation. This allowed us to realize stabilization using the central component of the hyperfine structure

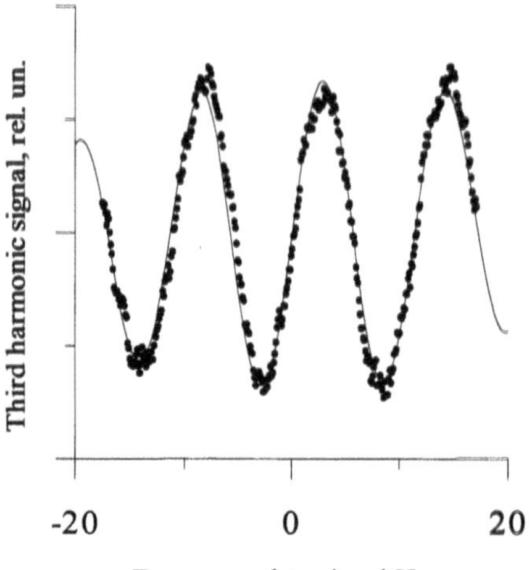

Fig. 2. An experimental record (*solid circles*) and the calculated form (*continuous line*) of the hyperfine structure of the $F_2^{(2)}$ methane line. The laser-3 output power is $\sim 30\,\mu W$

of the $F_2^{(2)}$ methane absorption line. The band of the servo-loop AFC2 was 2×10^{-3} Hz, and the static gain was about 500.

The narrow laser radiation spectrum of the reference laser-1 was obtained due to stabilization to the nonlinear resonance in methane. This spectrum did not exceed 10 Hz (Fig. 3).

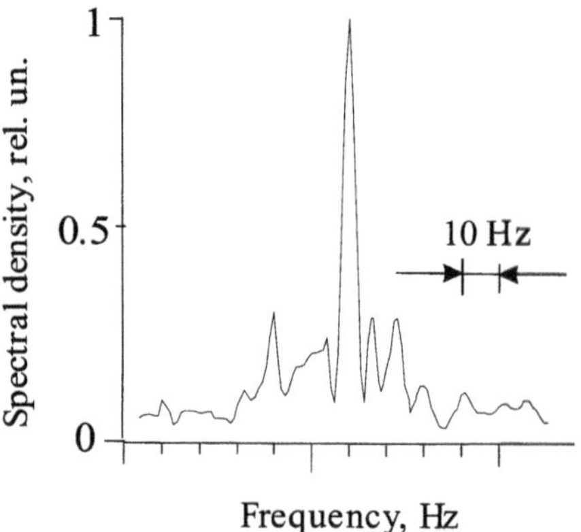

Fig. 3. An experimental record of the He–Ne/CH_4 laser standard radiation spectrum. The dependence of the spectral density as a function of frequency in Hz

To observe this we used the stationary He–Ne/CH_4 standard as a local oscillator. This standard was also used to measure the frequency stability of our transportable He–Ne/CH_4 laser.

Allan's variance is given in Fig. 4. For the averaging time $\tau < 10\,\mathrm{s}$, the experimental points were marked on the basis of the experimental dependencies obtained with the averaging time up to 10^{-2} s. The long-term frequency stability of $\sim 5 \times 10^{-15}$ of the transportable laser obtained in the experiments was at the level of the best H-masers.

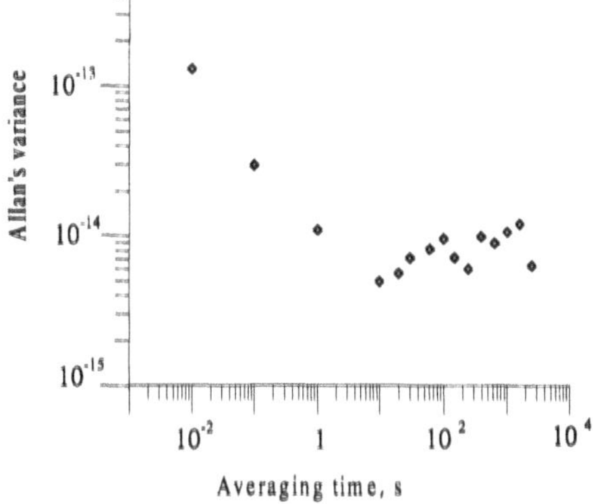

Fig. 4. Allan's variance of the transportable He–Ne/CH_4 frequency standard

The measurement of the absolute frequency value of the transportable standard was performed in PTB (Braunschweig) using the radio-optical chain [10] and the hydrogen standard previously calibrated on the cesium standard. By direct absolute frequency measurements, the frequency of the optical standard reproduced under optimal conditions was determined as follows: $\nu_2 = 88\,376\,181\,599\,937(23)\,\mathrm{Hz}$. First experiments on the measurement of the frequency of the 1S–2S transition of the hydrogen atom and the Rydberg constant performed with the help of the He–Ne/CH_4 frequency standard made it possible to decrease the error by more than an order of magnitude, up to 1.8×10^{-11} and 3.8×10^{-11}, respectively [11].

Long-term frequency measurements of the transportable He–Ne/CH_4 standard with respect to the frequency of the 7–6 transition were also carried out in Novosibirsk. The scheme of these measurements included a stationary laser spectrometer [12] and the error in the measurements of the shifts did not exceed 5 Hz. The measured frequency values of the He–Ne/CH_4 standard obtained in certain years are shown in Fig. 5 [13]. The obtained root-mean-square deviation of 18 Hz was equal to 2×10^{-13} in relative units.

A further increase in the long-term stability and reproducibility of the laser frequency standards can be obtained in the case of narrower optical references.

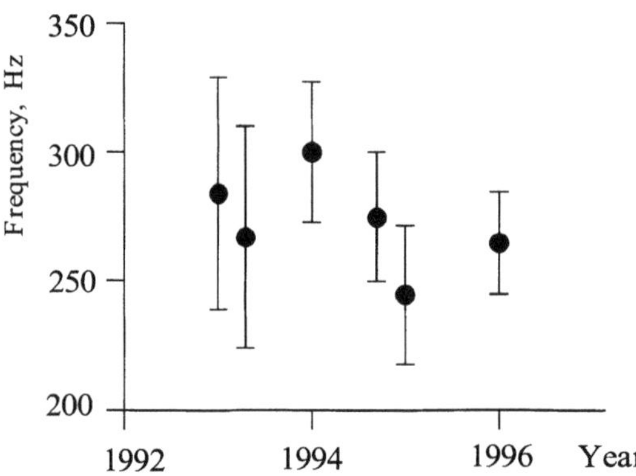

Fig. 5. Measured frequency values of the He–Ne/CH_4 standard for certain years

3 Laser Frequency Standards with an Intracavity Mirror Telescopic Light Beam Expander

A laser with a telescopic cavity is used for this purpose. The large cross-section of the light beam in the absorption cell makes it possible to decrease the transit broadening and to obtain narrow optical resonances that can be used for optical frequency standards [14].

The optical scheme of the telescopic laser is usually a four-mirror one (Fig. 6a). A short-focus mirror 2 and a long-focus mirror 3 form a telescopic beam expander. In the wide part of the light beam with a plane wave front between spherical mirror 3 and plane mirror 4 there is an absorption cell, and the gain tube occupies the narrow part between mirror 1 and short-focus mirror 2.

A shortcoming of using a telescopic expander is the influence of the angular disalignment of the cavity mirrors on the laser power. The main contribution to these fluctuations is from mirrors that touch the wide part of the light beam, namely, the third and the fourth mirrors. This effect is increased when the light section in the absorption cell is extended.

The situation is essentially changed when the W-like scheme is used (Fig. 6b) [15]. To obtain the plane wave front in the cell, mirror 4 should be located near the focus of mirror 3.

To obtain a narrow intensive resonance, it is necessary to have the largest possible cross-section of the light beam in the absorption cell proportional to $d_{12}(f_3/f_2)^2$. Here, f_i is the focusing length of the ith mirror, and d_{12} is the distance between mirror 1 and mirror 2. At the same time, as was shown experimentally and theoretically, the influence of the angular misalignment of mirror 3 and mirror 4 on the fluctuation of the laser radiation intensity in the W-like scheme is reduced by at least a factor of $f_3(d_{12}-f_2)/f_2^2$ in comparison to the V-like scheme [15]. Therefore, it will be more advantageous to use the

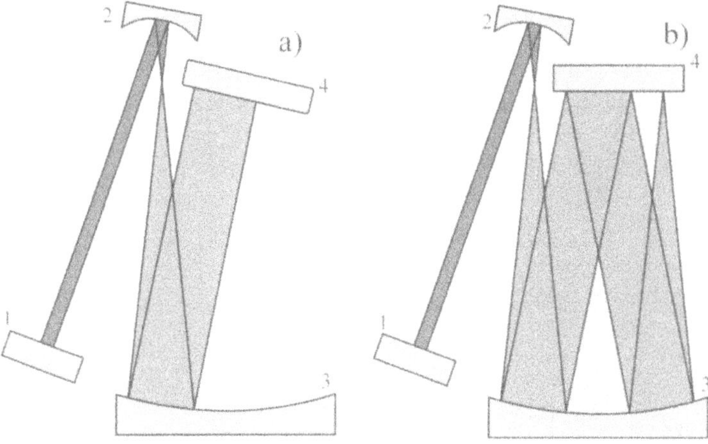

Fig. 6. Optical schemes of telescopic lasers: (**a**) V-like; (**b**) W-like

W-like scheme instead of the V-like one as the cross-section of the light beam in the absorption cell becomes wider.

An investigation was carried out using a He–Ne four-mirror laser at $\lambda = 3.39\,\mu m$. Mirror 1 and mirror 4 were plane and a telescopic expander of the light beam was formed using a mirror 2 with a focal length of 8 cm and an 80-cm mirror 3. The distance between mirror 2 and mirror 3 was equal to the sum of the focal lengths f_2 and f_3. In this case the cavity is most tolerant. The diameter of the light beam in the absorption cell was about 24 mm, and the wave front was plane. Mirror 4 was situated in the focus of mirror 3. The distance between mirror 1 and mirror 2 was equal to 80 cm. The gain tube occupied this plane. The maximum of the light beam section in the gain tube was near mirror 2, and its diameter was equal to 2.4 mm.

The nonlinear resonances obtained under the identical parameters for different schemes are shown in Fig. 7. The methane pressure in the absorption cell was equal to 100 μTorr, the output power was equal to 100 μW, the modulation frequency was equal to 169 Hz, and the deviation was equal to 1.6 kHz. With the V-like scheme (Fig. 7a), the noise is much higher than in the W-like scheme. Moreover, the resonance in the W-like scheme is higher, which is due to the double path of the light beam in the absorption cell.

The measurements of the frequency shifts of the telescopic laser stabilized by the 7–6 transition of the $F_2^{(2)}$ methane line made it possible to estimate the reproducibility at the level of 5 Hz [16].

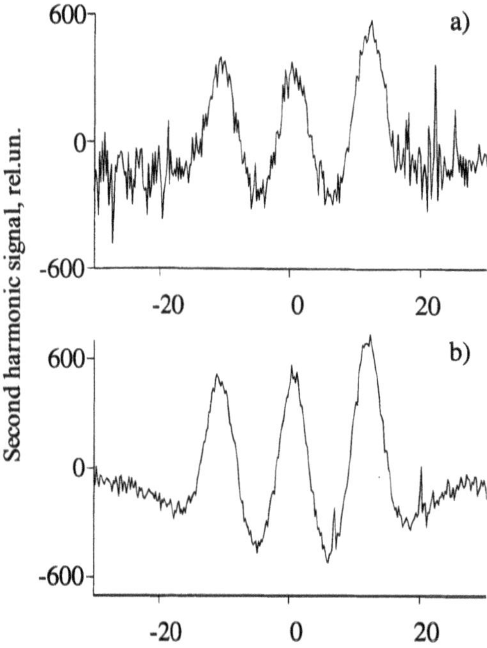

Fig. 7. A record of the saturated absorption methane resonances for different optical schemes: (**a**) V-like scheme; (**b**) W-like scheme

4 Prospects for Improving the Methane Standard

The potential for the methane standard and most other standards, is mainly limited by the second-order Doppler effect. Its influence can be reduced by both decreasing the gas temperature and using selection of cold particles. Extremely low temperatures, which can be used to observe nonlinear absorption resonances in a gas, are determined by the saturated vapor pressure. This pressure is $10^{-6}-10^{-4}$ Torr for methane at temperatures of 45 – 50 K. In such conditions, the second-order Doppler shift for molecules with the thermal velocity is less than 25 Hz, and the population of the main rotational level increases by more than an order of magnitude in comparison with the room temperature. This makes it possible to consider that recording of resonances at the R(0) transition with a transit parameter of 0.1 is quite plausible. In this case, the shift of the resonance maximum will be less than 0.25 Hz. If the temperature in the methane cell has an accuracy of 5%, the frequency reproducibility can reach 10^{-16}.

Acknowledgements

The authors would like to thank P.V. Pokasov and A.A. Lugovoy for their help in performing experiments.

References

1. R.L. Barger, J.L. Hall, Phys. Rev. Lett. **22**, 4 (1969)
2. S.N. Bagayev, A.K. Dmitriyev, Sov. J. Opt. Spectrosc. **34**, 337 (1973)
3. R.G. Beausoleit et al., Phys. Rev. A **35**, 1878 (1987)
4. S.N. Bagayev et al., Pis'ma Zh. Eksp. Teor. Fiz. **23**, 399 (1976)
5. S.N. Bagayev et al., Pis'ma Zh. Eksp. Teor. Fiz. **45**, 371 (1987)
6. G. Kramer et al., Appl. Phys. Lett. **37**, 354 (1980)
7. N.G. Basov et al., Sov. J. Quantum Electron. **14**, 866 (1987)
8. S.N. Bagayev et al., In: *Lasers Systems*, ed. by V.P. Chebotayev (Nauka, Novosibirsk 1980) (in Russian) p. 122
9. S.N. Bagayev, A.K. Dmitriyev, P.V. Pokasov, Laser Physics **7**, 4, 989 (1997)
10. C.O. Weiss et al., IEEE J. Quantum Electron. **24**, 1970 (1988)
11. T. Andreae et al., Phys. Rev. Lett. V **69**, 1923 (1992)
12. S.N. Bagayev et al., Appl. Phys. B **52**, 63 (1991)
13. S.N. Bagayev et al., in: Digest 2nd Intern. Symp. on Modern Problems of Laser Physics, Novosibirsk (Russia), July 28–August 2, 1997, PI-58
14. S.N. Bagayev, V.P. Chebotayev, Uspekhi Fizicheskikh Nauk **148**, 143 (1986)
15. S.N. Bagayev et al., Proc. 2nd Intern. Symp. on Modern Problems of Laser Physics **2**, 195 (1997)
16. S.N. Bagayev, A.K. Dmitriyev, A.A. Lugovoy, V.M. Semibalamut (in press)

The Parametric Frequency-Interval Divider

Harald R. Telle and Burghard Lipphardt

1 Introduction

During the last twenty years, two techniques have been developed to measure the frequency of visible radiation with respect to a Cs primary clock: harmonic frequency chains [1], and frequency chains operating in the difference frequency domain [2]. The latter technique has turned out to be much more flexible, since its spectral coverage can be restricted to the near-infrared and visible range, where tunable light sources are readily available. In particular, with the advent of optical frequency-comb generators based on Kerr-lens mode-locked lasers [3] this technique permits the measurement of any unknown frequency in this spectral range.

The key building blocks of the difference-frequency-domain chains are optical frequency-interval dividers which phase-coherently divide the interval between two optical signals by an integer. This integer may be a small number, two or three, in the case of a divider consisting of discrete elements [2] or of the order of 10^6 in the case of a frequency-comb with external self phase-modulation [4].

Both methods, dividers based on discrete components and comb generators, have already been combined, demonstrating that the spectral lines of such a frequency-comb generator based on femtosecond laser are equally spaced within a relative uncertainty of 3×10^{-17} [5].

One disadvantage of active optical frequency-comb generators is obvious: the method is restricted to frequency regions, where broadband laser gain media are available. In other spectral regions, one has to rely on frequency-interval divider stages based on discrete components. Usually, these schemes employ one or two fast nonlinear elements and an active electronic servoloop. A typical set-up comprises three lasers, two optical nonlinear crystals and an electronic phase-locked-loop (PLL) (see Fig. 1a).

The phase of the field E_3 emitted by laser 3 is controlled to satisfy $\nu_3 = (\nu_1 + \nu_2)/2$, where ν_1, ν_2 and ν_3 are the emission frequencies of lasers 1, 2 and 3, respectively. The same effect can be obtained by optical four wave mixing, i.e. by substitution of the two crystals with $\chi^{(2)}$-nonlinearity by one nonlinear element with strong $\chi^{(3)}$ (see Fig. 1b).

Although $\chi^{(3)}$ nonlinearities are usually small in most transparent media, they can approach considerably high values in semiconductors if inter- or

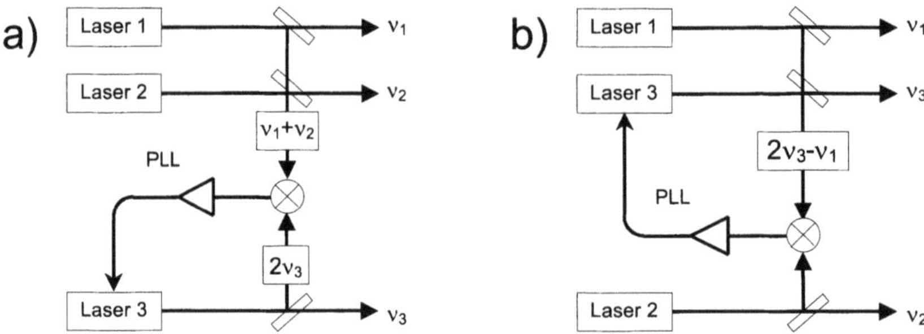

Fig. 1. Frequency interval divider using (**a**) $\chi^{(2)}$ and (**b**) $\chi^{(3)}$ nonlinearities

intra-band relaxation processes are involved. Then, frequency intervals with a width of several terahertz can be divided with this technique [6]. Furthermore, such a semiconductor can be simultaneously operated as a laser diode, i. e. as an optical oscillator. If its free-running output frequency ν_{LD} is tuned sufficiently close to the frequency ν_3 which bisects the initial frequency interval between ν_1 and ν_2, it will spontaneously phase-lock to a four-wave mixing product at $\nu_1 + \nu_2 - \nu_{\mathrm{LD}}$ [7], yielding $\nu_{\mathrm{LD}} = (\nu_1 + \nu_2)/2$. Thus, the device acts as a self-locking frequency-interval divider which does not require electronic servocontrol at all. Frequency intervals up to the sub-terahertz range have been divided with this technique [7]. However, the lock-in range of such a resonant frequency-interval divider is usually small, thus requiring additional electronic pre-stabilization of ν_{LD} for stable long-term operation.

In this paper, we introduce a novel concept that avoids the drawbacks of the methods mentioned above. It allows for broadband, self-locked interval bisection without requiring control loops or resonators. Although we demonstrate this method in the radio-frequency range for the sake of simplicity, it is straightforward to extend the method to the optical range using the optical Kerr effect in isotropic media.

2 Principle

Our approach follows the route of regenerative dividers for absolute frequencies, which is a well-known technique for frequency synthesis in the rf and microwave range [8].

In this scheme (Fig. 2a) the input signal at frequency ν is mixed with the $\nu/2$ filtered and amplified output of the same mixer yielding a signal at $\nu/2$. Thus, the self-consistency condition is satisfied for proper phase delay and sufficiently large gain of the feedback loop.

The regenerative frequency-interval bisection scheme proposed here employs two phase modulators PM1, PM2 for both input signals and a square-law detector DET (see Fig. 2b). Its operating principle can be most easily understood with the help of the self-consistency condition. Assuming a signal

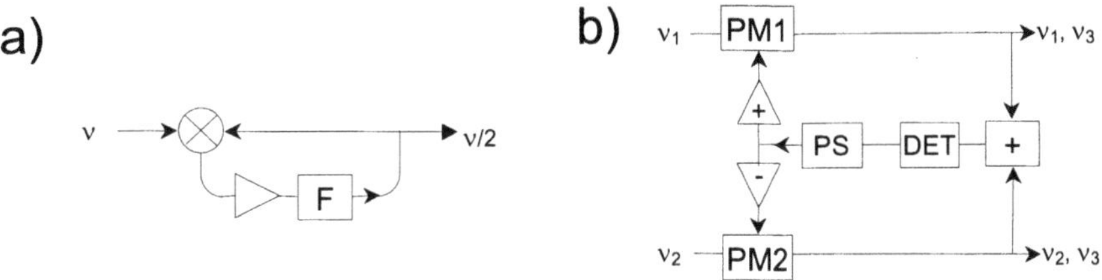

Fig. 2. Regenerative dividers for (**a**) absolute frequencies and (**b**) frequency intervals. F denotes the $\nu/2$ filter, PM the phase modulators, PS the variable phase shifter, DET the square-law detector

at $\nu_3 = (\nu_1 + \nu_2)/2$ with proper phase angle, the input signals at ν_1 and ν_2 can be considered as its AM side bands. This AM is demodulated by the square-law detector yielding a signal at $(\nu_1 - \nu_2)/2$. It is amplified and fed back to the two phase-modulators, which superimpose PM side bands onto both input signals, whose amplitude is given by the first order Bessel function $J_1(\Delta\phi)$, where $\Delta\phi$ is the modulation amplitude given by the overall loop gain. Two of these side bands coincide with the frequency of E_3, i.e. $\nu_3 = (\nu_1 + \nu_2)/2$. Thus, the self-consistency condition for the amplitude is satisfied for sufficiently high gain of the loop amplifiers. Once the threshold is surpassed, the amplitude of the bisecting signal grows until saturation sets in owing to the nonlinearity of the Bessel functions. Such type of gain saturation is known from electro-optical optical-parametric oscillators [9].

The self-consistency requirements can be visualized most easily in a frame rotating at frequency $\nu_3 = \nu_1 + \nu_2$ (see Fig. 3).

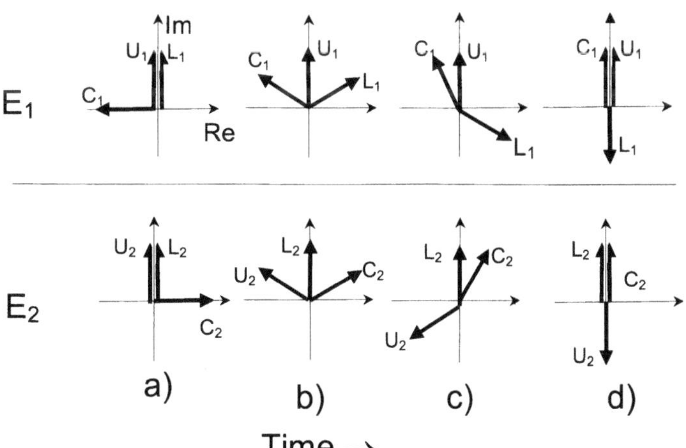

Fig. 3a–d. Parametric frequency-interval divider: phasor picture

Defining a counter-clockwise rotation as a positive frequency offset, E_1 as the lower-frequency and E_2 as the higher-frequency input signals, the phasors in Fig. 3a represent the time position of maximum phase deviation. Both the upper side band U_1 and the lower side band L_1 of E_1 are retarded with respect to the carrier C_1, i.e. E_1 shows a maximum negative phase deviation. Vice versa, U_2 and L_2 are advanced with respect to the carrier C_2 of E_2, yielding a maximum positive phase deviation. Since ν_1 is lower than ν_3, which defines the frame in Fig. 3, C_1 rotates clockwise in Fig. 3b to Fig. 3d, while C_2 rotates counter-clockwise. Consequently, the upper side band U_1 of E_1 and the lower side band L_2 of E_2 remain fixed in this frame, whereas L_1 and U_2 rotate with twice the frequency offsets of the corresponding carriers C_1 and C_2. The square-law detector, however, sees the superposition of E_1 and E_2: U_1 and L_2 form a new carrier, while C_1 and C_2 act as its AM side bands. The maximum AM deviation occurs in Fig. 3d, i.e. at a quarter period after the maximum phase deviation (Fig. 3a). Hence, self-consistency requires phase-modulations of PM1 and PM2 which are phase shifted by $+\pi/2$ and $-\pi/2$ (or vice versa) with respect to the phase angle of the AM as demodulated by DET. Obviously, a rotation of all phasors by π in Fig. 3 also satisfies the self-consistency condition. Thus, the phasor of the bisecting signal, i.e. the sum of U_1 and L_2, shows a π-ambiguity which is characteristic for all frequency-by-two dividers.

It should be noted, that both phase modulators are required for a proper interval bisection. If only one modulator is used, the self-consistency condition is satisfied for any frequency between ν_1 and ν_2 after two round-trips. Hence, such a system is expected to generate broadband noise but no distinct line at $(\nu_1 + \nu_2)/2$.

To demonstrate the operation of the proposed parametric frequency-interval divider, we have set up an electronic version of the scheme in Fig. 2b. Two input signals with $\nu_1 = 9.9$ MHz and $\nu_2 = 10.1$ MHz were phase-modulated in counter-phase by varicap-phase-modulators. The square-law detector was approximated by a power combiner and a double-balanced mixer. The round-trip phase-shift of the loop was set to $\pi/2$ at 100 kHz with the help of an all-pass filter PS.

Fig. 4 shows the measured spectra of the signals in front of DET. In Fig. 4a the bisection threshold was not yet reached, i.e. only the two input signals at 9.9 MHz and 10.1 MHz are found. In Fig. 4b the amplifier gain was increased by 2 dB and a very weak signal appeared at the bisection frequency $\nu_3 = 10.0$ MHz. The bisection threshold was surpassed in Fig. 4c: 1 dB more gain enlarged the 10.0 MHz signal by about 20 dB. With further increased gain (13 dB in Fig. 4d), saturation sets in, thus reducing the slope efficiency of the bisecting signal. At even higher gain values, the system passed the well known route to chaotic behavior, generating subsequently higher sub-harmonic orders.

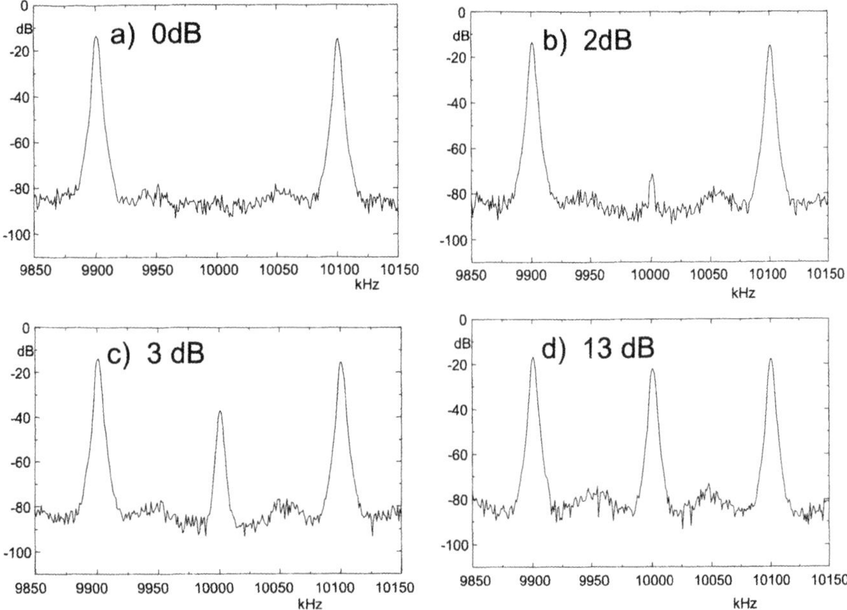

Fig. 4a–d. Parametric frequency-interval divider: experimental output spectra

3 Division of Optical Frequency Intervals

The model system presented here is of minor practical interest for the rf range, where digital frequency dividers and numerically controlled oscillators allow for much more sophisticated frequency-synthesis schemes. In the optical range, however, where those element are not available, one has to rely on simpler methods like the proposed scheme.

Although it can be extended to the optical region in several ways, the most instructive technique is depicted in Fig. 5. Both input signals, E_1 and E_2 are phase modulated during their forward transit through the Kerr phase modulator KPM by cross-phase modulation. The driving force of this modulator is the modulation of the polarization direction of the fields which are reflected by the mirror M and polarization controlled by the two retardation plates P1 and P2. Here, the loop delay time is given by the geometrical dimensions of possibly thin optical components and can be kept of the order of 1 mm or below. As a result, the response time of such an all-optical frequency divider can be much faster than an electronic device. Operation bandwidths of hundreds of GHz or even THz seem to be possible. The two essential features of a parametric interval divider,

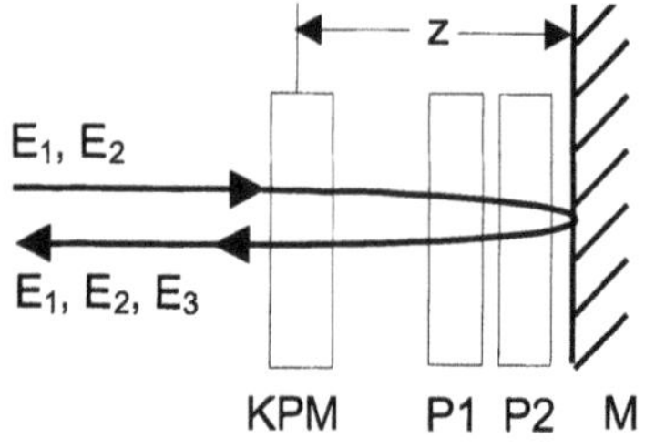

Fig. 5. Proposed optical parametric frequency-interval divider. KPM denotes the Kerr-lens phase modulator, P1 and P2 the retardation plates, and M the mirror

1. counter-phase-modulation of the two input signals, and
2. $\pi/2$ phase shift between the instantaneous phase of the phase modulation and the phase of the resulting intensity modulation,

are realized as follows: the two input signals E_1 and E_2 are orthogonally polarized, for example in x- and y-direction, respectively, as depicted in Fig. 6a. The interval-bisecting field E_3 is assumed to be already present after the Kerr modulator in the forward direction. Its polarization direction is assumed to be rotated by 45° with respect to the Cartesian coordinates of Fig. 6.

The thicknesses of the two retardation plates P1 and P2 are chosen to change the polarization of only one of the three signals, whereas the other two polarizations are left unchanged. This is possible with conventional birefringent crystals due to the dispersion of the birefringence. Estimates show that a crystal thickness of less than 1 mm is sufficient to satisfy this condition for interval widths of a few 10 THz in moderately birefringent crystals like $LiNbO_3$.

The upper part of Fig. 6 shows the processing of E_1. Plate P1 acts as a $n\lambda/4$-plate for E_1 and as a $m\lambda/2$-plate for E_2 and E_3, where n and m are integer numbers. Hence, E_2 and E_3 are not affected in a double pass. The polarization of E_1, however, is rotated due to the resulting $\lambda/2$-retardation

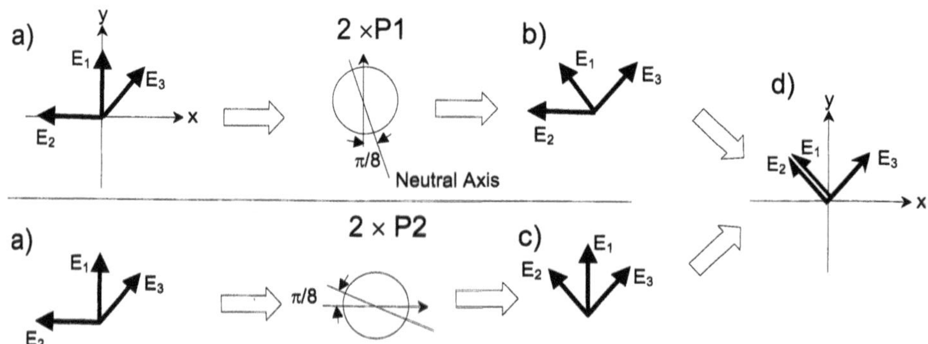

Fig. 6a–d. Proposed optical parametric frequency-interval divider: polarization management

by an angle which is twice the rotation angle α_1 of the neutral axis of P1 with respect to the input polarization. Choosing α_1 as $\pi/8$, E_1 is rotated by $\pi/4$ in Fig. 6b. The lower part of shows the same procedure for E_2. P2 has a thickness that yields a retardation of $n\lambda/4$ for ν_2 and $n\lambda/2$ for ν_1 and ν_3. It is rotated by $\alpha_2 = -\pi/8$ with respect to the coordinate axes. Thus, in a double pass, the polarization of E_2 is rotated by $-\pi/4$ (Fig. 6c). As a result of the two processes, the backward-running field comprises the two components E_1 and E_2 with a parallel polarization. It is rotated by $\pi/4$ with respect to the coordinate system. E_3 is orthogonally polarized with respect to E_1 and E_2. The superposition of all three backward-running fields leads to an AM at $\nu_1 - \nu_2$, which is not important here, and a periodic switching between linear polarization states in the x- and y-directions with a frequency of $(\nu_1 - \nu_2)/2$. This polarization-switched field drives the Kerr-effect phase modulator KPM, which may consist, for example, of a thin plate of semiconductor material, an organic crystal with strong $\chi^{(3)}$ nonlinearity or a thin cell filled with CS_2. It converts the polarization switching into a counter-phase-modulation of the orthogonally polarized input fields E_1 and E_2, since the Kerr-effect-induced phase deviation of a probe field which is parallel polarized with respect to the pump is three times larger as compared to orthogonal pump-probe polarizations (in the case of a purely electronic Kerr effect) [10]. Thus, requirement 1 mentioned above, phase-modulation in counter-phase, is satisfied. Requirement 2, $\pi/2$ phase delay of the feedback loop, can be achieved by choosing the proper separation z between the modulator KPM and the retro-mirror M.

Although we have assumed cw input fields so far, the proposed scheme can be extended to mode-locked lasers which emit a frequency-comb. Then, the generated field E_3 shows a comb spectrum as well. This might be useful, for example, to fill up the dark fringes [11] sometimes observed in the output spectrum of microstructure fibers used for spectral broadening of the comb by self-phase modulation.

4 Conclusion

To conclude, we have introduced a novel concept for the parametric division of frequency intervals. Our first electronic model system showed the expected behavior: a sharp threshold for interval bisection and low-noise operation over a dynamic range of $> 10\,\mathrm{dB}$ for both input signals. We have proposed a simple scheme to extend the concept to optical frequencies. Its compact size allows for broadband, almost non-resonant, bisection of large optical frequency intervals. This property and the sharp threshold may open up new perspectives for ultra-fast optical signal processing, gating and pulse shaping.

References

1. H. Schnatz, B. Lipphardt, J. Helmcke, F. Riehle, G. Zinner, Phys. Rev. Lett. **76**, 18 (1996)
2. H.R. Telle, D. Meschede, T.W. Hänsch, Opt. Lett. **15**, 532 (1990)
3. Th. Udem, J. Reichert, R. Holzwarth, T.W. Hänsch, Phys. Rev. Lett. **82**, 3568 (1999)
4. J. Reichert, R. Holzwarth, Th. Udem, T.W. Hänsch, Opt. Commun. **172**, 59 (1999)
5. Th. Udem, J. Reichert, R. Holzwarth, T.W. Hänsch, Opt. Lett. **24**, 881 (1999)
6. Ch. Koch, H.R. Telle, Opt. Commun. **91**, 371 (1992)
7. H.R. Telle, Ch. Koch, Proc. SPIE **1837**, 426 (1992)
8. R.L. Miller, Proc. IRE **27**, 446 (1939)
9. A. Wolf, H.R. Telle, Opt. Lett. **23**, 1775 (1998)
10. see, for example, G.P. Agrawal, in *Nonlinear Fiber Optics* (Academic Press, Boston 1989) p. 181
11. J.K. Ranka, R.S. Windeler, A.J. Stentz, Opt. Lett. **25**, 25 (2000)

External Laser Stabilization

John L. Hall

Perhaps a more public-relations-sensitive title would be "Sabbatical in a Silicon Valley Paradise." In fact, 1984 was a great year: George Orwell's view of that future did appear and it did touch the lives of a significant fraction of mankind – but only as an Apple Computer advertisement during television coverage of the Super Bowl football game. My wife Lindy was taking advanced classes in the Education Department at Stanford, exploring an idea for an internal "Peace Corps" kind of public service for young teachers. Support for college expenses could be provided for promising college students who wanted to make a difference – perhaps I mean to say, could make THE difference – by going into teaching in a public schools setting. One thus could hope to nudge the median teacher performance in a good direction, provide a reliable supply of upper-end young teaching staff, and enjoy a recurring transient of a year or two of public service by teachers in rural and/or some urban settings in the US not accustomed to participation by the "Best and Brightest."

1 The Perpetual Quest for Better Resolution and the Required Narrow Laser Linewidths

For Jan, the Spring at Stanford was a great joy, with the opportunity to interact with Prof. Theodor Hänsch, the late and beloved Prof. Arthur Schawlow, and with their many students, postdocs and visiting scholars. Theodor's hydrogen program was seriously preparing for cw spectroscopy on the 1S–2S two-photon transition. The idea was to use cavity build-up at 243 nm of the frequency-doubled output from a 486-nm frequency-stabilized dye laser. The coumarin 480 dye did not offer much gain, and it needed pumping by the very expensive violet photons from a Kr^+ ion laser. But this was the only game available. From JILA studies of locking to a stable high-finesse reference cavity using the FM sideband method, it was already possible to project success in locking a cw dye laser to the 1-Hz linewidth level of interest for hydrogen spectroscopy. But there was one serious problem: this calculation depended on having enough servo gain that the laser's intrinsic noise was reduced below the measurement noise. But how could we ever obtain such a quick response that the servo system could overpower the dye laser's intrinsic frequency noise? Using the free-flowing jetstream for the laser's active medium led to a servo speed requirement of 2–4-MHz unity gain. With less gain, there would still remain such large phase excursions at 100 – 300 kHz that residual phase excursions would usually exceed 1 radian, which would lead to phase modulation sidebands of significant strength at these frequency offsets from the nominal center frequency of the laser. If we could just get a little more bandwidth and high-frequency gain, the phase excursions would be servo-reduced well below the 1-radian level and thus would extract only a little power into their unwanted sidebands. In brief, with just a little more gain, the laser FWHM linewidth would fall from ~500 kHz to something more like 1 Hz, as set by the S/N obtained from the cavity-based frequency discriminator. It had been demonstrated earlier in work [1] with Ron Drever that this all worked fine, as just described. To get such a wide servo bandwidth, the PZT approach clearly was not going to provide a suitably fast response. Instead Ron and I used a Pockels cell with sufficiently low loss (~ 1%) that it could be used as an intra-cavity phase modulator. Changing the cavity round-trip phase is equivalent to a frequency shift of the laser cavity and hence the laser output. So just a few volts were needed to give the 1-MHz dynamic range for this fast transducer. A customary (~50 kHz BW) PZT channel provided the medium-term corrections, while for still larger and slower corrections we used an intracavity Brewster plate mounted on a galvo (angular) motor. Nearly everything was perfect for Theodor's hydrogen experiment. The dye laser cavity could be servo-stabilized with the needed speed, the output could be doubled with BBO, and everything looked good. Except that the Pockels cell optical insertion loss approximated or exceeded the dye laser gain, so the dye laser output power was somewhere between pitiful and nonexistent.

Factors like these can become depressing. It's something like saving capital gains tax when your estate's net worth is calculated: yes, we won something, but it's really not so interesting anymore if the laser is below the threshold. Theodor kept coming back to probing the basis on which one could be certain the laser response needed to be so quick. For one thing, certainly the time delay of the response needed to be short. If the servo loop could have a dead-time τ before it responded, then an error signal at the Fourier frequency $f_{\mathrm{max}} = 1/(2\pi\tau)$ could already have a 1-radian phase shift, indicating that a modest-gain correction would be only mildly helpful in reducing the error excursions. Let's have a look at the physics that sets the timescale.

2 The Dye Laser's Fast-Changing Frequency

The active medium of the jet stream dye laser is actually a free-flowing curtain of thick viscous solvent, such as ethylene glycol, with a few mg/l of dissolved dye in it. Inside the laser cavity, filled with Watts-level circulating power, certainly the optical phase is well-enough defined for a short time, say < 100 ns, within the cavity photon storage time. For longer times, the troubles arise because the dye liquid needs to be flowing very rapidly to deal with the thermal issues. Indeed, fast flow (5 – 10 m/s) is needed to carry away the heat associated with the $\sim 10\,\mathrm{MW/cm^2}$ optical intensity which is being dumped into the dye by the $\sim 4\,\mathrm{W}$ pumping beam. Under these flow conditions, the entire illuminated dye sample is exchanged in about 1 µs. Even so, the temperature is raised near to the boiling point in this brief time. With such a thermal track written along the length of the flowing dye jetstream, no wonder it is difficult to remove all traces of astigmatism from the dye laser's output beam. However, for frequency stabilization, the real killer is the roughness of the surface. Surely it won't be smoother than a monolayer. This small a length change, say 0.4 nm out of a 1-m cavity length, already represents a 200-kHz frequency shift. Important averaging occurs of course as the illuminated spot is perhaps 40 000 molecules in diameter. But important aerodynamic issues also work to make the surface unstable and rough. Consider the molecules flying along inside the jet-forming nozzle, very near to the stationary wall. Polishing defects on this inner surface are going to perturb the dyestream surface mightily. Special nozzles have been fabricated from polished sapphire plates, well polished, and bonded together to form the little tube. Going downstream from the nozzle nibs helps the surface roughness effect substantially, as the solvent's surface tension can work to make the surface smooth again. But now a new "gotcha" turns on: surface roughness is generated by the unstable boundary between the air and the fast moving jet. You'll agree that the tangential speed of the air boundary layer will eventually approximate that of the liquid, and the required momentum input begins abruptly when the liquid exits the nibs of the nozzle. The associated turbulence leads to a surface roughness that is expected to grow quadratically along the flow direction. Operationally, 1 mm to a few mm is the right range of distance between the nibs and the pump laser beam, and a 500-kHz fast linewidth is the typical result of the optical phase modulation process. So, as noted before, clearly a loop unity gain frequency of $\sim 2\,\mathrm{MHz}$ is the slowest one that will produce the gain needed at $\sim 300\,\mathrm{kHz}$ such that the intrinsic intra-cavity phase variations at this frequency can be servo-reduced $\sim$ 4- or 5-fold, and thus reduce the laser's phase modulation index to so small a value that little power is converted into the FM sidebands.

3 Theodor Buys the Speed Requirement – the a Posteriori External Laser Correction Scheme

Eventually Theodor appeared to buy my story that great speed was necessary for the frequency servo. But certainly he didn't want to accept the intracavity losses that a modulator would bring. Now he seized on some earlier JILA work [2] where we had been frequency-stabilizing and amplitude-stabilizing lasers using only external means, such as an Acousto-Optic Modulator (AOM). In such a way, one would not introduce any lossy intracavity phase modulator as was used in my work with Ron Drever, but still one would have a versatile and powerful external frequency control possibility. However, there was one important problem: while the AOM can exhibit rather larger modulation bandwidths (by focusing the light beam to a small diameter so the acoustic beam transits it quickly), there still remains an irritating time delay until the electrical error signal effects an optical frequency change. First we have to change the frequency of the oscillating RF energy in the transducer's RF "tank" circuit. Then the thin $LiNbO_3$ transducer itself needs to begin vibrating at a new frequency to launch the new acoustic waves. But most troublesome is the acoustic propagation delay between the transducer and the optical interaction region. With normal AOMs it is hard to get much below a 150 – 200 ns delay time. Moving the light beam closer to the transducer leads to excessive light absorption and optical beam distortion.

4 Adding the External EOM

I don't remember how the concept came up to use an external *phase* modulator, in addition to the AOM *frequency* shifter. For sure, this Electro-Optic Modulator (EOM) wouldn't be as handy as the frequency shifter, but it could be really fast. To integrate these two kinds of beasts within one servo system, some larger way of thinking about the servo things would be necessary. Of course this was a perfect challenge for Theodor: he is never worried about too large a project! One could understand in a direct way that – to combine our two transducers in a simple way – it would be sweet if a 1-V input into each channel produced the same 1-MHz frequency-shift result. Perhaps 1 V could create a 1-MHz shift from the AOM. But to have the external phase modulator produce a *frequency*-step rather than a *phase*-step, we needed to have a step input to our amplifier result in a ramping voltage output. In short, the driver for the EOM needed to be an electrical integrator. Of course, eventually, large voltages will arise and saturation may be expected. So this EOM would need to be tightly lashed with the AOM so that the AOM would take up the heavy lifting of producing a continuously present frequency shift. Actually, we also needed to get the laser's PZT into the act, so that the AOM would also only be dealing with transient corrections. Otherwise a long-term drift would lead to a frequency shift, which with an AOM would correspond to an angular shift and thus a changed Bragg diffraction efficiency, and a lot of other troublesome technical issues. (Later I and others came up with a double-passed AOM configuration where the frank angular deviation was suppressed optically.)

5 Doubly Correcting

Actually, there is a rather curious issue that comes up in this combined use of a fast transducer (the EOM channel) and a time-delayed one (the AOM channel). Consider what happens when a fresh error signal is presented to the system. The EOM channel is very quick and in just a few units of its unity gain time, the error signal will begin to be mainly suppressed. During this time there will be an error signal presented to the input of the AOM driver system, and so an appropriate AOM response will be assembled and launched out as an ultrasonic wave. After a few hundred ns of acoustic travel time, this "correction" information will arrive at the interaction point and introduce this channel's approximation to a perfect correction/response for the original perturbation. This would be great, except that the original perturbation had long ago been tracked to zero by the fast EOM loop. So here we are introducing a new perturbation after a time delay, which will elicit a fresh response quickly from the EOM channel and later – only after another acoustic delay time in the AOM channel – will the AOM response to this secondary effect arrive at the light beam. These transducers sound now more like part of an oscillator team, rather than a stabilizer team. Of course, for the electro-optic designer, there needs be no surprise – we know for sure that the error integral will show up in the mainline output, thanks to the AOM channel. Somehow, it is as if at long times we are "correcting" twice for the same perturbation. What we really need is a "hand-off" strategy. Let's have the EOM quickly do its best job of suppressing the error input but then, when the AOM signal begins to arrive, we want the EOM channel to cease its activity completely. Just forget about that correction! Actually, the EOM is capable of rather speedy response, so we could generate this "drop-the-action, Jack" command by simply stopping the input to its amplifier.

6 A Delay Line Can Speed Things Up?

Some discussions like this with Theodor thus led to one of the most counter-intuitive schemes I've heard – you can get a faster servo system if you build an interferometer with an added delay inside the control loop. Then the fast signal is presented through the short path to the fast error amplifier and this leads to a suppression of the error input. The signal also is traveling in the acousto-optic material, but now its electrical image is also traveling in the delay line within the electrical interferometer. Both signals arrive at their respective destinies at the same time, the EOM system's drive quickly goes to zero because of the designed-in destructive electrical interference. But the AOM is now making the needed corrections! A perfect hand-off. At least, it *could be* a perfect hand-off.

7 Analysis Needed

Very quickly we were at a bad place in the design: the system was too complex to just see how it would work. We needed some kind of computer-based analysis scheme. Of course, knowing Theodor and Art, you will already realize this means that it also has to be smart coding, since they would only be interested if it could run on the then state-of-the-art microcomputers, something like an Apple II with a MOS Technologies 65C02 8-bit CPU. Or one of the curious XT machines from a major manufacturer, with a genuine Intel 8086 processor inside. (Or was it an 8088?) These things came with the most inhomogeneous software. Theodor and Art even had some pioneering text-processing software on one of their machines, maybe it was a Radio Shack Z80 system. The program was called "Electric Pencil", or some such title. Lindy could type many times faster than this machine could handle – *that* was most distressing for both parties. The "making interesting software"-disease was even caught by a serious laser physicist, Prof. Murray Sargent, co-author of Sargent, Scully and Lamb, the laser and optical interactions principal reference book for many generations of laser students. I think Murray wanted to see his beautiful density matrix equations on the screen, not just after being worked over by a post-processor such as TEX (a contemporary Stanford project by Prof. D.E. Knuth). But big programs led to the need for robust operating system software. I have heard that Murray developed a crash-proofing system that prevented (his) buggy code from overwriting another program's sacred data and instructions. One recent book suggested that this optical and theoretical physicist had a related major connection with the Redmond Operating Systems in the Win 3.1 epoch.

8 Theodor Gets "Programming Fever"

So now our Theodor was caught by the programming fever. Not to program stacked blocks on one another, or watch bricks getting "broken out" by flying slightly-controllable missiles. *This* software should be able to analyze servo systems! Of course there was nothing commercially available. Theodor had the insight that the charge on a capacitor in a circuit, after being affected by currents flowing in the past, was very much like a bank balance, after it had been affected by past financial transactions. Now we are talking about something that is often attractive to bright people, even programmers. I speak of course about MONEY. So there was a marvelous program, VisiCalc, and its later competition, Lotus 1-2-3 late, which were organized in a way almost ideally suited to the analysis of servo systems in the time domain. At one moment, we have an error signal. It is amplified, with a delay. The output is derived from this signal, augmented by a bit of some accumulated value from this signal at previous times. Perhaps a term sensing the rate-of-change would later be useful also. I can remember talking with Theodor one Friday about how the now-ubiquitous spreadsheet could be used – could *perhaps* be used – to analyze electronic servo systems. The next time I came to the lab, Monday morning, there were many photocopy pages of a T.W. Hänsch document showing basic op-amp and RLC circuits, low-pass, high-pass, integrators, lag–lead filters, leaky integrators ... Building blocks for a spreadsheet servo design. Just add these processing steps one by one, and work out the time behavior of the open loop system. Want the loop closed? Make it recursive?

Well then, just let this output error signal now go back to the input ... Of course these financial-analysis dudes didn't want to invoke periodic boundary conditions or recursive self-sustained transient performance analysis. So their tools were slightly limiting. But Theodor made this scheme really sing. So here was this powerful and famous laser guy – now turned electronician – telling ME, an imported electronics guy of nominal guru rating, just which topologies would be interesting. Even suggesting component values, in a well-meaning and mostly innocent way, of course.

9 The Summary of this Epoch is Two-Fold

First, there are now real professional tools such as SimuLink that allow one to study electronic and other control systems in basically just the way forseen by this young Stanford professor. OK, modern software has a graphical user interface. But the base idea is the same: what you have now depends on what you had before and what you did in between. The second concrete output of this Spring quarter's work was a nice circuit box that I built at Stanford to

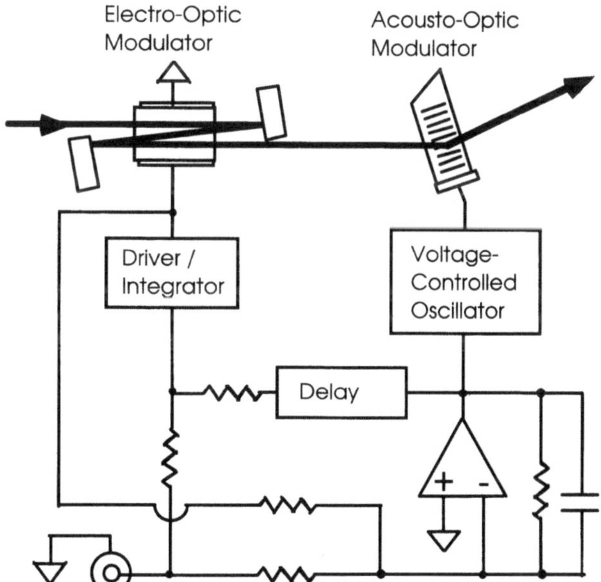

Fig. 1. External laser phase/frequency stabilizer. A servo correction signal from a phase/frequency discriminator is promptly applied to an Electro-Optic Modulator, via an integrator which produces a rate-of-change output for a *given* steady input, thus producing a *frequency* shift of laser. This input signal is cancelled after the delay time, chosen to match the ultrasonic delay of the Acousto-Optic Modulator. Total control authority is thereby "handed-off" from the EO to the AO channel ... Slow feedback from an integrator output gradually resets the DC levels

be used for the associated experiments that led to the joint publication [3]. But the particular chips chosen, while fast, were new and quite tweaky, and prone to oscillation. (This reminds me of the vast storeroom of "good old stuff" maintained at Stanford by Prof. Schawlow's technical assistant, Ken Sherwin. Ken was a genius. He's the one responsible for Art's ruby laser in the plastic "phaser" shell that appears in all the exploding Mickey Mouse balloon photos. Ken was an avid radio amateur, with the consequence that he needed one of those gigantic "land-shark" vehicles from Detroit, the better to house his mobile ham transmitters for the 14-m and 7-m amateur bands.

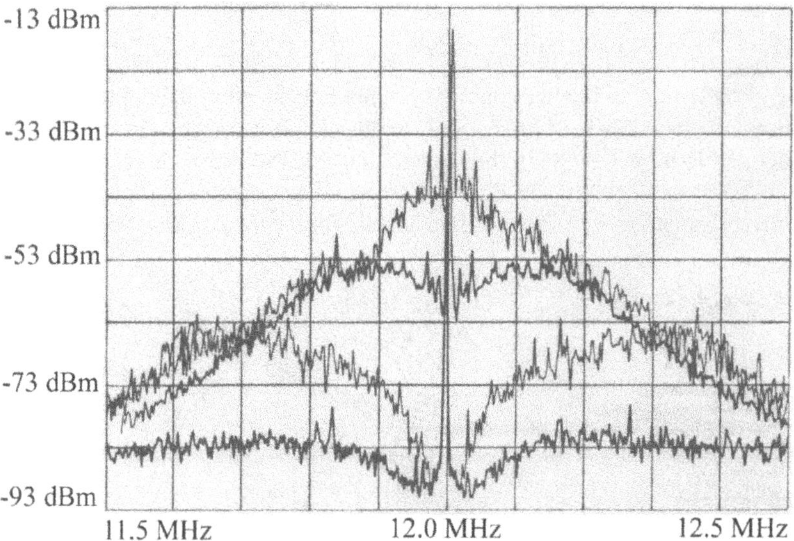

Fig. 2. External laser stabilization with AOM/EOM frequency servo system. A Dye laser is locked to a reference interferometer using the FM sideband technique. A spectrum analyzer display shows noise around the 12.0-MHz modulation frequency due to low-frequency laser FM noise. The top curve shows noise when the 699 dye laser is locked only to its own cavity. Correcting for (deliberately introduced) saturation, a 12-MHz coherent signal of -3 dBm would result when the laser is tuned to the half-height of the external reference cavity. With the laser locked by the external AOM/EOM system (v.1988) the noise is drastically reduced. The next-to-top curve has some servo gain ($\sim$14 dB). The next two curves have 20-dB successive gain increases, until the FM noise level is reduced to a near the measurement noise level (~ -83 dBm at a 1 kHz bandwidth). With the cavity FWHM = 250 kHz, this noise floor corresponds to a laser linewidth below 1 Hz. Below $\sim$50 kHz, measurement shot noise is written as a low-level FM onto the laser, so that the "error point" shows less apparent noise than the measurement noise level. The non-zero 12-MHz residual carrier shows that a small DC error signal is needed to match an input offset error in this early version of our fast servo amplifier. This work was done at JILA in 1987–88 in collaboration with Miao Zhu, and Fujio Shimizu, and Kazuko Shimizu. See [4, 5]

And to support the colossal unfurling mobile antennas. Even I, a visitor to the Stanford community, contributed to the general progress by giving him a jump start one night in the Varian Physics Department's parking lot since all of his several car batteries had been run down by extended radio broadcasting. But Ken *did* have a grid dip meter to check my circuits for oscillation. In fact, he had several.) So eventually the famed External Stablizer came into operation: The "Laser-Person's Magic Noise-Eating Box." Later improvements allowed one to obtain sub-Hz linewidth from a standard commercial dye laser [4, 5]. See Figs. 1 and 2. Now – some 17 years later, and with Jun Ye's expert help – the noise of cw Ti:sapphire lasers can be squashed with this external approach, down to noise levels of $20\,\mathrm{mHz}/\sqrt{\mathrm{Hz}}$, corresponding to a laser linewidth of 1.2 mHz relative to the reference cavity.

10 A Still Better Way?

The last point, and one which shows my friend's creative power quite clearly, is that none of the modern generations of cw two-photon spectroscopy of hydrogen in Garching have in fact made use of an External Stabilizer. Theodor just helped some clever optical engineers make an Electro-Optic Modulator of such low loss it can be inserted into the dye laser with nearly no deleterious losses. Now the servo system is simpler and can work even better.

Let me conclude by wishing you, dear Theodor, a long and continuing top creativity during the next few decades. Now I've got to get back to work designing more cool stabilizer circuits to go with your latest and greatest fad, the one-step phase-coherent optical comb connection between RF and optical domains. Theodor, congratulations and best wishes!

Acknowledgements

I am grateful to and thank many colleagues who have participated in the JILA work, and particularly Jun Ye in whose hands the labs continue. The research discussed here was funded by NIST, the National Science Foundation and the Office of Naval Research.

References

1. R.W.P. Drever, J.L. Hall, F.V. Kowalski, J. Hough, G.M. Ford, A.J. Munley, H. Ward, Appl. Phys. B **31**, 97 (1983)
2. J.L. Hall, H.P. Layer, R.D. Deslattes, 'An Acoustooptic Frequency and Intensity Control System for CW Lasers', 1977 IEEE/OSA CLEA, Digest p. 46, IEEE Cat. No. 77CH-1207-0 Laser
3. J.L. Hall, T.W. Hänsch, Opt. Lett. **9**, 502 (1984)
4. J.L. Hall, M. Zhu, F. Shimizu, K. Shimizu, 'External Frequency Stabilization of a Commercial Dye Laser at the Hertz Level', Proc. 1988 IQEC, Tokyo
5. M. Zhu, J.L. Hall, J. Opt. Soc. Am. B **10**, 802 (1993)

Miniaturized Laser Magnetometers and Clocks

Robert Wynands

It is with great pleasure that I contribute to this volume honoring Theodor Hänsch on the occasion of his 60th birthday (or, in the present context of precision measurements, shouldn't we call it his 61st, really?). As a Ph.D. student constructing the original Garching frequency chain in his lab I was infected with a "passion for precision" that has influenced my subsequent career in laser spectroscopy and atomic physics. Here I would like to report on the application of some of the techniques we developed in Garching for frequency measurements on atomic hydrogen to the spectroscopy of thermal alkali atomic vapors, leading to very sensitive optical magnetometers and to compact cesium atomic clocks.

1 Introduction

Man has always had a great interest in the passage of time. The change of the seasons is of obvious importance for hunting and agriculture, and calendar sticks dating back to the stone age have been found. Today we count time in seconds on our wrist watches and in femtoseconds for modern laser chemistry experiments. State-of-the-art atomic clocks have reached fractional inaccuracies [1] reaching 10^{-15} [2, 3], corresponding to one second offset in thirty million years. Optical clocks, based on electronic transitions in trapped atoms or ions [4] stand poised to improve on this by a few orders of magnitude.

Impressive as this progress is, in many cases picosecond or even nanosecond accuracy is adequate, for instance satellite navigation [5] and telecommunications [6]. Here it is more important to have available a robust, inexpensive, and power-efficient device. Meeting all those specifications is difficult with today's devices: a quartz crystal suffers from long-term drift, and conventional optically pumped rubidium clocks are too power-hungry for some applications. Here we describe a small atomic clock based on laser spectroscopy in a thermal cesium vapor whose better than 10^{-11} fractional inaccuracy, lack of deformable parts, and power efficiency should make it perfect for mobile applications.

Although we can now determine position with GPS receivers, the compass is still the most common device to determine the direction "north" by showing the direction of the magnetic field of the Earth. While the geostatic field is relatively large and easy to measure, for other fields more effort is required. For instance, geologists monitor small changes of the geomagnetic field precisely for seismic studies, archeologists measure local magnetic field variations in search of ancient habitation sites and other buried objects. In material science alternating magnetic fields are used for material testing and even our own body produces tiny but measurable magnetic fields. The strongest of these, the field of the human heartbeat, hardly reaches a millionth of the Earth's magnetic field. It is commonly believed that its detection is possible only with SQUIDs, superconducting quantum interference devices. In the second part of this contribution we will present experiments on a compact all-field magnetometer. It allows us to measure magnetic fields down to a fifty-millionth of the geomagnetic field strength and maybe even the field of the beating human heart. An important ingredient is the use of a magnetic gradiometer in order to reduce the influence of external stray magnetic fields.

2 Operating Principle: Dark-Line Resonances

Dark-line resonances can be observed when an atomic three-level system (Fig. 1) is illuminated by a bichromatic light field [7, 8]. If the difference frequency of the two light fields is exactly equal to the splitting between the two lower states, the atomic population can be transferred (optically pumped) into a coherent superposition of these two states, the so-called coherent dark state, or coherent population trapping (CPT) state. It no longer couples to the excited state because of destructive interference between transition pathways. The absorption of light is reduced, resulting in electromagnetically induced transparency (EIT).

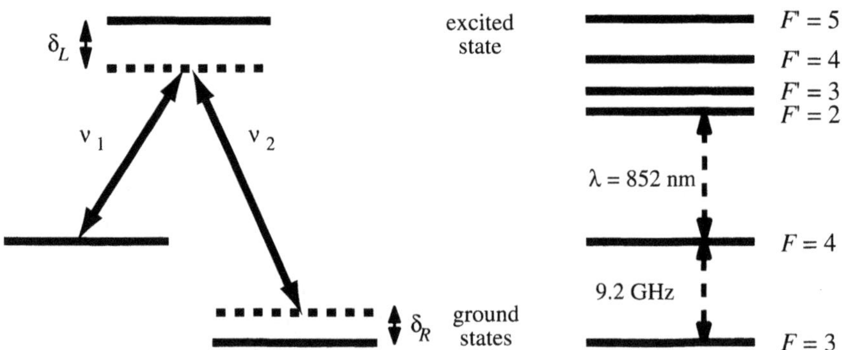

Fig. 1. *Left:* The three-level Λ-system with detuning δ_L from optical resonance and Raman detuning δ_R from the center of the dark resonance. *Right:* Approximation of the three-level Λ-system by the cesium D_2 line. In the experiments described here the Doppler and pressure broadened optical linewidth is larger than the hyperfine splitting in the excited state

Using the cesium D_2 line, for example, Λ-systems can be formed with the hyperfine components $F = 3$ and 4 of the $6S_{1/2}$ ground state and the hyperfine-split excited states $6P_{3/2}$, $F' = 3$ and 4. Such a CPT resonance absorption signal is shown in Fig. 2, where the detector voltage is normalized to the value at maximum Doppler absorption. The difference frequency between the two optical fields (horizontal axis) is varied near the ground state's hyperfine splitting, and the total laser power through the cell is monitored. Although the absorption is changed by less than 1 %, the resulting Q-values are greater than 2×10^8, since linewidths below 50 Hz are possible [9, 10]. For this reason dark-line resonances have been considered as frequency references [11, 12] or for other types of precision applications [13] for many years now.

An important ingredient for small dark resonance linewidths is the addition of a buffer gas like neon or argon to the cesium vapor. Without buffer gas the cesium atoms only spend a few microseconds interacting with the laser beam, resulting in time-of-flight broadened linewidths of about 10 kHz. The frequent collisions of the cesium atoms with the buffer gas atoms impede the free motion and thus keep the cesium atoms in the laser beam for several milliseconds so that the narrow resonance lines become possible.

A second key ingredient is the use of a bichromatic light field with very stable relative phases. During my time in Garching I implemented a technique (a so-called optical divider stage, one of Theodor Hänsch's many ingenious ideas) that allowed us to phase-lock two lasers with a very wide frequency separation, up to hundreds of THz, in fact [14]. So the 9.2 GHz frequency difference required for the cesium dark resonance spectroscopy was surely no problem in principle. Using a fast photodiode and suitable rf electronic components we could phase-lock two "Hollberg" diode lasers [15] with the

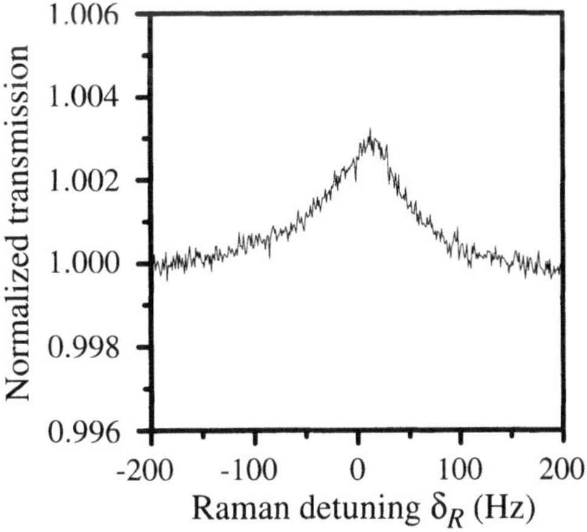

Fig. 2. Spectrum of the dark resonance absorption (CPT) signal with dc detection of the laser power transmitted through the cell

required frequency splitting [16]. After superposing their output beams with the help of a single-mode optical fiber a bichromatic light field with extremely good phase stability was obtained. Later we modified and extended the technique to digital phase-locked loops, another Garching development [17]. This allowed us to use extended-cavity diode lasers which are much easier to handle than the "Hollberg" type, leading to extremely narrow linewidths of the dark resonance [9]. Nowadays we use an experimentally simpler technique that became practical with the advent of new types of laser diodes: vertical-cavity surface-emitting lasers (VCSELs). These lasers can have a modulation bandwidth of more than 10 GHz [18] so that direct modulation of the laser's injection current produces modulation sidebands with the proper frequency difference and perfect relative phase [19]. The dark resonance in cesium can then be prepared using the two first-order modulation sidebands created by 4.6 GHz modulation or by modulating at the full hyperfine frequency and using the carrier plus one of the sidebands.

In a magnetic field the dark resonance splits into several Zeeman components (Fig. 3). Their number and relative strengths can be determined from the Clebsch–Gordan coefficients for the optical transitions between Zeeman sublevels of ground and excited states [20]. The central component is affected by magnetic fields only in second order and corresponds to the coupling of the $m = 0$ sublevels in the two ground states, the same ones as in a standard cesium clock. The principle of operation of the dark resonance atomic clock is to control the rf source that determines the laser difference frequency such that it always stays right at the center of the $0-0$ dark-resonance component. This rf frequency is the output signal of the clock/frequency standard.

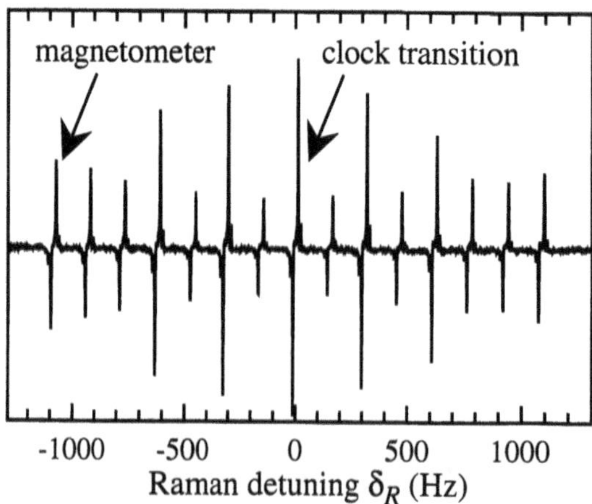

Fig. 3. In a magnetic field (here 45 μT flux density at 45° to the light propagation direction) the dark resonance absorption signal splits into several Zeeman components. The substructure in the line shape of each component is due to the frequency modulation spectroscopy employed

The outer Zeeman components (corresponding to the coupling of ground states $|F = 3, m = 3\rangle$ and $|F = 4, m = 4\rangle$, for instance) react most strongly to magnetic field changes, at a rate of 24.5 Hz/nT. Monitoring their position therefore allows for the sensitive detection of tiny flux density changes.

3 Atomic Clocks

Traditionally, an atomic clock is based on the transition between the states $|F = 3, m = 0\rangle$ and $|F = 4, m = 0\rangle$ in a Cs atomic beam, excited by a 9.2 GHz microwave field in two Ramsey interaction zones [21]. In this setup, Q-values of $1 \ldots 2 \times 10^8$ are typically obtained. "Fountain clocks", made possible by recent advances in laser cooling techniques, use a cloud of cold atoms launched against gravity, resulting in long interaction times, narrow linewidths, and substantially higher Q-values [22, 23, 2, 3].

In order to provide more compact and lower-cost frequency references for applications with less stringent stability requirements alternative techniques have been investigated over the years, including optically pumped microwave double resonances [24, 11, 25] and Raman scattering [26]. Relatively compact (shoe-box sized) and inexpensive ($\geq$ \$1500) frequency references are available commercially using lamp-pumped alkali-vapor cells and external microwave fields to probe the clock transitions [21]. We will now describe a small clock relying on coherent population trapping resonances (CPT dark-line resonances) [27–30] in a thermal cesium atomic vapor.

3.1 Experimental Setup for the Clock

In the experimental setup, a single-mode VCSEL is used because of the simplicity of its operation, for instance without any external optics for mode control. With a threshold current of 0.82 mA at 31.2 °C and a linewidth of approximately 50 MHz it emits 870 μW in a single spatial and spectral mode at 852 nm. The linearly polarized output from the VCSEL is sent through a quarter-wave plate to create a circular polarization and is attenuated by neutral density filters, leaving $\sim$ 10 μW in a beam 4 mm in diameter at the entrance to the vapor cell (Fig. 4). When the VCSEL's injection current is modulated at 4.6 GHz, half the ground-state hyperfine splitting of Cs, the two first-order sidebands can be used to excite the resonance. Because of the large modulation bandwidth of the VCSEL (about 6 GHz in this case) the modulation requires only 12.6 mW of rf power in order to obtain as much as 47% and 19% of the optical power in the blue and red first-order sidebands, respectively, leaving only 8% in the carrier. The remaining power is in the sidebands of higher order. The sideband asymmetry can be explained by additional AM modulation that occurs when the laser current is modulated.

The light is sent through a Cs vapor cell 2 cm long and 2.5 cm in diameter (although much smaller cells can also be used). A photodiode detects the

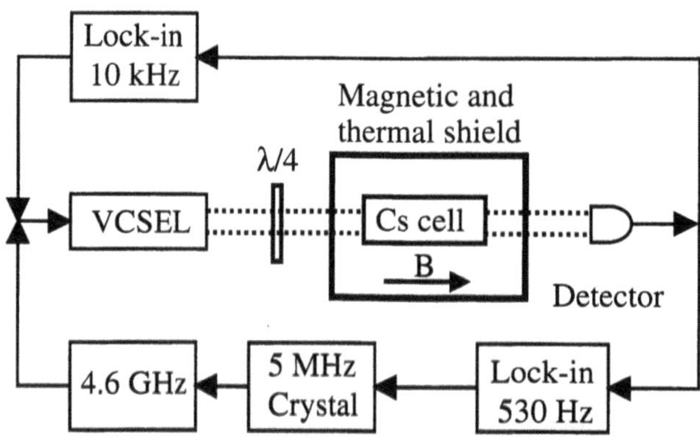

Fig. 4. Experimental setup of the compact dark resonance clock. The 4.6 GHz signal modulates the laser injection current to produce two sidebands separated by 9.2 GHz which probe the dark-line resonance in the Cs cell. The lock-in amplifier at 10 kHz is used to lock the laser to the Cs resonance and the lock-in amplifier at 530 Hz is used to lock the 5 MHz quartz crystal that in turn controls the frequency of the 4.6 GHz source

transmitted intensity. The laser is tuned such that the two first-order sidebands are resonant with the transitions from the two $6S_{1/2}$ ground-state components to the excited $6P_{3/2}$ state. Due to the large tuning rates of the laser frequency with injection current and temperature (of the order of -300 GHz/mA and -25 GHz/K), the laser current and temperature must be carefully controlled. The laser is locked to the center of the Doppler-broadened optical resonance line by sinusoidally modulating its current at 10 kHz. The absorption signal on the photodiode is demodulated with a lock-in amplifier.

A second servo-loop locks the rf modulator to the dark resonance signal. This is done by modulating the 5 MHz reference crystal for the 4.6 GHz synthesizer at 530 Hz and detecting in turn the resulting modulation that appears on the optical power transmitted through the cell. Phase modulation indices as high as 0.7 were obtained with these parameters. When the 4.6 GHz source is tuned over the dark resonance a dispersive error signal is produced with a full-width at half-maximum (FWHM) around 100 Hz. The Cs vapor cell is thermally isolated and temperature stabilized to within 100 μK of 32 °C. It sits inside a mu-metal shield that reduces stray transverse magnetic fields by a factor of $\sim$2000. To isolate the dark resonance on the $m_3 = 0 \leftrightarrow m_4 = 0$ Zeeman component, a small homogeneous longitudinal magnetic field of 10 μT flux density is applied. Neon at a pressure of 9.7 kPa was added to the Cs vapor to reduce transit-time and Doppler effects. While the CPT resonance is narrowed compared to the case of pure Cs vapor, the optical resonance is broadened by 86 MHz/kPa Ne [31]. This means that the homogeneous width of the optical transition exceeds the laser linewidth. In addition, the CPT resonance frequency is shifted upwards by $\sim 46\,\mathrm{kHz}$ by Cs-Ne collisions at this pressure [9].

With the rf modulation applied, the laser is locked near the position of maximum optical absorption and the 4.6 GHz frequency is scanned over the dark-line resonance. The dark-line height and width are recovered by fitting the error signal with an FM lineshape of a Lorentzian line [32]. In addition, the 4.6 GHz signal can be locked to the resonance by feeding an error signal to the tuning port on the quartz crystal of reference. The thus-stabilized 4.6 GHz signal can then be compared to a reference frequency derived from a hydrogen maser. In this way the frequency of the small clock can be measured as a function of time and its stability optimized through appropriate choices of experimental parameters.

3.2 Atomic Clock Results

Laser and rf power, beam diameter, buffer-gas pressure, vapor cell temperature, and modulation frequencies and indices for the two servo-loops are chosen to obtain the optimal signal-to-noise ratio and minimum sensitivity to external disturbances. As an example, only the noise level as a function of detuning from optical resonance will be discussed. All the details can be found in [33, 34].

3.2.1 Noise Level

In this atomic clock, the conversion of laser FM noise to AM noise by the slope of the atomic absorption profile contributes substantially to the clock's short-term instability. While this noise source is well known in laser-pumped atomic clocks, we have discovered an additional FM-AM noise source which arises because of optical pumping between the atomic ground state hyperfine levels.

The overall noise at the 530 Hz lock-in detection frequency strongly depends on the detuning of the laser frequencies from resonance with the excited state. The measured noise power spectral density at 530 Hz is shown in Fig. 5. The squares are the experimental values, normalized to the power on the detector. Using a Fabry–Perot resonator, we have independently measured the frequency (FM) and the amplitude (AM) noise of the laser. When the laser frequency jitters back and forth, the total light intensity changes due to the change in Doppler-broadened absorption (linear FM-AM conversion). This change is proportional to the absolute value of the slope of the optical absorption line. The shape of the Doppler profile was measured and numerically differentiated. Using the numerically determined slope and the measured FM noise the dashed line in Fig. 5 was calculated. In addition, detector shot noise and laser AM noise were added as constant offsets. Near the maximum of the Doppler absorption another process becomes important: nonlinear FM-AM conversion. Whenever the shape of the absorption profile

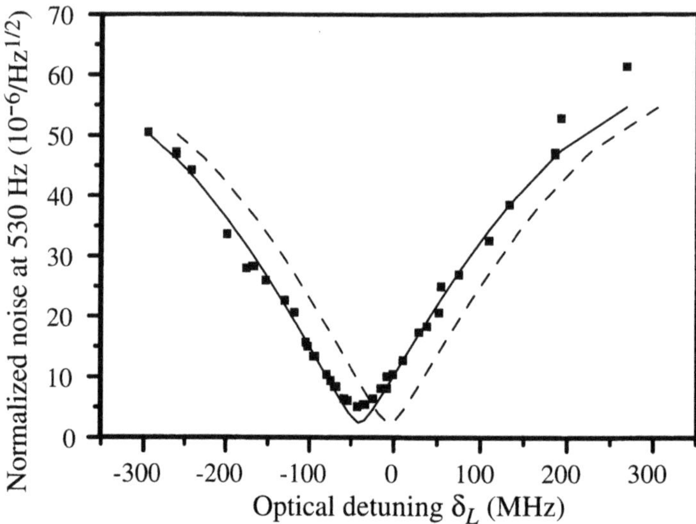

Fig. 5. Normalized noise power spectral density at the lock-in amplifier's detection frequency of 530 Hz. The *squares* are experimental values while the *dashed line* is calculated without free parameters, using the independently measured frequency and amplitude noise of the laser. The *solid line* is a copy of the *dashed line*, shifted by 40 MHz to lower frequencies in order to account for the effect of Fourier-frequency dependent optical hyperfine pumping

has a quadratic component (like near the maximum), noise components at all pairs of Fourier frequencies whose frequency sum or difference is 530 Hz are mixed to 530 Hz where they are picked up by the lock-in amplifier. This process was also included in the calculation of the dashed line in Fig. 5. When this curve is shifted by 40 MHz to lower frequencies (solid line in the figure) one finds very good agreement, except at the very bottom where the measured values are higher than the calculated ones.

Both the shift and the excess noise can be explained by a dependence of the efficiency of optical hyperfine pumping on the frequency of the laser jitter [34]. The basic mechanism is the following. Because of the resonant nature of the atom–light interaction, the absorption of light by the atoms depends strongly on the laser frequency. Since the absorption transfers population between the ground state hyperfine levels (optical pumping), the populations of these levels depend on the laser frequency. The level populations in turn affect the transmitted intensity since atoms in a particular level absorb only light from one of the two incident optical fields. As a consequence, frequency variations of the laser that are slower than the optical pumping rate cause fluctuations in the atomic ground state populations which then produce AM noise on the transmitted beam through the cell. This process can be described qualitatively by a simple theoretical model.

Part of this additional noise can be cancelled at certain optical detunings by the normal FM-AM conversion due to the slope of the effective atomic absorption profile. But even for the optimum detuning, there remains a component of the noise out-of-phase with the original FM noise on the laser which cannot be cancelled out. This excess FM-AM conversion is an important new noise source originating from optical pumping and may well limit the performance of future microwave frequency references.

In summary, one finds that the agreement between the dots and the solid line in Fig. 5 is very good. This indicates that noise processes are well understood and that it is the laser's FM noise that dominates.

3.2.2 Clock Performance

The clock performance after an initial round of optimization is shown by the black symbols in Fig. 6. For short averaging times τ the instability drops at a rate of $9.3 \times 10^{-12}/\sqrt{\tau/\mathrm{s}}$, comparable to good commercial rubidium frequency standards. Although the cell is thermally isolated and temperature stabilized, residual temperature drifts still limit the long-term stability of the frequency reference. A linear shift rate with temperature of 10 Hz/K at 9.2 GHz for 9.7 kPa of Neon was measured in the range between 25 °C and 50 °C. This is consistent with the measurements for direct microwave excitation of the $0-0$ resonance [35]. For temperature instabilities less than 100 μK this shift would cause frequency instabilities in the 10^{-13} region, far below the measured values for long times. The residual temperature dependence

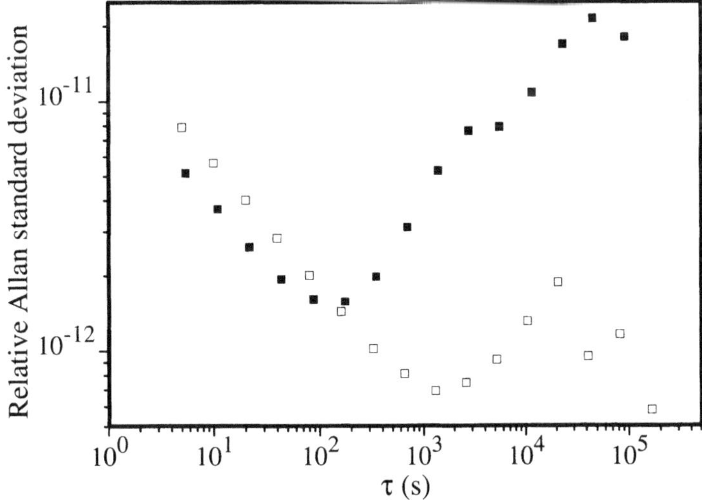

Fig. 6. Clock performance in terms of the relative Allan standard deviation as a function of averaging time τ. *Black symbols:* pure neon as a buffer gas; *open symbols:* neon–argon mixture

which limits the Allan deviation could be caused by inadequate temperature stabilization of the cell windows. We observed a correlation between the room temperature and the center frequency of the dark-line resonance, even though the temperature sensor near the center of the cell did not show any residual temperature fluctuations.

Argon as a buffer gas shifts the resonance in the opposite direction as a function of temperature. In a cesium cell with a suitable mixture of neon and argon, therefore, the effective temperature-dependent shift rate was reduced below 0.4 Hz/K for a total pressure of 5.3 kPa. This improved the long-term stability of the clock significantly at the expense of a degraded short-term stability by a factor of two, due to additional collisional broadening of the resonance line (open symbols in Fig. 6). The Allan standard deviation now reaches its minimum of 7×10^{-13} near $\tau = 1000$ s.

4 Magnetometry

Sensitive optical magnetometry has a long history. Already in 1969 a sensitivity of about 100 fT in 3 s of averaging time was achieved in a combined optical and radio frequency setup based on the Hanle effect [36], with a more recent result of 1.8 fT/$\sqrt{\mathrm{Hz}}$ [37] which is comparable to the best SQUID detectors. The nonlinear Faraday effect has also been employed in laboratory setups, where under the assumption of shot-noise limited operation a sensitivity of 0.3 fT/$\sqrt{\mathrm{Hz}}$ was extrapolated [38].

As for a dark resonance magnetometer, it was suggested several years ago that the index of refraction near such a dark resonance would not suffer from power broadening and thus would be superior to conventional optically pumped magnetometers by several orders of magnitude [39, 40]. It turns out that the freedom from power broadening predicted in [39, 40] is a general feature of any resonance that shows the same power broadening behavior as a simple two-level system, as long as a sample with an optical length of one absorption length is considered. Furthermore, in later work it was pointed out that such a magnetometer would instead suffer from an ac Stark shift-induced reduction of sensitivity [41]. Nevertheless it is interesting to investigate where the experimental sensitivity limit of a dark resonance magnetometer lies, in particular because recent technological advances [19] allow for the construction of very compact and robust devices that are suitable for applications outside a laser research laboratory. The sub-pT performance for the detection of acmagnetic field components in the kHz range has been reported in [42], as well as its ability to provide the numerical value of the strength of a static field. The influence of magnetic field direction was considered in [20] and also in [43]. Here we concentrate on the sensitivity of a cesium dark resonance magnetometer to small changes of a nominally static field.

4.1 Experimental Setup for the Magnetometer

The experimental setup (Fig. 7) is similar to that in [19]. A vertical-cavity surface-emitting laser diode (VCSEL) at 852 nm wavelength provides a beam of 7 mm diameter and powers of typically a few μW at the sample cell. The VCSEL drive current is modulated at 9.2 GHz, the ground state hyperfine splitting in cesium atoms. The carrier and one of the two first-order modulation sidebands provide the bichromatic light field for the preparation of the dark state. This modulation frequency is itself frequency-modulated at 4.5kHz in order to facilitate lock-in detection. The laser carrier frequency is locked to the absorption maximum of the transition from one of the ground states to the $6\,^2P_{3/2}$ state (the D_2 line) using the DAVLL scheme [44]. The laser light is circularly polarized (a dark resonance cannot be prepared with linear polarization in a Cs cell with buffer gas [20]) and sent through a thermal vapor cell inside a magnetic shield. The cell is only 2 cm long and contains 5.1 kPa of neon as a buffer gas. Both longitudinal and transverse magnetic fields can be applied to the cell. The transmitted light is detected on a slow photodiode and demodulated with a dual-phase lock-in amplifier, thus making available in real-time both the amplitude and the phase response of the medium without the need for an interferometric setup with its mechanical stability problems.

In order to monitor only one Zeeman component at a time a constant transverse offset field of a few μT is applied to the cell. In this case the outermost Zeeman components shift with total flux density at a rate of 24.5 Hz/nT. By keeping the 9 GHz modulation frequency fixed at the center of the chosen resonance line a change in magnetic field strength can be detected through a change in the index of refraction, monitored by the in-phase signal. The sensitivity δB_{min} of this scalar magnetometer is determined by the smallest detectable change (with signal-to-noise ratio of unity) in signal amplitude S,

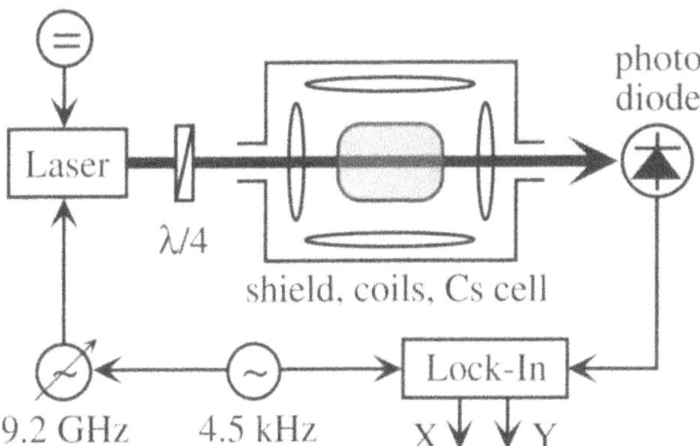

Fig. 7. Experimental setup of the dark resonance magnetometer

given a noise level ΔS:

$$\delta B_{\text{min}} = \frac{1\,\text{nT}}{24.5\,\text{Hz}} \frac{\Delta S}{|\text{d}S/\text{d}f|} \tag{1}$$

where f is the 9 GHz modulation frequency. The noise consists of the same contributions mentioned before for the clock and therefore will not be discussed here. The only difference is that due to a different stabilization technique (which does not require a second, low-frequency modulation) the shift of the noise curve with respect to the low-frequency absorption line does not occur.

The dependence of the slope of the in-phase signal on optical detuning was found to be rather flat over the range of interest here so that the optimum sensitivity is reached when the laser is locked to the frequency of the noise minimum. Furthermore, the sensitivity according to (1) is strongly dependent on laser power. For higher power the signal increases, but since the resonance width and all noise contributions increase as well there is an optimum intensity for best magnetometric sensitivity [9]. In a series of measurements for various laser intensities and lock-in amplifier time constants resonance curves were recorded, their slope at line center evaluated by fitting a straight line, and the noise level measured by subtracting a quadratic background from the outer wings of the measured line shape and then taking the root-mean-square value. Using (1) one can then extrapolate a value for the magnetometric sensitivity of the dark resonance magnetometer.

In Fig. 8 the resulting sensitivity is plotted as a function of lock-in time constant τ where each data point corresponds to the sensitivity at the opti-

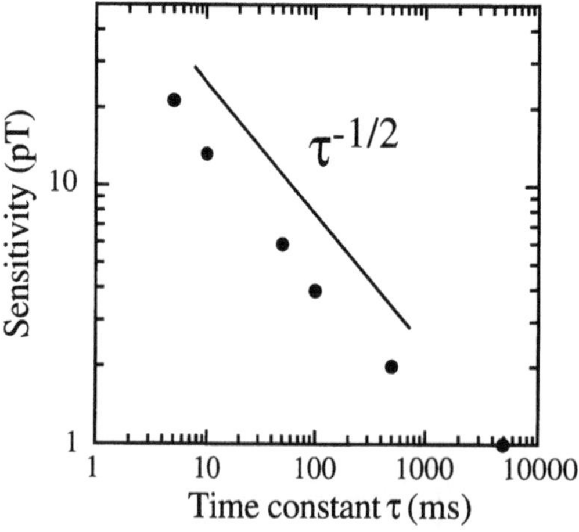

Fig. 8. Extrapolated sensitivity of the dark resonance magnetometer as a function of lock-in amplifier time constant. Note that these results correspond to a three-fold improved sensitivity or a ten-fold improved time resolution as compared to the data published in [45]

mum laser intensity for this particular time constant. This optimum intensity changes slightly from time constant to time constant due to the specific shape of the detector frequency noise spectrum which, unfortunately, also contains sharp peaks from electrical interference. The sensitivity improves from around 13 pT at $\tau = 10\,\mathrm{ms}$ to 1 pT at $\tau = 5\,\mathrm{s}$. The sensitivity for longer times is partly limited by drifts of the current producing the offset field and by thermal drifts of the vapor cell: a temperature rise of 5 mK causes the same shift of the dark resonance as a 1 pT flux density change. This temperature dependence can be eliminated with a suitable mixture of buffer gas species [21], as also demonstrated above for the clock.

The improvement in sensitivity with increasing time constant cannot be expected to follow the simple white-noise behavior proportional to $\tau^{-1/2}$ because all points are taken with slightly different laser intensities. Nevertheless, in order to compare the sensitivity with that obtained in previous work [36–38] a $\tau^{-1/2}$ dependence was fitted to the data points for $\tau < 1\,\mathrm{s}$. For the second-order low-pass filter used in the lock-in amplifier the effective bandwidth is $1/8\,\tau$, and one obtains a sensitivity of $4\,\mathrm{pT}/\sqrt{\mathrm{Hz}}$. This is about the same as the noise level of the best commercial flux-gate magnetometers.

In addition to the influence of laser noise a systematic limitation of sensitivity is given by the fact that in our experiments on the D_2 line there is substantial absorption at the dark resonance due to the additional excited state hyperfine levels. This results in a reduction of absorption by less than 1% compared to the Doppler-broadened optical absorption line and thus the advantages of a non-absorbing resonance in an ideal Λ-system are only partly exploited. It should be possible to overcome these problems and thus improve the sensitivity by switching to the D_1 line, where due to the level structure of the excited state the experimental realization of an ideal Λ-system is much easier. We are also exploring various other ways of increasing the signal contrast, for instance by optical pumping [46].

5 Gradiometer

In many practical cases the field to be measured is well localized but weak as compared to the fluctuations of the ambient magnetic field. In such a situation a gradiometer is very helpful: two identical sensors are employed, one close to the signal source, the other one a short distance away. While the first sensor detects signal and background the second one is too far away to be influenced by the signal proper but still sees the background fluctuations. If this background is homogeneous it will cancel when the difference of the two sensor signals is taken.

The advantage of such a gradiometer setup is that no magnetic shielding is needed in the ideal case. The improvement that can be reached with such a gradiometric measurement becomes apparent from the comparison of the two traces in Fig. 9. By applying carefully calibrated test gradients we were

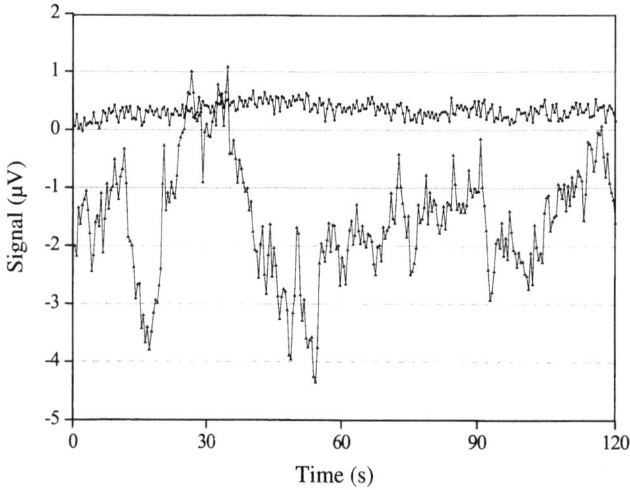

Fig. 9. *Bottom trace:* temporal variations of the output signal measured by a single dark resonance magnetometer standing freely in the laboratory. *Top trace:* difference signal of two magnetometers 1.2 cm apart, showing a 20-fold reduction in field noise sensitivity

able to establish a detection limit of 14 pT/mm for 0.1 s of averaging time in a completely unshielded environment (i.e., inside a standard laser laboratory).

6 Conclusion

We have shown that coherent population trapping resonances (dark resonances) can be used as compact atomic frequency references, with a performance level already comparable to commercial rubidium clocks. The dark resonance magnetometer is also just as good in terms of sensitivity as the best commercial room-temperature magnetometers. The long-term stability/sensitivity is limited by the temperature dependence of the resonance position in a buffered cesium vapor cell while the frequency noise of the laser diode determines the short-term stability/sensitivity. An important effect is the noise-frequency-dependence of the optical hyperfine pumping due to unequal optical transition rates on the two legs of the Λ system. This noise process should also be important in other types of optically pumped frequency standards.

We are confident that the dark resonance clock can be integrated into a very small package. Even when using commercial components, all the optics fit into a box of volume $16 \times 18 \times 66\,\mathrm{mm}^3$ (Fig. 10 shows a first prototype). Eventually the electronics could be integrated into a box of similar size. Since there are no movable, acceleration-sensitive parts (unlike in clocks using microwave resonators) and the estimated total power consumption is below

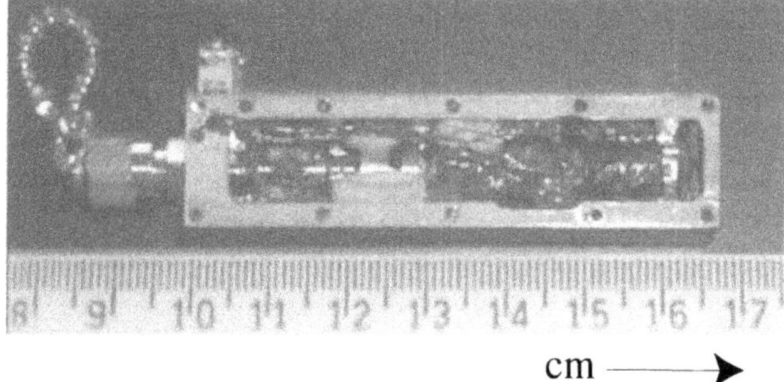

Fig. 10. Photograph of the prototype physics package for the compact atomic clock with cover removed. The case is made from mu-metal and shields its contents from external magnetic fields. It contains the laser, collimator, polarizer, quarter-wave plate, attenuator, cesium vapor cell with heating and thermal insulation and coils for a longitudinal magnetic offset field, and a photodiode

100 mW the clock would be ideal for mobile applications, for instance in portable GPS receivers. In outlying repeater stations for telecommunications networks the reduced power and size requirements should also be advantageous, as is the estimated cost of about $200.

As for the magnetometer, even in the face of measured [36, 37] or predicted [38] femtotesla sensitivity for other magnetometer schemes the picotesla sensitivity of the dark resonance magnetometer is very interesting. First of all, the sensor volumes of the Hanle and Faraday devices quoted above are about a hundred times larger than the 0.8 cm^3 active volume in our case, which alone enhances their sensitivity more than tenfold but at the same time reduces spatial resolution and increases sensitivity to field inhomogeneities. Secondly, Faraday magnetometers are sensitive to the longitudinal field component only, so a small change in field direction relative to the sensor axis (or vice versa in a mobile application) produces a large signal fluctuation. The dark resonance magnetometer, on the other hand, can be operated as a scalar magnetometer, i.e., it measures the total magnetic field strength $|\boldsymbol{B}|$ independent of direction. Because of selection rules [20] one might have to choose a different Zeeman component of the dark resonance in order not to lose sensitivity, depending on field direction. Optical alignment and mechanical stability are completely uncritical: the laser beam only has to hit the photodiode behind the vapor cell. This should make the sensor insensitive to acceleration. In comparison to flux-gate magnetometers a big advantage is that there are no magnetic parts near the sensor, so the sensor head does not distort the field to be measured.

The all-optical setup allows for fiber coupling of the sensor head and for its operation far from the base station with laser and power supply. Unlike today's Faraday magnetometers that only work near zero field strength the dark resonance sensor can be operated at the geostatic field strength and much higher without losing its picotesla sensitivity. In fact, we cannot see changes in the signal-to-noise ratio for flux densities from microteslas to several milliteslas, a limit dictated by the largest permissible current through our field coils. It would be interesting to explore the sensitivity at flux densities of tens of milliteslas where the selection rules change due to recoupling of atomic angular momenta [47].

Acknowledgements

Several students and visitors have made contributions to the work described here. Let me just mention my most recent students, Svenja Knappe, Christoph Affolderbach, and Markus Stähler. For the cesium atomic clock we closely collaborate with Leo Hollberg and John Kitching at NIST in Boulder. Svenja and John also contributed to an early version of this text. Some of the results were obtained using Hugh Robinsons's wonderful "hand-crafted" alkali vapor cells or the high-modulation-bandwidth VCSEL prototypes kindly provided by the group of Karl-Joachim Ebeling at Ulm University. We also thank the DAAD, the Land NRW (Bennigsen-Foerder Prize), the DFG, the US Army, and NIST for funding these projects. And last, but not least, I would like to thank Theodor Hänsch for introducing me to the fascinating world of precision laser spectroscopy.

References

1. J. Barnes, A. Chi, L. Cutler, D. Healey, D. Leeson, T. McGunigal, J. Mullen, W. Smith, R. Sydnor, R. Vessot, G. Winkler: IEEE Trans. IM-**20**, 105 (1971)
2. A. Clairon, P. Laurent, G. Santarelli, S. Ghezali, S. Lea, M. Bahoura: IEEE Trans. IM-**44**, 128 (1995)
3. S. Jefferts, D. Meekhof, L. Hollberg, D. Lee, R. Drullinger, F. Walls, C. Nelson, F. Levi, T. Parker: *NIST Cesium fountain frequency standard: preliminary results*, 1998 IEEE Int. Freq. Contr. Symp. 2–5 (1998)
4. L. Hollberg, C.W. Oates, E.A. Curtis, S.A. Diddams, Th. Udem, J.C. Bergquist, R.E. Drullinger, W.M. Itano, D.J. Wineland: *Optical frequency standards – the clocks of the future*, Int. Conf. on Coherent and Nonlinear Optics (ICONO) 2001, Minsk (June 2001)
5. B. Hoffmann-Wellenhof, H. Lichtenegger, J. Collins: *GPS: Theory and Practice* (Springer, Berlin, Heidelberg 1994)
6. J.A. Kusters, C.A. Adams: RF Design, 28–38 (1999)
7. G. Alzetta, A. Gozzini, L. Moi, G. Orriols: Il Nuovo Cim. B **36**, 5 (1976)
8. E. Arimondo: Prog. Opt. **35**, 257 (1996)
9. S. Brandt, A. Nagel, R. Wynands, D. Meschede: Phys. Rev. A **56**, R1063 (1997)
10. M. Erhard, S. Nußmann, H. Helm: Phys. Rev. A **62**, 061802(R) (2000)
11. P. Kumar, J.H. Shapiro: Opt. Lett. **10**, 226 (1985)
12. J.E. Thomas, S. Ezekiel, C.C. Leiby Jr., R.N. Picard, C.R. Willis: Opt. Lett. **6**, 298 (1981); J.E. Thomas, P.R. Hemmer, S. Ezekiel, C.C. Leiby Jr., R.N. Picard, C.R. Willis: Phys. Rev. Lett. **48**, 867 (1982); P.R. Hemmer, S. Ezekiel, C.C. Leiby Jr.: Opt. Lett. **8**, 440 (1983)

13. R. Wynands, A. Nagel: Appl. Phys. B **68**, 1 (1999); Erratum: Appl. Phys. B **70**, 315 (2000)
14. R. Wynands, T. Mukai, T.W. Hänsch: Opt. Lett. **17**, 1749 (1992)
15. B. Dahmani, L. Hollberg, R. Drullinger: Opt. Lett. **12**, 876 (1987)
16. O. Schmidt, R. Wynands, Z. Hussein, D. Meschede: Phys. Rev. A **53**, R27 (1996)
17. M. Prevedelli, T. Freegarde, T.W. Hänsch: Appl. Phys. B **60**, S241 (1995)
18. R. King, R. Michalzik, C. Jung, M. Grabherr, F. Eberhard, R. Jäger, P. Schnitzer, K.J. Ebeling: *Oxide confined 2D VCSEL arrays for high-density inter/intra-chip interconnects*; in Vertical-cavity Surface-Emitting Lasers II, R.A. Morgan, K.D. Choquette (eds.), Proc. SPIE **3286**, 64–71 (1998)
19. C. Affolderbach, A. Nagel, S. Knappe, C. Jung, D. Wiedenmann, R. Wynands: Appl. Phys. B **70**, 407 (2000)
20. R. Wynands, A. Nagel, S. Brandt, D. Meschede, A. Weis: Phys. Rev. A **58**, 196 (1998)
21. J. Vanier, C. Audoin: *The Quantum Physics of Atomic Frequency Standards* (Adam Hilger, Bristol 1989)
22. J.R. Zacharias: unpublished (1953), described in N.F. Ramsey: Rev. Mod. Phys. **62**, 541 (1990)
23. M. Kasevich, E. Riis, S. Chu, R. De Voe: Phys. Rev. Lett. **63**, 612 (1989)
24. M. Poelker, P. Kumar, S.-T. Hoe: Opt. Lett. **16**, 1853 (1991)
25. C. Rahman, H.G. Robinson: IEEE J. QE–**23**, 452 (1987)
26. N. Vukičević, A.S. Zibrov, L. Hollberg, F.L. Walls, J. Kitching, H.G. Robinson: IEEE Trans. Ultrason.Ferroel. **47**, 1122 (2000)
27. P.R. Hemmer, M.S. Shahriar, H. Lamela-Rivera, S.P. Smith, B.E. Bernacki, S. Ezekiel: J. Opt. Soc. Am. B **10**, 1326 (1993)
28. N. Cry, M. Têtu, M. Breton: IEEE Trans. IM **42**, 640 (1993)
29. J. Vanier, A. Godone, F. Levi: Phys. Rev. A **59**, R12 (1999); F. Levi, A. Godone, J. Vanier: IEEE Trans. Ultrason. Ferroel. **46**, 609 (1999); A. Godone, F. Levi, J. Vanier: IEEE Trans. IM **48**, 504 (1999)
30. J. Kitching, S. Knappe, N. Vukičević, L. Hollberg, R. Wynands, W. Weidemann: IEEE Trans. IM **49**, 1313 (2000)
31. N. Allard, J. Kielkopf: Rev. Mod. Phys. **54**, 1103 (1982)
32. G.C. Bjorklund, M.D. Levenson, W. Lenth, C. Ortiz: Appl. Phys. B **32**, 145 (1983)
33. S. Knappe, R. Wynands, J. Kitching, H.G. Robinson, L. Hollberg:J. Opt. Soc. Am. B, in press
34. J. Kitching, H.G. Robinson, L. Hollberg, S. Knappe, R. Wynands: J. Opt. Soc. Am. B, in press
35. N. Beverini, F. Strumia, G. Rovera: Opt. Comm. **37**, 394 (1981)
36. C. Cohen-Tannoudji, J. DuPont-Roc, S. Haroche, F. Laloë: Phys. Rev. Lett. **22**, 758 (1969)
37. E.B. Alexandrov, M.V. Balabas, A.K. Vershovskii, A.E. Ivanov, N.N. Yakobson, V.L. Velichanskii, N.V. Senkov: Opt. Spectrosc. **78**, 292 (1995)
38. D. Budker, D.F. Kimball, S.M. Rochester, V.V. Yashchuk, M. Zolotorev: Phys. Rev. A **62**, 043403 (2000)
39. M.O. Scully, M. Fleischhauer: Phys. Rev. Lett. **69**, 1360 (1992)
40. M. Fleischhauer, M.O. Scully: Phys. Rev. A **49**, 1973 (1994)
41. M. Fleischhauer, A.B. Matsko, M.O. Scully: Phys. Rev. A **62**, 013808 (2000)

42. A. Nagel, L. Graf, A. Naumov, E. Mariotti, V. Biancalana, D. Meschede, R. Wynands: Europhys. Lett. **44**, 31 (1998)
43. H. Lee, M. Fleischhauer, M.O. Scully: Phys. Rev. A **58**, 2587 (1998)
44. K.L. Corwin, Z.-T. Lu, C.F. Hand, R.J. Epstain, C.E. Wieman: Appl. Opt. **37**, 3295 (1998)
45. M. Stähler, S. Knappe, C. Affolderbach, W. Kemp, R. Wynands: Europhys. Lett. **53**, 323 (2001)
46. N.D. Bhaskar: IEEE Trans. Ultrason. Ferroel. Frequ. Control **42**, 15 (1995)
47. R. Höller, F. Renzoni, L. Windholz, J.H. Xu: J. Opt. Soc. Am. B **14**, 2221 (1997)

Part II

At the Limits of High Resolution Spectroscopy

Theodor W. Hänsch and Arthur Schawlow during the 1980s at Stanford University

High-Noise, Low-Resolution Spectroscopy

Steven Chu

For over three decades, Theodor Hänsch has invented numerous elegant lasers and laser methods that have defined and redefined "laser physics at the limit". With these methods, he has extended the reach of physics with experiments of great precision and sensitivity.

As an example, Fig. 1 summarizes the billionfold increase in the resolution of hydrogen spectroscopy in the past forty years. This exponential improvement is marked by two dramatic changes in the slope, both due to Theodor. The first revolution occurred in 1972, when he used a tunable dye laser to measure the Balmer-α line of hydrogen [1]. The second quantum jump occurred in 1992, when he made an absolute frequency measurement of the 1S–2S transition relative the a CH_4-stabilized HeNe laser with a chain of phase-locked lasers [2].

Beginning with saturation spectroscopy, he has invented numerous methods of probing the sharp atomic lines hidden beneath the Doppler-broadened transitions [3]. Ultimately, Theodor and Art Schawlow were able to devise a method of essentially eliminating Doppler broadening by laser cooling the atoms [4].

Theodor's discovery of the extraordinary phase stability of widely separated frequencies in a mode-locked laser allows us to directly count optical frequencies [5]. This discovery has begun to revolutionize laser spectroscopy and the entire field of metrology.

Stimulated by Theodor's beautiful application of two-photon spectroscopy to the 1S–2S transition of hydrogen, Allen Mills and I applied this technique to the purely leptonic atoms, positronium and muonium. Rather than try to match the beautifully precise measurements of Theodor and his co-workers, we decided to make advances in the field of low-resolution, high-noise laser spectroscopy with our excitation of the 1S–2S transition in muonium [6]. After a considerable amount of hard work at the muon facility at KEK in Japan, we were able to obtain the spectra shown in Fig. 2 with a grand total of ~ 8 atomic excitations, slightly over one excitation per collaborator! These atoms were difficult to produce on demand and, admittedly, most of our efforts were spent developing an intense source of thermal muonium. Our source was eventually able to produce an unprecedented one atom per laser pulse in a volume of $\sim 20\,\mathrm{cm}^3$, corresponding to a time-averaged density of $\sim 2 \times 10^{-6}$ atoms/cm^3. Needless to say, suppression of spurious background counts also occupied much of our time.

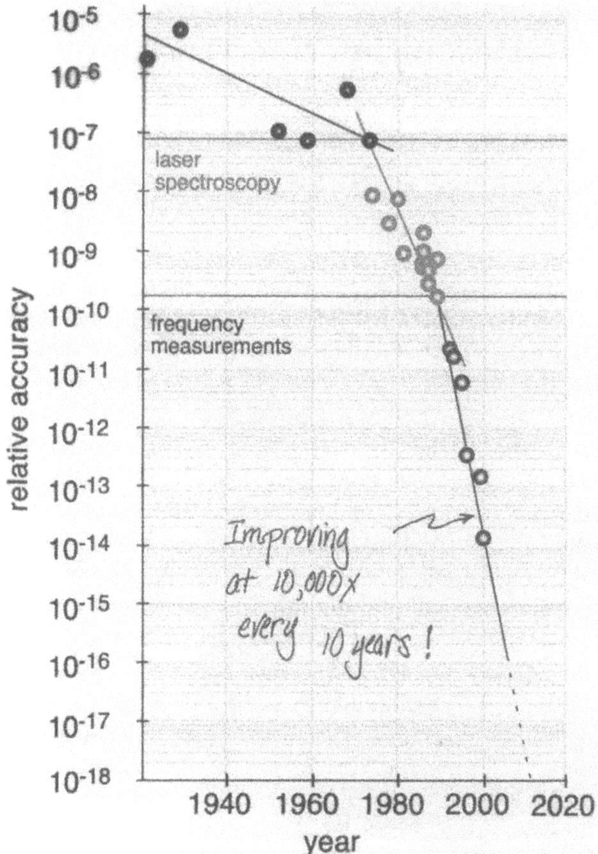

Fig. 1. The recent history of hydrogen sprectroscopy

Ironically, during this same period, we also demonstrated the magneto-optic trap (MOT) [7]. In contrast to the muonium work, the MOT worked effortlessly. The impact of this almost trivial-to-do experiment provided me with an important lesson: do the easy, important experiments first and save the less important, harder experiments for later. The challenge is how to identify the easy experiments that will have great impact.

In this respect, I had the good fortune to work with Art Ashkin. He was a master at identifying these types of experiments, as exemplified by his beautiful experiments trapping glass and polystyrene beads, viruses and bacteria [8]. I came to Stanford realizing that there was an opportunity to explore the molecular domain between trapping atoms and bacteria. Besides, Art's optical tweezers experiments looked like great fun.

I set about to trap a long piece of DNA with the intent of looking at enzymes such as RNA polymerase walk along the DNA. Working with Steve Kron, then an MD-Ph.D. student at Stanford, we began to develop the molec-

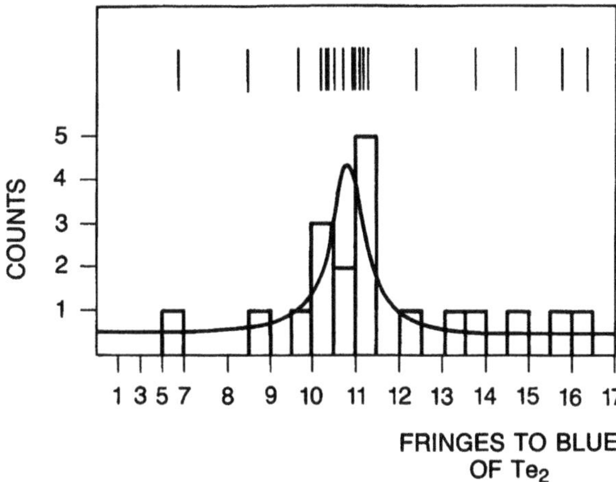

Fig. 2. The 1S–2S resonance in muonium (μ^+e^-). The data is the sum of all scans totaling 19 hours of integration time

ular biology protocol needed to glue micron-sized polystyrene spheres to the ends of the molecule. By 1990, we were able to stretch out and simultaneously image a single fluorescently labeled DNA molecule in water [9].

Once we began to play with the molecules, we noticed that our DNA molecules would spring back like a rubber band when released from the stretching forces. This accidental observation of a single molecule rubber band led me to realize that DNA could be used as a model polymer, large enough to visualize and simultaneously manipulate and yet small enough for the basic equation of motion (the Langevin equation) to still be valid [10]. We had stumbled onto a new way of addressing long standing questions in polymer dynamics, and were able to complete a number of relatively easy but significant experiments [11].

During the course of this work, we discovered something very new. Identical molecules placed in the same external conditions would take several distinct pathways to a new equilibrium state, as shown in Fig. 3. This "molecular individualism" was never anticipated in a half a century of theoretical and experimental work. In hindsight, our discovery turned out to be a simple example of non-equilibrium statistical mechanics [12], but it stimulated us to conjecture whether other macromolecules, particularly biomolecules, would act in a similar fashion.

Virtually all of our knowledge in chemistry and biology has been gleaned from experiments that measure the average properties of a large ensemble of molecules. We asked ourselves whether biological processes such as protein folding and enzyme activity could also have a number of different paths from one equilibrium state to another. If these paths existed, would they have

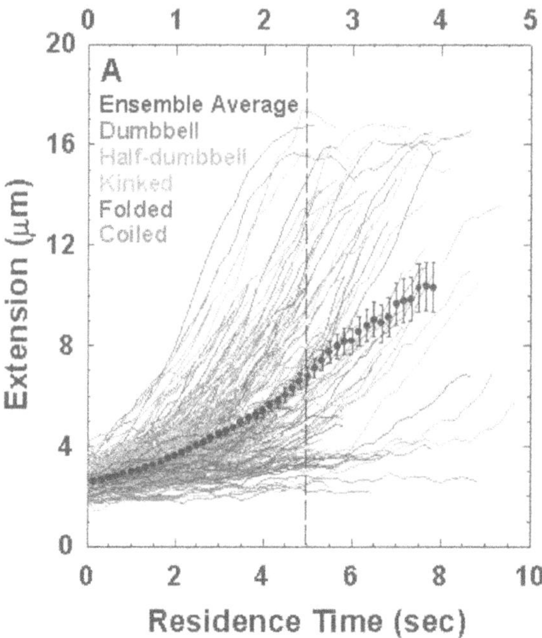

Fig. 3. The extension of a randomly coiled polymer in an elongational flow

been discovered by experiments that looked at bulk samples? Could these ensemble experiments have missed rare transient molecular states?

In our polymer experiments, we simply scaled up the size of the DNA until we could image the polymer in an optical microscope. Unfortunately, the size of biologically interesting molecules was fixed by nature, and we needed a new trick to follow the motion of an individual protein.

The method we decided to use for our single-molecule studies was actually an old trick: fluorescence resonance energy transfer (FRET) [13]. Donor and acceptor dyes attached to two sites of a biological molecule can measure the distance between the two dyes: donor fluorescence emission is strongly quenched in a distance-dependent manner by the acceptor, while the acceptor emission increases due to the energy transfer. Thus, a measurement of a change in fluorescence from the two dyes can be used as an indicator of a change in the conformation of the host molecule. Since the two fluorophorers are on different parts of the biomolecule, intramolecular motion can be measured free from the Brownian motion of the molecular center of mass.

Thus, we embarked on a research direction that is the antithesis of the ultra-high resolution spectroscopy made possible by Doppler-free methods, laser cooling and atomic fountains. FRET required us to differentiate between donor and acceptor emission frequencies separated by $\sim$100 nm. Furthermore, the fluorescent signals of individual molecules (that would photo-bleach after emitting $\sim 10^6$ photons) did not give gloriously large signals.

When we entered the field, the FRET technique had recently been extended to the single-molecule domain [14]. Although interesting fluctuations in energy transfer were seen as arising from photophysical events, orientation and distance changes, their relation to changes in the overall shape of a protein was uncertain. Also, in these earlier experiments, the molecules diffused in and out of an observation region defined by a confocal microscope.

In our first work, we chose to study a model RNA system consisting of three short RNA segments in a planar geometry with $\sim 120°$ angles between each pair of helices as shown in Fig. 4 [15]. Upon the binding of either a ribosomal protein (S15) or Mg^{2+} ions, one of the helices was believed to rotate by 60°. In this work, we showed that the conformational change of individual surface immobilized RNA molecules can be unambiguously observed with FRET. Since the molecules were immobilized, the change could be induced by changing the buffer conditions. We were able to make the RNA wave its arms on command by rapidly exchanging solutions with and without Mg^{2+} ions. Furthermore, the changes in the fluorescence signal were in good agreement with the predicted energy-transfer efficiency based on the known structures of the RNA system and the properties of the dye molecules. The time resolution of our single molecule measurement was a few milliseconds, limited by the shot noise in our photon counting system.

After we demonstrated the feasibility of working with immobilized single molecules with a model system, we turned our attention to protein and RNA folding. There were presumed to be many folding pathways that an amino acid chain or RNA strand can take en route to its native, biologically active

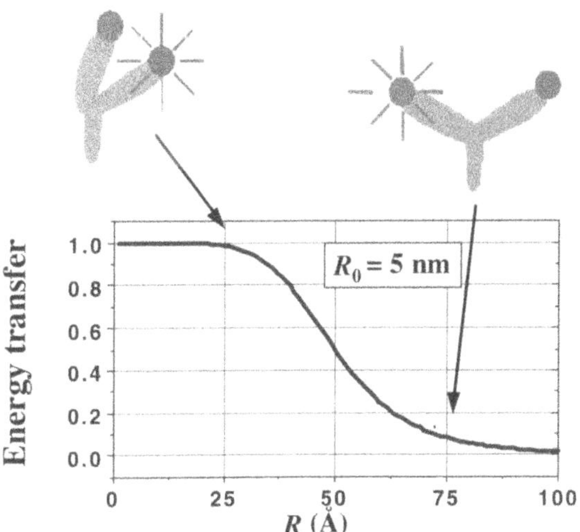

Fig. 4. Sketch on the RNA 3-junction helix and the conformational change induced by Mg^{2+} ions or the S15 Protein. The curve shows the energy-transfer efficiency: $1/[1 + (R/R_0)^6]$

state. Furthermore, it is not obvious that these molecules would have a unique native state. For example, thermal fluctuations may induce a particular enzyme to flip between biologically active and inactive states. We felt that there was a good chance of observing such behavior if we could follow the behavior of individual molecules.

Protein folding to the native state can be represented by a "folding landscape" as depicted in Fig. 5. Most proteins fold rapidly into their native states in a time scale of milliseconds or less and generally avoid partially folded, metastable states. Thus, single molecule methods based on fluorescence would have a hard time resolving this motion.

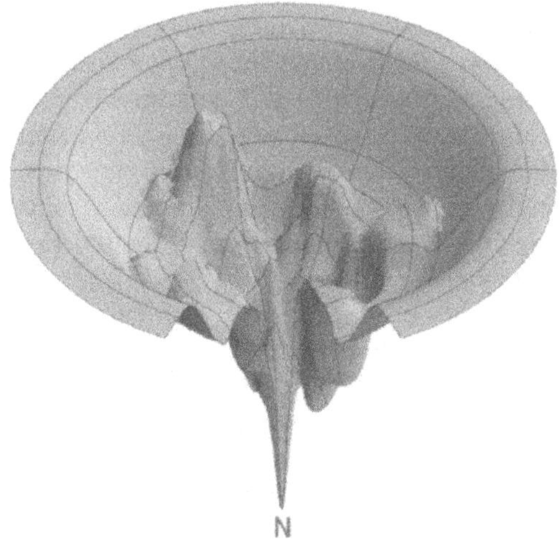

Fig. 5. Schematic of the folding landscape of a protein where free energy is plotted vertcally and the other axes are used to represent a multi-dimensional set of reaction coordinates relevant to folding. The native state N defines the free energy minimum. Random starting ponts on the rim of the folding funnel could result in different paths on the native state N

On the other hand, the folding time of RNA enzymes, referred to as "ribozymes", could be easily resolved with the single-molecule methods we were using. Also, the molecules were expected to fold in a more rugged folding landscape with long-lived metastable states that could be detected with our methods. In short, the study of RNA folding was the equivalent of "low hanging fruit", begging to be plucked first.

In our next work, we studied the catalysis and folding of individual *Tetrahymena* ribozyme molecules [16]. This enzyme has historical importance as the first RNA discovered that could fold, on its own, into an active enzyme. The discovery ribozymes resulted in the Nobel Prize for Tom Cech and Sidney Altman.

Our study of the *Tetrahymena* ribozyme demonstrated, for the first time, that single-molecule fluorescence could be used to observe the biological activity of RNA. We showed that dye-labeled and surface-immobilized ribozymes used in the experiment were functionally indistinguishable from the unmodified ribozyme free in solution. Our approach allowed us to observe a reversible folding step in which a portion of the RNA "docks" and "undocks" from the ribozyme core in real time. A rarely populated docking state, not measurable by ensemble methods, was also observed.

We went on to follow the unfolding and refolding of this immobilized molecule back to its full biological activity. In the overall folding process, intermediate folding states and multiple folding pathways were observed. Previously folding rates, measured in solution, were confirmed, and a new folding pathway with an observed folding rate constant of $1\,s^{-1}$ was discovered. We then performed a bulk experiment that verified our single-molecule result.

In subsequent work, we followed folding of the *Tetrahymena* ribozyme beginning from distinct regions within the folding landscape by varying the starting solution conditions [17]. By following the folding of individual ribozyme molecules as a function of initial Na^+ concentration, we found that folding proceeds through channels in the landscape, with the initial solution conditions determined the probability of populating a particular channel (see Fig. 6). These results suggest that the molecular contacts formed at the onset of folding and the order in which additional contacts are formed have profound effects on the overall folding kinetics for RNA.

We have also followed the enzymatic pathway of the Hairpin ribozyme, perhaps the simplest RNA enzyme known [18]. Previous studies of the Hairpin ribozyme have suggested that the enzyme follows a complex set of steps that result in the cleavage of an RNA substrate. Using both FRET and a novel triple-probe combination of FRET and a fluorescence quencher, we were able

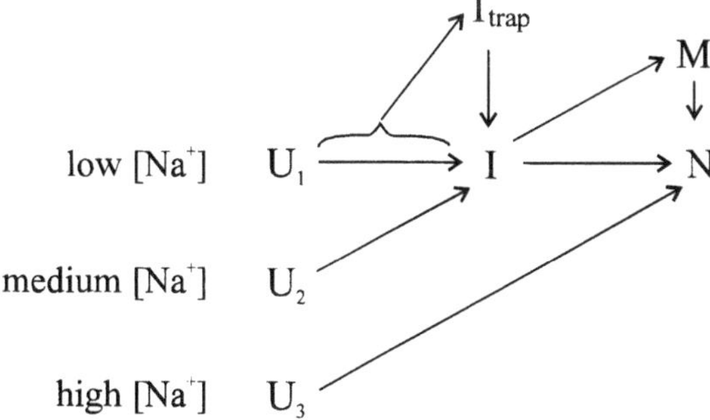

Fig. 6. By changing the initial Na concentration, differnent secondary structures U_1, U_2 and U_3 cause the enzyme to fold along different pathways. The time spent in $I_{trap} \sim 60\,s$. M is a long-lived misfolded state that lasts for $> 100\,h$ at room temperature. N is the native state an I is a new state postulated because of these measurements

to map out the sequence of multiple steps in the complex reaction pathway of the hairpin ribozyme and directly measure many of the rate constants associated with the function of this enzyme.

Measurements on bulk samples were unable to unravel the enzymatic pathways, in part because of thermal fluctuations between a "docked" state, where cleavage occurs, and an "undocked" state where the enzyme is inactive. Our ability to measure the docking and undocking rates of this enzyme showed that the docking/undocking reactions are the rate-limiting steps for this enzyme, and not the cleavage and ligation reactions, as was previously believed.

The most surprising result was that one of the reaction sub-steps exhibited a type of dynamic disorder. An undocking step in the reaction showed at least four distinct reaction rates. Furthermore, there was a strong correlation between events: if the time to undock was fast, there was a high probability that the time for the next undocking step would also be fast (see Fig. 7). This

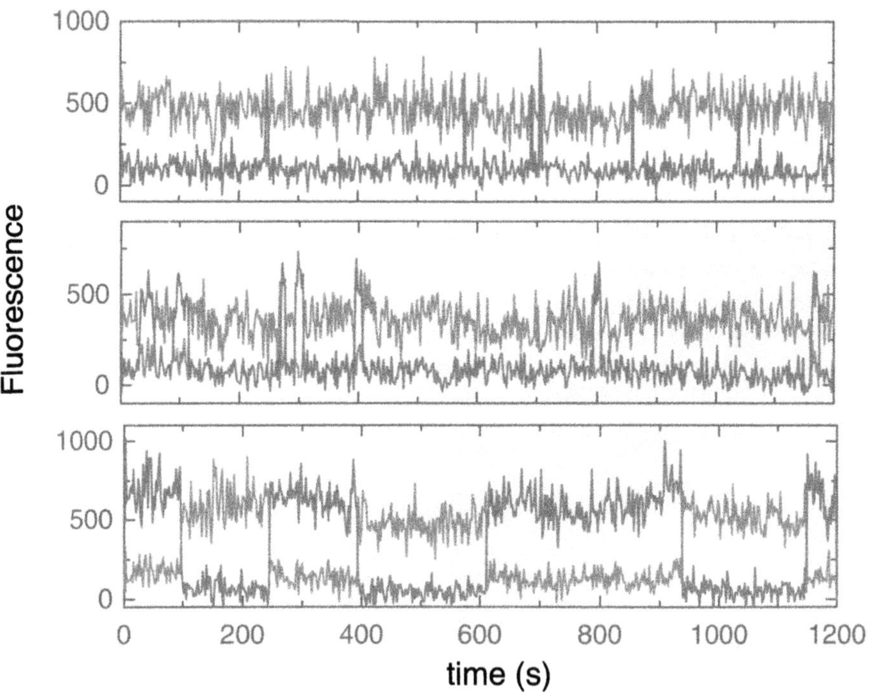

Fig. 7. FRET time traces of the docking and undocking of single Hairpin ribozyme molecules. High acceptor flourescence (lower traces in top two figures) indicates the docked state. Note that short, medium and long docking events have a high probability of being followed by a similar event. Occassionally, we see time traces that show transitions that go between the various docked states (traces "switch" in bottom figure)

work revealed that this simple enzyme exhibits a type of molecular memory. We are now searching for the molecular origin of this effect.

We have also used single-molecule FRET to study the interaction of *E. coli* Rep helicase with DNA [19]. Helicase enzymes unwind double-stranded DNA into its single-stranded components, and play a critical role in replication and DNA repair. We were able to observe several distinct conformational changes on individual strands of DNA induced by the enzyme. Moreover, we were able to associate specific FRET states with known molecule structures obtained through X-ray diffraction. Thus, our single-molecule studies could now link known X-ray structures with specific FRET signals. We found rapidly fluctuating transient states in the presence of ATP which were not detected with traditional molecular biology techniques. This work showed that single molecule studies can provide a low-resolution recording that sheds light on the transitions between molecular structures that can be captured by X-ray diffraction.

Our experience with relatively simple molecular systems has encouraged us to study more complex biological systems. We have also begun long-term studies on three other biological systems. In collaboration with Jodi Puglisi, in the Structural Biology Department at Stanford, we are developing methods to follow the manufacture of proteins by the prokaryotic ribosome as shown in Fig. 8. (Prokaryotes are cells, such as bacteria, that do not have a nucleus.) Since many of our antibiotics kill bacteria by binding to this molecular machine, a detailed understanding of how these mechanisms work will help us design improved antibiotics.

The following is an abbreviated outline of how proteins are manufactured in bacteria. The manufacture of proteins begins with a number of initiation proteins that causes a transfer RNA (upper tRNA) to bind to the ribosome.

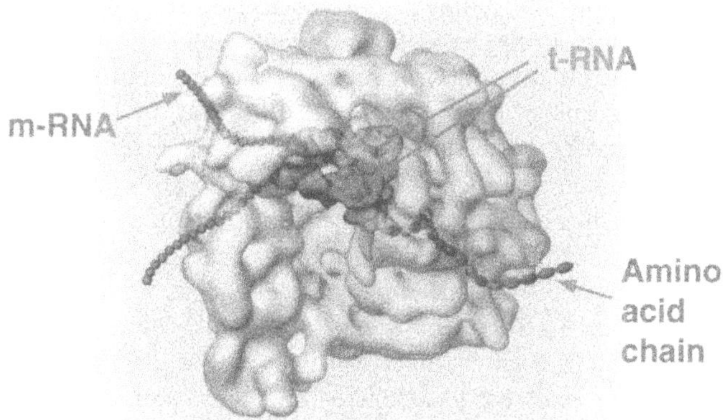

Fig. 8. A low-resolution electron density map of the ribosome shwoing the location of tRNA in the upper P- and lower A-sites. The MRNA and amino acid chain are sketches

Another tRNA (lower) binds to the adjacent site with the aid of additional proteins. The particular species of tRNA is determined by the 3-bases presented by the messenger RNA (mRNA) at the foot of the binding sites. After the tRNA is properly positioned, the two amino acids attached to the tRNA are linked together by another enzyme. After the formation of this bond, the green tRNA falls off the ribosome and the purple RNA moves to the green site, freeing up its binding site for another tRNA.

We hope to follow the steps of elongation at the single-molecule level. By placing fluorescence probes on the tRNA, on the amino acids attached to the tRNA, and on suitable proteins that are part of the ribosome, we hope to explore the kinetic behavior of this molecule.

In collaboration with Axel Brunger in the Neurology and Neurological Sciences Departments at Stanford, we have begun single-molecule studies of the dynamics of the SNARE family of proteins. This family of proteins are involved in signaling of nerve cells at the synapse junction.

Communication between nerves begins with a set of voltage pulses that propagate down the axon of a nerve cell. This voltage pulses causes a synaptic vesicle containing neurotransmitters to fuse with the cell membrane as shown in Fig. 9.

The fusion event, believed to be caused by the SNARE proteins, allows the neuro-transmitters contained within the vesicle to be ejected from the cell where they can diffuse rapidly across the synaptic junction to receptor sites on the other nerve cell. Additional proteins regulate the overall rates of vesicle fusion.

In experiments now in progress, we hope to observe the details of vesicle fusion between artificially formed vesicles attached to an artificial membrane deposited on a quartz substrate. These lipid structures contain the SNARE proteins believed to cause fusion. Single-molecule fluorescence microscopy methods will be used to follow the dynamics of confirmation of individual proteins and how their motion is related to fusion events.

Finally, we have also begun a collaboration with James Nelson in the Molecular and Cellular Physiology Department at Stanford to study adhesive molecules that bind cells together. A major binding mechanism is due to a family of proteins known as cadherins. Cadherins are cell-surface proteins that bind to identical molecules on adjacent cells. Besides their role in maintaining the architecture of mature tissue, cadherins are involved in the formation and differentiation of tissue during development of an organism. Normal cells that do not bind to other cells will undergo programmed cell death (cell suicide); so it is suspected that these molecules also affect the regulatory proteins in the nucleus. Deficiencies and mutations of certain cadherins correlate with the onset and metastasis of cancer.

Though the interactions between cadherins have been extensively studied, little is known about the orientation and molecular interactions of the protein in its bound complex. Our initial goal is to determine the molecular

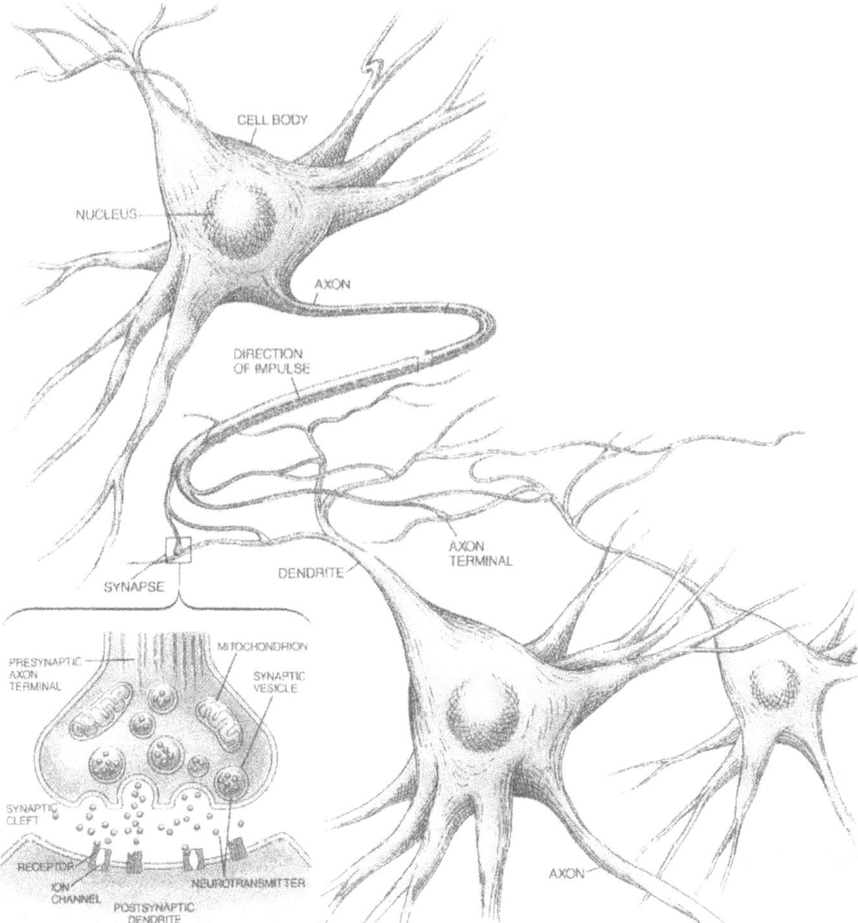

Fig. 9. Illustration of neurotransmission at synapse

mechanisms for cadherin-mediated adhesion using a combination of force measurements using an atomic force microscope (AFM) with single-molecule FRET.

Currently we are developing AFM methods to study cadherin binding. We are also engineering recombinant cadherins with mutations that will allow us to label the protein with fluorescent dyes. In this way, we should be able to measure the force of cadherin binding with molecular-scale distance measurements made possible with FRET. If these initial physics measurements are successful, we will try to relate the external physical binding of cadherin molecules to the regulatory mechanisms in the nucleus of the cell.

The work I have described is a far cry from my roots in atomic physics: parity violation in atomic transitions, positronium and muonium spectroscopy,

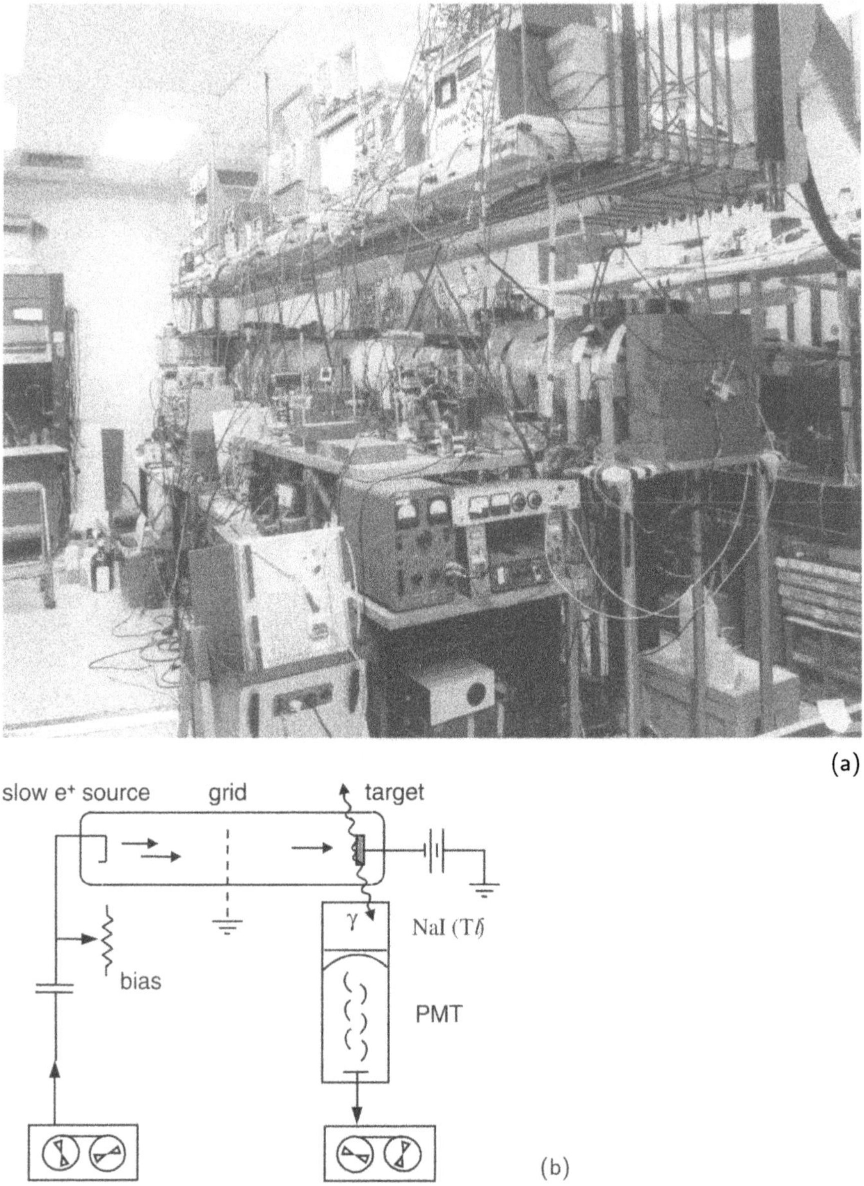

(a)

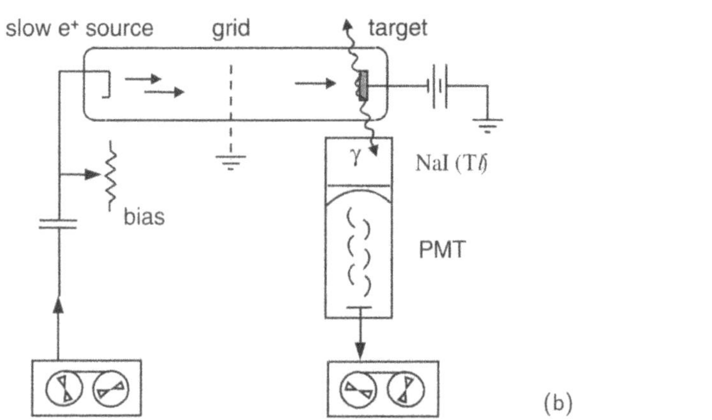

Fig. 10. (**a**) A view of our positronium apparatus at Bell Labs, circa 1983. (**b**) Schematic of the first anti-matter vacuum tube

laser cooling and trapping, and atom interferometry. I still use laser spectroscopy, albeit at very low resolution. Because we are trying to observe the workings of individual molecules in water solutions, the signals are plagued with significant noise. However, these applications of physically based methods, when coupled to molecular biology methods, provide a new window into the world of molecular biology.

In closing, I return to atomic physics to discuss an application to high noise, low resolution spectroscopy.

Figure 10a is a photograph of the apparatus used to generate the thermal positronium we used in an experiment to measure the 1S–2S transition. In the background to the right is the laser table holding the cw dye laser/pulsed amplifier system and associated metrology equipment.

What may not be apparent to the casual observer is that this apparatus was also the world's first positron vacuum tube. A schematic of Allen Mills' invention is shown in Fig. 10b. He made a grounded-grid, PNP vacuum tube amplifier with amplitude-modulated positron transmitter and gamma-ray receiver. Since the gamma ray detector, consisting of a NaI (Tl) scintillator and photomultiplier tube, had an average counting rate of a few kilohertz, audio-frequency modulation of the positron source was possible. Surprisingly, the Bell Labs patent attorneys were not wildly enthusiastic about this invention.

Fig. 11. Theodor as a young professor at Stanford

The reason I resurrect him this invention is to pay homage to Theodor´s approach to science. He introduced us to optical wavemeters pulled by toy railroad trains [20], incr(edible) (Jell-O) lasers [21], and bits of aluminium held in a Paul Trap made out of a ladies earring, and dancing to the tune of the Blue Danube Waltz [22].

In Fig. 11, we see Theodor, as an assistant professor at Stanford, looking at the world with child-like joy and wonder. That spirit still remains in all that he does.

References

1. T.W. Hänsch, I.S. Shahin, A.L. Schawlow, Nature **235**, 63 (1972)
2. T. Andrease, W. König, R. Wynands, D. Leibfried, F. Schmidt-Kaler, C. Zimmermann, D. Meschede, T.W. Hänsch, Phys. Rev. Lett. **69**, 1923 (1992)
3. P.W. Smith, T.W. Hänsch, Phys. Rev. Lett. **26**, 740 (1971)
4. T.W. Hänsch, A.L. Schawlow, Opt. Commun. **13**, 68 (1975)
5. M. Bellini, T.W. Hänsch, Appl. Phys. B **65**, 677 (1997); M. Bellini et al., Phys. Rev. Lett. **81**, 297 (1998)
6. S. Chu, A.P. Mills Jr., A.G. Yodh, K. Nagamine, H. Miyake, T. Kuga, Phys. Rev. Lett. **60**, 101 (1988)
7. E.L. Raab, M. Prentiss, A.E. Cable, S. Chu, D.E. Pritchard, Phys. Rev. Lett. **59**, 2631 (1987)
8. For a review, see A. Ashkin, Proc. Natl. Acad. Sci. USA **94**, 4853 (1997)
9. S. Chu, S. Kron, Int. Quantum Electronics Conf. Tech. Digest (Optical Soc. Am. Washington D.C. 1990) p. 202
10. S. Chu, Science **253**, 861 (1991)
11. As a partial sampling, see: T. Perkins, D.E. Smith, S. Chu, Science **64**, 819 (1994); T.T. Perkins, S.R. Quake, D.E. Smith, S. Chu, Science **264**, 822 (1994); T.T. Perkins, D.E. Smith, R.G. Larson, S. Chu, Science **268**, 83 (1994); D.E. Smith, T.T. Perkins, S. Chu, Phys. Rev. Lett. **75**, 4146 (1995); D.E. Smith, T.T. Perkins, S. Chu, Macromolecules **29**, 1372 (1996); S.R. Quake, S. Chu, Nature **388**, 151 (1997); T.T. Perkins, D.E. Smith, S. Chu, Science **276**, 2016 (1997); D.E. Smith, S. Chu, Science **281**, 1335 (1998); D.E. Smith, H.P. Babcock, S. Chu, Science **283**, 1724 (1999); H.P. Babcock, D.E. Smith, J. Hur, E. Shaqfeh, S. Chu, Phys. Rev. Lett. **85**, 2018 (2000); H. Babcock, R. Teixeira, J. Hur, E. Shaqfeh, S. Chu, submitted to Science (2001); J. Hur, E. Shaqfeh, H. Babcock, S. Chu, submitted to Phys. Rev. Lett. (2001)
12. T.T. Perkins, D.E. Smith, S. Chu in *Flexible Chain Dynamics in Elongational Flow*, ed. by H.K. Kausch, T.Q. Nguyen (Springer, Berlin, Heidelberg 1999)
13. L. Stryer, Science **162**, 526 (1968)
14. T.J. Ha. T. Enderle, D.F. Ogletree, D.S. Chennla, P.R. Selvin, S. Weiss, Proc. Natl. Acad. Sci. USA **93**, 6264 (1996); G.J. Schutz, W. Trabesinger, T. Schmidt, Biophys. J. **74**, 2223 (1998); T.J. Ha, A.Y. Ting, J. Liang, W.B. Coldwell, A.A. Deniz, D.S. Chennla, P.G. Schultz, S. Weiss, Proc. Natl. Acad. Sci. USA **96**, 893 (1999)
15. T. Ha, X. Zhuang, H.D. Kim, J.W. Orr, J.R. Williamson, S. Chu, Proc. Natl. Acad. Sci. USA **96**, 9077 (1999)

16. X. Zhuang, L.E. Bartley, H.P. Babcock, R. Russell, T. Ha, D. Herschlag, S. Chu, Science **288**, 2048 (2000)
17. R. Russell, X. Zhuang, H.P. Babcock, I.S. Millett, S. Doniach, S. Chu, D. Herschlag, Proc. Natl. Acad. Sci. USA (2001) (submitted)
18. X. Zhuang, H.D. Kim, H.P. Babcock, N. Walter, S. Chu, to be submitted to Science, (2001)
19. T. Ha, H.P. Babcock, W. Cheng, T.M. Lohman, S. Chu, unpublished (2001)
20. H-R. Xia, S.V. Benson, T.W. Hänsch, Laser Focus **17**, 54 (1981)
21. T.W. Hänsch, in *Lasers, Spectroscopy and New Ideas: A Tribute to Arthur L. Schawlow*, ed. by. W.M. Yen, M.D. Levenson (Springer, Berlin, Heidelberg 1987) pp. 3–16
22. An after-dinner speech given by Theodor Hänsch

Two-Photon Spectroscopy of Hydrogen

François Biraben and Lucile Julien

1 Introduction

Among the various research themes where Theodor Hänsch has made a decisive contribution, that of the hydrogen atom has a special place in our hearts. One reason is that for over more than than 15 years our Paris group has been in hard but friendly competition with his group, which was a quite stimulating though difficult experience for us.

As pointed out in a well-known paper of Hänsch, Schawlow and Series [1], the hydrogen atom has a central position in the history of 20th-century physics. As it is the simplest of atoms, it has played a key role as a test of fundamental theories, and hydrogen spectroscopy is associated with the successive major advances in the understanding of atomic structures. During the 1970s, the development of tunable lasers opened the way to considerable improvements in resolution, thanks to the possibility of eliminating the Doppler broadening of the lines. Never at a loss for new ideas, Theodor Hänsch has made very essential contributions to this field, both in single-photon [2] and in two-photon spectroscopy. Nowadays, when one gives students a lecture on high-resolution spectroscopy, one is led inevitably to illustrate it using recordings obtained on atomic hydrogen by his group.

Because of the limited length of this paper, we will discuss here only work performed on hydrogen by two-photon spectroscopy.

2 The 1S–2S Transition

The name of Theodor Hänsch has appeared ever since the beginning of two-photon spectroscopy. One of us (F. B.) has directly experienced the birth of this field, because his thesis, led by Bernard Cagnac, dealt with this subject. At the time, using a flashlamp-pumped dye laser, we succeeded in observing a two-photon transition in sodium with a linewidth of a few hundred MHz and a signal-to-noise ratio of about one [3]. A few months later, Theodor Hänsch published the results of a similar experiment with a cw dye laser: the linewidth was 10 MHz and the signal to noise ratio about 100 [4]. Our group in Paris was for the first time in competition with Theodor Hänsch and, obviously, it was very difficult!

The idea to apply this two-photon spectroscopy to the 1S–2S transition in hydrogen was immediately proposed by several authors: Cagnac [5], Baklanov and Chebotayev [6], and T. Hänsch [4]. It was a very attractive proposal, because of the very small natural width (1.3 Hz), but also very difficult, due to the wavelength of this transition (243 nm) and, since there is no intermediate state, the relatively low transition probability. For these reasons, we gave up this idea in Paris at the time. Nonetheless, these difficulties did not deter Theodor Hänsch from accepting this challenge.

Very quickly, Theodor Hänsch obtained in 1975 the first results, using a pulsed dye laser and a flowing hydrogen cell. Table 1 shows the features of the successive 1S–2S experiments since this pionnering work. For comparison, we have also included the work of several other groups in Southampton [10], Oxford [14–16], Yale [20] and MIT [21]. The progress since 1975 is illustrated by the impressive reduction of the linewidth, from 800 MHz to 2 kHz. The crucial improvements were cw excitation, UV production with a BBO crystal, and the use of either a magnetic trap or a cold atomic beam colinear with the UV laser beams. In the more recent design, the Garching experiment uses this longitudinal geometry. Atomic hydrogen produced in a discharge outside the vacuum chamber is emitted collinearly to the axis of the UV beams by a nozzle which may be cooled at liquid helium temperature. At the end of the atomic beam, the 2S atoms are detected by applying a dc electric field to mix the 2S and 2P states, after which the Lyman-α fluorescence is monitored. Moreover, to reduce the second-order Doppler effect, a time-of-flight technique is employed to select the slowest atoms.

Table 1. Main steps of the study of 1S–2S transition in hydrogen since 1975.The indicated linewidths are the fullwidth in term of atomic frequency (two times the optical frequency at 243 nm)

Year	Reference	UV production	H production	Linewidth
		Pulsed laser experiments		
1975	Stanford [7]	lithium formate crystal	flowing cell	800 MHz
1980	Stanford [8]	lithium formate crystal	flowing cell	300 MHz
1986	Stanford [9]	urea crystal	atomic beam	120 MHz
1986	Southampton [10]	urea crystal	flowing cell	200 MHz
		CW laser experiments		
1985–1989	Stanford [11–13]	SFG in KDP	flowing cell	16 MHz
1987–1989	Oxford [14]	SHG in BBO	flowing cell	9 MHz
1990–1992	Oxford [15, 16]	SHG in BBO	atomic beam	10 MHz
1990–2000	Garching [17–19]	SHG in BBO	cold atomic beam	120→2 kHz
1994	Yale [20]	SHG in BBO	low pressure cell	10 MHz
1996	MIT [21]	SHG in BBO	magnetic trap	6 kHz

Table 2. Determination of the 1S Lamb shift from direct comparison of hydrogen frequencies

Year	Reference	Transitions	L_{1S} (MHz)
1975	Stanford [7]	1S–2S and 2S–4P	8600 (800)
1980	Stanford [8]	1S–2S and 2S–4P	8151 (30)
1992	Oxford [16]	1S–2S and 2S–4P	8168 (8)
1992	Garching [22]	1S–2S and 2S–4S/4D	8172.82 (11)
1995	Garching [24]	1S–2S and 2S–4S/4D	8172.878 (51)
1995	Yale [20]	1S–2S and 2S–4P	8172.834 (48)
1996	Paris [25]	1S–3S and 2S–6S/D	8172.825 (47)

The main results from these experiments are more and more accurate determinations of the 1S Lamb shift and the Rydberg constant. Two kinds of experiments must be distinguished: (i) direct frequency comparison with another hydrogen transition, (ii) absolute wavelength or frequency measurements. In the first method, pioneered by Theodor Hänsch, the 1S–2S frequency is compared with transitions whose energies lie namely in a ratio 4, the 2S–4P [7, 8, 16, 20], or 2S–4S/D transitions [22–24]. This last measurement gave us the opportunity to collaborate with Theodor Hänsch, because the 2S–4S/D set-up in Garching was similar to the two-photon experiment in Paris (see Sect. 3). In our group, we have also performed an experiment based on the same idea, but by comparing the 1S–3S and the 2S–6S/D frequencies [25]. As a result of all this work, the accuracy of the measurement of the 1S Lamb shift has been improved from 800 MHz in 1975 (Hänsch's first value) to a mere 50 kHz today (see Table 2; in this table, the three last results have been updated by taking into account a better determination of the 2S Lamb shift [26]).

The absolute determination of the 1S–2S interval (see Table 3) used first an interferometric method, with a calibrated absorption line of $^{130}Te_2$ at 486 nm serving as reference [9, 10, 12–14]. In these experiments, the accuracy was limited by that of the $^{130}Te_2$ reference (about 2.7×10^{-10}). In the last decade, this limitation has been overcome thanks to optical frequency measurements. In a first experiment, Theodor Hänsch used a frequency chain which linked the 1S–2S frequency (2466 THz) to a transportable CH_4-stabilized He–Ne frequency standard at 88 THz. In this chain, a residual frequency splitting of 2 THz was initially measured with a Fabry–Pérot interferometer [27], then with an optical divider [28]. Right now, this complex frequency chain has been superseded by a femtosecond laser-frequency comb, which links in one swoop the Cs clock at 9 GHz to the optical frequency. Thanks to this technique, pioneered in Garching, Hänsch's group has recently succeeded in measuring the 1S–2S interval with respect to a transportable Cs atomic fountain clock from the Laboratoire Primaire du Temps et des

Table 3. Measurements of the 1S–2S interval. The frequency $\nu_0 = 2\,466\,061$ GHz is subtracted from the frequency ν_{1S-2S} of the 1S–2S interval

Year	Reference	$\nu_{1s-2S} - \nu_0$
	Tellurium line reference	
1986	Southampton [10]	397 (25) MHz
1986	Stanford [9]	395.6 (4.8) MHz
1987	Oxford [14]	414.13 (79) MHz
1989	Stanford [13]	413.2 (1.8) MHz
	Frequency chain	
1992	Garching [27]	413 182 (45) kHz
1997	Garching [28]	413 187.34 (84) kHz
2000	Garching [29]	413 187.103 (46) kHz

Fréquences (LPTF) in Paris. This last measurement reduces the uncertainty to 46 Hz (i.e. a relative uncertainty of about 2×10^{-14}) [29]. This 1S–2S splitting, which was in 2000 the most precise value of any optical frequency, has been recently superseded by the Hg^+ clock transition [30].

3 Spectroscopy of the 2S–nS and 2S–nD Transitions

In Paris, the hydrogen experiment was begun in 1983. At this time, we chose to study the two-photon transitions between the 2S metastable level and the upper nS and nD levels. In our opinion, these transitions had several advantages: (i) they lie in the near infrared range (for instance 778 nm for $n = 8$), where the dye lasers are very efficient, (ii) the levels involved are relatively long lived, leading to small linewidths, and (iii) the Lamb shift of the 2S level had been measured very accurately by microwave spectroscopy, and it was easy to extract the Rydberg constant from the 2S–nS/D interval. Consequently, these measurements were complementary to those on the 1S–2S transition.

The principle of these experiments is described in [31–33]. In order to obtain very narrow signals, we use the folllowing experimental conditions: (i) metastable atoms are produced in an atomic beam to avoid collisional broadening, (ii) the atomic beam is collinear with the two counterpropagating laser beams, so that the transit time broadening is reduced. The metastable atomic beam is formed by electronic excitation of a 1S hydrogen atomic beam. Due to the inelastic collision with the electron, the atomic trajectory is deviated by an angle of about 20°. The metastable yield is monitored at the end of the atomic beam: an electric field quenches the metastable state and two photomultipliers detect the Lyman-α fluorescence. The light source, in the range 730–820 nm, was previously a dye laser and is now a titanium–sapphire laser. To enhance the two-photon transition probability, the whole apparatus is placed inside a Fabry–Pérot cavity, where the optical power can be as much as 100 W in each direction. When the laser frequency is in resonance with the 2S–nS/D transition, the atoms in the nS or nD states undergo a radiative cascade towards the 1S state in a proportion of about 95 %. There occurs an optical quenching of the metstable level before the detection region and the optical excitation can be detected via the corresponding decrease of the 2S beam intensity. Figure 1 shows a typical signal obtained in the case of the $2S_{1/2}(F = 1)$–$8D_{5/2}$ transition of deuterium. In this recording, the decrease of the metastable intensity is 18% and the linewidth is 2 MHz (in terms of atomic frequency). By comparison with the natural width of the 8D level (572 kHz), there is a large broadening which is mainly due to the inhomogenous light shift experienced by the atoms passing through the Gaussian profile of the laser beams. To evaluate this effect, the signal is recorded for several laser intensities and, for each recording, a theoretical profile is fitted to the experimental curve. An extrapolation to zero laser power gives the exact position of the line.

In our first experiments, we used an interferometric method to compare the hydrogen wavelengths to an iodine-stabilized He–Ne laser. With this method, we determined the frequencies of the 2S–nD transitions in hydrogen and deuterium for the levels $n = 8$, 10 and 12 [34, 35]. From these measurements, we deduced the Rydberg constant with a relative accuracy of 1.7×10^{-10}, limited mainly by the standard laser. Later, in collaboration with the group of A. Clairon from LPTF, we have made optical frequency

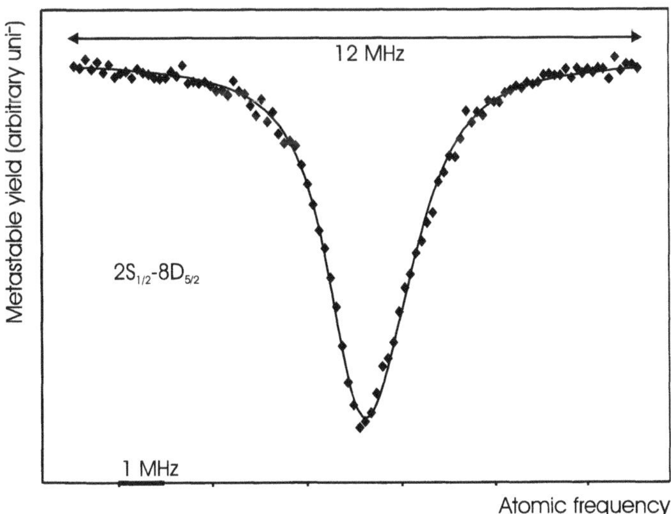

Fig. 1. Fit of the experimental line profile with the theoretical one for the $2S_{1/2}(F = 1)$–$8D_{5/2}$ transition in deuterium. The light power deduced from the fit is 90.6(2.2) W and the decrease of the metastable yield is 18%

measurements. In 1993, the optical frequencies of the 2S–8S/D hydrogen transitions were measured with a frequency chain using two standard lasers (the iodine-stabilized and the methane-stabilized helium–neon lasers). The precision was in the range of 10^{-11} [36, 37]. In 1996, these measurements were remade in hydrogen and deuterium with an accuracy better than one part in 10^{11} [38]. A new frequency chain was used, with a new standard laser, namely a diode laser at 778 nm stabilized on the $5S_{1/2}$–$5D_{5/2}$ two-photon transition of rubidium (LD/Rb laser). The frequency of this standard was measured with a frequency chain at LPTF [39]. More recently, in order to check these 2S–8S/D frequency measurements, a new chain has been built to measure the frequencies of the 2S–12D transitions in hydrogen and deuterium [40]. These experiments are described in detail in [26].

4 The Rydberg Constant Today

The values of the Rydberg constant deduced from the above measurements are summarized in Table 4. In the first part of this table, the interferometric method limits the accuracy to about 10^{-10}. Thanks to optical frequency

Table 4. Determination of the Rydberg constant

Method and transitions involved	$(R_\infty - 109\,737)$ cm^{-1}
2S–Lamb shift and interferometric measurements	
CODATA 1986 [41]	0.315 34 (13)
2S–8D and 2S–10D (Paris, 1986 [34])	0.315 692 (60)
2S–3P (Yale, 1986 [42])	0.315 689 (71)
2S 4P (Yale,1987 [43])	0.315 731 (29)
1S–2S (Stanford, 1987 [12])	0.315 710 (70)
1S–2S (Oxford, 1987 [14])	0.315 731 (31)
2S–8D, 2S–10D and 2S–12D (Paris, 1989 [35])	0.315 709 (18)
2S–Lamb shift and optical frequency measurements	
1S–2S and 2S–4S/D (Garching, 1992 [27])	0.315 684 1 (42)
2S–8S/D (Paris, 1992 [36])	0.315 683 0 (31)
2S–8S/D (Paris, 1993 [37])	0.315 683 4 (24)
1S–2S and 2S–4S/D (Garching, 1995 [24])	0.315 684 9 (30)
2S–8S/D (Paris, 1997 [38, 26])	0.315 686 1 (13)
2S–8S/D and 2S–12D (Paris, 1999 [38, 40, 26])	0.315 685 5 (11)
Lamb shift scaling law and optical frequency measurements	
1S–2S, 2S–8S/D and 2S–12D (Garching and Paris, [28, 38, 40, 26])	0.315 685 4 (10)
General least-squares adjustment	
2S–2P, 1S–2S, 2S–8S/D and 2S–12D	0.315 685 50 (84)

measurements, the uncertainty has been reduced by an order of magnitude (second part of Table 4). A large part of this uncertainty is due to the 2S Lamb shift value used in the data analysis. To overcome this difficulty, one can use the scaling law of the Lamb shift [44] and obtain the Rydberg constant without requiring the measurements of the 2S Lamb shift (third part of Table 4). Finally, it is possible to make an average of these different determinations of R_∞ by performing a least squares adjustment which takes into account all the precise measurements [26]. The result ($R_\infty = 109\,737.315\,685\,50(84)\,\mathrm{cm}^{-1}$) is similar to that of the 1998 adjustment of the fundamental constant [45], with a relative uncertainty of 7.7×10^{-12}. By comparison with the 1986 adjustment [41], the uncertainty is reduced by a factor of about 150. Moreover, this general adjustment provides us a very precise determination of the 1S Lamb shift $L_{1\mathrm{S}} = 8172.840(22)\,\mathrm{MHz}$ [26].

As is clear from this table, the competition between the Paris and Garching groups has actually turned out to be complementary, the best determination of fundamental data in hydrogen being obtained by a joint analysis of our experimental results.

5 Conclusion and Prospects

This rapid review shows clearly the major part played by Theodor Hänsch in the story of hydrogen spectroscopy over the past 25 years or so. As we have limited this paper to two-photon spectroscopy, we have not cited his work on the production of cw coherent Lyman-α radiation. The recent observation of the Lyman-α line in the cw regime opens the way to laser cooling of hydrogen or to simple laser spectroscopy of antihydrogen [46]. The precision of the Rydberg constant and Lamb shift is now limited by the uncertainties in the 2S–nS/D frequencies, which in the Paris experiment are mainly due to the light shifts. To obtain more accurate values of these frequencies, a first possibility is to use ultracold hydrogen to increase the interaction time and decrease the light shifts [47]. In Paris, we intend to measure the optical frequency of the 1S–3S transition. In this case, as the number of atoms in the 1S atomic beam is about 10^8 times larger than in the metastable atomic beam, we can observe the transition with a very low light power and, consequently, with negligible light shifts. Using such experiments, the uncertainties in the Rydberg constant and the Lamb shift should be further reduced.

References

1. T.W. Hänsch, A.L. Schawlow, G.W. Series, Sci. Am. **240**, 72 (1979)
2. T.W. Hänsch, M.H. Nayfeh, S.A. Lee, S.M. Curry, I.S. Shahin, Phys. Rev. Lett. **32**, 1336 (1974)
3. F. Biraben, B. Cagnac, G. Grynberg, Phys. Rev. Lett. **32**, 643 (1974)

4. T.W. Hänsch, K.C. Harvey, G. Meisel, A.L. Schawlow, Opt. Commun. **11**, 50 (1974)
5. B. Cagnac, G. Grynberg, F. Biraben, J. Phys. **34**, 845 (1973)
6. E.V. Baklanov, V.P. Chebotayev, Opt. Commun. **12**, 312 (1974)
7. T.W. Hänsch, S.A. Lee, R. Wallenstein, C. Wieman, Phys. Rev. Lett. **34**, 307 (1975); S.A. Lee, R. Wallenstein, T.W. Hänsch, Phys. Rev. Lett. **35**, 1262 (1975)
8. C. Wieman, T.W. Hänsch, Phys. Rev. A **22**, 192 (1980)
9. E.A. Hildum, U. Boesl, D.H. McIntyre, R.G. Beausoleil, T.W. Hänsch, Phys. Rev. Lett. **56**, 576 (1986)
10. J.R.M. Barr, J.M. Girkin, J.M. Tolchard, A.I. Ferguson, Phys. Rev. Lett. **56**, 580 (1986)
11. C.J. Foot, B. Couillaud, R.G. Beausoleil, T.W. Hänsch, Phys. Rev. Lett. **54**, 1913 (1985)
12. R.G. Beausoleil, D.H. McIntyre, C.J. Foot, E.A. Hildum, B. Couillaud, T.W. Hänsch, Phys. Rev. A **35**, 4878 (1987)
13. D.H. McIntyre, R.G. Beausoleil, C.J. Foot, E.A. Hildum, B. Couillaud, T.W. Hänsch, Phys. Rev. A **39**, 4591 (1989)
14. M.G. Boshier, P.E.G. Baird, C.J. Foot, E.A. Hinds, M.D. Plimmer, D.N. Stacey, J.B. Swan, D.A. Tate, D.M. Warrington, G.K. Woodgate, Nature **330**, 463 (1987), Phys. Rev. A **40**, 6169 (1989)
15. C.J. Foot, P. Hannaford, D.N. Stacey, C.D. Thompson, G.H. Woodman, P.E.G. Baird, J.B. Swan, G.K. Woodgate, J. Phys. B **23**, L203 (1990)
16. C.D. Thompson, G.H. Woodman, C.J. Foot, P. Hannaford, D.N. Stacey, G.K. Woodgate, J. Phys. B **25**, L1 (1992)
17. C. Zimmermann, R. Kallenbach, T.W. Hänsch, Phys. Rev. Lett. **65**, 571 (1990)
18. F. Schmidt-Kaler, D. Leibfreid, S. Seel, C. Zimmermann, W. König, M. Weitz, T.W. Hänsch, Phys. Rev. A **51**, 2789 (1995)
19. A. Huber, B. Gross, M. Weitz, T.W. Hänsch, Phys. Rev. A **59**, 1844 (1999)
20. D.J. Berkeland, E.A. Hinds, M.G. Boshier, Phys. Rev. Lett. **75**, 2470 (1995)
21. C.L. Cesar, D.G. Fried, T.C. Killian, A.D. Polcyn, J.C. Sandberg, I.A. Yu, T.J. Greytak, D. Kleppner, J.M. Doyle, Phys. Rev. Lett. **77**, 255 (1996)
22. M. Weitz, F. Schmidt-Kaler, T.W. Hänsch, Phys. Rev. Lett. **68**, 1120 (1992)
23. M. Weitz, A. Huber, F. Schmidt-Kaler, D. Leibfreid, T.W. Hänsch, Phys. Rev. Lett. **72**, 328 (1994)
24. M. Weitz, A. Huber, F. Schmidt-Kaler, D. Leibfreid, W. Vassen, C. Zimmermann, K. Pachucki, T.W. Hänsch, L. Julien, F. Biraben, Phys. Rev. A **52**, 2664 (1995)
25. S. Bourzeix, B. de Beauvoir, F. Nez, M.D. Plimmer, F. de Tomasi, L. Julien, F. Biraben, D.N. Stacey, Phys. Rev. Lett. **76**, 384 (1996)
26. B. de Beauvoir, C. Schwob, O. Acef, J.-J. Zondy, L. Jozefowski, L. Hilico, F. Nez, L. Julien, A. Clairon, F. Biraben, Eur. Phys. J. D **12**, 61 (2000)
27. T. Andreae, W. Köning, R. Wynands, D. Leibfried, F. Schmidt-Kaler, C. Zimmermann, D. Meschede, T.W. Hänsch, Phys. Rev. Lett. **69**, 1923 (1992)
28. Th. Udem, H. Huber, B. Gross, J. Reichert, M. Prevedelli, M. Weitz, T.W. Hänsch, Phys. Rev. Lett. **79**, 2646 (1997)
29. M. Niering, R. Holwarth, J. Reichert, P. Pokasov, Th. Udem, M. Weitz, T.W. Hänsch, P. Lemonde, G. Santarelli, M. Abgrall, P. Laurent, C. Salomon, A. Clairon, Phys. Rev. Lett. **84**, 5496 (2000)

30. Th. Udem, S.A. Diddams, K.R. Vogel, C.W. Oates, E.A. Curtis, W.D. Lee, W.M. Itano, R.E. Drullinger, J.C. Bergquist, L. Hollberg, Phys. Rev. Lett. **86**, 4996 (2001)
31. J.C. Garreau, M. Allegrini, L. Julien, F. Biraben, J. Phys. (France) **51**, 2263 (1990)
32. J.C. Garreau, M. Allegrini, L. Julien, F. Biraben, J. Phys. (France) **51**, 2275 (1990)
33. J.C. Garreau, M. Allegrini, L. Julien, F. Biraben, J. Phys. (France) **51**, 2293 (1990)
34. F. Biraben, J.C. Garreau, L. Julien, Europhys. Lett. **2**, 925 (1986)
35. F. Biraben, J.C. Garreau, L. Julien, M. Allegrini, Phys. Rev. Lett. **62**, 621 (1989)
36. F. Nez, M.D. Plimmer, S. Bourzeix, L. Julien, F. Biraben, R. Felder, O. Acef, J.J. Zondy, P. Laurent, A. Clairon, M. Abed, Y. Millerioux, P. Juncar, Phys. Rev. Lett. **69**, 2326 (1992)
37. F. Nez, M.D. Plimmer, S. Bourzeix, L. Julien, F. Biraben, R. Felder, Y. Millerioux, P. de Natale, Europhys. Lett. **24**, 635 (1993)
38. B. de Beauvoir, F. Nez, L. Julien, B. Cagnac, F. Biraben, D. Touahri, L. Hilico, O. Acef, A. Clairon, J.J. Zondy, Phys. Rev. Lett. **78**, 440 (1997)
39. D. Touahri, O. Acef, A. Clairon, J.J. Zondy, R. Felder, L. Hilico, B. de Beauvoir, F. Biraben, F. Nez, Opt. Commun. **133**, 471 (1997)
40. C. Schwob, L. Jozefowski, B. de Beauvoir, L. Hilico, F. Nez, L. Julien, F. Biraben, O. Acef, A. Clairon, Phys. Rev. Lett. **82**, 4960 (1999)
41. E.R. Cohen, B.N. Taylor, Rev. Mod. Phys. **59**, 1121 (1987)
42. P. Zhao, W. Lichten, H.P. Layer, J.C. Bergquist, Phys. Rev. A **34**, 5138 (1986)
43. P. Zhao, W. Lichten, H.P. Layer, J.C. Bergquist, Phys. Rev. Lett. **58**, 1293 (1987)
44. S.G. Karshenboim, J. Phys. B **29**, L29 (1996), Z. Phys. D **39**, 109 (1997)
45. P.J. Mohr, B.N. Taylor, Rev. Mod. Phys. **72**, 351 (2000)
46. K.S.E. Eikema, J. Walz, T.W. Hänsch, Phys. Rev. Lett. **86**, 5679 (2001)
47. L. Willmann, D. Kleppner, in *The Hydrogen Atom*, ed. by S.G. Karshenboim, F.S. Pavone, F. Bassani, M. Inguscio, T.W. Hänsch, Springer Lecture Notes in Physics, Vol. 570 (Springer, Berlin, Heidelberg 2001) pp. 42–56

Precision Spectroscopy on the Lyman-α Transitions of H and He

Kjeld S.E. Eikema, Wim Ubachs, Wim Vassen, and Wim Hogervorst

1 Introduction

The development of well-controlled, wavelength-tuneable lasers has had a tremendous impact on atomic and molecular physics. New methods of investigating the properties of atoms and molecules became available, with major applications in the field of metrology. Also exciting new possibilities to manipulate the motion of atoms were introduced. Hänsch and coworkers over the years played, and continue to play a major role in this field, developing new techniques for high-resolution laser spectroscopy and applying them, e.g. to ultra-precise studies of the hydrogen atom, as well as initiating innovative research in the field of cold atoms and Bose–Einstein condensation. Already in 1971 Hänsch et al. introduced the now well-known technique of laser-saturation spectroscopy in an experiment on the I_2 molecule [1] and the Na atom [2], later improved and refined with the introduction of various polarization schemes [3, 4]. This method was applied by Hänsch et al. to study the Balmer-α line of the H atom in high resolution [5]. The technique to eliminate first-order Doppler effects in a two-photon excitation process, for the first time experimentally applied to investigate the $3S$–$5S$ transition in Na by Biraben et al. [6], was subsequently used by Hänsch et al. [7] to study the $1S$–$2S$ transition in H with a pulsed laser system. This immediately resulted in improved data on Lamb shifts and Rydberg constant and pointed the direction for future work, involving sophisticated CW laser light sources at 243 nm and the use of cooled atoms to reduce second-order Doppler-shifts. Using a highly stable CW laser source at 243 nm in combination with a cooled beam of hydrogen atoms the frequency of the $1S$–$2S$ transition has been determined with an accuracy of 1.8 parts in 10^{14} [8]. When this transition frequency is used together with frequencies for other transitions (e.g. $2S$–$8D$ measured by Biraben et al. [9]) values for Lamb shifts and the Rydberg constant with unprecedented accuracies can be established.

In 1975 Hänsch and Schawlow already pointed out that laser radiation pressure could be applied to cool atoms [10], suggesting future applications on cooling atomic hydrogen. This requires a source of, preferably CW, vacuum ultraviolet Lyman-α laser radiation at 121.5 nm, a source that was recently built and applied by Eikema et al. in the Hänsch group [11, 12]. This source and its first operation in a study of the Lyman-α $1S$–$2P$ transition will be discussed in Sect. 2 of this contribution. In Sect. 3 a similar measurement on the $1\,^1S$–$2\,^1P$ Lyman-α transition in neutral helium, at 58.4 nm, will be presented [13–15]. In Sect. 4 new possibilities to measure the $1\,^1S$–$2\,^1S$ transition in helium in a two-photon experiment with high accuracy will be discussed. The proposed approach is based on recent, exciting developments in the Hänsch group to measure large optical frequency differences with a train of femtosecond laser pulses, recognizing that the regularly spaced comb of frequency modes of a mode-locked femtosecond laser can be used as a precise ruler [16]. With this optical ruler technique not only has the most accurate value for the frequency of the $1S$–$2S$ transition in atomic hydrogen been determined [8], but it also made an unprecedented accuracy of 1.1 parts in 10^{14} possible for an optical transition in the mercury ion [17].

2 The $1S$–$2P$ Lyman-α Transition in Hydrogen

The $1S$–$2P$ transition in hydrogen at 121.56 nm is rather special. Apart from being the strongest optical dipole transition in hydrogen, it is also the one required to perform closed-cycle laser cooling from the $1S$ ground state. Application of laser cooling can reduce the second-order Doppler effect and enhance the number of atoms at low velocities in precision spectroscopy on a beam of hydrogen. It is therefore remarkable that even after almost 40 years of laser spectroscopy, the $1S$–$2P$ transition had not been observed up to now with its natural linewidth of 100 MHz (1.6 ns lifetime of the $2P$ state). There is however a more pressing reason to finally harness the $1S$–$2P$ transition: the antihydrogen atom. Creation of antihydrogen seems imminent given the tremendous progress in production and handling of elementary antiparticles. If optical transitions in hydrogen and antihydrogen could be compared to high precision, exciting new tests of CPT and possibly gravitation can be performed. One of the challenges of antihydrogen is its rapid annihilation on contact with normal matter. Nondestructive optical techniques such as laser cooling and optical detection using the $1S$–$2P$ transition (assumed to be the same in antihydrogen) will be vital.

To get the most out of $1S$–$2P$ laser cooling (Doppler limit 2.4 mK) and spectroscopy, a narrow band continuous source of Lyman-α is required. Although several *pulsed* (broadband $>$ 100 MHz) sources of Lyman-α were demonstrated over the last two decades, not a single coherent *continuous* source was realized. One of the difficulties is that Lyman-α is at the edge of the vacuum ultraviolet (VUV) spectral region. Not only is this radiation readily absorbed in air, optics has to be made from single crystal MgF_2 or LiF material, which typically absorbs 50% of Lyman-α in just a few mm. To produce VUV, high power pulsed lasers have been used to drive third-order nonlinear interactions. Lyman-α can be generated by third harmonic conversion of 365 nm in krypton, or by frequency mixing using up to three different wavelengths in, e.g. mercury. Simply replacing multimegawatt pulsed lasers

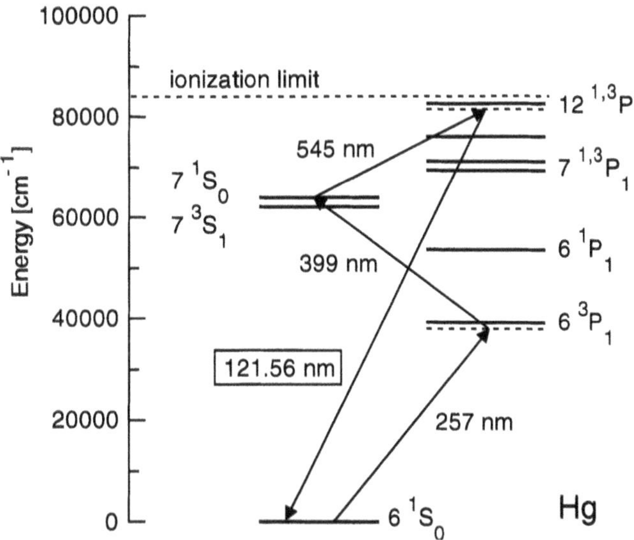

Fig. 1. Four-wave-mixing scheme in mercury for producing continuous coherent Lyman-α

by continuous lasers with only a few watt optical power would reduce the VUV yield to below a photon per second. For detection and laser cooling of antihydrogen at least 10 nW ($\approx 10^{10}$ ph/s) is required. Strong focusing of the laser beams in the nonlinear medium does improve VUV yield for continuous lasers, but it is not sufficient.

The solution is to choose all wavelengths involved in the mixing process close to, or in coincidence with atomic resonances to enhance the nonlinear susceptibility. Mercury has the appropriate level structure to implement this approach. As can be seen in Fig. 1, the combination of a frequency doubled single-mode argon laser (257 nm), and a frequency-doubled Ti:sapphire laser (399 nm) can be made exactly resonant with the $7s\,^1S_0$ state. A dye laser at 545 nm is then sufficient to reach the Lyman-α photon energy. The combination of exact and near-resonances, and the 100% duty-cycle of continuous lasers basically compensates the much lower peak power compared to pulsed lasers. In this way up to 20 nW Lyman-α radiation has been generated in mercury vapor. At a slightly longer wavelength of 122.1 nm, up to 200 nW was produced [12]. Both the short wavelength and the yield are unprecedented for continuous four-wave-mixing in the VUV. As all the fundamental lasers have a bandwidth below a few MHz, the generated VUV radiation has a bandwidth well below 10 MHz. The result is a tuneable source with an average Lyman-α power comparable to the pulsed source previously used for cooling and spectroscopy of magnetically trapped hydrogen [18], but with an almost two orders of magnitude narrower bandwidth. Several improvements, such as changing to a pure solid-state laser system or using isotopically pure

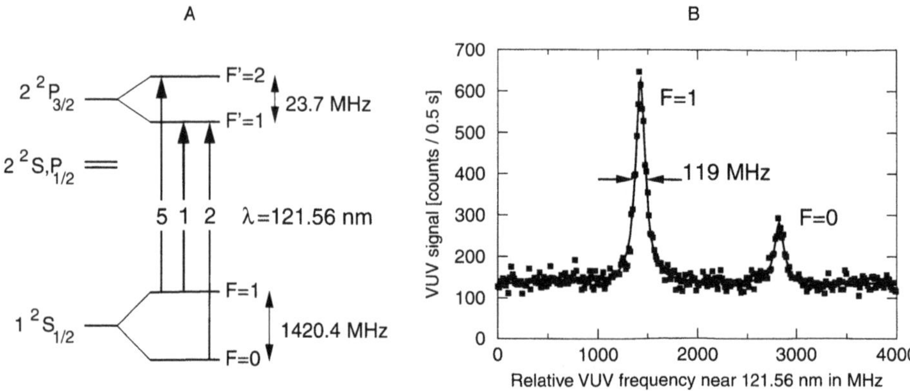

Fig. 2. The $1s\,^2S_{1/2}$–$2p\,^2P_{3/2}$ excitation scheme, and recorded resonance in atomic hydrogen using continuous coherent Lyman-α. The numbers in the arrows indicate the relative strengths of the transitions

mercury, could improve the Lyman-α yield even further by at least one or two orders of magnitude.

To demonstrate the capabilities of this source, and its prospects for antihydrogen, $1S$–$2P$ spectroscopy in a beam of normal atomic hydrogen was performed. The natural resonance width is 100 MHz, but in hydrogen already at a transverse temperature of 1 K a Doppler broadening of 1 GHz occurs. Therefore a liquid-nitrogen-cooled hydrogen source was used in combination with a narrow slit-skimmer to obtain a beam with a temperature of $\approx 80\,$K and a divergence of 1:400. This beam intersected the carefully collimated Lyman-α light perpendicularly. Due to the optics in the Lyman-α beam for spatial filtering and collimation about $1-2\,$nW reached the hydrogen interaction zone. Excitation of the $1S$–$2P$ was then monitored by photon-counting the resulting fluorescence with a solar-blind photomultiplier. In Fig. 2 the excitation scheme and the experimental result are shown. Most measurements were performed on the $1s\,^2S_{1/2}$–$2p\,^2P_{3/2}$ transition that can be used for laser cooling. A resonance width of 119(2) MHz was measured, based on 24 recordings. This is the first $1S$–$2P$ recording with a near natural linewidth (the excess 19 MHz is ascribed to residual Doppler broadening). The 23.7 MHz hyperfine structure of the $2p\,^2P_{3/2}$ is less than the natural width of 100 MHz and causes little extra broadening (2 MHz) as mainly the $F = 1$ to $F' = 2$ is excited. It does, however, change the double peak separation between excitation from the $F = 0$ and $F = 1$ by effectively 20 MHz. The resulting theoretical value of 1400.5 MHz is in good agreement with the experimental value of 1396(6) MHz. From the results presented it is clear that the continuous coherent Lyman-α source can play an important role in the future for detection, spectroscopy, and laser cooling of antihydrogen.

3 The $1\,^1S$–$2\,^1P$ Lyman-α Transition in Helium

For many years atomic hydrogen was the only atomic system in which QED theory could be tested and this work was primarily pushed by Hänsch and coworkers in Garching. Progress in theoretical calculations also made multi-electron atoms and ions of interest for QED studies. Energy level calculations without QED and higher-order relativistic effects in, e.g. helium are nowadays so accurate that experimental transition frequencies can be used as a test for Lamb-shift calculations, just like in atomic hydrogen. Helium is particularly interesting because of two-electron contributions to the Lamb shift. These contributions are best tested in ^{1}S states where they contribute up to 10% to the total Lamb shift. Of course the Lamb shift is largest in the $1\,^1S$ ground state. However, the ground state is difficult to access with laser radiation due to the large energy difference with excited states. In 1993 our group in Amsterdam showed in a preliminary experiment on the $1\,^1S$–$2\,^1P$ transition at 58.4 nm that this energy difference can be bridged using high-power pulsed laser systems and harmonic upconversion [13]. This encouraged us to start a dedicated experiment using a CW dye laser at 584 nm and pulsed amplification techniques. A schematic of the setup is shown in Fig. 3. A pulse-dye amplifier (PDA), pumped by a Nd:YAG laser, amplifies light

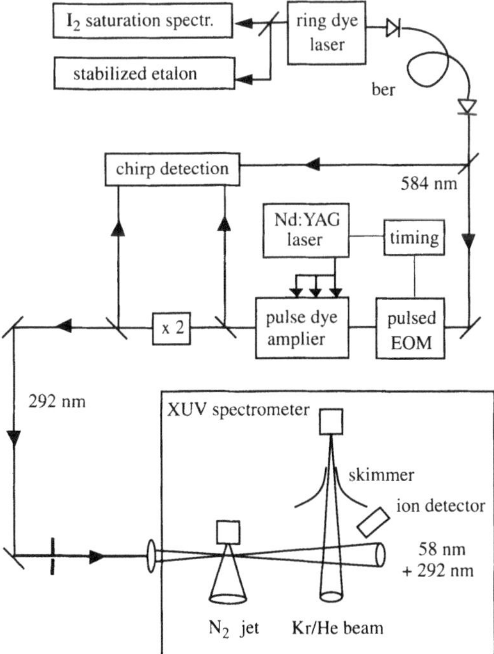

Fig. 3. Overview of the experimental setup to measure the $1\,^1S$ Lamb shift in helium

at 584 nm from a ring dye laser in 6.5 ns pulses. This output is frequency doubled in a nonlinear crystal and subsequently focused in a pulsed jet of N_2 for fifth harmonic generation. In this process $10^5 - 10^6$ photons per pulse at 58.4 nm are generated. To reduce Doppler effects the $1\,^1S$–$2\,^1P$ transition is induced in a skimmed and pulsed beam of 10% helium seeded in 90% krypton in a crossed-beam arrangement. A few percent of the atoms excited to $2\,^1P$ is ionized by the UV light and detected. The primary calibration is performed at 584 nm by saturation spectroscopy on the P88(15-1)-o transition in I_2.

Precision spectroscopy with such a pulsed laser system is hampered by unwanted phase modulation effects in the PDA. Phase modulation results in time-dependent frequency excursions, also called chirp, which lead to a calibration error for the pulsed output relative to the frequency standards based on CW laser saturation spectroscopy in the visible. This chirp is a result of phase modulation due to time-dependent gain in the amplification process. To counteract this chirp a $LiTaO_3$ EOM was developed to modify the phase of the CW seed beam prior to amplification in the PDA. A scan under antichirp conditions is shown in Fig. 4. The experimental linewidth is only a factor 2 larger than the natural linewidth of the upper $2\,^1P$ state, which is 300 MHz. Included in the figure is a saturated absorption spectrum of the relevant I_2 line (the calibrated one indicated with an asterisk) as well as an etalon spectrum. The $1\,^1S$–$2\,^1P$ transition frequency could be determined with an accuracy of 9 parts in 10^9: 5130495083(45) MHz [15]. A Lamb shift for the $1\,^1S$ ground state of $-41224(45)$ MHz was deduced, in excellent agreement with a theoretical value of 41233 MHz. The accuracy of the theoretical value was recently re-evaluated by Drake [19] leading to an uncertainty in theory

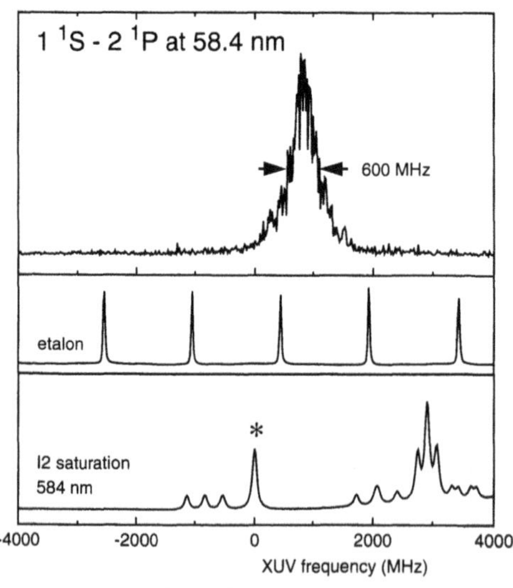

Fig. 4. The $1\,^1S$–$2\,^1P$ resonance transition for anti-chirped PDA pulses

of 91 MHz. As in the case of atomic hydrogen, the experimental accuracy is better than the theoretical one.

4 Outlook

The recent use of a mode-locked laser as a frequency ruler is a prime example of the creativity in the Hänsch group. This work is not only important for frequency synthesis and optical interval division. Already in 1978 Hänsch proposed and performed precision spectroscopy using mode-locked pulses from a dye laser to *excite* Na [20]. Although only small intervals could be measured in this experiment (1 GHz), modern mode-locked Ti:sapphire lasers can vastly improve this situation. The principle is that repetitive excitation with phase-controlled pulses results in Ramsey-like resonance features that are much narrower than the bandwidth of the individual optical pulses. This type of spectroscopy does, however, require better control of the phase reproducibility of the pulses than frequency metrology. In frequency metrology, phase variations can be averaged out by extended counting times as long as no cycles are missed. In the case of direct spectroscopy with pulses, phase noise results in a broadening of the Ramsey resonances. The chirp effect discussed in the previous section can be a significant source of phase distortion. Although this influences the excitation by a single pulse, to first order it merely influences the envelope and not the sharp resonances in multipulse spectroscopy. If all sources of phase noise can be sufficiently controlled, then spectroscopy with mode-locked lasers can combine high excitation probability with a high resolution. However, use of short and therefore broad-bandwidth optical pulses limits the Doppler broadening reduction that can be obtained in two-photon spectroscopy. For atoms that are difficult to excite otherwise (e.g. low signals on two-photon transitions with UV or VUV light), but that can be trapped and cooled, this is fortunately not a severe limitation. Antihydrogen ($1S$–$2S$) and helium ($1s^2\,{}^1S$–$1s2s^1S$) are therefore interesting candidates for this type of excitation. Also nonlinear optics with ultra-short pulses may greatly benefit from the use of phase-controlled pulses. The generation of X-ray pulses with attosecond duration for studies of ultrafast phenomena may be envisioned [21]. Both aspects of phase-controlled optical pulses are currently being investigated in our group.

References

1. T.W. Hänsch, M.D. Levenson, A.L. Schawlow, Phys. Rev. Lett. **26**, 846 (1971)
2. T.W. Hänsch, I.S. Shahin, A.L. Schawlow, Phys. Rev. Lett. **27**, 707 (1971)
3. C. Wieman, T.W. Hänsch, Phys. Rev. Lett. **36**, 1170 (1976)
4. T.W. Hänsch, D.R. Lyons, A.L. Schawlow, A. Siegel, Z-Y. Wang, G-Y. Yan, Opt. Commun. **37**, 87 (1981)
5. T.W. Hänsch, I.S. Shahin, A.L. Schawlow, Nature **235**, 63 (1972)

6. F. Biraben, B. Cagnac, G. Grynberg, Phys. Rev. Lett. **32**, 643 (1974)
7. T.W. Hänsch, S.A. Lee, R. Wallenstein, C. Wieman, Phys. Rev. Lett. **34**, 307 (1975)
8. M. Niering, R. Holzwarth, J. Reichert, P. Pokasov, T. Udem, M. Weitz, T.W. Hänsch, P. Lemonde, G. Santarelli, M. Abgrall, P. Laurent, C. Salomon, A. Clairon, Phys. Rev. Lett. **84**, 5496 (2000)
9. C. Schwob, L. Jozefowski, B. de Beauvoir, L. Hilico, F. Nez, L. Julien, F. Biraben, O. Acef, A. Clairon, Phys. Rev. Lett. **82**, 4960 (1999)
10. T.W. Hänsch, A.L. Schawlow, Opt. Commun. **13**, 68 (1975)
11. K.S.E. Eikema, J. Walz, T.W. Hänsch, Phys. Rev. Lett. **83**, 3828 (1999)
12. K.S.E. Eikema, J. Walz, T.W. Hänsch, Phys. Rev. Lett. **86**, 5679 (2001)
13. K.S.E. Eikema, W. Ubachs, W. Vassen, W. Hogervorst, Phys. Rev. Lett. **71**, 1690 (1993)
14. K.S.E. Eikema, W. Ubachs, W. Vassen, W. Hogervorst, Phys. Rev. Lett. **76**, 1216 (1996)
15. K.S.E. Eikema, W. Ubachs, W. Vassen, W. Hogervorst, Phys. Rev. **A55**, 1866 (1997)
16. J. Reichert, M. Niering, R. Holzwarth, M. Weitz, Th. Udem, T.W. Hänsch, Phys. Rev. Lett. **84**, 3232 (2000)
17. T. Udem, S.A. Diddams, K.R. Vogel, C.W. Oates, E.A. Curtis, W.D. Lee, W.M. Itano, R.E. Drullinger, J.C. Bergquist, L. Hollberg, Phys. Rev. Lett. **86**, 4996 (2001)
18. I.D. Setija, H.G.C. Werij, O.J. Luiten, M.W. Reynolds, T.W. Hijmans, J.T.M. Walraven, Phys. Rev. Lett. **70**, 2257 (1993)
19. G.W.F. Drake, in: The Hydrogen Atom, Precision Physics of Simple Atomic Systems, ed. by S.G. Karshenboim, F.S. Pavone, G.F. Bassani, M. Inguscio, T.W. Hänsch, (Springer 2001) pp. 57–80; G.W.F. Drake, W.C. Martin, Can. J. Phys. **76**, 679 (1998)
20. J.N. Eckstein, A.I. Ferguson, T.W. Hänsch, Phys. Rev. Lett. **40**, 847 (1978)
21. A. Apolonski, A. Poppe, G. Tempea, Ch. Spielman, Th. Udem, R. Holzwarth, T.W. Hänsch, F. Krausz, Phys. Rev. Lett. **85**, 740 (2000)

Towards Laser Spectroscopy of Antihydrogen

Jochen Walz

Much of the fascination behind the ongoing effort to produce and to investigate *cold* antihydrogen atoms comes from ultrahigh-resolution laser spectroscopy of ordinary atomic hydrogen. Striking advances in Theodor Hänsch's group at the Max-Planck Institute for Quantum Optics (MPQ) and elsewhere [1] have long inspired the vision of measuring antihydrogen spectra to a similar level of precision. The comparison of hydrogen and antihydrogen would then represent one of the most stringent tests of the fundamental CPT (charge conjugation-parity inversion-time reversal) symmetry [2]. A foretaste has been given by the recent production and observation of *fast* antihydrogen atoms, first at CERN [3] and then at Fermilab [4].

Cold antihydrogen atoms might also open up a new field of experimentally investigating antimatter gravity [5]. Antihydrogen atoms are neutral and thus immune to stray electric fields that plague experiments in gravitation with charged particles [6]. Although possible – at least in principle – experiments in gravitation using antihydrogen represent a formidable challenge [7].

The starting point for antihydrogen experiments is low-energy antiprotons. CERN's new Antiproton Decelerator (AD) is the only source of low-energy antiprotons in the world. Section 1 gives a brief account of the AD. In order to produce cold antihydrogen atoms both antiprotons and positrons can be trapped, cooled, and made to interact using Penning traps. The experiment of the ATRAP collaboration, led by Gerald Gabrielse at Harvard, is described in Sect. 2. The generation of laser radiation at 121.56 nm wavelength (Lyman-α) will most likely be a pivotal technique for both CPT experiments and for experiments on antimatter gravity. The development of the first source of continuous coherent Lyman-α radiation and future laser experiments with antihydrogen are discussed in Sect. 3.

1 The Antiproton Decelerator (AD) at CERN

The initial plans for experiments with cold antihydrogen atoms faced a dramatic situation: The Low Energy Antiproton Ring (LEAR) at CERN was to be shut down at the end of 1996 in order to free resources for the LHC. At that time LEAR was the only source of low-energy antiprotons in the world. Thus it seemed that future antihydrogen experiments would not be possible because of the lack of low-energy antiprotons.

However, following a recommendation of the SPS and LEAR Committee in 1995, Dieter Möhl and Stephan Maury from the CERN PS Division studied the possibility of a simplified scheme to provide low-energy antiprotons. The former antiproton complex at CERN involved four rings, the Antiproton Collector (AC), the Antiproton Accumulator (AA), the Proton Synchrotron (PS) and LEAR [8]. The new simplified scheme uses the AC alone with appropriate modifications to function as an antiproton cooling and decelerating ring, now called the Antiproton Decelerator (AD) [9]. Simplifying the former antiproton complex to just one machine comes at the expense of reducing the number of antiprotons by one to two orders of magnitude. Most important, however, the simplified scheme reduces cost and that made the project possible in times when CERN was struggling to find support for the LHC.

The AD project was approved by the CERN management early in 1997 based on the condition that the cost and manpower are largely supported by external laboratories who were also required to help with the operation. Rolf Landua at CERN had the task of coordinating the initial AD physics programme and the efforts of the User Community to provide the external funding. Institutions from several countries (Japan, Germany, Italy, Denmark, and the United States) supported the project. Theodor Hänsch organized a substantial contribution from the Max-Planck Society in Germany.

The AD produced the first slow antiprotons for experiments at the end of 1999. It was fully commissioned and begun routine operation in July 2000 [10]. Currently the AD delivers 4×10^7 antiprotons with a momentum of $100\,\mathrm{MeV}/c$ (corresponding to 5.3 MeV kinetic energy) every 110 seconds in bunches with 180 ns length.

Three experiments in the AD hall share the antiproton beam. ASACUSA investigates exotic "atomcules" produced in collisions of slow antiprotons with matter [11, 12]. The two other experiments, ATHENA [13] and ATRAP [14], have the common goal of producing and investigating cold antihydrogen atoms.

2 The Antihydrogen Trap (ATRAP) Experiment

Figure 1 shows a section of the present ATRAP experiment at CERN. The experiment is located on a platform three meters above the floor of the AD hall. There are two experimental zones on this platform. The first zone is currently used and accommodates a superconducting solenoid with a 6 T magnetic field in the vertical direction. The magnet bore has a 10 cm diameter and is at liquid nitrogen temperature. Positrons come from a 150 mCi ^{22}Na source. The source is hanging on a fishing line and can be lowered from a shielding lead blockhouse into the magnet. Radiation safety requires that there are shielding walls around the zone. Besides the experimental zone is a laser cabin which is built as a Faraday cage. Laser beams can be sent into the zone through pipes in the concrete shielding wall.

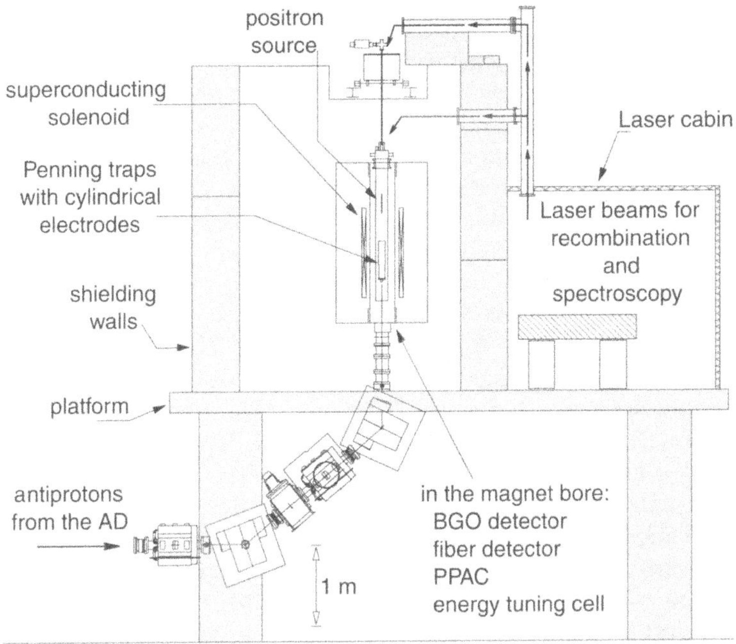

Fig. 1. The present ATRAP setup in the AD hall

The antiproton pulse from the AD is bent upwards using dipole magnets and comes through a hole in the platform. The antiprotons pass through a 10 μm titanium window and two gas detectors. The anodes of these two PPACs (parallel plate avalanche counter) [15] are segmented into five parallel strips, 2 mm wide with 0.5 mm gaps. The two anodes are oriented perpendicular to each other, thus providing two-dimensional information which is used to steer the antiproton beam. The antiproton pulse is centered on the trap axis for optimum catching efficiency.

The energy of the antiprotons can be varied using a 15 mm long gas cell containing a mixture of helium and sulfur-hexafluoride. He and SF_6 have significantly different stopping powers. Adjusting the mixture of the two gases provides an energy tuning range of 570 keV at a beam energy of 4.6 MeV at the entrance of the gas cell. Fine energy tuning is important in order to maximize the number of trapped antiprotons.

After passing another 10 μm titanium window the antiprotons are slowed to keV energies in a 125 μm thick beryllium degrader. The slowed antiprotons leaving the degrader enter a trap region composed of stacked cylindrical electrodes. The antiprotons are reflected by a negative high voltage on the top of the trap. Before they can return to the degrader, another negative high voltage on the bottom is applied quickly to the degrader. In this way the antiprotons are trapped. A cold cloud of electrons in a small additional potential well is then used for sympathetic cooling of the antiprotons. The antiprotons accumulate in the small central well as they cool down to 4 K. After a while the outer high-voltage well can thus be opened again to catch antiprotons from another AD pulse without losing the antiprotons that are already in the trap. Many antiproton pulses can be stacked in this manner.

The crucial techniques of trapping, cooling, and stacking antiprotons were developed by the TRAP collaboration at CERN's former Low Energy Antiproton Ring (LEAR). These experiments, led by Gabrielse at Harvard, were recently reviewed [16].

Charged pions coming from antiproton annihilation events are detected using scintillators with spatial resolution. 384 scintillating fibers in the magnet bore surround the vacuum enclosure of the trap. These fibers are organized in three layers. The fibers in the first two layers are offset to close the gaps and are arranged to form a 150° helix. The fibers in the outer layer are vertical. In addition to the fiber detector, there are two layers of scintillator "paddles" on the outside of the magnet. They are used in coincidence with the fiber detector to discriminate antiproton annihilation events against background.

Positrons from the radioactive source enter the trap vacuum through a 10 µm titanium window. Some of them slow in single crystals of tungsten. A 2 µm foil is used as a transmission moderator and a 2 mm crystal is used as a reflection moderator. While leaving the transmission moderator, slow positrons can pick up electrons thus forming Rydberg positronium atoms. These atoms can move into the trapping region until they are ionized by the electric field of the trap. The resulting positrons can be accumulated [17].

The electrode stack contains a unique rotatable trap electrode that acts as a ball valve separating the positron accumulation region from the antiproton trap section below. The tungsten moderator crystals thus cannot be struck by antiprotons when the rotatable electrode is in its closed position. This leaves undisturbed the essential layer of adsorbed gas on the transmission moderator, without which positron loading ceases [18].

Positron annihilations will be detected using twelve BGO crystals in a cylindrical ring with a height of 12 cm, an inner radius of 26 mm, and 12.8 mm thickness. This detector, too, will be inserted into the magnet bore from below and will be located between the vacuum enclosure of the trap and the fiber detector. Plexiglass pipes will be used to guide the light out of the magnet bore and onto photomultipliers.

Positrons and antiprotons can both be accumulated at the same time, the two trapping regions being separated by the ball valve in its closed position. The ball valve can then be opened and positrons can be transfered into the trap section below. There, the positrons can be made to interact with the antiprotons. A recent experiment used a nested potential well structure based on a long shallow potential well with steep edges to confine the antiprotons. Cold positrons are preloaded into a short inverted potential well in the center. Antiprotons are launched into the longer outer potential well with enough kinetic energy so that they can pass through the positron cloud in the inverted well. It has been observed that antiprotons rapidly lose energy and cool. This positron cooling of antiprotons [19] represents the closest approach to cold antihydrogen so far. It shows that antiprotons and positrons can be made to interact in the same trap structure at low energies.

The next milestone is clearly to produce cold antihydrogen atoms. Several recombination schemes are currently being investigated: three body recombination [20], pulsed field recombination [21], laser stimulated recombination [22], and antihydrogen formation in charge-exchange collisions of excited positronium atoms with antiprotons [23]. Cold antihydrogen atoms should afterwards be confined using a Ioffe trap [24]. This next-generation apparatus will be set up in the second zone on the ATRAP platform.

3 Source of Continuous Coherent Lyman-α Radiation

In addition to the challenges of producing and trapping cold antihydrogen atoms, one basic problem is anticipated: the number of atoms for future experiments with antihydrogen will be extremely small. This is rather obvious considering the following numbers. Current ultrahigh-resolution $1S$–$2S$ laser spectroscopy uses 10^{15}–10^{17} atoms/s in a beam [25] or 10^{10}–10^{13} atoms in a magnetic trap [26]. And these atoms are at temperatures of less then a few degrees kelvin. On the other hand, the AD delivers 4×10^7 antiprotons every 110 s at 5.3 MeV energy. The efficiency of all the subsequent processes to form antihydrogen will add several orders of magnitude to the discrepancy. Thus future antihydrogen experiments have to be far more efficient than current state of the art hydrogen experiments. Antihydrogen has the nice feature that atoms can be detected with high efficiency using annihilation detectors. Nevertheless, new laser techniques are urgently needed.

Radiation at 121.56 nm wavelength (Lyman-α) can be used for laser-cooling of antihydrogen on the strong $1S$–$2P$ transition. This will be essential for experiments to study the gravitational acceleration of antimatter [7]. Cooling antihydrogen atoms in a magnetic trap from 1 K to the Doppler limit of 2.4 mK reduces their ability to climb up against gravity from 840 m to 2 m. This clearly helps a lot but even so additional cooling techniques might be necessary in order to make experiments in antimatter gravity possible.

Laser cooling with Lyman-α radiation will also be most important for high-resolution laser spectroscopy experiments, since it provides a means to center antihydrogen atoms in a Ioffe trap. This increases the overlap with laser-beams at 243 nm wavelength for Doppler-free $1S$–$2S$ spectroscopy thereby enhancing the excitation rate. In addition, accumulating the atoms in the minimum of a magnetic trap also reduces systematic effects [27].

Unfortunately, producing Lyman-α laser radiation is a huge technological difficulty. At the time of the ATRAP proposal, only *pulsed* Lyman-α sources were available. The obvious disadvantage of pulses put aside, these sources typically have a broad emission bandwidth. This is undesirable since it can cause spurious optical pumping to untrapped magnetic substates of antihydrogen.

A new project was therefore started in the fall of 1996 at the MPQ together with Kjeld Eikema, joining as a postdoc from Amsterdam. The goal was to develop a source of *continuous* coherent Lyman-α radiation. The first scheme attempted was based on four-wave mixing in magnesium vapor. It turned out, though, that the magnesium scheme is very inefficient. It was scarcely possible to detect radiation at 123 nm produced by four-wave mixing [28]. Following a suggestion by Wim Hogervorst the setup was converted for four-wave mixing based on mercury vapor. Producing continuous Lyman-α radiation with the mercury scheme worked quite well [29]. The new Lyman-α source has since been used to observe the 1S–2P transition in atomic hydrogen with almost natural linewidth [30]. Anette Pahl, Birgit Schatz, and Peter Fendel are now trying to further improve the Lyman-α source and possibly even to demonstrate its usefulness for laser-cooling.

A source of continuous Lyman-α laser-light might also solve the basic problem for antihydrogen spectroscopy – that the number of atoms will be extremely small. The strong Lyman-α transition can ultimately be used to detect single excitation events on the weak $1S$–$2S$ transition with unit efficiency in a shelving scheme. This method for high-resolution spectroscopy requires – in principle – just one single antihydrogen atom in a trap [27, 28].

4 Conclusion

Recent progress towards laser spectroscopy of antihydrogen includes both the continuous Lyman-α source at the MPQ and the trap experiment with interacting antiprotons and positrons at CERN. The ATRAP experiment is now well under way and I would like to thank Theodor Hänsch for letting me participate in this project from the beginning. Much remains to be done before cold antihydrogen is observed and precise laser spectroscopy is performed. But then, again, "perhaps the greatest surprise would be none at all" [31].

Acknowledgements

This work is supported by the German BMBF and the Max-Planck Society.

References

1. F. Biraben, T.W. Hänsch, M. Fischer, M. Niering, R. Holzwarth, J. Reichert, T. Udem, M. Weitz, B. de Beauvoir, C. Schwob, L. Jozefowski, L. Hilico, F. Netz, L. Julien, O. Acef, J.-J. Zondy, A. Clairon, 'Precision Spectroscopy of

Atomic Hydrogen'. In: *The Hydrogen Atom: Precision Physics of Simple Atomic Systems.* ed. by S.G. Karshenboim, F.S. Pavone, G.F. Bassani, M. Inguscio, T.W. Hänsch (Springer, Berlin, Heidelberg 2001) pp. 17–41
2. R. Bluhm, V.A. Kostelecký, N. Russel, Phys. Rev. Lett. **82**, 2254 (1999)
3. G. Baur, G. Boero, S. Brauksiepe, A. Buzzo, W. Eyrich, R. Geyer, D. Grzonka, J. Hauffe, K. Kilian, M. LoVetere, M. Macri, M. Moosburger, R. Nellen, W. Oelert, S. Passagio, A. Pozzo, K. Röhrich, K. Sachs, G. Schepers, T. Sefzick, R.S. Simon, R. Stratmann, F. Stinzing, M. Wolke, Phys. Lett. B **368**, 251 (1996)
4. G. Blanford, D.C. Christian, K. Gollwitzer, M. Mandelkern, C. T. Munger, J. Schultz, G. Zioulas, Phys. Rev. Lett. **80**, 3037 (1998)
5. M.M. Nieto, T. Goldmann, Phys. Rep. **205**, 221 (1991), Phys. Rep. **216**, 343 (1992)
6. T.W. Darling, F. Rossi, G.I. Opat, G.F. Moorhead, Rev. Mod. Phys. **64**, 237 (1992)
7. G. Gabrielse, Hyperfine Interact. **44**, 349 (1988)
8. J. Eades, F.J. Hartmann, Rev. Mod. Phys. **71**, 373 (1999)
9. S. Maury, Hyperfine Interact. **109**, 43 (1997)
10. A. Hellemans, Science **288**, 788 (2000)
11. T. Yamazaki, N. Morita, R.S. Hayano, E. Widmann, J. Eades, To appear in Phys. Rep.
12. ASACUSA collaboration, *Atomic Spectroscopy and Collisions Using Slow Antiprotons* (Proposal CERN/SPSC 97-19, 1997)
13. ATHENA collaboration, *Antihydrogen Production and Precision Experiments* (Proposal CERN/SPSC 96-47, 1996)
14. ATRAP collaboration, *The Production and Study of Cold Antihydrogen* (Proposal CERN/SPSC 97-8, 1997)
15. J. Gub, C. Gund, D. Pansegrau, G. Schrieder, H. Stelzer, Nucl. Instrum. Methods A **453**, 522 (2000)
16. G. Gabrielse, Adv. At., Mol., Opt. Phys. **45**, 1 (2001)
17. J. Estrada, T. Roach, J.N. Tan, P. Yesley, G. Gabrielse, Phys. Rev. Lett. **84**, 859 (2000)
18. G. Gabrielse, D.S. Hall, T. Roach, P. Yesley, A. Khabbaz, J. Estrada, C. Heimann, H. Kalinowsky, Phys. Lett. **B 455**, 311 (1999)
19. G. Gabrielse, J. Estrada, J.N. Tan, P. Yesley, N.S. Bowden, P. Oxley, T. Roach, C.H. Storry, M. Wessels, J. Tan, D. Grzonka, W. Oelert, G. Schepers, T. Sefzick, W.H. Breunlich, M. Cargnelli, H. Fuhrmann, R. King, R. Ursin, J. Zmeskal, H. Kalinowsky, C. Wesdorp, J. Walz, K.S.E. Eikema, T.W. Hänsch, Phys. Lett. B **507**, 1 (2001)
20. M.E. Glinsky, T.M. O'Neil, Phys. Fluids **B 3**, 1279 (1991)
21. C. Wesdorp, F. Robicheaux, L.D. Noordam, Phys. Rev. Lett. **84**, 3799 (2000)
22. A. Wolf, Hyperfine Interact. **76**, 189 (1993)
23. E.A. Hessels, D.M. Homan, M.J. Cavagnero, Phys. Rev. **A 57**, 1668 (1998)
24. J.T.M. Walraven, Hyperfine Interact. **76**, 205 (1993)
25. A. Huber, B. Gross, M. Weitz, T.W. Hänsch, Phys. Rev. A **59**, 1844 (1999)
26. C.L. Cesar, D.G. Fried, T.C. Kilian, A.D. Polcyn, J.C. Sandberg, I.A. Yu, T.J. Greytak, D. Kleppner, J.M. Doyle, Phys. Rev. Lett. **77**, 255 (1996)
27. T.W. Hänsch, C. Zimmermann, Hyperfine Interact. **76**, 47 (1993)
28. J. Walz, A. Pahl, K.S.E. Eikema, T.W. Hänsch, Hyperfine Interact. **127**, 167 (1999)

29. K.S.E. Eikema, J. Walz, T.W. Hänsch, Phys. Rev. Lett. **83**, 3828 (1999)
30. K.S.E. Eikema, J. Walz, T.W. Hänsch, Phys. Rev. Lett. **86**, 5679 (2001)
31. T.W. Hänsch, A.L. Schawlow, G.W. Series, Sci. Am. **240**, 72 (1979)

Ramseyfication of the Resonant Nonlinear Faraday Effect

Antoine Weis

1 Introduction

In 1846 Michael Faraday observed that the polarization of light traversing a glass slab is altered by a longitudinal magnetic field. In 1898 Macaluso and Corbino were the first to study the Faraday effect (FE) of an atomic sodium vapor using resonance radiation. The resonant FE was quantitatively explained by Voigt in the early 20th century in terms of the Zeeman effect on a Lorentz oscillator. In the vicinity of absorption lines both absorptive and dispersive effects contribute to the FE, so that the emerging light is elliptically polarized and the Faraday rotation angle ϕ_{F} is defined as the rotation of the polarization ellipse with respect to the incoming polarization.

In 1987 we used the resonant FE to measure the magnetic field inside a hot cesium vapor and found that the magneto-rotation angles were several orders of magnitude larger than estimated by Voigt's theory [1]. Moreover, the dependence of the effect on laser frequency and on the magnetic field did not obey the anticipated behavior. This marks our (re)discovery of the nonlinear Faraday effect (NLFE) which had been studied in parallel or before in Sm by Mlynck in Hannover and in Na by Gawlik in Cracow. The theoretical interpretation of the NLFE at that time was far from satisfactory, in particular for multilevel atoms. At MPQ (Max-Planck-Institut für Quantenoptik) developed a quantitative model of the NLFE and extended the NLFE to atomic beams. The scope of this paper is not to review the field, but to give an account of our own contributions to the understanding and applications of the NLFE in multilevel atoms.

2 The Resonant Linear Faraday Effect

A magnetic field breaks the rotational symmetry of an atomic medium and leads to both circular birefringence and circular dichroism. In transparent media the FE is due to circular birefringence and the Faraday rotation angle ϕ_{F} is given by

$$\phi_{\mathrm{F}} = \frac{\varphi_{+} - \varphi_{-}}{2} = \left(N'_{+} - N'_{-}\right) \frac{\pi L}{\lambda}, \tag{1}$$

where $\varphi_\pm$ are the phases acquired by the circularly polarized eigenmodes and $N'_\pm$ the real parts of the complex indices of refraction $N_\pm$ of these eigenmodes. L is the length of the medium and λ the vacuum wavelength of light. The equation holds even in the presence of absorption, when ϕ_F represents the orientation of the emerging polarization ellipse.

In a multilevel atom the Zeeman interaction does not only shift the level energies but also induces a mixing of hyperfine wave functions

$$|\widetilde{F,M}\rangle = |F,M\rangle + \sum_{f,m} \frac{\langle f,m|-\boldsymbol{\mu}\cdot\boldsymbol{B}|\,F,M\rangle}{E_{F,M}-E_{f,m}}\,|f,m\rangle\;, \tag{2}$$

which has to be taken into account for the correct description of the experimental signals. We have investigated the influence of wave function mixing on the FE of Doppler-broadened absorption lines in Cs [2], of Doppler-free selective reflection resonances in Cs [3], and of Doppler-free lines in Rb [4] and Cs [5] beams. All spectra clearly exhibit the influence of hyperfine mixing, which is shown for the case of a Cs beam in Fig. 1.

If the laser is tuned to the center of one hyperfine line and the magnetic field is scanned, $\phi_\mathrm{F}(B)$ shows a typical dispersive-shaped resonance, whose

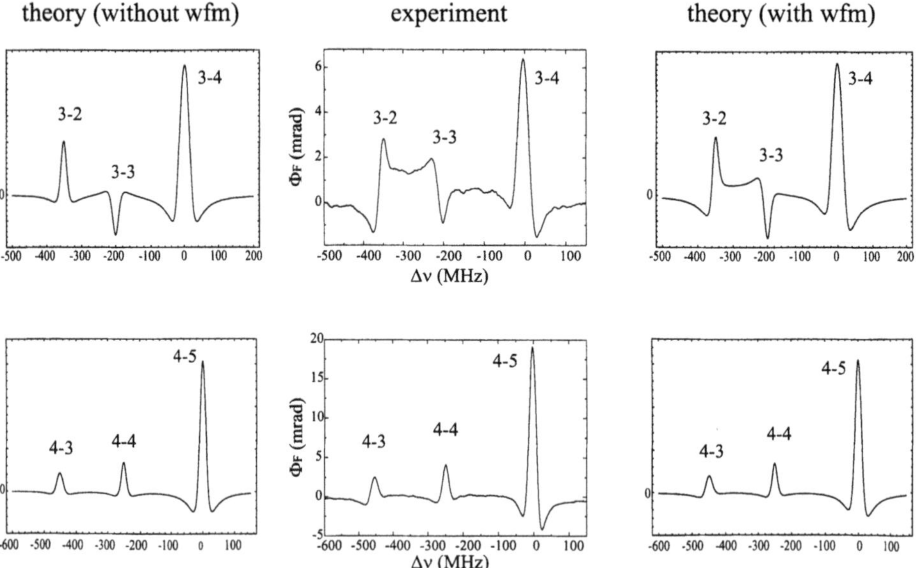

Fig. 1. Linear Faraday effect ($B = 7.64\,\mathrm{G}$) of the $F = 4 \to F' = 3,4,5$ (*bottom*) and $F = 3 \to F' = 2,3,4$ (*top*) transitions of the Cs $\mathrm{D_2}$-line in a thermal beam. Experimental results (*center*) and theoretical calculations neglecting (*left*) and taking hyperfine wave function mixing (2) into account (*right*). The lineshapes were calculated assuming Lorentzian-shaped resonances (30 MHz FWHM), while the experimental line shapes have a residual Doppler broadening of 30 MHz

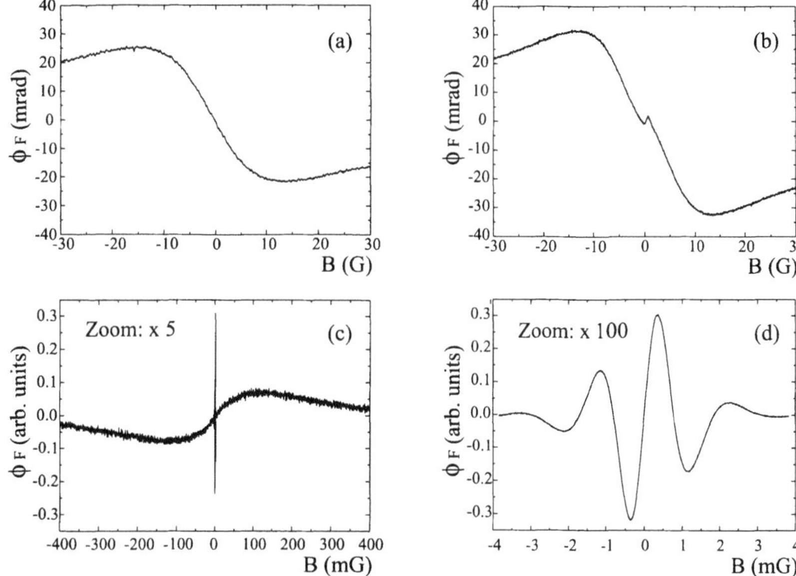

Fig. 2a–d. Experimental magneto-optical rotation on the D_2-line of a Cs beam (geometry of Fig. 3a): LFE (**a**) and onset of NLFE (**b**) of probe beam only (no pump beam); (**c**) expanded NLFE of probe beam and Faraday–Ramsey spectrum (central structure) induced by pump beam; (**d**) expansion of Faraday–Ramsey spectrum

width is determined by the optical linewidth (Fig. 2a). This resonance structure is typical for the *linear Faraday effect* (LFE) and can only be observed with light intensities far below optical saturation. The *nonlinear Faraday effect* (NLFE) appears as a very narrow dispersive resonance centered on the linear FE structure when the light intensity is increased (Fig. 2b).

3 The Resonant Nonlinear Faraday Effect

The origin of the resonant NLFE is the spin alignment created by resonant optical pumping with linearly polarized light. In the 1980s several theoretical models of the NLFE of simple transitions were developed. Besides a single theoretical work regarding open transitions in multilevel atoms there was, in 1990, no theoretical model describing partially closed transitions as they occur in the alkalis. In an effort to find a lowest-order treatment of the NLFE for such transitions I developed, together with Sergej Kanorsky, a simple but effective three-step model of the NLFE [6, 7]:

1. Creation of spin alignment in a hyperfine ground state by optical pumping with linearly polarized light. This alignment breaks the rotational symmetry of the medium which acquires a linear dichroism/linear birefringence whose axis is determined by the orientation of the alignment.

2. Precession and transverse relaxation of the induced alignment in the magnetic field. The steady state magnitude and orientation of the alignment (angle α with respect to the light polarization) can be calculated in a way similar to the treatment of the Hanle effect.
3. Detection of the rotated alignment based on the linear birefringence/ linear dichroism that it induces. The same light beam probes the alignment and emerges elliptically polarized with an orientation

$$\phi_{\mathrm{F}} = (N"_{\parallel} - N"_{\perp}) \frac{\pi L}{\lambda} \sin 2\alpha \,, \tag{3}$$

where $N"_{\parallel/\perp}$ are the imaginary parts of the index of refraction of light polarized parallel/perpendicular to the alignment axis.

According to (3) the NLFE is thus a linear dichroism, while the LFE (1) is a circular birefringence. The three steps occur in general simultaneously when a single laser produces and probes the alignment. Under well chosen experimental conditions however, the description in terms of sequential steps is nonetheless an excellent approximation as we have shown in our study of the NLFE of cesium vapor [6].

3.1 Line Shape and Strength of the NLFE

The relative strengths of the NLFE of different hyperfine components is determined by $N"_{\parallel} - N"_{\perp}$, which is proportional to the alignment

$$N"_{\parallel} - N"_{\perp} \propto A_{zz} \propto \sum_{M} p_M \left\langle F, M \left| 3F_z^2 - F^2 \right| F, M \right\rangle \tag{4}$$

produced in the hyperfine ground state $|nL_J,\, F\rangle$ in step 1 (p_M=level populations). The optical spectral line shape of the NLFE thus coincides with the optical excitation line shape, i.e. an absorptive Lorentzian (Gaussian) in Doppler-free (Doppler-broadened) spectra.

The magnetic field dependence of ϕ_{F} is determined by time averaging $\sin 2\alpha$ in (3) with weights determined by the relaxation process. In the case of an exponential isotropic relaxation of the alignment (at the rate γ_2) this time average yields a dispersive Lorentzian:

$$\phi_{\mathrm{F}}(x) \propto \frac{x}{1+x^2} \,, \text{ where } x = \frac{2\,\omega_{\mathrm{L}}}{\gamma_2} = \frac{B}{B_0} \,, \text{ with } B_0 = \frac{\hbar\,\gamma_2}{2\,g_{\mathrm{F}}\,\mu_{\mathrm{B}}} \,. \tag{5}$$

In general the line shape is determined by the distribution of the magnetic interaction times, i.e. the distribution of times between the pumping and the probing process. In a collision-free vapor of atoms with a Maxwellian velocity distribution (average velocity $\overline{v}$) and a Gaussian TEM_{00}-laser beam (waist

w) the line shape is no longer Lorentzian [8], but can be approximated (for an infinitely long cell) by

$$\phi_{\mathrm{F}}(x) \propto x \ln |x| \ , \text{ where } \ x = \omega_{\mathrm{L}} \tau_{\mathrm{TOF}} = \frac{B}{B_0}, \text{ with } B_0 = \frac{\hbar \overline{v}}{g_{\mathrm{F}}\, \mu_{\mathrm{B}}\, w}\,, \tag{6}$$

where τ_{TOF} is the average time of flight through the beam. In (5) and (6) the scaling fields B_0 determine the linewidths of the NLFE resonances.

Long spin-precession times thus lead to narrow NLFE resonances, which can be obtained using buffer gases or spin-preserving wall coatings. We have reported the first observation of a narrowing of the NLFE resonance in a small paraffin-coated cell [9], a technique which has been further refined by Budker and collaborators [10]. In a coated cell, atoms pumped inside a narrow laser beam can re-enter this beam after hundreds or thousands of wall collisions, thus leading to a considerable narrowing of the NLFE resonance. Based on the pump-precess-probe interpretation the extension of NLFE spectroscopy to atomic beams with spatially separated light fields [4, 11] was quite natural. We have called this technique Faraday–Ramsey spectroscopy (FRS).

3.2 Faraday–Ramsey Spectroscopy

The geometry of FRS experiments is sketched in Fig. 3. A thermal beam of cesium atoms traverses two linearly polarized laser beams from a single-mode diode laser locked to a hyperfine transition of the Cs D_2-line. These optical interaction zones are separated by 30 cm and the whole arrangement is contained in two mu-metal cylinders. The interzone spacing contains a solenoid, transverse coils and a pair of electrodes, which allow us to apply homogeneous magnetic and electric fields. The probe beam detects the rotated alignment by a polarimetric technique yielding a signal proportional to ϕ_{F}.

FRS is closely related to Ramsey's method of magnetic resonance with separated oscillatory fields, where two phase-locked radio-frequency fields each provide a $\pi/2$-pulse to particles initially polarized along the magnetic

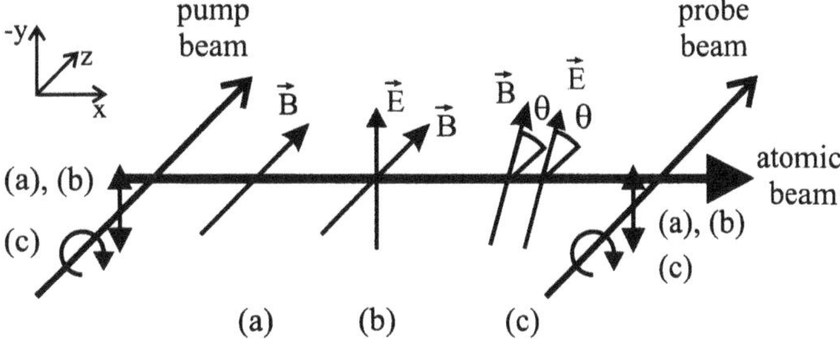

Fig. 3. Experimental arrangement of light polarizations and static fields for the recording of Faraday–Ramsey signals: pure magnetic interaction (a); measurement of the Aharonov–Casher phase shift (b), and of electric tensor polarizabilities (c)

field. In the interzone region the transverse magnetization evolves in the magnetic field and the second zone completes the spin-flip process in a phase-sensitive way. In FRS the two r.f. zones are replaced by the laser beams which create and probe the transverse spin polarization (alignment with linear polarized light and orientation with circularly polarized light). This technique does not rely on the use of oscillatory fields, and the relative phase of the r.f. fields in the conventional Ramsey zones corresponds in FRS to the relative orientations of the pump and probe polarizations.

The line shape of the FRS signal is determined by the time, or equivalently the velocity average of $\sin 2\alpha$. In a monochromatic beam, $\phi_{\mathrm{F}}(B)$ has a sinusoidal field dependence, while a velocity distribution damps these oscillations and yields a typical Ramsey fringe spectrum (Fig. 2d) given by

$$\phi_{\mathrm{F}}(x = \frac{B}{B_0}) \propto \int_o^\infty \sin\left(\frac{x}{u}\right) u^2 \,\mathrm{e}^{-u^2} \mathrm{d}u, \text{ with } B_0 = \frac{\hbar\, v_0}{2\, g_{\mathrm{F}}\, \mu_{\mathrm{B}}\, L}$$

for a Maxwellian velocity distribution. B_0 determines the fringe width.

4 Applications of Faraday–Ramsey Spectroscopy

The narrow fringes and the very good signal-to-noise ratio make FRS a very sensitive tool for the investigation of spin coherence perturbations in the interaction region. Although electric fields have no direct influence on spin polarization, the combination of electric interactions and spin-orbit and/or hyperfine interactions does affect spin polarization. FRS has proven to be well suited for the investigation of linear and quadratic Stark effects.

4.1 Linear Stark Effect: the Aharonov–Casher Phase Shift

In a static electric field E applied at right angles to the magnetic field and the beam direction (Fig. 3b), the Ramsey fringes experience a phase shift due to the motional magnetic field $\boldsymbol{B}_m = \frac{\boldsymbol{v}}{c^2} \times \boldsymbol{E}$ seen by the atoms in their rest frame. The line shape is then given by

$$\phi_{\mathrm{F}}(x, y = \frac{E}{E_0}) \propto \int_0^\infty \sin\left(\frac{x}{u} + y\right) u^2 \,\mathrm{e}^{-u^2} \mathrm{d}u, \text{ with } E_0 = \frac{\hbar\, c^2}{2\, g_{\mathrm{F}}\, \mu_{\mathrm{B}}\, L}. \quad (7)$$

By suitable topological transformations one can show [12] that the electric field induced phase is equivalent to the Aharonov–Casher (AC) effect, a topological phase that a magnetic moment acquires when carried on a closed loop around a charged wire. In Garching we have measured this AC phase shift on a Rb beam with a precision of 1% [12] and recently we improved the precision to 0.3% in a Cs beam [5]. Comparison with theory is limited by the precision ($\approx 1\%$) with which the electric field integral over the interaction zone can be calculated. The higher experimental precision thus allows us to use the AC effect for the absolute calibration of such electric field integrals [5].

4.2 Measurement of Electric Tensor Polarizabilities in the Alkalis

The change of energy of an atomic state $|nL_j; F, M\rangle$ exposed to a static electric field E oriented along the quantization axis is given by

$$\Delta E(F,M) = -\frac{1}{2}\,\alpha\, E^2, \text{with } \alpha = \alpha_0 + \frac{3M^2 - F(F+1)}{F(2F-1)}\,\alpha_2(F)\,, \tag{8}$$

where the static electric polarizability α can be decomposed into scalar (α_0) and tensor (α_2) parts. The scalar polarizability arises in second-order perturbation theory. Because of the rotational symmetry of the $^2S_{1/2}$ alkali ground states their tensor polarizabilities vanish in second order, but get small finite values in third-order perturbation theory due to the combined hyperfine and Stark interactions. In alkali ground states α_2 is 6 to 7 orders of magnitude smaller than α_0, and leads to typical level shifts in the Hz range for electric fields of 10 kV/cm. The tensor polarizabilities of the alkalis were measured in the sixties in atomic beam experiments using conventional Ramsey spectroscopy and showed large discrepancies between experiment and theory (see references in [5]). There is a manyfold interest in remeasuring the tensor polarizabilities: on one hand they serve as a sensitive test for high-precision atomic physics calculations, and on the other hand they are a source of systematic effects in experiments searching for permanent electric dipole moments in atoms and in atomic clocks. We also plan to measure the tensor polarizability of Cs atoms trapped in solid He matrices in order to get more insight into He matrix-induced perturbations of atomic properties and need a reliable reference value of α_2 for free atoms.

We propose [5] to remeasure the tensor polarizabilities using FRS and the arrangement shown in Fig. 3c. Pumping is done with circularly polarized light, while the static fields E and B are parallel to one another and oriented at an angle θ with respect to the pumping direction. Detection proceeds via the measurement of the circular dichroism of the probe laser using a double (magnetic field and probe polarization) modulation technique. At MPQ we were able to show [13] that this technique allows us to measure α_2. These results were not published as the analysis of systematic effects could not be completed before I left Garching. We have now set up a similar experiment using a Cs beam and observed the first quadratic Stark signals. The experiments and their analysis are still underway and will be extended to the Rb and K isotopes after completion of the work on Cs.

As a final remark we want to mention that the measurement of these tiny quadratic effects is dominated by a linear Stark effect background from the AC phase shift, which is yet another example of the famous "yesterday's sensation, is today's calibration and tomorrow's background".

5 Personal Remarks and Acknowledgements

In 1990 I accepted a job offer by Theodor Hänsch and I worked at MPQ, where most of the work described in this paper was carried out. I am grateful to Theodor for giving me the freedom to devote part of my activities to the deeper understanding of the nonlinear Faraday effect, which we had discovered by accident a few years earlier at ETH-Zurich. It is thus with great pleasure that I dedicate this paper to Theodor on the occasion of his 60th birthday and in memory of the fascinating years that I spent working in his team.

I also express my warmest thanks to the students and visiting scientists whose contributions have led to the results presented above: Xiao-hong Chen, Jörg Skalla (born Wurster), Eva Pfleghaar, Sergej Kanorsky, Vladimir Sautenkov, Birgit Schuh, Axel Görlitz, Markus Niering, Ulrich Rasbach and Christian Ospelkaus.

References

1. X. Chen, V.L. Telegdi, A. Weis, Opt. Commun. **74**, 301 (1990)
2. X. Chen, V.L. Telegdi, A. Weis, J. Phys. B. **20**, 5653 (1987)
3. A. Weis, V.A. Sautenkov, T.W. Hänsch, J. Phys. II (France) **3**, 263 (1993)
4. B. Schuh, Magneto-optische Ramsey-Spektroskopie. Ph.D. Thesis, LMU München (1994) unpublished
5. U. Rasbach, C. Ospelkaus, A. Weis, 'Linear and quadratic Stark effect studies of the Cs ground state using Faraday-Ramsey spectroscopy'. In: XVIIth Int. Conf. Coherent and Nonlinear Optics, Minsk, June 26–July 1, 2001 (to be published as SPIE Proceedings)
6. S.I. Kanorsky, A. Weis, J. Wurster, T.W. Hänsch, Phys. Rev. A **47**, 1220 (1993)
7. A. Weis, J. Wurster, S.I. Kanorsky, J. Opt. Soc. Am. B **10**, 716 (1993)
8. E. Pfleghaar, J. Wurster, S.I. Kanorsky, A. Weis, Opt. Commun. **399**, 303 (1993)
9. S.I. Kanorsky, J. Skalla, A. Weis, Appl. Phys. B **60**, 165 (1995)
10. D. Budker, V. Yashchuk, M. Zolotorev, Phys. Rev. Lett. **81**, 5788 (1998)
11. B. Schuh, S.I. Kanorsky, A. Weis, T.W. Hänsch, Opt. Commun. **100**, 451 (1993)
12. A. Görlitz, B. Schuh, A. Weis, Phys. Rev. A **51**, R4305 (1995)
13. M. Niering, Messung der elektrischen Tensorpolarisierbarkeit des $5S_{1/2}$-Grundzustands von ^{85}Rb und ^{87}Rb. Diploma thesis, LMU München (1996) unpublished

Sensitive Detection Techniques of Laser Overtone Spectroscopy

Wolfgang Demtröder and Thorsten Platz

This paper is dedicated to Prof. Theodor W. Hänsch on the occasion of his 60th birthday. I first met him 33 years ago in Heidelberg, where he worked as a graduate student of Prof. Toschek on Doppler-free saturation spectroscopy. The discussion with him was very enlightening and I was impressed by his deep understanding of any detail and by his imagination. Eight years later I spent some months with him in his labs at Stanford University, where I learned from him about the beauties of polarization spectroscopy and admired his experimental skills, building a laser wavemeter with a toy train. His delight in trying new experimental techniques and following his imagination in his private lab has brought about many beautiful ideas which turned out to be realizable and enabled considerable progress in high-resolution and high-precision laser spectroscopy.

The present paper reports on linear Doppler-free spectroscopy in another spectral range to that in which Theodor Hänsch is working. It shows how much can be learned from sub-Doppler spectroscopy, a field where Theodor Hänsch has made essential and pioneering contributions.

1 Introduction

High vibrational levels in electronic ground states of molecules play an important role for enhancing chemical reactions. For all reactions with potential barriers a minimum amount of excitation energy is necessary to overcome this barrier and to start the reaction. The excitation energy can be supplied, for instance, by absorption of photons, where a specific excited level is populated. Due to the increasing anharmonicity of the ground state potential and the strongly rising density of states with increasing energy, couplings between high-lying vibrational levels become more and more important. This may result in the redistribution of the primary excitation energy selectively pumped into a specific high-lying rovibronic level of the molecule into many vibrational modes (intramolecular vibrational redistribution = IVR) [1, 2].

The investigation of molecular structure and dynamics and in particular of such IVR processes is one of the goals of overtone spectroscopy. Since the optical transition probability strongly decreases with increasing number of excited vibrational quanta, spectroscopic techniques of measuring overtone transitions have to be sufficiently sensitive and demand detection limits for transitions with absorption coefficients $\alpha < 10^{-9}$.

Besides optoacoustic spectroscopy [3], cavity-ring down spectroscopy [4] or intracavity laser absorption spectroscopy [5], the recently developed optothermal spectroscopy with bolometric detection [6] can fulfill this requirement. It has the further advantage of sub-Doppler spectral resolution, which makes it with regard to high resolution superior to most of the other sensitive techniques.

The present paper describes the experimental setup of optothermal spectroscopy and illustrates its sensitivity and spectral resolution by some examples of sub-Doppler overtone spectra of acetylene, ethylene and chloroform.

2 Experimental Setup

The experimental setup is shown schematically in Fig. 1. The molecules under investigation are mixed in a ratio 1:10–1:30 with helium at a total pressure of up to 1 bar. The gas mixture flows through a nozzle with 50 μm diameter into the vacuum chamber and is collimated by a conical skimmer. The collimated molecular beam is crossed perpendicularly by a single mode tunable cw laser (color center or Ti:sapphire laser) which excites the molecules selectively into high-lying rovibronic levels.

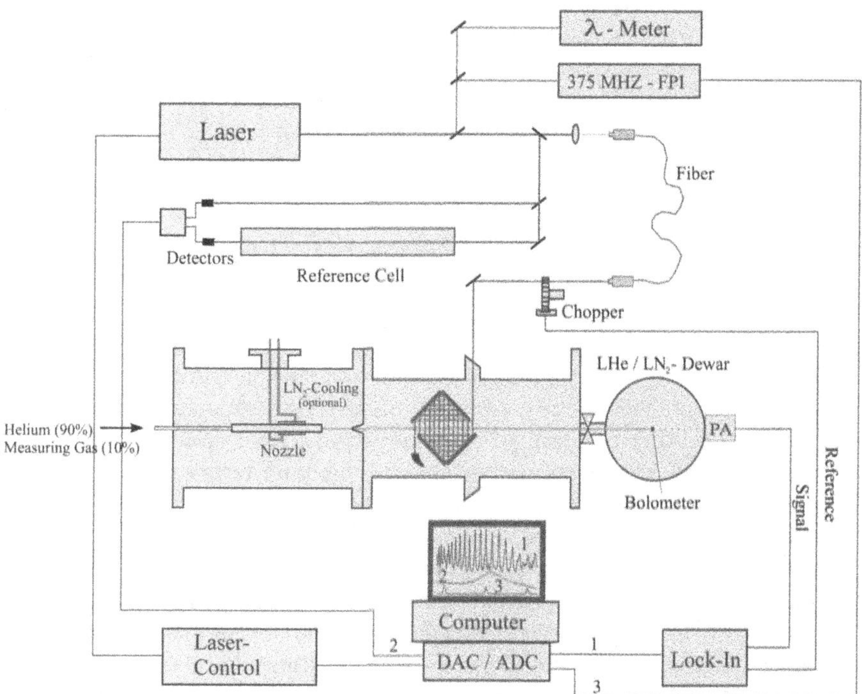

Fig. 1. Experimental arrangement for optothermal spectroscopy in a collimated molecular beam

Since the absorption path length for a single crossing with the collimated molecular beam is only 1 mm, it would be desirable to increase the number of crossings to reach a higher absorption probability. This can be achieved in different ways. One solution is based on a multiple-reflection device (Fig. 2a) formed by a pair of 90° mirrors. The number of crossings can be adjusted by the shift Δ between the upper and lower mirror pair. The effective total absorption length is, however, limited by the Rayleigh length of the Gaussian laser beam. In order to obtain the optimum overlap between the laser beam and the molecular beam, the laser beam diameter should be matched to the molecular beam diameter. This restricts the effective number of crossings to $N = L_R/L_C$ where L_R is the Rayleigh length and L_C is the pathlength between two successive crossings. Typical values for N reach 15 – 20.

A better choice is an optical enhancement cavity (Fig. 2b) consisting of two spherical mirrors with equal radius r at a distance d. The beamwaist $w_0 = [(\lambda/2\pi)\sqrt{d(2r-d)}]^{1/2}$ of the fundamental cavity mode [7], located at the crossing with the molecular beam , can be optimized by the proper radius r of the spherical mirrors. The laser beam is guided into the vacuum chamber through a single-mode optical fiber and is mode-matched by two lenses to the fundamental Gaussian mode of the enhancement cavity. With mirror reflectivities of $R = 99\%$ a theoretical enhancement factor of $F = 1/(1-R) = 100$ should be possible, if ideal mode matching can be achieved. The measured factor was $F = 50$, due to reflections by the mode-matching lenses and a slight mode mismatch. With an incoming laser power of 200 mW this means an intracavity laser power of 10 W, resulting in an intensity of $3 \times 10^3\,\mathrm{W/cm^2}$ in the beam waist. However, for higher overtone transitions this high intensity might still be below the saturation intensity.

The commercial bolometer consists of a small doped silicon crystal ($0.3 \times 0.3\,\mathrm{mm^2}$) cooled down to 1.6 K by pumping on the liquid helium. The excited molecules impinging on the cold surface transfer their translational and their excitation energy (and eventually also their condensation energy) to the crystal, thus increasing its temperature by a small amount ΔT. This results in a decrease

$$\Delta R = \frac{\partial R}{\partial T}\Delta T$$

of its electrical resistance, and a corresponding signal $\Delta V = I \cdot \Delta R$ across the bolometer when a small constant current I is sent through the silicon crystal. The temperature increase

$$\Delta T = (P_{\mathrm{trans}} + P_{\mathrm{exe}} + P_{\mathrm{cond}})/G$$

is proportional to the total energy per second that the molecules deposit at the detector and is inversely proportional to the heat conductivity G of the detector. The time constant τ of the system $\tau = C/G$ equals the ratio of the heat capacity C and heat conductivity G and has for our bolometer a value of $\tau \approx 1$ ms.

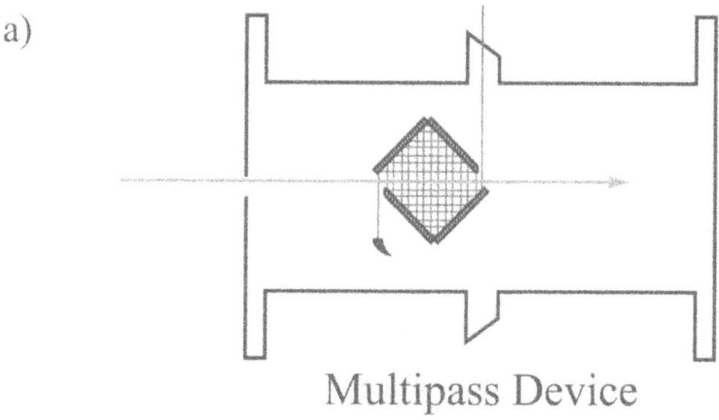

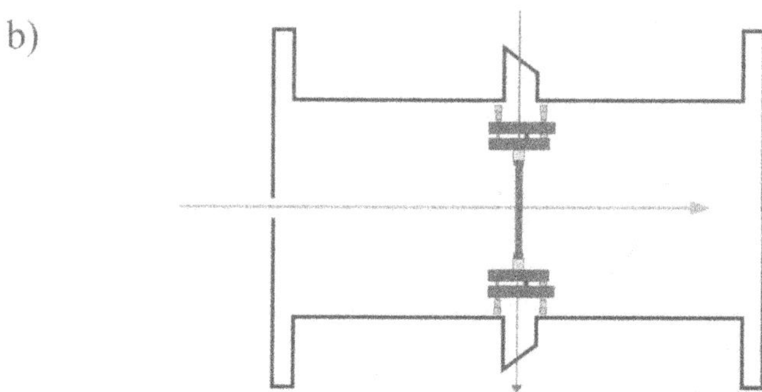

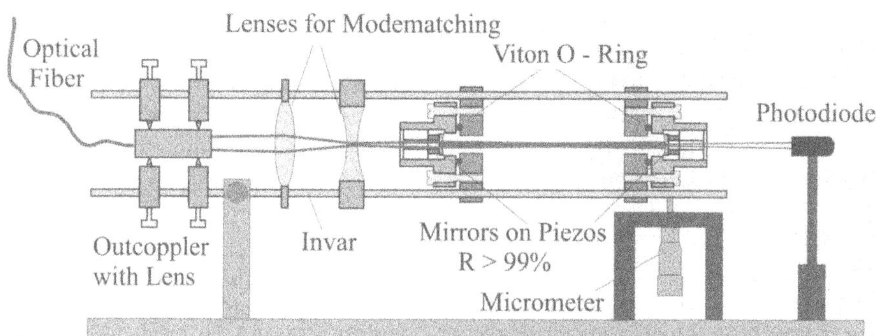

Fig. 2. Multiple crossing of laser and molecular beams: (**a**) multireflection arrangement, (**b**) enhancement cavity with mode-matching optics

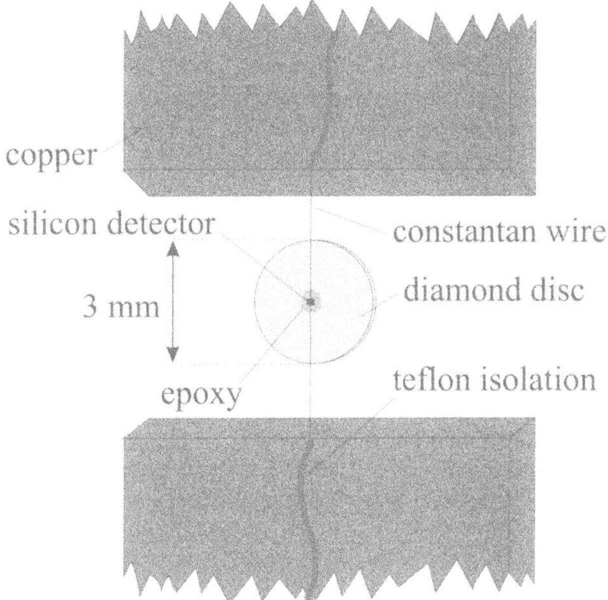

Fig. 3. Doped silicon detector with diamond disc for increasing the detection area

In order to match the detector surface to the geometrical cross-section of the collimated molecular beam, a thin diamond disc with 3 mm diameter is epoxied to the front surface of the silicon plate (Fig. 3). Because of the high Debeye temperature of the diamond, its specific heat at $T = 1.6\,\mathrm{K}$ is extremely small and the diamond therefore does not significantly increase the total heat capacity of the system. The lower limit for the detectable power is $8 \times 10^{-14}\,\mathrm{W}$ where the signal-to-noise ratio becomes S/N = 1 at a detection bandwidth of $\Delta\nu = 1\,\mathrm{Hz}$. With an incident laser power of 500 mW this means a relative absorption $\alpha \cdot L \approx 1.6 \times 10^{-13}$. For the enhancement cavity with an intracavity power of 10 W the lower detection limit is even $\alpha \cdot L \geq 8 \times 10^{-15}$. With an effective absorption path length $L = 2\,\mathrm{cm}$ this means a minimum detectable absorption coefficient of $\alpha \geq 10^{-13}\,\mathrm{cm}^{-1}$ for the multipass arrangement. With an excitation energy of $\Delta E = 1\,\mathrm{eV}$ per molecule the minimum detectable rate of excited molecules is then $\mathrm{d}N/\mathrm{d}t \geq 5 \times 10^5\,/\mathrm{s}$.

3 Measurements

The signal-to-noise ratio S/N achieved with this bolometer using the multiple crossing configuration of Fig. 2a is illustrated by Fig. 4, which shows the rotational line P(3) of the overtone transition $(1, 0, 3, 0^0, 0^0) \leftarrow (0, 0, 0, 0^0, 0^0)$ in acetylene. The linewidth of 5.5 MHz (FWHM) is limited by the residual Doppler-broadening due to the finite collimation of the molecular beam. The

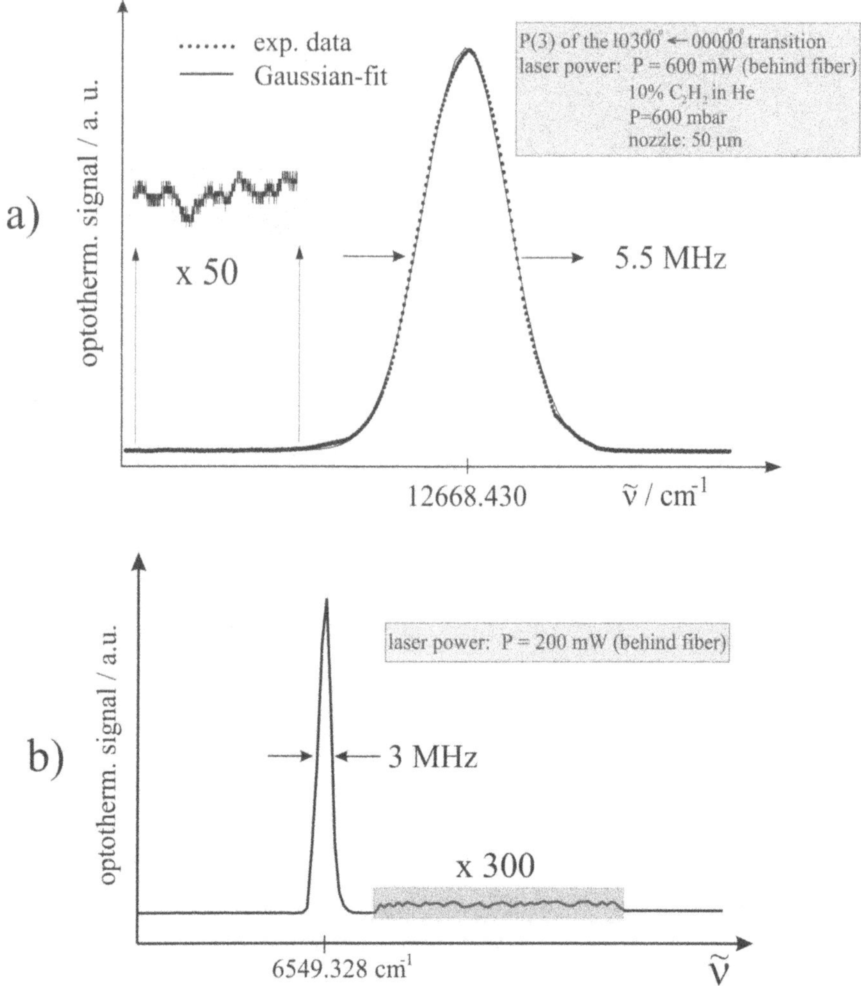

Fig. 4. (**a**) P(3)-rotational line of the $(1, 0, 3, 0^0, 0^0) \leftarrow (0, 0, 0, 0^0, 0^0)$ overtone transition in acetylene; (**b**) P(3) line of the $(1, 0, 1, 0^0, 0^0) \leftarrow (0, 0, 0, 0^0, 0^0)$ transition

noise background has been magnified by a factor of 50 in order to quantify the achieved S/R=10^3. The P(3) line of the first overtone transition $(1, 0, 1, 0^0, 0^0) \leftarrow (0, 0, 0, 0^0, 0^0)$ of acetylene C_2H_2 (Fig. 4b) can be even detected with S/N = 70 000.

In Fig. 5a a comparison is shown between identical small sections of the $(\nu_5 + \nu_9)$ excitation in ethylene C_2H_4. The upper spectrum was taken with a Bruker 4 m Fourier spectrometer which nearly reaches the Doppler limit in its spectral resolution. The second spectrum was measured with optoacoustic spectroscopy in an acoustic resonance cell placed inside an optical multipass

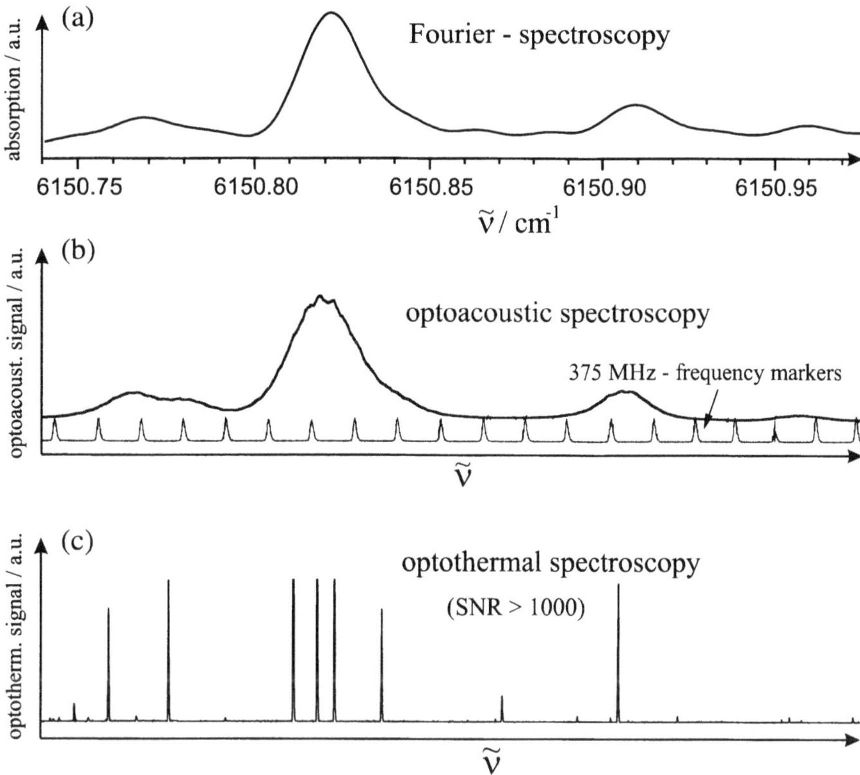

Fig. 5a–c. Comparison between identical sections of the ethylene ($\nu_5+\nu_9$) overtone spectra around 6150.8 cm^{-1}. (**a**) Fourier spectrum, (**b**) optoacoustic, (**c**) optothermal spectrum

Herriot cell. The lower spectrum represents the Doppler-free optothermal spectrum, which shows that within the Doppler-broadened profiles of the two upper spectra several overlapping transitions are hidden which can be completely separated in the Doppler-free optothermal spectrum.

The analysis of the C_2H_4 overtone spectrum in the spectral range between 6147 cm^{-1} to 6170 cm^{-1} was based on the method of combination differences. Since the molecular constants of the vibrational ground state are well known from infrared spectra of fundamental transitions [8], the energy separation of two transitions sharing a common upper level (e.g. P(3), R(1) and Q(2) transitions) allow in most cases the unambiguous determination of the rotational quantum number N' and the projection quantum numbers K'_a and K'_c of the upper level.

The ethylene overtone spectra are of particular interest, since ethylene plays an important role in agriculture. Plants under stress emit ethylene and unripe fruits experience shortened riping times when stored in an ethylene atmosphere [9]. These processes are enhanced by the excitation of ethylene into higher vibrational levels. The question of which molecular reactions take place in these processes is still open. It is therefore necessary to investigate the structure and dynamics of ethylene in higher vibrational states more thoroughly.

Although many upper levels are not perturbed and their spectra could be readily assigned, the analysis showed that quite a lot of transitions deviate from their expected positions, indicating perturbations by other nearby rovibronic levels. Although a systematic deperturbation procedure allowed the identification of several perturbing states, not all of the measured perturbed levels could be completely reproduced by this procedure within the experimental accuracy of $2 \times 10^{-3}\,\mathrm{cm}^{-1}$. The reason is, that up to now not much information has been available about all the vibronic states that can interact with the excited $(\nu_5 + \nu_9)$ levels.

More success was achieved for the overtone spectra of chloroform, $CHCl_3$, which occurs as several isotopomers, due to the natural relative abundances $^{35}Cl{:}^{37}Cl = 75.4\%{:}24.6\%$ of the Cl isotopes. While the $CH^{35}Cl_3$ (42.9%) represents a symmetric top (C_{3v}), $CH^{35}Cl_2^{37}Cl$ (41.9%) must be described as an asymmetric top (point group C_s).

Figure 6a shows the high-resolution optothermal spectrum of the Q-branches in the $2\nu_1$ overtone transition of chloroform around $5941\,\mathrm{cm}^{-1}$. The insert shows for comparison the Doppler-limited Fourier spectrum. This clearly demonstrates the increase in spectral resolution which yields much

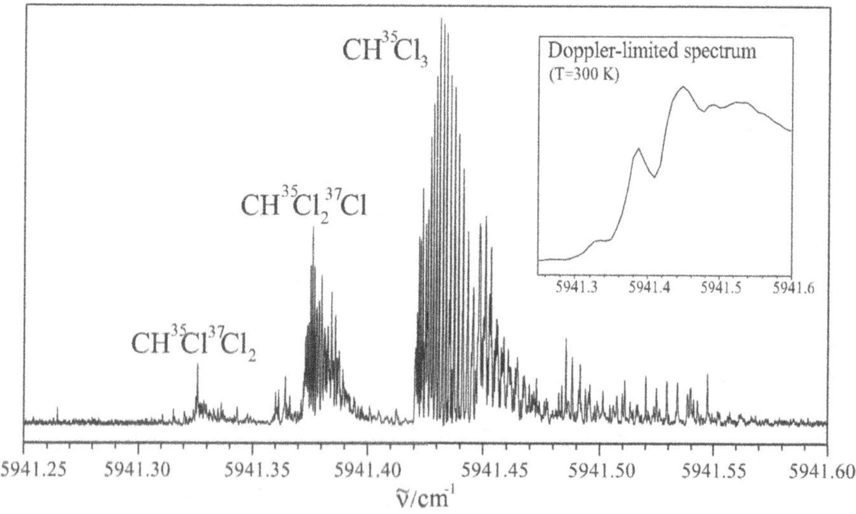

Fig. 6. Optothermal spectrum showing the Q-branch in the $2\nu_1$ overtone transition of the three strongest chloroform isotopomers. The insert shows for comparison the Doppler-limited Fourier spectrum

more details and thus more information on the molecular structure. It turns out that the spectrum of the asymmetric rotor $CH^{35}Cl_2^{37}Cl$ is easier to assign than that of the symmetric rotor $CH^{35}Cl_3$. The reasons for this are severe perturbations of the upper level of $CH^{35}Cl_3$ by nearly coincident other vibronic levels. The assignment of the $CH^{35}Cl_2^{37}Cl$ spectrum was possible with the expert cooperation of Dr. Hollenstein, ETH Zürich. Figure 7 illustrates for the K-substructure of two rotational transitions the comparison between experiment and simulation, based on the molecular constants which had been derived from the analysis of the experimental spectra. The agreement is quite satisfactory.

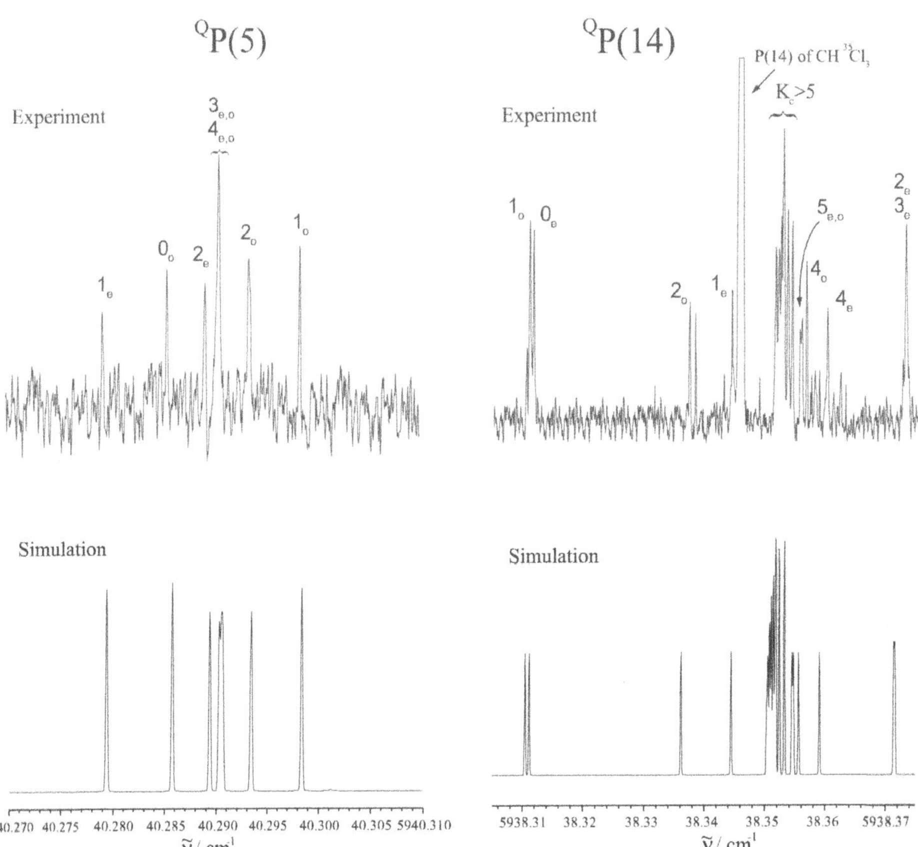

Fig. 7. Two rotational transitions measured with Doppler-free optothermal spectroscopy compared with simulations based on molecular constants derived from the measured spectra

4 Conclusion

These few examples have shown that Doppler-free optothermal spectroscopy represents a very sensitive technique for measuring even very weak overtone transitions. Besides these excitations into bound high-vibrational levels, transitions into predissociating states can also be detected. The dissociating fragments leave the collimated beam, thus reducing the number of molecules reaching the detector. This results in negative signals which, therefore, can be readily distinguished from transitions into bound levels. Such transitions into predissociating levels are of great interest for the study of molecular clusters and van der Waals dimers, since they allow information about coupling between excitations of the monomer and the van der Waals bound. Several research groups are performing active research in this area [6].

References

1. K. Lehmann, G. Scoles, Annual Rev. Phys. Chem. **45**, 241 (1994), J. Chem. Phys. **95**, 3891 (1991)
2. X. Luo, T. Rizzo, J. Chem. Phys. **93**, 862 (1990), J. Chem. Phys. **94**, 889 (1990)
3. V. Zharov, V. Letokhov, *Laser Optoacoustic Spectroscopy*, Springer Ser. Opt. Sci., Vol. 37 (Springer, Berlin, Heidelberg 1986)
4. D. Romanini, K. Lehmann, J. Chem. Phys. **99**, 6287 (1993), V. Baer, Appl. Phys. **B55**, 463 (1992)
5. A. Campargue, F. Stoeckel, M. Chevenier, Spectrochim. Acta Rev. **13**, 69 (1990)
6. R. Miller, in *Atomic and Molecular Beam Methods*, ed. by G. Scoles (Oxford Univ. Press, Oxford 1992), pp. 192–211
7. A. Siegmann, *Lasers* (Oxford Univ. Press, Oxford 1986)
8. J. Legrand, M. Herlemont, A. Fayt, J. Mol. Spectrosc. **171**, 13 (1995)
9. H. de Vries, M. van Lieshout, F. Harren, J. Reuss, Infrared Phys. **36**, 483 (1995)

Multiple-Beam Atom Interferometry: An Overview

Michael Mei, Sile Nic Chormaic, Sebastian Fray, and Martin Weitz

A review of recent experiments in multiple-beam atom interferometry is presented. In contrast to the sinusoidal interference pattern obtained in two-beam interferometry, the signal in this work is a sharply peaked Airy function similar to that obtained in multiple-beam optical interferometry. Apart from this fringe sharpening, collapse and revival of the interference signal – a phenomenon that is characteristic of multiple-beam interferometry – is observed. We present studies of a generalized Aharanov–Bohm phase shift and of controlled decoherence in multiple-beam atomic interference. Finally, recent work on multiple-beam atom interferometry using cold rubidium in a fountain configuration is discussed.

1 Introduction

Since the early 1990s, as a result of technological advances made in laser physics and microstructure fabrication, several designs of atom interferometers have been developed [1]. The specific interest in using atoms for interferometry rather than neutrons [2] or electrons [3] is due to the small atomic de Broglie wavelength, thus permitting such interferometers to reach a very high sensitivity and gain increasing significance in precision measurements of rotational or gravitational effects. One such measurement is the recoil energy shift of an atom on absorbing a photon, since this yields an accurate value of the fine structure constant [4]. Other fundamental uses of atom interferometers include tests for general relativity, tests of topological phases and atom coherences. Possibly the most significant development that influenced atom interferometry is the ability to manipulate the velocity of a beam of atoms using laser cooling techniques as proposed by Hänsch and Schawlow [5]. Since then, the field of laser cooling has advanced significantly and atom temperatures as low as a few μK can be routinely obtained [6]. As laser cooling improved, so too did the methods used for realizing atom interferometers.

In this work, we present experimental results that were obtained using multiple-beam atom interferometry. While for conventional two-beam atom interferometers sinusoidal fringes are observed, the interference signal here shows a more sharply peaked Airy function-like fringe pattern, such as from optical Fabry–Pérot interferometers. The sharper fringes translate to a higher resolution compared with that of the corresponding two-beam interferometer. A multiple-beam atom interferometer was demonstrated by Weitz, Heupel and Hänsch using a series of laser pulses projecting atomic wavepackets onto velocity-dependent dark states [7] and subsequently in other laboratories using different methods [8–10]. Besides fringe sharpening, further interesting effects occur in multiple-beam interference experiments when the phase difference between adjacent paths is nonconstant for all paths. It has long been known that nonlinear phase terms are the cause of the Talbot images observed in near field optics [11], where quadratic phase terms occur in the Fresnel approximation of the wave equation. In a multiple-beam interferometer, nonlinear phase terms cause collapse and revival effects of the fringe pattern. Nonlinear phase terms can appear due to photon recoil in an atom interferometer [12, 9, 10] or an additionally applied Aharonov–Bohm type phase shift [13]. Using multiple-beam interference with atoms, decoherence studies have been performed [14]. In the final section of this chapter, we describe work that is being undertaken on rubidium atoms in a fountain configuration using intense pulses of laser light as atomic beamsplitters.

2 Multiple-Beam Atom Interferometry: Initial Experiments

In the first experiments, a multiple-beam atom interferometer was built with thermal atoms from a caesium beam. The atomic wavepackets were coherently split and later recombined by laser pulses, which project the atoms onto a multilevel velocity-dependent dark state. Such dark quantum superpositions with entangled internal and external states yield novel possibilities for manipulating matter waves. A two-beam atom interferometer, where two spatially separating components of a dark state were overlapped, was first demonstrated [15]. If a multilevel dark state is used instead, atomic wavepackets can be divided into more than two components.

For the multiple-beam atom interferometer, we use the $6S_{1/2}(F = 4)$–$6P_{1/2}(F' = 4)$ transition of the caesium D1 line near 894 nm and two counter-propagating laser beams in a σ^+–σ^- configuration. During the first pulse, caesium atoms are projected onto a dark superposition of the five even magnetic ground state sublevels with magnetic quantum numbers $m_F = \pm 4, \pm 2, 0$, where each sublevel is associated with a distinct momentum as shown in Fig. 1. The momenta of two neighboring paths differ by $2\hbar k$. The paths then separate spatially for a time T after which a second laser pulse is applied. In general, the atom at this time will be in a superposition of dark and coupled states. The second pulse projects the atom once more onto a dark state by optically pumping all other components into the $F = 3$ hyperfine ground state, which is not detected. Thus, the second pulse splits each of

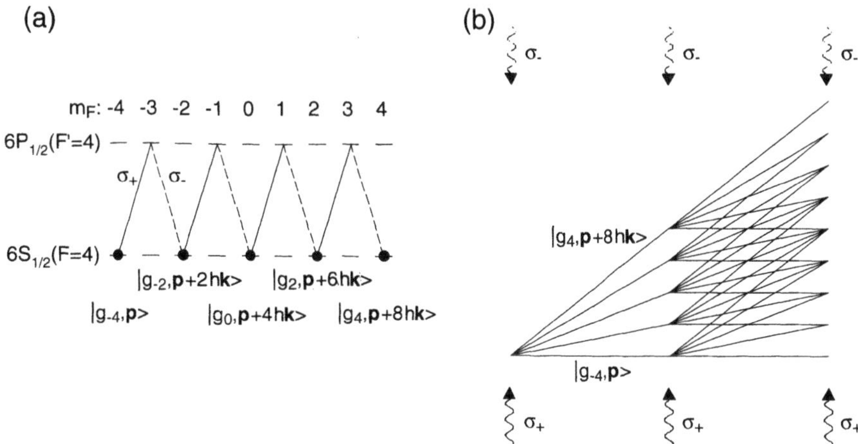

Fig. 1. (**a**) Level scheme and (**b**) recoil diagram for an atomic multiple-beam interferometer

the paths into five further paths. A third pulse recombines the paths and interference is observed.

The experimental set-up has been described in detail elsewhere [7, 12]. A beam of thermal caesium atoms enters an optical interaction region and is irradiated by the laser beams. This region is magnetically shielded and has a small homogeneous magnetic field ($\sim$ 10 mG) along the direction of the laser pulses. Typical interference signals are shown in Fig. 2 for different drift times, T. For $T = 5\,\mu$s good contrast is obtained, and the signal shows the fringe sharpening expected for multiple-beam interference. For larger drift times T, the fringe signal collapses. This collapse is not observed for many common optical interference experiments, such as e.g. a Fabry–Pérot resonator. In this experiment, the optical pulses acting as atomic beam-splitters transfer momentum to an atom in units of $2\hbar k$. Besides a linear phase term, a phase connected to the photon recoil energy and quadratic in transverse atomic momentum appears. This is intuitively understood by considering the kinetic energy differences between the paths. The interference pattern collapses for larger times T, when the accumulated quadratic phase between a central and an outermost path approaches unity. However, once the accumulated quadratic phase between all adjacent paths reaches an integral multiple of 2π, the interference pattern is revived. This revival occurs for pulse spacings T of integer multiples of π/ω_{r} (corresponding to 66.8 µs here) with $\omega_{\mathrm{r}} = 2\hbar k^2/m$. The first revival is clearly visible in the experimental data (Fig. 2), though with a reduced contrast. Besides this effect, the possibility of realizing a multiple-beam atom interferometer with larger drift times, as required for precision measurements, is demonstrated. For example, the photon recoil energy can be determined from a precise measurement of the revival time.

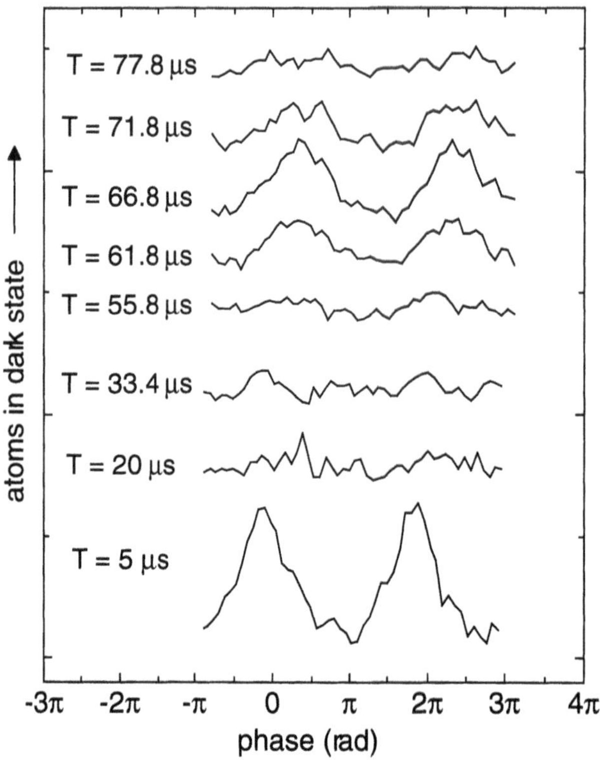

Fig. 2. Experimental fringe signals for the atom interferometer of Fig. 1 as a function of the phase of the third optical pulse for different pulse spacings T

3 Generalized Aharonov–Bohm Phase Shifts

In 1959 Aharanov and Bohm showed that potentials are of great physical significance in quantum mechanics [16] because phase shifts can be induced in regions of nonzero potential even where no classical force is present. This is in stark contrast to the role played by potentials in classical mechanics, where they are viewed as being purely convenient mathematical tools. Aharanov and Bohm predicted two effects – vector and scalar – and these have been experimentally demonstrated using electrons [17], neutrons [18] and atoms [19] in two-beam interference experiments. More recently, we have demonstrated the presence of a generalized Aharanov–Bohm phase shift using multiple-beam atomic interference [13]. The phase shifts studied here were associated with only the internal atomic degrees of freedom. By pulsing an off-resonant light field, a potential and thence an accumulated phase shift was generated that scales quadratically with the interfering path number. This can be interpreted as a scalar Aharanov–Bohm type effect, whereby the nonlinear phase shift is achieved even though no classical forces act on the atoms. One can read this tensorial phase by using an interferometer in which the different paths consist of different magnetic sublevels. One should note that while the original work relied on interfering paths with the same charge passing through regions of differing potential, our experiment uses the idea of a path-dependent atomic polarizability such that the same electric field is used for all the paths (with different magnetic numbers) and this removes the requirement for spatially separated paths.

The experiment is similar to that described previously, only this time two *copropagating* optical Ramsey pulses are applied to the atoms in the interaction region [7]. This results in no net momentum transfer to the atoms and each atom is in a superposition of purely internal Zeeman states. The off-resonant light field that generates the nonlinear phase shift (via a tensor contribution to the atomic polarizability) is typically 400 MHz blue detuned from the $6S_{1/2}(F = 4)$–$6P_{1/2}(F' = 4)$ line. Fig. 3 shows typical interference

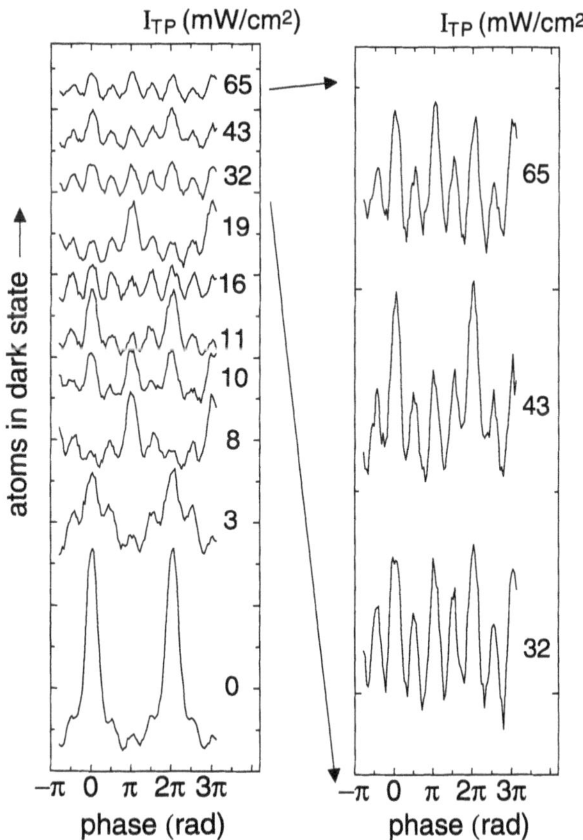

Fig. 3. Multiple-beam Ramsey interference signals for different intensities I_{TP}. The three uppermost spectra are also shown on the right-hand side on an expanded scale

signals. The signals are plotted as a function of the phase of the second Ramsey pulse for different off-resonant light intensities, I_{TP}. When $I_{\mathrm{TP}} = 0$, an Airy-function-like signal is observed. As I_{TP} increases a phase shift quadratic in the path number is introduced resulting in collapse and revival of the fringes for different intensities. Linear phases would result in a mere shifting of the interference fringes and it is the nonlinear contribution that leads to the observed collapse and revival.

4 Studies of Controlled Decoherence

In the experiments discussed in the last section, the observed loss and regain of the interference signal is caused by (ideally) completely coherent processes. We soon became interested in examining the effect of a partial loss of coherence on the fringe pattern. Here, we describe an experiment in which photons are scattered off one of the four interfering paths in a multiple-beam atomic interferometer [14].

The experiment again is based on a multiple-beam Ramsey set-up. During the first optical pulse, caesium atoms are pumped into a dark coherent superposition of the four magnetic sublevels $m_{\mathrm{F}} = \pm 3, \pm 1$ of the $F = 3$ hyperfine ground state, which is probed with a second optical pulse that projects the atoms onto a dark state. Between these Ramsey interactions, the $m_{\mathrm{F}} = 3$ path can be observed by applying a sequence of microwave and optical pulses designed to scatter photons selectively off this path (see [14] for details). With no photon scattering, we again observe an Airy-function-like fringe pattern as a function of the phase of the second Ramsey pulse, as shown by the solid line in Fig. 4a. When scattering photons off the path in $m_{\mathrm{F}} = 3$, the fringe contrast decreases as shown by the dashed line. The data shown in Fig. 4b were recorded with an applied phase shift of π for the $m_{\mathrm{F}} = 3$ path. With no scattering of photons (solid line), the fringe contrast is significantly lower than that in (Fig. 4a), as the fourth path now is severely out of phase with respect to the other paths. When scattering photons off this path, the interference contrast *increases*. This is contrary to the situation in two-beam interferometers, where decoherence inevitably leads to a loss of fringe contrast. It is important to note that in all cases the scattering of photons leads to a loss of coherence in the atomic quantum state as 'which-path' information is carried away by the photon field. When only detecting the atomic degrees of freedom, this missing information – corresponding to a nonzero entropy – shows up as decoherence. Our results suggest that in multiple-beam interference a single Michelson fringe contrast is not sufficient to quantify decoherence. In [14] we describe an approach to quantify decoherence in this experiment using the possible path guessing likelihood and show that in all cases this quantity increases with the scattering of photons off a path.

We wish to point out that this experiment can be seen as a model system for the study of controlled decoherence in quantum systems, as e.g. quantum

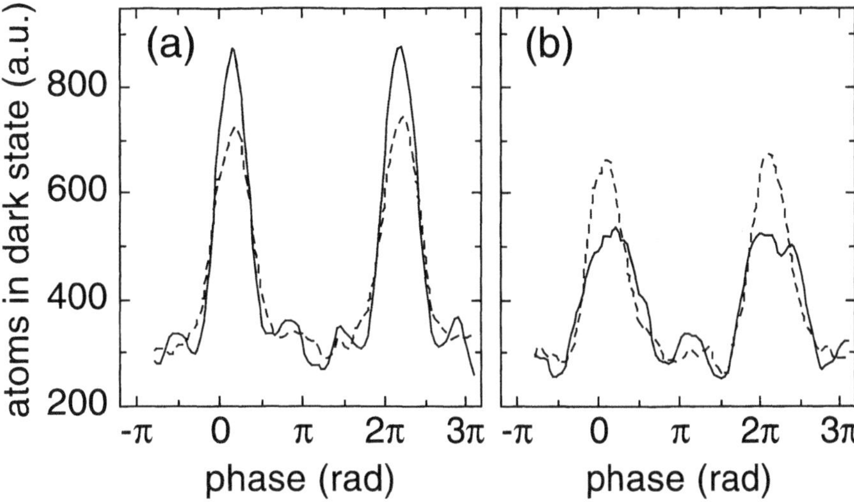

Fig. 4. (**a**) Multiple-beam Ramsey interference signals without (*solid*) and with (*dashed*) an applied 9 µs long optical pulse scattering photons off the path in $m_F = 3$. (**b**) Interference spectra recorded with an additional phase shift of π applied to the path in $m_F = 3$. Again, both spectra without (*solid*) and with (*dashed*) scattering of photons off this path are shown

logic gates. In quantum computation science, the study of decoherence is of immense interest, as such experiments aim towards the realization of complex entangled quantum systems. In this experiment, the four paths of the Ramsey interference set-up could represent two quantum bits. From a quantum information science point of view, the π phase shift of the $m_F = 3$ path corresponds to the operation of a phase gate, which is an elementary quantum gate. The photon scattering corresponds to coupling to an engineered reservoir. For large couplings, this results in an output state that is independent of the input parameters of the quantum gate.

5 Multiple-Beam Atom Interferometry with Cold Atoms and Intense Pulses of Light

In this section we report on work in progress. An atom interferometer for laser-cooled ^{85}Rb atoms in a fountain configuration is being developed. *Ultracold* atoms have an advantage over thermal atoms due their smaller velocity. This yields a longer atom–light field interaction time, thereby increasing the sensitivity of the apparatus. We plan to further increase the sensitivity by implementation of a novel atom interferometer with a larger number of interfering paths. In our previous scheme, the number of paths was limited by the internal atomic structure. These limitations are removed when the different interfering paths have the same internal quantum state and only differ in their momentum. This can be achieved by irradiating the atoms with a pulsed optical standing wave that is resonant with an open transition from a stable ground state $|g\rangle$ to an excited state $|e\rangle$. Spontaneous emission during the pulse is avoided if the pulse duration is less than the upper state natural lifetime. As shown in Fig. 5, coherent partial beams of different momenta are generated by multiple absorption and stimulated emission of photons [20]. After the pulse, all components left in the upper electronic state spontaneously decay and, assuming a favorable branching ratio, no longer contribute to the signal. The number of remaining ground state components is not limited by the internal atomic structure, but rather by the available laser power. In a different picture, the beamsplitting process can be understood in terms of diffraction off a transient phase grating.

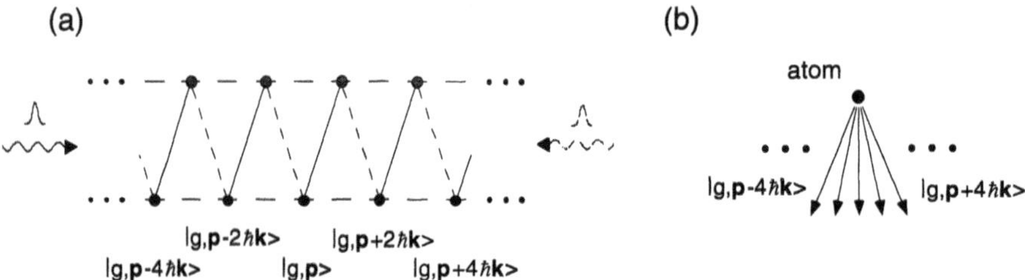

Fig. 5. (**a**) Schematic of populated internal and momentum states for a two-level atom initially in state $|g, \boldsymbol{p}\rangle$ that is subject to a pulsed, resonant standing wave. After the pulse, the components left in the upper electronic state (*dashed*) spontaneously decay and, for an open transition, no longer contribute. (**b**) Spatial picture of the achieved coherent beamsplitting

A multiple-beam interferometer can be realized by a sequence of three pulsed standing waves (with equidistant pulse spacing), where the first two should have a pulse duration less than the natural lifetime, as mentioned. For the final standing wave, it seems advantageous to use a pulse time longer than the upper state natural lifetime, so that only atoms passing precisely through the antinodes are transmitted. By scanning the position of this final pulse and detecting the number of transmitted atoms, the atomic interference pattern can be plotted. The 'slit width' of this optical absorption mask depends on the optical pulse area. This detection scheme relies on saturation of the transition. We wish to remark that such optical masks have been experimentally demonstrated in the context of atom lithography [21]. One can show that the expected fringe width of the described interferometer for, e.g. gravitation or rotation increases linearly with the number of paths, while it increases quadratically for measurement of the photon recoil energy. On the other hand, too many paths may decrease the fringe contrast. We note that Cahn et al. [9] have recently observed multiple-beam atom interference by scattering off a nonresonant standing wave. Although the number of paths was not limited by internal atomic structure, the possible sensitivity for the atomic recoil energy increased linearly with the path number. This is attributed to the detection scheme used, which was based on backscattering of light on atomic density gratings and was sensitive to sinusoidal density modulations only.

An alternative solution could be the replacement of the first two beam-splitting pulses by comparatively long pulsed standing waves. This experiment would be analogous to a sequence of three atomic absorption gratings. The only drawback would be the reduced number of transmitted atoms, but one expects high fringe contrast.

The current status of the experiment is as follows. We have constructed an atomic fountain with laser-cooled ^{85}Rb atoms (Fig. 6a), which are first precooled in a standard magneto-optical trap. Cooling to temperatures of 10–15 μK is obtained by applying a 30 ms optical molasses phase after the MOT has loaded. During the final 1.5 ms of the molasses phase, the atoms are launched vertically upwards by shifting the two upper trapping beams slightly to the red and the two lower beams by the same amount to the blue [6]. Once the atoms have reached the selected launch height, they turn around, fall under gravity and are finally detected below the original trapping region. Due to the geometric design of the experiment a maximum time of 300 ms between launch and detection is attainable.

When operating the atomic fountain, an optimum SNR of 250 was obtained by using frequency modulated fluorescence detection. This technique is similar to that described in [22]. As in ordinary transmission FM spectroscopy, the principle is based on the idea that sharp resonance structures can be separated from a slowly varying background by frequency modulation. A frequency modulated resonant standing wave probe laser beam passes vertically through the cold atomic cloud. Light emitted by the atoms is

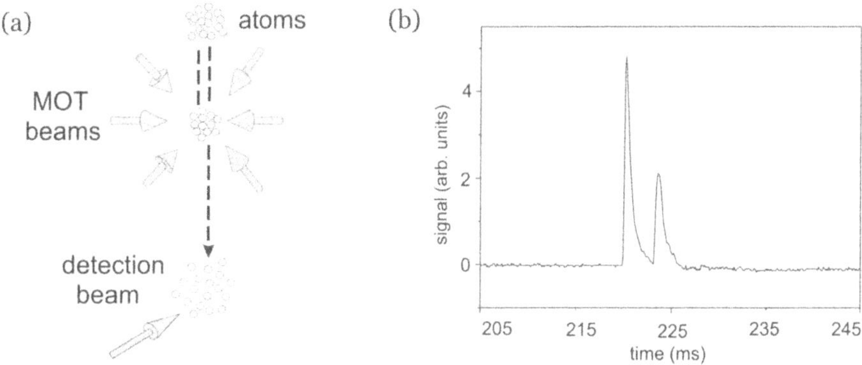

Fig. 6. (**a**) Experimental setup of the rubidium atomic fountain. (**b**) Demodulated fluorescence signal from cold atoms in the fountain. The first (second) peak gives a measure for the number of atoms detected in the upper (*lower*) hyperfine state

detected by a PMT. The signal from the PMT is RF amplified and then mixed with the modulation frequency. This modulation frequency (2.93 MHz) must be smaller than the natural linewidth of the transition (5 MHz) so that the sidebands contribute to the overall fluorescence signal. Typical signals obtained are shown in Fig. 6b. The background signal from hot atoms is negligible due to the frequency modulation detection.

To realize the novel multiple-beam atom interferometer, we plan to apply a series of three optical interferometer pulses along the direction of atom propagation, and then apply the described frequency modulated detection. It is essential to work with atoms in the magnetic field insensitive $m_F = 0$ Zeeman level. The optical interferometer pulses, each consisting of two counterpropagating beams in a $\sigma^+ - \sigma^+$ configuration, are tuned to the $F = 3$, $m_F = 0 - F' = 4$, $m_F = 1$ component of the rubidium D2 line. It is more probable for the upper state to spontaneously decay to states of higher magnetic quantum number than the $m_F = 0$ ground state. These can be easily distinguished from those that do decay to the $m_F = 0$. The three pulses for the interferometer are generated in a high-finesse cavity [23]. Light from a Ti:sapphire laser is stored and enhanced in the resonator. Three RF pulses, which can be as short as 10 ns, are applied to an intracavity AOM and the entire optical energy within the cavity is extracted. Each pulse has about 150 W on exiting the cavity, with which we expect up to 10 interfering paths. The maximum allowable drift time between the first (second) and second (third) pulse is of the order of 100 ms, depending on the launch height used.

To conclude, using this novel scheme we plan on removing the internal structure's influence from the interferometer, thereby increasing the number of interfering beams. Using the cold atom fountain source it should be possible to make precision measurements of gravitational effects and the photon recoil with multiple-beam atom interferometry. A precision measurement of the photon recoil energy of the rubidium atom would allow for a new determination of the fine structure constant α. As current tests of quantum electrodynamic theory (QED) are limited by the accuracy to which α is known, a sufficiently precise measurement of the photon recoil would allow for a more critical test of QED [4]. Alternatively, by measuring the Earth's fractional acceleration between two atomic species (e.g. rubidium and potassium), the weak equivalence principle could be tested on the atomic scale [24]. Limits better than 10^{-10} are expected.

6 Conclusion

The principle of multiple-beam interferometry as applied to atoms has been clearly demonstrated, and some first applications of this technique have been exploited. In the future, it is anticipated that this method will add a new tool for precision measurements based on atomic interferometry. Atom interferometers allow for more precise tests of quantum electrodynamic theory and general relativity. Along different lines, interferometry with systems of increased size may help us to exploit the boundary between the microscopic (quantum) and the macroscopic (classical) world. It is a pleasure for the authors to write an article in a book honouring Theodor Hänsch. We very much appreciate his teaching in scientific taste and the benefit of using intuitive physical pictures, and his generous support. We all have good memories of our time working in the division of Theodor Hänsch at the MPI für Quantenoptik, whether this work be ongoing or in the past.

References

1. See, for example, articles in *Atom Interferometry*, ed. by P. Berman (Academic Press, San Diego 1997)
2. H. Maier-Leibnitz, T. Springer, Z. Phys. **167**, 386 (1962); H. Rauch, W. Treimer, U. Bonse, Phys. Lett. A **57**, 369 (1972)
3. L. Marton, J.A. Simpson, J.A. Suddeth, Phys. Rev. **90**, 490 (1954); G. Möllenstedt, Naturwiss. **42**, 41 (1955)
4. B. Young, M. Kasevich, S. Chu, in [1], p. 363; J.L. Hall, Ch.J. Bordé, K. Uehara, Phys Rev. Lett. **37**, 1339 (1976)
5. T.W. Hänsch, A.L. Schawlow, Opt. Commun. **13**, 68 (1975)
6. See, e.g., J. Opt. Soc. Am. B **6**, 2020 ff (1989)
7. M. Weitz, T. Heupel, T.W. Hänsch, Phys. Rev. Lett. **77**, 2356 (1996)
8. H. Hinderthür et al., Phys. Rev. A **56**, 2085 (1997)
9. S.B. Cahn et al., Phys. Rev. Lett. **79**, 784 (1997)
10. L. Deng et al., Phys. Rev. Lett. **83**, 5407 (1999)
11. K. Patorski, *Progress in Optics XXVII*, ed. by E. Wolf (North-Holland, Amsterdam 1989)
12. M. Weitz, T. Heupel, T.W. Hänsch, Europhys. Lett. **37**, 517 (1997)
13. M. Mei, T.W. Hänsch, M. Weitz, Phys. Rev. A **61**, 020101-1 (2000)
14. M. Mei, M. Weitz, Phys. Rev. Lett. **86**, 559 (2001)
15. M. Weitz, B.C. Young, S. Chu, Phys. Rev. Lett. **73**, 2563 (1994)
16. Y. Aharonov, D. Bohm, Phys. Rev. **115**, 485 (1959)
17. R.G. Chambers, Phys. Rev. Lett. **5**, 3 (1960)
18. B.E. Allman, A. Cimmino, A.G. Klein, G.I. Opat, Phys. Rev. Lett. **68**, 2409 (1992); G. Badurek et al., *ibid* **71**, 307 (1993)
19. S. Nic Chormaic et al., Phys. Rev. Lett. **72**, 1 (1994)
20. C.J. Bordé et al., Phys. Rev. A **30**, 1836 (1984)
21. K.S. Johnson et al., Science **280**, 5369 (1998)
22. M.J. Snadden, R.B. Clarke, E. Riis, Opt. Commun. **152**, 283 (1998)
23. T. Heupel, M. Weitz, T.W. Hänsch, Opt. Lett. **22**, 1719 (1997)
24. C. Lämmerzahl, Class. Quant. Grav. **15**, 13 (1998)

Part III

Precision Investigations of Fundamental Physical Problems

Pursuing Fundamental Physics with Novel Laser Technology

Carl E. Wieman

One theme that has run through Theodor Hänsch's career has been his development of new laser technology that he then uses to explore interesting fundamental physics. In this paper I would like to discuss a little about how I got started following this approach to doing physics as Theodor's student, and then how it led me to go on to measure parity violation in atoms to test the current theories of elementary particle physics. As an undergraduate at MIT, I read this paper about a new design of laser that produced intense very narrowband tunable light. Someone named Hänsch, whom I had never heard of, had written it. I still remember being quite excited by the thought that now one could get enough photons of any color in a narrow bandwidth to saturate virtually any visible atomic transition. Even as an undergraduate I realized that this represented a tremendous advance, and that there must be lots of interesting physics that could be done with such a tool. This sentiment was a major factor in my ending up at Stanford a few years later with Theodor Hänsch as my Ph.D. advisor. Although I did not realize it at the time, I was one of his first graduate students (the second after Siu Au Lee?), and I started working with him on the effort to observe the $1S$–$2S$ two-photon transition in hydrogen. Theodor Hänsch had already used his narrowband laser to demonstrate Doppler-free saturated absorption spectroscopy in hydrogen, and thereby greatly improved the resolution of hydrogen spectroscopy and the determination of the Rydberg constant. Now he was working on a much more challenging and audacious goal: the observation of the very exotic (at that time) process of a two-photon transition between the simplest two levels in hydrogen. This was the only atom where spectroscopy allowed one to readily test the fundamental theory of quantum electrodynamics. So it was exciting to be a part of this effort. In the same spirit as most of his other work, the success of this experiment was entirely dependent on the invention of new and better laser technology. So in the early days we were working on pulsed oscillators followed by multiple amplifier stages and frequency doubling in very aggravating unreliable crystals. Like many things Theodor Hänsch developed, most of this multistage amplifier technology is now standard stuff, but then it had not yet been invented and there was lots of work and trial and error necessary to get it all to function properly. Although I assume that we must have actually worked during the day at some times, all my memories of that period are only of working with Theodor very late at night. He always seemed to be at his happiest and most enthusiastic when he was working in the lab, particularly if it was about midnight.

Eventually, this new narrowband high-power laser technology was working well, and the doubling crystal survived barely long enough for us to see the $1S$–$2S$ transition and get what, at the time, we thought were impressively narrow lines. Over the next few years, we then extended this work to higher precision by developing more laser technology – a CW single-mode blue dye laser that went through the pulsed amplifiers. The CW blue dye laser may have been a technology that was a bit ahead of its time. I had to mix up the dye in five gallon batches because it wore out so quickly, and we had to replace the large-frame UV argon-ion laser tubes two or three times a month for a year while they perfected the window technology. Fortunately the Spectra-Physics factory was only about a mile away and all the tubes were free. It must have cost the company a fortune, but they were very anxious to have all those visitors to Hänsch's lab see that he was using a Spectra-Physics laser (even if it did not work very well). By the time I finished graduate school, we had obtained hydrogen spectra that were good enough that they were able to confirm a previously untested QED contribution to the hydrogen energy levels. For me, this really proved that the Hänsch approach of building better laser tools to do fundamental physics was both a lot of fun, and really did work. Of course, Theodor has pursued this up to the present time in developing ever-better laser technology in order to push the precision of the hydrogen $1S$–$2S$ transition to the almost ridiculous levels of precision that he has now obtained. This has reached resolutions and has tested QED to a level that, when I was working on these experiments, I would never have believed possible.

After graduate school, I continued in this spirit and went looking for problems where advances in laser technology could allow one to study basic physics questions. The unification of the weak and electromagnetic interactions that had been predicted shortly before seemed exactly such a problem. The Bouchiats [1] had pointed out that if this theory of unification was correct, it would result in a tiny amount of parity violation (PV) in atoms that, in principle, could be detected using atomic spectroscopy. Rather than trying to further test QED, this offered the exciting potential of testing the structure of atoms "beyond QED". Having learned from Theodor Hänsch that building better lasers and niftier laser technology could solve any atomic spectroscopy problem, I applied that philosophy to this challenging problem. Ultimately it ended up proving successful, although the lessons I had learned as a student about lots of late nights being necessary to get hard experiments to work also turned out to be very important! In parity violation measurements in atoms, one is doing sensitive measurements of atomic line strengths, rather than precision measurements of line centers as in the hydrogen work. However, the same fundamental feature of laser light is important for the success of both, namely the ability to get a great many photons that one can control to be at exactly the desired frequency. Another similarity to the measurement of the $1S$–$2S$ transition (although not quite as bad) was that in both cases many years elapsed in the pursuit of ever-higher precision. The parity violation work [2, 3] lasted nearly two decades and involved numerous students and postdocs. The primary motivation of these parity violation experiments was to test what is now known as the standard model of electroweak unification. Initially, the interest was just to see whether parity was violated in atoms, but later it became clear that it was interesting to measure this violation as precisely as possible. Although we learned over the years that the standard model worked very well, there have always been compelling reasons to think that it remains an incomplete description of nature. So there have been many extensions or "improvements" to the standard model proposed. Much of elementary particle physics research over the past two decades has been aimed at finding evidence of a deviation from the predictions of the standard model that matches the predictions of one of these extensions. In most cases the "new" physics would manifest itself as a small change in the experimental observable, and so the more precise the measurement in either atoms or at high energy, the more sensitive the test. Atomic parity violation measurements are uniquely sensitive to several types of possible new physics such as extra Z bosons, leptoquarks, or compositeness to the Fermions. A secondary motivation for this work grew out of a discovery that was made in our second-generation experiment that managed to measure atomic PV to a few percent uncertainty. We found that there was a small component to atomic parity violation that depends on the nuclear spin: a sort of hyperfine structure to atomic PV. This component of atomic parity violation is called the "anapole moment" contribution. Although we were unaware of it at the time, the parity-violating interactions between hadrons had been predicted to give such an effect. (Because we were unaware of this, we tried very hard to make it go away, but it stubbornly persisted.) Our later more precise measurements of the anapole moment effect gave a unique and important probe of parity violation in purely hadronic nuclear interactions, for which there are very few good measurements.

Although the details are complicated, the basic concepts behind PV in atoms and its measurement are quite simple. The neutral current interaction arising from the exchange of a Z boson between the electrons and the nucleons in an atom leads to a mixing of the S and P states in the atoms. The amount of mixing scales with the atomic number (Z) as Z^3, but even for large Z it is only about one part in 10^{11}, which is painfully small! This dependence forced me to abandon my beloved hydrogen, the atom that I had come to know and love as a graduate student, and turn to cesium. (After a serious 20 year relationship with cesium, I recently dumped it to devote myself to yet a third atom, rubidium. I feel downright fickle when compared with Theodor Hänsch, who has never wavered in his long attachment to hydrogen!) This parity-violating mixing is detected by observing the electric dipole transition amplitude between two S (or two P) states in an atom. In the case of cesium it is the $6S$ to $7S$ transition at 540 nm. Such an E1 amplitude can only exist if parity is violated. Because the amplitude is so small, it can only be observed by using interference techniques. In such a technique the nS to $n'S$ transition rate is given by

$$R = |A_0 \pm A_{\mathrm{pv}}|^2 = A_0^2 \pm 2A_0 A_{\mathrm{pv}} + A_{\mathrm{pv}}^2, \tag{1}$$

where A_{pv} is the desired PV amplitude and is proportional to the S–P mixing, and A_0 is a larger parity conserving amplitude. Because the interference term is linear in A_{pv}, it can be large enough to measure; however, it must be distinguished from the large background due to the A_0^2 contribution. For there to be a nonzero interference term, the experiment must have an inherent "handedness", and if the handedness is reversed, the interference term will change sign, and can thereby be distinguished.

In our experiments the A_0 is a "Stark-induced" E1 amplitude due to an applied DC electric field. This field mixes S and P states giving rise to an E1 amplitude that can interfere with A_{pv} under the proper conditions. The handedness of the experiment arises from the combination of orthogonal electric, magnetic, and laser fields that define a handed coordinate system of the apparatus. The Stark-induced $6S$–$7S$ transition is excited in an intense cesium atomic beam by a 540 nm beam from a dye laser that intersects the atomic beam at right angles. The PV interference term changes sign, causing a modulation in the excitation rate, as this coordinate system is reversed back and forth between right- and left-handed. There are four such reversals: changing the sign of each of the E and B fields, reversing the laser light from left to right circular polarization, and reversing the sign of the m level that is being excited. This use of multiple redundant reversals greatly suppresses possible systematic errors. Although it is simple in principle, there are many technical challenges in achieving an adequate signal-to-noise ratio for the measurements, and the elimination of potential systematic errors is even more difficult. It is hard to convey how difficult and time-consuming this and similar PV experiments are to those who have not worked on such projects. To provide a little perspective, I can compare the third-generation PV experiment that ended in 1998 with the progress in my laser cooling and trapping program. The laser trapping and cooling program grew out of our efforts to develop diode laser technology that was the beginning of our third-generation PV experiment, and from then on both programs proceeded in parallel. Both of these involved roughly the same number and quality of people and monetary support (although the parity-violation group tended to work somewhat harder, and for the last several years the cooling and trapping group had Eric Cornell). Over the subsequent 1.5 decades, the cooling and trapping group produced 18 Physics Review Letters articles, as well as many other publications on a variety of topics including the attainment of Bose–Einstein condensation (BEC), while the parity violation group was carrying out a single measurement! This somewhat illustrates the difference between PV measurements and "normal" atomic physics. The $6S$ and $7S$ levels in cesium are each split into $F = 3$ and $F = 4$ hyperfine levels. We drive transitions from a single m state of either the $F = 3$ ground state up to a single m state of the $F = 4$ excited state, or the $6S4$ to $7S3$ transition. In the first two generations of the experiment we applied a magnetic field that was sufficient to allow us to resolve the individual m states, and then used the frequency of the dye laser to select which level was excited. In the third generation, we improved the signal size by optically pumping all of the atoms into the m state that we were exciting, rather than using the $1/16$ of the atoms that normally populates a single m state. This optical pumping also allowed us to more efficiently detect the $7S$ excitation using a technique similar to the "atom shelving" used in ion trapping. About half of the time the $7S$ atoms decayed back down into the empty $6S$ F level. We excited atoms that refilled this "empty" F level on a cycling transition to the P state, and therefore scattered and detected many (200) photons for each $6S$–$7S$ excitation. These changes improved the signal size and detection efficiency, but at a cost of requiring four additional high-performance lasers. Three of these were used in the optical pumping and detection and the fourth was needed to measure the exact degree of optical pumping for calibration.

There are a number of aspects of this apparatus that have involved extensive development of laser technology, as I alluded to at the beginning of this article. First, it requires five lasers (four diode, one dye) that must be stabilized to nearly state-of-the-art performance. This means frequency stabilities on the order of one part in $10^{14}/\mathrm{s}^{1/2}$ and intensity stabilities of one part in $10^{6}/\mathrm{s}^{1/2}$. Getting this level of performance in a dye laser operating in the green at high power (400 mW) was difficult, but was helped considerably by the expert help of Jan Hall just down the hall. More surprises and more work were involved in getting all those diode lasers to work. This required quite a lot of development work, as we discovered that simply controlling the central frequency of the diode laser to this kind of stability was not nearly good enough; we also had to narrow up the linewidth. That required a fairly extensive program to figure out how to combine optical and electronic feedback to get well-behaved narrowband very stable diode lasers. That technology provided a much cheaper and simpler laser source for cooling and trapping atoms. That was what got me interested in that field and ultimately resulted in our obtaining BEC many years later. One of the biggest technological advances that made the PV experiment possible traces its origins back to my student days. This is the power-buildup cavity, a resonant Fabry–Perot cavity that is used to build up the optical power in order to enhance the excitation of the weak $6S$–$7S$ transition. In my work with Theodor Hänsch, we used a power-buildup cavity to make the power a few times higher. In the PV work we were exciting an extremely weak transition and so needed as much narrowband power at 540 nm as possible.

That led us to vigorously pursue power buildup technology. Over the years, we progressed from buildup factors of a few hundred up to nearly 100 000. However, we did eventually find that it was possible to get too much power, because it started to ionize the atoms and distort the mirror coatings. So we backed off a little (to 30 000) for most of the data. With each increase in buildup, concomitant increases in laser-frequency stability and cavity-mirror stability were required. Our final cavity had the spacing between the mirrors held constant to better than 1/100 of an atomic diameter per $s^{1/2}$. Actually, that was not the hardest part. The hardest part was that this stability had to be achieved while holding the mirrors extremely gently to avoid inducing tiny amounts of birefringence that could give a systematic error. So quite a few years were involved developing the dye laser, diode lasers, power-buildup cavity, and a number of other technologies needed for this experiment. Meeting any one of this list of technical requirements is not easy, but by far the most difficult aspect of the technology for this experiment was that all these things had to work at the same time, and they had to do that for hours, days, and months in order to acquire the necessary statistics.

After achieving this necessary level of performance from the apparatus, we then looked at the signal representing the $6S$–$7S$ excitation rate as we performed the five parity reversals (E, B, polarization, and there are two ways to reverse m in the optical pumping process). The quantity of primary interest is the fraction (about 6 parts in 10^6) of the signal that modulates with all five reversals. However, there are many other modulation channels that we monitored to measure the modulation with different subsets of these five reversals. These channels provided a great deal of information about the experiment; for example, the size of many stray nonreversing and misaligned field components. This information is essential for dealing with possible systematic errors.

Although a lot of time was required to develop the laser technology, probably as much time and more suffering was involved in the study and elimination of systematic errors. This got worse with each generation of the experiment as the target precision improved. For the final generation that achieved a fractional uncertainty of 0.3%, dealing with the systematics required about 7000 hours of data, or roughly 95%, of that acquired. The approximately five years spent on this was a lot less fun than developing the technology! The general approach that was used was that first we calculated all the possible ways that combinations of stray (nonreversing) and misaligned electric and magnetic fields (either oscillating or DC fields) could mimic the PV signal. This involved looking at all field components. Then we found a way to measure and control all the relevant field components to the necessary level. One of the painful aspects of this experiment was that it was necessary to consider all the possible combinations of gradients of these fields as well as the fields themselves. The measurement and control of all these fields and gradients was done by the combination of appropriate construction of the apparatus, monitoring 32 different modulation channels, and using 31 servosystems to stabilize everything about the experiment. Many of the modulation channels needed to come directly out of the same data as was used for the parity measurement, but there were several others that came from a set of auxiliary experiments that were interspersed with the acquisition of the PV data. The second stage in eliminating systematic errors was based on extensive statistical tests. Basically, these were quantitative ways to decide whether anything about the experiment was changing on any timescale by more than what one would expect just due to the shot noise. Ultimately we reduced the systematic uncertainty to the 0.1% level.

The final result for the parity-violating amplitude on the two different hyperfine lines is:

$$\begin{aligned} \mathrm{Im}(E1_{\mathrm{pv}}/\beta) = {} & 1.6349(80)\,\mathrm{mV/cm} \quad \text{for } F = 4 \text{ to } F' = 3 \\ & 1.5576(77)\,\mathrm{mV/cm} \quad \text{for } F = 3 \text{ to } F' = 4\,. \end{aligned} \tag{2}$$

Since we are detecting the PV amplitude as a fractional modulation in an electric field-induced rate, it is given in units of the equivalent electric field required to give the same mixing of S and P states.

It was a relief to see that the results of all of our three generations of this experiment agreed with each other, as well as with the earlier less-precise measurement of Bouchiat. It was also exciting to see that the tantalizing hint of a dependence of the PV on hyperfine transition (the nuclear spin dependence) that was seen in our 1988 result and hotly debated was confirmed by the later work. Our final result is that this difference is 0.077 mV/cm, with a fractional uncertainty of 14%. This is primarily due to the nuclear anapole moment. This is the first (and still only) measurement of an anapole moment, some forty years after its existence was first proposed by Zeldovich. It has provided new information on parity violation in hadronic interactions that does not agree very well with the previous theoretical analysis and very limited experimental data. This has resulted in a re-examination of this entire subject and has stimulated considerable new work.

If we take an appropriately weighted average of the two lines, all of the nuclear-dependent parts cancel out, and we are left with a quantity with a fractional uncertainty of 0.3% that can be used to compare to the predictions of the standard model. That comparison is complicated by the need for atomic structure calculations that relate the measured PV value to the fundamental electron–quark coupling constants of the standard model. When our experimental result was first published, the best calculations, made nearly a decade earlier, indicated that we were lower than the Standard-model (SM) prediction by about the uncertainty of those atomic calculations, which was stated to be 1%. However, this uncertainty was based on the comparison of the calculated and measured values for many other properties of the low-lying states of cesium, such as hyperfine splittings and oscillator strengths. The calculations of those quantities are quite similar to that of the parity violation and were done using the same highly sophisticated techniques. After the completion of the 1998 PV measurement, we went back and checked some of those experimental numbers and carried out an updated analysis [3] of the theoretical uncertainty using the improved experimental numbers obtained by ourselves and others. Remarkably enough, it turned out that in every case where there was an improved experimental measurement, the agreement with these sophisticated atomic structure calculations also improved. So the best estimate of the atomic calculation error based on this updated comparison of measured and calculated observables was 0.4%, rather than 1% [3]. At that level there was then a 2.5 sigma discrepancy between the atomic PV results and the standard-model prediction. Naturally, this attracted considerable attention. On the elementary-particle side it was pointed out how this could be consistent with the existence of an additional high-mass Z boson of a certain type that would also fix up a smaller standard-model difference in the results of a high-energy experiment. On the atomic side it resulted in renewed attention to the atomic calculations. A number of possible problems with the calculations were checked and eliminated. Then a bright young theorist discovered that a correction that had previously been thought to be insignificant, the second-order Breit correction, was surprisingly large and reduced the discrepancy with the standard model down to a modest 1.5 sigma, if one used the same uncertainty [4]. However, this same Breit correction also considerably improved the comparison between other calculated and measured properties of cesium, and hence arguably should have reduced the uncertainty. In the process of a couple of other groups checking and confirming the accuracy of this calculation of the Breit correction, it was realized that there is also a nonnegligible vacuum polarization correction that was neglected. This has very recently been calculated and found to shift the result by 0.4% away from the standard-model value, so that there is again a larger than two sigma discrepancy. It is a tantalizing situation that whispers of a possible problem with the standard model, but cries out for further work on the atomic theory and its experimental confirmation at the 0.1% level. In any case, at worst this is confirming the standard model at the fraction of a percent level in atoms, and thereby setting the best available constraints on a variety of possible new physics such as extra Z bosons, leptoquarks, and composite fermions. At best, it is providing us with indications of new physics beyond the standard model.

This work has epitomized the Hänsch approach to physics of developing novel laser technology that allows one to probe fundamental physics through atomic spectroscopy.

References

1. M.A. Bouchiat, C. Bouchiat, Phys. Lett. **48B**, 111 (1974a), J. Physique **35**, 899 (1974b), J. Physique **36**, 493 (1975)
2. S.L. Gilbert, M.C. Noecker, R.N. Watts, C.E. Wieman, Phys. Rev. Lett. **55**, 2680 (1985); M.C. Noecker, B.P. Masterson, C.E. Wieman, Phys. Rev. Lett. **61**, 310 (1988); C.S. Wood, S.C. Bennett, D. Cho, B.P. Masterson, J.L. Roberto, C.E. Tanner, C.E. Wieman, Science **275**, 1759 (1997)
3. S.C. Bennett, C.E. Wieman, Phys. Rev. Lett. **82**, 2484 (1999)
4. A. Derevianko, Phys. Rev. Lett. **85**, 1618 (2000)

Precision Optical Measurements and Fundamental Physical Constants

Savely G. Karshenboim

A brief overview is given of precision determinations of values of the fundamental physical constants and the search for their variation with time by means of precision spectroscopy in the optical domain.

1 Introduction

The electron and proton are involved in various phenomena in different fields of physics. As a result, fundamental physical constants related to their properties (such as the electron/proton charge (e), the electron and proton masses (m_{e} and m_{p}), the Rydberg constant (Ry), the fine structure constant (α), etc.), can be traced in various basic equations of the physics of atoms, molecules, solid state, nuclei and particles etc. Until recently most accurate measurements came from radio-frequency experiments only, which supplied us with precise values of most of the fundamental constants and accurate tests of the quantum theory of simple atoms (bound-state QED). The optical measurements delivered to us a value for the Rydberg constant, but were not accurate enough to provide any competitive test of the QED.

During the last decade the status of the optical measurements changed dramatically because of

- advances in atomic spectroscopy, for example, Doppler-free two-photon spectroscopy of atomic hydrogen, with significantly increased resolution;
- advances in the technology for measuring optical frequencies. New frequency chains routinely deliver the high accuracy of the microwave cesium radiation (related to the definition of the *second*) to the optical domain.

Two-photon methods were developed for hydrogen spectroscopy at Stanford, Oxford, LKB (Laboratoire Kastler Brossel), Yale and MPQ (Max-Planck-Institute für Quantenoptik) during two last decades (see a review in [1]). Improved accuracy of the Rydberg constant by few orders of magnitude is supplying us with a precision test of the theory of the Lamb shift in hydrogen and deuterium atoms (see [2] and references therein). The accuracy of these tests is even higher than that of traditional microwave measurements. We discuss that in detail in Sect. 2.

The progress with the spectroscopy of the $1s$–$2s$ transition [3] and with the development of a new type of frequency chain [4] was so large that the transition offered also an opportunity to search for a variation of the fundamental constants with time. The appearance of the new frequency chain developed at MPQ [4] and successfully applied at several laboratories (MPQ, JILA [5], NIST, PTB) has greatly changed the situation for the optical measurements (see a review in [6]). This is important for metrology and the design of new frequency standards, for the search of variations of the fundamental constants and for numerous other applications. We consider the application to the search for such variations in Sect. 3.

2 Rydberg Constant and the Lamb Shift in the Hydrogen Atom

The discovery of the Lamb shift in the hydrogen atom was a starting point of the most advanced quantum theory – quantum electrodynamics (QED). This theory predicts a number of quantities with a great accuracy. There are only a few examples where an accurate theory also allows precise measurements. Those are [7, 8] the anomalous magnetic moment of the electron and muon and some transition frequencies in simple atoms (e.g. see Fig. 1). Most success was obtained with the study of the hydrogen atom. However, the progress of traditional microwave measurements has been quite slow and reached the 10 ppm level of accuracy only for the hydrogen Lamb shift (Fig. 2).

It turned out that the two-photon Doppler-free spectroscopy of gross structure transitions (such as $1s$–$2s$, $1s$–$3s$, $2s$–$4s$, etc.) allowed access to narrower levels and could deliver very accurate values sensitive to QED effects. But to interpret those values in terms of the Lamb shift, two problems had to be solved:

- the Rydberg constant determines a dominant part of any optical transition and has to be known itself;
- a number of levels are involved and it is necessary to be able to find relationships between the Lamb shifts ($E_{\rm L}$) of different levels. Otherwise the experimental data would be of no use because the number of unknown quantities exceeds the number of measured transitions.

The former problem has been solved by comparison of two transitions determining both: the Rydberg and the QED contributions. Presently the two best results to combine are the $1s$–$2s$ frequency in hydrogen and deuterium [3, 9] and the $2s$–$8s/d$ transition in the same atoms [10]. The latter problem has been solved with the help of a specific difference [11]

$$\Delta(n) = E_{\rm L}(1s) - n^3 E_{\rm L}(ns) , \tag{1}$$

which can be calculated more accurately than the Lamb shift of the individual levels.

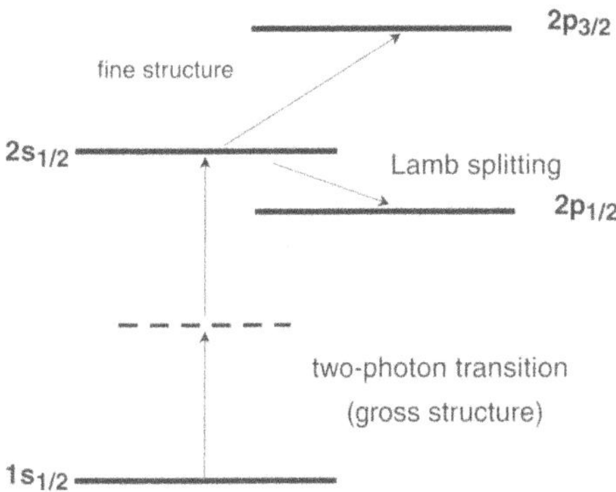

Fig. 1. Example of different transitions in the hydrogen atom. Transitions within the fine structure and Lamb splitting are in the microwave range, while the $1s$–$2s$ two-photon transition lies in ultraviolet domain. The Lamb splitting is the difference of the Lamb shift of the $2s_{1/2}$ and $2p_{1/2}$ levels

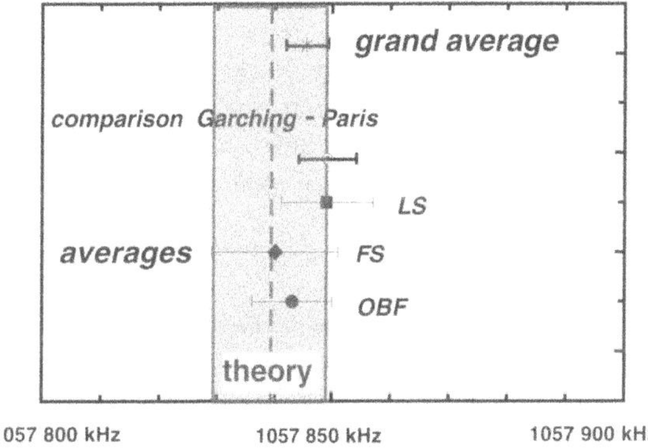

Fig. 2. Present status of the Lamb splitting in hydrogen ($2s_{1/2}$–$2p_{1/2}$). The figure contains values derived from various experiments. *LS* stands for the Lamb splitting measurements (see Fig. 1), *FS* is for the fine structure, and *OBF* stands for optical beat frequency (simultaneous measurement of two optical transitions). The theoretical estimation is taken from [14]

A successful deduction of the Lamb shift (Fig. 2) in the hydrogen atom provides us with a precision test of bound-state QED and offers an opportunity to learn more about the proton size. Bound-state QED is quite different from QED for free particles. The bound-state problem is complicated itself even in the case of classical mechanics. The hydrogen atom is the simplest atomic system; however, a theoretical result for the energy levels is expressed as a complicated function (often a perturbative expansion) of a number of small parameters [7] (see review [13] for a collection of theoretical contributions):

- α, which counts the QED loops;
- the Coulomb strength $Z\alpha$;
- electron-to-proton mass ratio;
- ratio of the proton radius to the Bohr radius.

Indeed, in hydrogen $Z = 1$, the origin of the correction is very important and the behaviour of expansions in α and $Z\alpha$ differs from each other. In particular, the latter involves large logarithms $[\ln(1/Z\alpha) \sim 5]$ [7, 12] and big coefficients. It is not possible to do any exact calculations and one must at least use expansions in some parameters. In such a case the hardest theoretical problem is not to make the calculation, but rather to estimate the uncalculated terms related to higher-order corrections of the expansion.

The other problem is due to the proton size [14], which leads to a major uncertainty of the theory. The finite size of the proton leads to a simple expression, but to obtain its numerical value one needs to determine the value for the proton charge radius. Unfortunately, no such measurements are available at the moment. It happens now that the most accurate value can be obtained from a comparison of the experimental value of the Lamb shift derived from the optical measurements of the hydrogen and QED theory [14].

The other quantity deduced from the optical measurements on hydrogen and deuterium is the Rydberg constant. The recent progress is clearly seen from the recommended CODATA values of 1986 and 1998 [15]:

$$\mathrm{Ry}_{86} = 10\,973\,731.534(13)\ \mathrm{m}^{-1} \text{ and } \mathrm{Ry}_{98} = 10\,973\,731.568\,549(83)\ \mathrm{m}^{-1}\,.$$

The former value was derived from one-photon transitions (Balmer series) and was slightly improved later (by a factor of 4.5), but all further progress

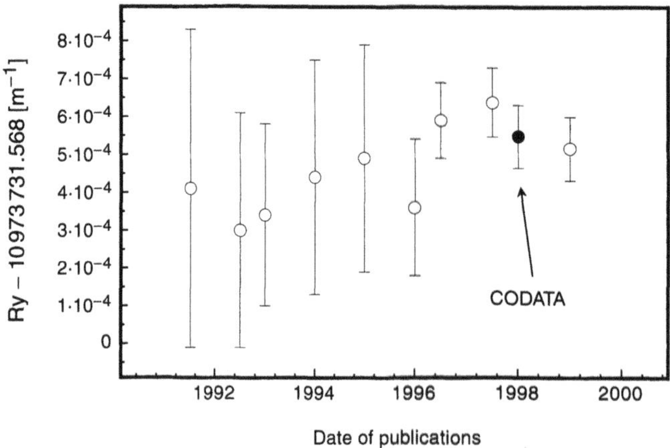

Fig. 3. Recent progress in the determination of the Rydberg constant (mainly at MPQ and LKB)

that led to the 1998 value (30 times improvement) was a result of the study of two-photon transitions in hydrogen and deuterium. Fig. 3 shows a comparison of several recently published values for the Rydberg constant.

Some other fundamental constants can also be determined by means of laser spectroscopy. To complete the overview let us mention determination of

- the muon-to-electron mass ratio from three-photon ionization of the ground state of muonium at a resonance point of the two-photon excitation of the $2s$ state [16];
- the fine structure constant α derived from the helium fine structure;
- the fine structure constant deduced from Raman recoil spectroscopy of the cesium D_1 line [17]. For this evaluation a precise value of the absolute frequency of the D_1 line is also necessary.

3 Optical Measurements and Variation of the Fundamental Constants with Time

After the success in the understanding of the electromagnetic, weak and, in part, strong interactions we arrived at some sort of threshold of a new physics. Naturally, it is not clear on which front one is most likely to discover new physical phenomena. The present-day attempts to discover a new physics are rather a kind of a random search for a hidden treasure. Unfortunately, lacking theoretical predictions there seems to be no better method up to now. One of the few directions for such a search is related to the variation of the fundamental constants with time. There is no common model for such a phenomena, but the accepted picture of the evolution of our universe strongly implies that. 'Variation' means a slow drift or oscillation of parameters of the interaction and particle properties (like their masses). We commonly believe that during a small fraction of the very first second of its existence our universe came through a number of phase transitions and that the interactions and particles as we know them now did not exist before those transitions. So, philosophically speaking, we have to acknowledge the variation of constants as some kind of trace of the early great changes. Physically speaking, we understand that the critical question for a detection of the variation is the rate at which it occurs, because variation rates of 10^{-10} yr^{-1} and 10^{-20} yr^{-1} are not the same. Variation of, for example, the fine structure constant at the former limit would have already been detected a few decades ago, while the latter rate will rather be a challenge for physicists in a few decades. Present-day searches are related to a level of $10^{-13} - 10^{-15}$ yr^{-1}, which corresponds to a fractional shift of the constants smaller than $10^{-3} - 10^{-5}$ for the life time of the universe.

There are a number of possibilities for such a search, but the optical measurements play a specific role, because of their clear interpretation. However, the fact of variation is more important than the accurate interpretation. In the case of a negative result no limitation for a variation can be assigned without a reasonable interpretation. From the theoretical point of view we have to expect simultaneous variations of all coupling constants, masses and magnetic moments. If one wishes to find a solid interpretation, some reduction of all variable quantities to a very few is important. However, nuclear properties are the result of the strong interactions and involve effects that cannot be calculated. Only optical transitions are completely free from this problem.

Table 1. Comparison of different kinds of searches for the variation of fundamental constants. Here T is the half-period of the oscillations and Δt is the time separation for a comparison. In the case of the oscillations their presnt-day phase is crucially important. The *Laboratory search* contains all laboratory measurements, while *Optical transitions* are for comparison of only optical transitions. The limitations expected in *1-2 years* are also presented

	Geochemical study	Astrophysical observation	Laboratory search	Optical transitions
Oscillation	$\Delta t > T$	$\Delta t > T$	phase - ?	phase - ?
Space	–	important	–	–
Statistics	–	essential	–	–
Strong Int.	sensitive	not sensitive	sensitive	not sensitive
α	+	+	+	+
m_e/m_p	–	+	+	–
g_p	–	+	+	–
g_n	–	–	+	–
Limitations	10^{-17} yr^{-1}	10^{-15} yr^{-1}	10^{-15} yr^{-1}	10^{-13} yr^{-1}
In 1-2 years				10^{-14} yr^{-1}

We summarize a comparison of different searches in Table 1 [18]. Let us discuss briefly the specific features of the different methods.

- To detect a variation one needs to compare some quantity $A(t)$ measured at time t and $t+\Delta t$. However, the obvious estimation of the variation rate $\Delta A/\Delta t$ is valid only in the case of a slow drift. In the case of oscillations (such oscillations were suggested because of some astrophysical reasons) the estimate must rather be $\Delta A/T$, where T is the half-period of the oscillation ($\sim 10^8$ yr). That makes astrophysical ($\delta t \sim 10^{10}$ yr) and geochemical ($\delta t \sim 2\times 10^9$ yr) estimates much weaker, while in the case of any laboratory experiments we arrive at the question of the phase of such an oscillation. One sees that the laboratory search and a search over time (and space) are two different kinds of experiments and serve different purposes.
- Astrophysical data have two features different from other types of data. First, we can observe the object separated from us both in space and

time, as there might be a correlation between time- and space-variations. Secondly, the astrophysical data are hard to interpret on an event-by-event base. They used to be treated statistically and it is necessary to study the correlations of the data.

- Another important question for interpretations is involvement of the strong interactions. In the case of geochemical data that is the main effect and there is no reasonable interpretation of such data at all. In the case of laboratory measurements, the strong interaction is important because of the nuclear magnetic moments (see below).
- We can look only for variations of dimensionless quantities such as a ratio of two frequencies and we need to reduce them to a variation of a few dimensionless constants, which are the fine structure constant α, the electron-to-proton mass ratio, and the g-factors of the proton and neutron. This is possible because of two reasons, which are
  - known non-relativistic dependence of any transition frequency on the fundamental constants;
  - the Schmidt model predicts nuclear magnetic moments for odd Z;
- Non-relativistic theory in particular predicts (see [18] for details), that
  - any gross structure transition frequency depends only on the Rydberg constant ($\sim$ Ry);
  - any fine structure transition frequency is proportional to α^2Ry;
  - any hyperfine structure transition is proportional to $\alpha^2(\mu/\mu_{\mathrm{B}})$Ry, where μ is the nuclear magnetic moment and μ_{B} is the Bohr magneton.
- The Schmidt model predicts the magnetic moment of nuclei with odd mass number A as a result of the spin and orbit contributions of a single nucleon (the remaining protons and neutrons are coupled in pairs and do not contribute). Indeed this is a very rough model, but there is no other way to reduce all magnetic moments to few quantities ($m_{\mathrm{e}}/m_{\mathrm{p}}$, g_{p}, g_{n}). The problem of interpretation is now related to the inaccuracy of the Schmidt model because of the strong interaction, which is not under control.

The present-day level of limitations is different for various methods and the optical experiments do not look like a good choice in general and in comparison with other laboratory experiments. However, the discussion above explains in part that there is no clear interpretation of the geochemical data and there are some doubts in the astrophysical data. There is a single laboratory result in Table 1 that is better than 10^{-13} yr^{-1}: a comparison of the hyperfine splitting in Rb and Cs [19]. This experiment could be a search for a variation of g_{p}. Such an interpretation is valid only if the Schmidt model can be applied, but unfortunately there are essential deviations from the Schmidt model due to the strong interaction (mainly in Cs). The actual interpretation remains unclear.

We briefly discuss optical laboratory experiments below. The current limitations are quite weak ($\sim 10^{-13}$ yr^{-1}) because most of the results were obtained recently at a level of a fractional uncertainty of 10^{-14} (see Table 2). These limitations were obtained either after short-term monitoring or after a comparison with a previous less accurate result. Future limitations based on reproduction of the recent experiments in 2000–2001 must easily deliver limitations that are better by an order of magnitude.

Table 2. Some results of recent precision optical measurements (H, $1s-2s$; ^{40}Ca, $^3P_1-{}^1S_0$; ^{115}In$^+$, $5s^{2\,1}S_0-5s5p\,^3P_0$; ^{171}Yb$^+$, $6s^{2\,2}S_{1/2}-5d^{2\,2}D_{3/2}$; ^{199}Hg$^+$, $^2S_{1/2}-{}^2D_{5/2}$), their fractional uncertainty δ and their sensitivity to a variation of the fine structure constant $\kappa=\partial\ln F_{\rm rel}(\alpha)/\partial\ln\alpha$

Atom	Frequency [Hz]	Place and date of measurement	δ [10^{-14}]	κ [24]
H	2 466 061 413 187 103 (46)	MPQ, 1999, [3]	1.8	0.00
Ca	455 986 240 494 158 (26)	NIST, 2000, [20]	5.7	0.03
In$^+$	1 267 402 452 899 920 (230)	MPQ, 1999, [21]	18	0.21
Yb$^+$	688 358 979 309 312 (6)	PTB, 2001, [22]	0.9	1.03
Hg$^+$	1 064 721 609 899 143 (10)	NIST, 2000, [20]	0.9	−3.18

Since all the transitions in Table 2 are related to the gross structure, one could wonder how to obtain any information about a variation of the fine structure constant if all of them are proportional (in the non-relativistic approximation) to the Rydberg constant. The signal is due to the relativistic corrections. Their importance (for the hyperfine structure) was first pointed out in [23] and they were later calculated for a bunch of the optical transitions in [24]. The transition frequency is now equal to $c\,\mathrm{Ry}\,F_{\rm rel}(\alpha)$ and a non-trivial relativistic factor $F_{\rm rel}(\alpha)$ is a key point to look for in detecting the variation of the fine structure constant by optical means. The sensitivity $\kappa=\partial\ln F_{\rm rel}(\alpha)/\partial\ln\alpha$ to a variation of α is given in Table 2 accordingly to [24].

A comparison of two optical frequencies can be performed directly (e.g. as in an Hg–Ca comparison at NIST [20]) or indirectly (via a comparison of both frequencies to the cesium standard).

A direct comparison of two distinct optical frequencies is now possible using the newly developed femtosecond frequency chains. This approach has the advantage of a high short-term stability as compared with the cesium standard. This allows for a simple *time structure* of the experiment [18] with a few direct comparisons. A comparison with cesium often involves some secondary standard with a high short-term stability. Such a clock being an artefact that is not related to any transition should involve an unknown drift with time as the constants are drifting. Their short-term stability cannot be actually proved. A common reason for a statement on the good short-term stability of such a standard (like a hydrogen maser for example) is that the scattering of their frequencies with time is small. But there is no idea about a possible common mode rejection. The stronger reason to believe in a good short term stability is a comparison between different clocks. But still there might be a common mode rejection as it should be the case for a variation of the constants. A direct comparison of two optical transitions allows us to avoid this problem and offers a new opportunity to securely derive a limitation for a variation of the fine structure constant or maybe even to detect such a variation.

4 Summary

Nowadays, the optical spectroscopy of atoms provides an essential input for the determination of the fundamental physical constants, including the most accurately known fundamental constant, the Rydberg constant, which plays an important role in atomic physics. The so-called atomic unit of frequency and energy are related to this constant being $c\,\mathrm{Ry}$ and $hc\,\mathrm{Ry}$, respectively. A definition of the *second*, attractive from a general point of view, could be based on a fixed value of the Rydberg constant. This would be practically acceptable after the following steps are achieved:

- the completion of a frequency chain that connects the optical with the microwave domain and, at the same time, should allow for a comparison of any optical transition with the $1s$–$2s$ frequency in hydrogen (performed at MPQ [4] and now also at NIST, JILA, PTB);
- accurate measurement of the $1s$–$2s$ transition (performed at MPQ [3] with an accuracy compatible with any other optical transition but not yet competitive with cesium and rubidium fountain clocks);
- proper QED theory of the Lamb shift (essentially developed within the last decades, but it still needs more progress);
- determination of the proton charge radius (not known with sufficient accuracy, but a promising experiment is in progress [25]).

The hydrogen atom was a candidate for the primary frequency standard in the 1960s because of the hyperfine splitting, but this attempt failed for a number of reasons. Now it is time for a strong competition of new frequency standards, and the hydrogen atom has its second chance.

Actually there is one more fundamental constant, which is associated with the optical measurements. That is the speed of light, which is equal to

$$c = 299\,792\,458\ \mathrm{m/s} \tag{2}$$

because of the definition of the *meter*. Despite this constant being fixed by definition there is still an uncertainty related to the practical realization of the *meter*. The problem is that the *second* is defined with the help of the cesium hyperfine transition, while the *meter* is related to the optical domain. The accepted recommendations for its realization is based on optical transitions [26]. In the case of any direct application of (2) one meets three sources of uncertainty:

- a realization of the *second* in the microwave domain;
- a realization of the *meter* in the optical domain;
- a bridge between both realizations to the same transition (i.e. the frequency chain).

After recent progress in frequency chain metrology and a long-term development of the microwave standards, the present-day limitations come mainly from the optical standards, which however are developing in a very promising way.

The old-fashion frequency chains, which really presented the state of the art in the field just few years ago, appear now as some kind of dinosaur, which should very soon disappear. Those chains are much bigger, much more expensive, much more complicated in construction and in use, and at the same time less accurate than the new frequency-comb chains. The main disadvantage of the old technology was a limited possibility in its use. The old chains were designed as a single-problem chain and it was necessary to redesign it by adding some more items to adjust it to new transitions. The new chain provides us with the possibility of measuring any optical transitions, and some measurements like the cesium D_1 line, for example, needed for the determination of the fine structure constant [17] are now a routine problem. The chain offers new horizons for precision optical spectroscopy and we look forward for soon reproductions of recent experiments [3, 20, 22] which must deliver new secure limitations for a possible variation of the fine structure constant with time at a level of 10^{-14} per year.

Acknowledgement

I met Theodor Hänsch for the first time at ICAP'94 and became a frequent visitor to the MPQ. I have been always impressed by the free and fruitful atmosphere of his lab. Starting as a pure theorist I am now trying to be between theory and experiment and I enjoyed lots of discussions with Theodor Hänsch and his collaborators, which were really stimulating. During the 8 years around the MPQ, a highlight of our cooperation certainly was the organization of the *Hydrogen Atom, 2* meeting [8], in which both of us were involved. I always experienced his support, without which the project could not have been realized. Personally, he did only a few adjustments. But those were the crucial things that I had missed, and that was one more chance to see how efficient he is. I do really wish to thank him for the support and hospitality he extended in Garching and especially in Florence.

References

1. F. Biraben, L. Julien, *this volume*
2. M. Weitz, A. Huber, F. Schmidt-Kaler, D. Leibfried, W. Vassen, C. Zimmermann, K. Pachucki, T.W. Hänsch, L. Julien, F. Biraben, Phys. Rev. A **52**, 26641 (1995)
3. M. Niering, R. Holzwarth, J. Reichert, P. Pokasov, Th. Udem, M. Weitz, T.W. Hänsch, P. Lemonde, G. Santarelli, M. Abgrall, P. Laurent, C. Salomon, A. Clairon, Phys. Rev. Lett. **84**, 5496 (2000)
4. J. Reichert, M. Niering, R. Holzwarth, M. Weitz, Th. Udem, T.W. Hänsch, Phys. Rev. Lett. **84**, 3232 (2000);
   R. Holzwarth, Th. Udem, T.W. Hänsch, J.C. Knight, W.J. Wadsworth, P.St.J. Russell, Phys. Rev. Lett. **85**, 2264 (2000)
5. S.A. Diddams, D.J. Jones, J. Ye, S.T. Cundiff, J.L. Hall, J.K. Ranka, R.S. Windeler, R. Holzwarth, Th. Udem, T.W. Hänsch, Phys. Rev. Lett. **84**, 5102 (2000)
6. Th. Udem, A. Ferguson, *this volume*
7. S.G. Karshenboim, in *Atomic Physics* **17**, AIP Proc. **551**, ed. by E. Arimondo et al. (AIP, New York 2001) pp. 238–253
8. S.G. Karshenboim, F.S. Pavone, F. Bassani, M. Inguscio, T.W. Hänsch, *Hydrogen Atom: Precision Physics of Simple Atomic Systems* (Springer, Berlin, Heidelberg 2001)
9. A. Huber, Th. Udem, B. Gross, J. Reichert, M. Kourogi, K. Pachucki, M. Weitz, T.W. Hänsch, Phys. Rev. Lett. **80**, 468 (1998)
10. C. Schwob, L. Jozefowski, B. de Beauvoir, L. Hilico, F. Nez, L. Julien, F. Biraben, Phys. Rev. Lett. **82**, 4960 (1999)
11. S.G. Karshenboim, JETP **79**, 230 (1994); Z. Phys. D **39**, 109 (1997)
12. S.G. Karshenboim, JETP **76**, 541 (1993)
13. M.I. Eides, H. Grotch, V.A. Shelyuto, Phys. Rep. **342**, 63 (2001)
14. S.G. Karshenboim, Can. J. Phys. **77**, 241 (1999)
15. P.J. Mohr, B.N. Taylor, Rev. Mod. Phys. **72**, 351 (2000)
16. K. Jungmann, in [8], pp. 81–102
17. J.M. Hensley, A. Wicht, B.C. Young, S. Chu , in *Atomic Physics* **17**, AIP Proc. **551**, ed. by E. Arimondo et al. (AIP, New York 2001) pp. 43–57
18. S.G. Karshenboim, Can. J. Phys. **78**, 639 (2000)
19. C. Salomon, Y. Sortais, S. Bize, M. Abgrall, S. Zhang, C. Nicolas, C. Mandache, P. Lemonde, P. Laurent, G. Santarelli, A. Clairon, N. Dimarcq, P. Petit, A. Mann, A. Luiten, S. Chang, in *Atomic Physics* **17**,AIP Proc. **551**) ed. by E. Arimondo et al. (AIP, New York 2001) pp. 23–40
20. Th. Udem, S.A. Diddams, K.R. Vogel, C.W. Oates, E.A. Curtis, W.D. Lee, W.M. Itano, R.E. Drullinger, J.C. Bergquist, L. Hollberg, Phys. Rev. Lett. **86**, 4996 (2001)
21. J. von Zanthier, Th. Becker, M. Eichenseer, A.Yu. Nevsky, Ch. Schwedes, E. Peik, H. Walther, R. Holzwarth, J. Reichert, Th. Udem, T.W. Hänsch, P.V. Pokasov, M.N. Skvortsov, S.N. Bagayev, Opt. Lett. **25**, 1729 (2000)
22. J. Stenger, Ch. Tamm, N. Haverkamp, S. Weyers, H.R. Telle, e-print physics/0103040
23. J.D. Prestage, R.L. Tjoelker, L. Maleki, Phys. Rev. Lett. **74**, 3511 (1995)

24. V.A. Dzuba, V.V. Flambaum, J.K. Webb, Phys. Rev. A **59**, 230 (1999); V.A. Dzuba, V.V. Flambaum, Phys. Rev. A **61**, 034502 (2000).
25. R. Pohl, F. Biraben, C.A.N. Conde, C. Donche-Gay, T.W. Hänsch, F.J. Hartmann, P. Hauser, V.W. Hughes, O. Huot, P. Indelicato, P. Knowles, F. Kottmann, Y.-W. Liu, V.E. Markushin, F. Mulhauser, F. Nez, C. Petitjean, P. Rabinowitz, J.M.F. dos Santos, L.A. Schaller, H. Schneuwly, W. Schott, D. Taqqu, J.F.C.A. Veloso, in [8], pp. 454–466
26. T.J. Quinn, Metrologia **30**, 523 (1993) ; idem., ibid. **36**, 211 (1999)

Quantum Electrodynamics and All That

Krzysztof Pachucki

An overview of recent advances in quantum electrodynamics is presented.

1 Basics

It is the purpose of this small contribution to present an overview of the current status of quantum electrodynamics (QED), open issues and unsolved problems to the large audience not necessarily familiar with sophisticated calculations and experiments performed during the last few decades. The complexity in the formulation of bound state problems and the lack of modern textbooks form a difficult barrier. However, the beauty of bound state QED is acknowledged once you perform the calculations by yourself.

Quantum electrodynamics as a quantum field theory has not yet been formulated in a strictly mathematical manner. It only exists as a compound of prescriptions for the calculations of physical quantities. These prescriptions are however self-consistent. The apparent divergences which arise in the calculations are eliminated through the well-established regularization and renormalization procedure. It is the Poincaré and gauge symmetry which uniquely determine QED: no physical quantity may depend on the renormalization scheme. However, what is not well recognized, is the fact that the renormalization procedure works only when asymptotic states are involved, for example for the case of cross-sections, bound state energies, or g-factors. Other classes of problems, which involve time dependence, or for example, the shape of natural lines, are not yet well formulated within quantum electrodynamics. The renormalization procedure for calculations of processes starting and finishing at some definite time does not exist. There are still many issues to be considered such as infrared divergences or the notion of the wave function in QED. So-called infrared divergences are related to the fact that the state of a single electron is not physically distinguishable from the state of a single electron plus long wavelength photons. It affects the meaning of the electron wave function. Moreover the wave function loses the probabilistic interpretation when relativity is involved. For example, for two particles it would have to depend on two time variables. These problems do not prevent giving precise predictions for static properties of bound states. I will concentrate here on simple systems, such as hydrogen, positronium and anything formed with two particles of opposite charge.

2 Hydrogen

The spectrum of hydrogen played a crucial role in the development of quantum electrodynamic theory. As it is the simplest of stable atoms, theoretical calculations are the most accurate here. Precise theory has stimulated improved measurements and vice versa. Apart from the well-known $2S_{1/2}$–$2P_{1/2}$ splitting, two photon transitions between S-levels, especially the 1S–2S transition, are of main interest. This is because the 2S state is metastable, and the transition to the 1S state has a natural width of 1.3 Hz, as compared to the 2P state, where it is of order 10^8 Hz. The measurement of the 1S–2S transition by Hänsch's group in Garching [1] is the most precise of any optical measurement performed so far. Their result:

$$\nu(\mathrm{1S\text{–}2S}) = 2\,466\,061\,413\,187\,103(46)\,\mathrm{Hz} \tag{1}$$

with a precision of 46 Hz has almost reached the natural width of this transition. There is no hope that theory might be equally accurate, since any theoretical result must involve the electron mass and the fine structure constant. Both of them are known with much less accuracy. Currently this measurement is interpreted as a determination of the Rydberg constant. This is the constant which gives the nonrelativistic transition frequencies in hydrogen:

$$\nu(n \to m) = R\left(\frac{1}{m^2} - \frac{1}{n^2}\right). \tag{2}$$

A further measurement of a different transition, 2S–6S by the Paris group [2], or 2S–4S by the Garching group [3], leads to the result for such a specific combination, which cancels out nonrelativistic energy. The most known example is

$$\Delta = E(\mathrm{4S\text{–}2S}) - \frac{1}{4}\,E(\mathrm{2S\text{–}1S}). \tag{3}$$

It could be regarded as a test of QED, beside the well-known E(2S–2P). The high precision of these measurements has stimulated the intensive development of theoretical calculations. The most interesting were so-called recoil corrections, corrections which account for the finite mass of the nucleus. In the Dirac equation one assumes that the nucleus is a static source of the Coulomb field. Finite mass means that the nucleus can move. There does not exist an extension of the Dirac equation which incorporates finite nucleus mass. The use of a reduced mass, instead of the rest mass of the electron, in the Dirac equation is incorrect. For hydrogen-like systems it is convenient to perform an expansion in the electron–nucleus mass ratio, and consider each term of this expansion separately. Beautiful formulas were discovered by Shabayev [4] for the first term in the mass ratio, which are valid for any charge of the nucleus, even for hydrogen-like uranium. They have been applied for both analytical and numerical calculations of recoil corrections. Other types of interesting recoil corrections are related to the proton, or in general nucleus, self-energy. As the electron self-energy gives most of the Lamb shift, the nuclear self-energy is a tiny correction, diminished by the square of the electron–nucleus mass ratio. Although small, it is nevertheless important in view of high-precision hydrogen and deuterium spectroscopy [5]. Moreover it overlaps with the finite nuclear size effect, and a clear distinction between both effects is not possible. Currently, attention is directed toward two-loop corrections, as they give the largest uncertainty in the theoretical predictions. These corrections have very peculiar features, not encountered anywhere else. The power series in $Z\,\alpha$ has very slow convergence [6]. The presence of many logarithmic terms $\ln(Z\,\alpha)$, and the fact the higher-order coefficients in this series could be orders of magnitude larger then the leading ones, makes it difficult, if not impossible, to estimate the as yet uncalculated terms in the power series expansion. For that reason attempts to calculate the two-loop correction directly numerically are being made. Very recently, Yerokhin and Shabayev [7] have been able to complete the two-loop calculations started by Mallampali and Sapirstein a few years ago in [8]. However their result is valid only for heavy hydrogen-like ions, and no result exists for hydrogen, except for a few terms in the asymptotic expansion in powers of $Z\,\alpha$. Let us summarize this consideration by a comparison of the current theoretical predictions for the 1S Lamb shift in hydrogen ($Z = 1$) from [6]

$$E_{\rm L}(1{\rm S})_{\rm th} = 8\,172\,816\,(10)(32)\,{\rm kHz} \tag{4}$$

with the experimental result from [2] and [3]

$$E_{\rm L}(1{\rm S})_{\rm exp} = 8\,172\,837\,(22)\,{\rm kHz}\,. \tag{5}$$

The close agreement indicates that so far QED is a correct theory. The only doubt is with regard to the proton charge radius. It is not yet known with sufficient accuracy.

In the same way other so-called exotic atoms are being studied: muonium, positronium, muonic hydrogen or antiprotonic helium. Their theory does not differ significantly from that of hydrogen, except for positronium. The positronium system is the most beautiful and also the most difficult application of quantum electrodynamic theory. Beautiful, because there is no parameter other than the fine structure constant; difficult because the mass ratio of the electron and positron is equal to 1 as long as the CPT theorem is valid, and therefore an analog of the Dirac equation for positronium system does not exist. Relativistic corrections could only be calculated by considering the complete QED theory. We will not present details, because they are too difficult. Instead it is safer to refer to recent work on this subject [9, 10].

3 Beyond QED

Since the world is not completly described by QED, it is natural to investigate effects, usually very small at the atomic scale, coming from weak, strong and gravitational interactions. Weak interactions are clearly detectable because they break parity (P) symmetry; for a very good review see the book of

Khriplovich [11]. Left and right handed chemical molecules may have different vibrational frequencies. States of an atom do not have definite parity, and for example the electric dipole transition between $6S_{1/2}$ and $7S_{1/2}$ in cesium has been measured with high accuracy by the Colorado group [12]. In hydrogen, or better in deuterium, the close lying states of opposite parity $2S_{1/2}$ and $2P_{1/2}$ are mixed, so one hopes to observe the E1 transition $2S_{1/2} \rightarrow 1S_{1/2}$. Nevertheless, this effect in hydrogen is too small to be observed. Parity breaking effects are strongly enhanced with Z, the charge of the nucleus. Another form of symmetry violation, the combined charge and parity CP symmetry, is the source of asymmetry in the amount of matter and antimatter in our world. On the atomic scale, CP violation leads to the appearance of electric dipole moments for elementary particles or nuclei. These electric dipole moments, or so-called Schiff moments of nuclei, may induce dipole moments of the whole atom. It is hoped to observe such effects only for the heaviest atomic or molecular systems. Current predictions of the standard model, however, do not support this hope. Finally, the breaking of CPT symmetry, the combined charge, parity and time reversal symmetry, may lead to different transition frequencies for hydrogen and antihydrogen, the atom formed of a positron and antiproton. And once again, the 1S–2S transition, due to its very small width, is the natural candidate for the verification of CPT symmetry.

Acknowledgements

The author had the chance to work as a theoretician within the experimental group of Theodor Hänsch in Garching for three years after finishing his Ph.D. on the Lamb shift in hydrogen. Discussions with Martin Weitz, Dietrich Leibfried, Thomas Udem, Jochen Walz, Antoine Weiss and Theodor Hänsch have created an atmosphere for intensive scientific work. Let me thank them for their cooperation.

References

1. M. Niering et al., Phys. Rev. Lett. **84**, 5496 (2000)
2. S. Bourzeix, et al., Phys. Rev. Lett. **76**, 384 (1996)
3. M. Weitz, A. Huber, F. Schmidt-Kaler, D. Leibfried, T.W. Hänsch, Phys. Rev. Lett. **72**, 328 (1994)
4. V.M. Shabaev, Teor. Mat. Fiz. **63**, 394 (1985) [JETP **37**, 211 (1973)]
5. A. Huber et al., Phys. Rev. Lett. **80**, 468 (1998)
6. K. Pachucki, Phys. Rev. A **63**, 042503 (2001)
7. V.A. Yerokhin, V.M. Shabaev, physics/0107036
8. S. Mallampali, J.Sapirstein, Phys. Rev. A **57**, 1548 (1998)
9. K. Pachucki, Phys. Rev. A **56**, 297 (1997)
10. A. Czarnecki, K. Mielnikov, A. Yelkhovsky, Phys. Rev. Lett. **82**, 311 (1998)
11. I.B. Khriplovich, *Parity Nonconservationin Atomic Phenomena* (Gordon and Breach, Philadelphia 1991)
12. M.C. Noecker, B.P. Masterson, C.E. Wieman, Phys. Rev. Lett. **61**, 310 (1988); C.S. Wood, et al., Science **275**, 1759 (1997)

Lasers to Test Fundamental Physics in Space

Reinald Kallenbach

The state of the very early Universe, when it was sufficiently small to presumably appeal to our laureate, Prof. Theodor W. Hänsch, is yet unexplored. The forthcoming joint ESA-NASA mission LISA (Laser Interferometer Space Antenna) may probe this epoch. This mission will apply lasers with a frequency stability comparable to that developed in one of Theodor Hänsch's laboratories during the late 1980s. Relativistic length changes on the order of an atom's size due to the oscillating curvature of space in primordial gravitational waves may be measured over a distance of five million kilometers in space.

1 Fundamental Physics in Space

The Universe is the largest object to test fundamental laws of physics. Its structure is determined by four forces with very different coupling strengths. Each force acts on certain particle families: 1) the strong force on hadrons and quarks, 2) the electromagnetic force on charged particles, 3) the weak force on leptons, and 4) the gravity on mass or the equivalent energy.

Besides the electromagnetic interaction, only the gravitational force has infinite range. As the effects of positive and negative charges cancel at large distance, gravity, although the weakest of the four interactions, dominates the evolution of astrophysical objects and of the Universe as a whole. The Big Bang scenario (Fig. 1) is ruled by Einstein's equations of general relativity. The expansion rate H_0 of the Universe's scale factor S is derived as

$$H_0^2 = \left(\frac{\dot{S}}{S}\right)^2 = \frac{8\pi}{3}G\rho - \frac{kc^2}{S^2} + \frac{\Lambda}{3} \quad , \quad \rho_c = \frac{3H_0^2}{8\pi G} \quad (\text{if } \Lambda = 0) \quad , \tag{1}$$

with the gravitational constant G, the mean density of the Universe ρ, the cosmological constant Λ representing the energy density of the vacuum, and the parameter k characterizing the space curvature with $k = 1$ for a closed, $k = 0$ for a flat, and $k = -1$ for an open Universe. A critical density ρ_c is necessary to close the Universe if there is no vacuum energy. The radiation-dominated dense early Universe had an expansion timescale $S/\dot{S} \propto 1/\sqrt{\rho}$. It behaved like a relativistic gas with $\rho \propto T^4$, so that the temperature evolved

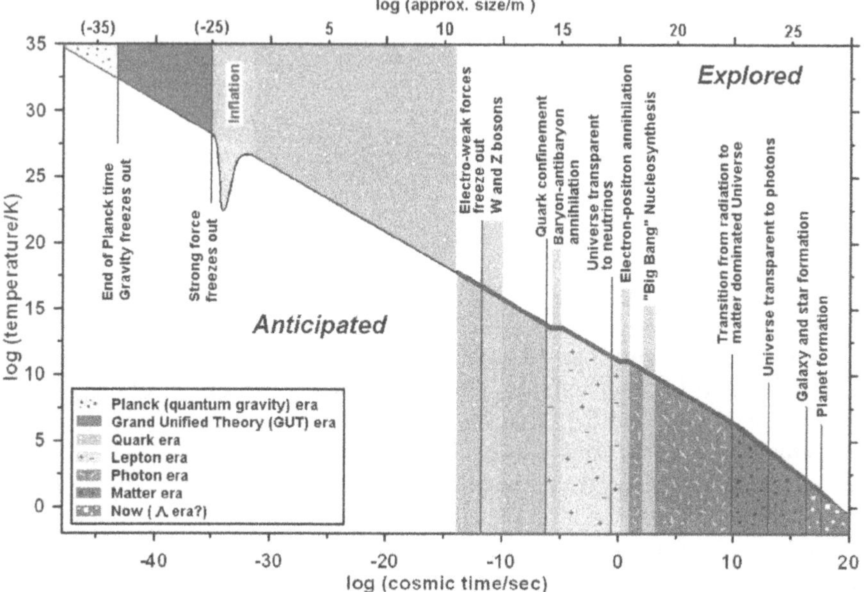

Fig. 1. Big Bang scenario

as $T \propto S^{-1} \propto t^{-1/2}$. While the Universe expanded, the temperature dropped by many orders of magnitude. Phase transitions of the Universe occured, when the mean thermal particle energy kT decreased below the rest energy of certain exchange particles or of specific particle–antiparticle pairs, so that their creation became unlikely. The different eras of the Universe are named after their dominant constituent. After a cosmic time of $\sim 10^{10}$ s, matter became dominant, and the Universe was ruled by the relations $\rho \propto S^{-3}$ or $T \propto t^{-2/3}$.

In daily life, we deal with baryonic matter. Interestingly, the weak interaction had a major impact on the primordial nuclei abundance ratios $^4\mathrm{He}/^1\mathrm{H}$ and D/H, which froze during the Big Bang nucleosynthesis (BBN, Fig. 1). The temperature-dependent rates of the weak reactions (nν, pe$^-$) and (ne$^+$, p$\overline{\nu}$) competed with the expansion rate H_0, which scaled with the sum of the Universe's photon and neutrino density [11]. Assuming three neutrino flavors, the ratio $n_\mathrm{p}/n_\mathrm{n}$ froze to $\sim$5.1 at temperatures below 10^{10} K. Within $\sim$260 s, about 70% of all neutrons were consumed to form primordial ^{4}He and D, while the rest suffered β-decay. The currently best observational value for primordial ^{1}H and ^{4}He abundances, $(^4\mathrm{He}/^1\mathrm{H})_\mathrm{pr} = 0.244 \pm 0.002$ [26], is well compatible with this BBN scenario. As outlined in Fig. 2, the contribution of baryonic matter to the total density of matter in the Universe can be derived from the present-day D/H ratio. Baryonic matter only contributes a few percent! This suggests that other ("dark") forms of energy or matter exist.

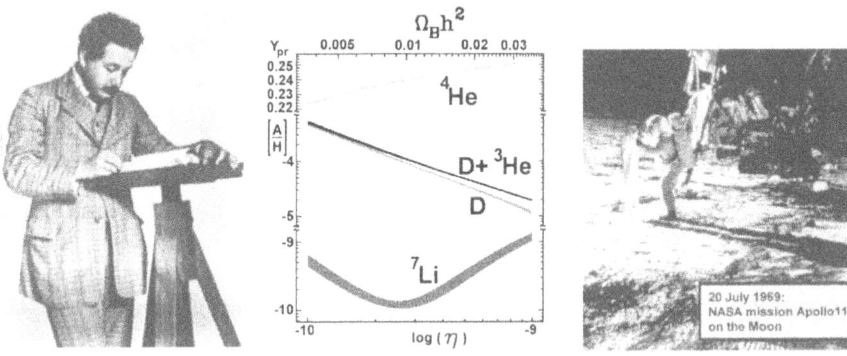

Fig. 2. To test the concepts of general relativity and cosmology by A. Einstein (here in Bern, 1904), E.E. Aldrin, and N. Armstrong installed an aluminum foil of the University of Bern on the lunar surface to collect solar wind ions. Solar wind [12] and meteoritic [8] (protosolar) ^{3}He/^{4}He ratios yield the protosolar D/H abundance ratio [10], because the Sun burns D to ^{3}He. Based on the present-day baryon to photon number ratio, $\eta = 4 \times 10^{-10}$ [20], and on nucleosynthesis models (e.g. [3]), which include the fact that D cannot be synthesized in stars, the protosolar D/H constrains the contribution, Ω_B, of baryonic matter to the total energy density of the Universe, to be less than 0.074 (if $H_0 = 55$ km/s/Mpc [25], i.e. $h = 0.55$)

The strong force predominantly leaves its imprints on baryonic matter by processes in stars, in particular in massive stars and supernovae. Current nucleosynthesis models, e.g., [22], fairly well describe galactic chemical evolution, leading to the protosolar composition of heavy isotopes. This composition is represented in the nearly unfractionated sample of the solar wind [16, 18]. With respect to its abundances, the processes that have formed Earth and Mars [1, 19] have fractionated heavy volatile isotopes, including the isotopes of C, N, and O, which form organic molecules.

The strong interaction also may have inflated the early Universe from microscopic to human scales (Fig. 1). The strong force freezes out at a cosmic time of $\sim 10^{-35}$ s, when the Universe's temperature is equivalent to 10^{16} GeV. At such particle energies the coupling strengths of the strong and electroweak interactions are predicted to be equal [14]. With decreasing temperature, the supersymmetry of this GUT (grand unified theory) state breaks. The Higgs boson acquires a mass and causes unusual changes in the vacuum energy [23]. Thus, with an exceptionally large parameter Λ, the expansion law (1) describes an exponential inflation of the Universe. This may have been the cause for the excess of matter over antimatter. The rapidly inflating Universe has not sufficient time to restore charge–parity (CP) symmetry, so that a proton excess – balancing the electron excess – can remain [14].

Rapid inflation also explains the observed isotropy of the cosmic microwave background radiation (CMBR). Although this electromagnetic "imprint" had decoupled much later, when the Universe became transparent to photons at a cosmic time $t = 300\,000\,\mathrm{y}$ [20], we only see a largely inflated originally small cosmic parcel with little temperature variation [14].

The very early Universe before the inflationary phase is completely unexplored. Presumably, it had microscopic scales and was extremely dense and hot, with temperatures above 10^{33} K. Thermal particles had energies of more than 10^{19} GeV, so that their equivalent mass was larger than the Planck mass, $m_\mathrm{P} = \sqrt{\hbar c/G} \approx 2.18 \times 10^{-8}$ kg. The latter marks the transition to the state of quantum gravity, where particles become black holes: the de Broglie wavelength, $2\pi\hbar/(mc)$, is shorter than the particle's Schwarzschild radius, $2Gm/c^2$. The Planck mass is associated with the Planck length, $l_\mathrm{P} = \hbar/(m_\mathrm{P} c) \approx 1.6 \times 10^{-35}$ m, and the Planck time, $l_\mathrm{P}/c = \sqrt{\hbar G/c^5} \approx 5.4 \times 10^{-44}$ s. After a few Planck times the Universe had expanded and cooled below 10^{33} K, so that the concept of space and time became applicable to describe the interaction between particles.

At this stage, the Universe became transparent to gravitational waves, which propagate freely with the speed of light as an oscillating curvature in space-time. Because of their weak interaction, they may have preserved imprints on very early phase transitions of the Universe. The primordial waves generated after a few Planck times may now be redshifted to frequencies corresponding to ~ 0.9 K (Fig. 3). Gravitational waves from the inflationary phase are expected to be scale-free. They are constrained by COBE results to have acquired a fraction of the critical energy density of $\Omega_\mathrm{GW} \leq 8 \times 10^{-14}$ [5]. Gravitational radiation may also arise from cosmic strings, i.e. from defects left over from a GUT-scale phase transition (Fig. 3). Colliding vacuum bubbles from the electroweak phase transition may generate gravitational waves peaking at 0.1 mHz with up to $\Omega_\mathrm{GW} \approx 3 \times 10^{-7}$ [5, 21]. According to theory and the indirect proof for gravitational radiation [13], stronger sources of gravitational waves in the "present" Universe may "disturb" the detection of the primordial background. However, the direct detection of any gravitational wave would be a breakthrough, because it is the first direct

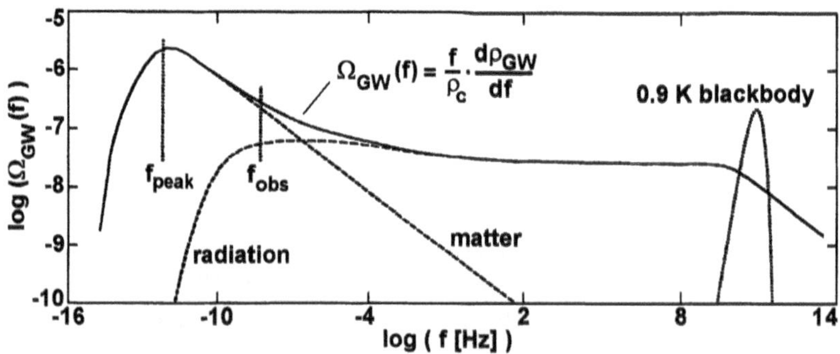

Fig. 3. Power spectrum of gravitational wave radiation from cosmic strings [4]

proof for the existence of the graviton, the gravity's exchange particle, in analogy to the photon of electromagnetic radiation.

2 Direct Detection of Gravitational Waves in Space

The most precise tool to probe relative length changes is a Michelson laser interferometer, located in space, where it is free of terrestrial Newtonian distortions. In the 1990s, the LISA team [5] designed a mission (Fig. 4) to directly detect gravitational waves. At that time, photons with a coherence length of at least 2000 km were generated [15]. For instance, the frequency width of a dye laser was ~ 140 Hz, when compared for 1 s to the 1S–2S two-photon transition of atomic hydrogen (Fig. 6; [27]). The coherence length of photons constrains the orbits at 1 AU of the three drag-free LISA spacecraft around the Sun. If they are separated by 5×10^6 km, the interferometer arm lengths vary by less than 10 000 km. That way, interference fringes from laser photons with a coherence length of 10 000 km are expected (Fig. 4) to indicate length changes on the order of the optical wavelength, corresponding to small deviations from the flat Minkowski-metric $h \approx 10^{-16}$. The precision may reach $\sim 10^{-23}$, if the bandwidth of the gravitational waves is much narrower than the laser's noise bandwidth, and if the signal is sampled for one year, increasing the signal-to-noise ratio by a factor $\sim \sqrt{3 \times 10^7}$.

Gravitational waves with an amplitude $h \approx 10^{-26}$ at Earth have been indirectly observed for years. They were emitted with a frequency of 36 μHz from the binary neutron star PSR1913+16 [13] at a distance of 8.3 kpc. These values calibrate the radiation strength of other astrophysical sources. In first order, they are described as a "dumbbell" with two point masses M separated by a distance $2R$. When rotating at an angular frequency ω with

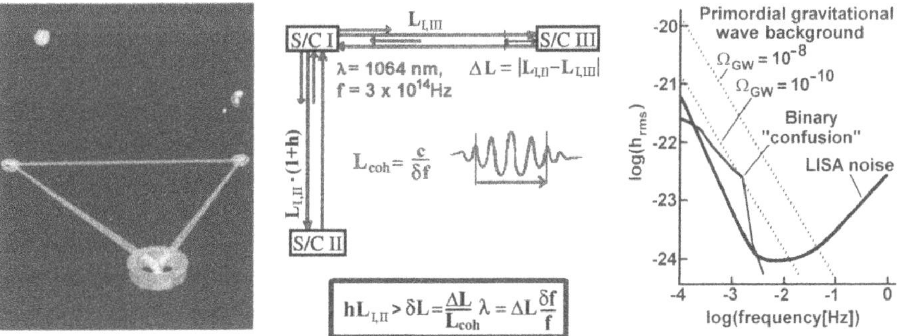

Fig. 4. The three LISA spacecraft [5] form a Michelson laser interferometer with arm lengths of 5×10^6 km and, for practical reasons, with 60° angles between its redundant arms. Depending on the direction of their propagation, gravitational waves cause different relative length changes h in different arms. The signals of close white dwarf binaries may "confuse" the detection of the primordial waves [5]

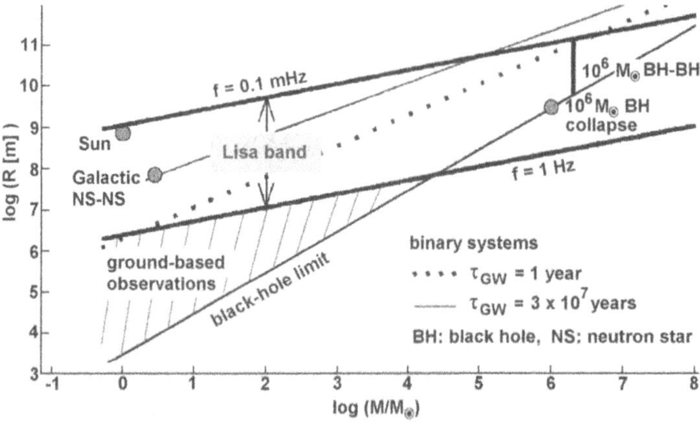

Fig. 5. Astrophysical sources of gravitational waves (after Danzmann et al. [5])

a nonaxisymmetric moment of inertia D, it emits to lowest order gravitational quadrupole waves along its axis of rotation. The total wave power is

$$P \approx \frac{G}{c^5} D^2 \omega^6 \ . \tag{2}$$

If the centrifugal and gravitational force balance each other, i.e. $R\omega^2 = GM/R^2$, the emission frequency ω, the damping rate $1/\tau$, and the amplitude h are

$$\frac{\omega}{\mathrm{Hz}} \approx 10^{5.3} \zeta^{3/2} \frac{M_\odot}{M} \ , \quad \tau\omega \approx \zeta^{-5/2} \ , \quad h \approx 10^{-14} \ \frac{\mathrm{kHz}}{f} \ \frac{\mathrm{kpc}}{r} \ \zeta^{5/2} \ , \tag{3}$$

with the solar mass $M_\odot$, the distance to the object r, and $\zeta = R_\mathrm{S}/R$, where R_S is the objects's Schwarzschild radius. There is a finite number of known Galactic binaries at 1–30 kpc and extragalactic objects (Fig. 5) that would generate well-separated identifiable signals in the LISA detection range [5].

A problem for the detection of primordial gravitational waves may arise from a continuous background of close white dwarf binaries (CWDBs) in the Galactic disk (Fig. 4). Perpendicular to the disk, however, the primordial background, even that from the Universe's inflationary phase, may be detectable, if the LISA sensitivity is further pushed. The lasers designed for LISA [24] are improved in frequency stability compared to earlier lasers by a factor 2 to 3 (Fig. 6), although problems with length drifts of the reference cavity seem to persist. It would be desirable to achieve an absolute flat noise level of $0.5\,\mathrm{Hz}/\sqrt{\mathrm{Hz}}$, as was demonstrated for the relative stability of a dye laser in 1990 [15] with respect to a reference cavity.

Currently, it is planned to measure the phase noise of an ultra-stable oscillator (USO) with an intrinsic stability o $\sim 7 \times 10^{-12}/\sqrt{\mathrm{Hz}(t/1\,\mathrm{s})}$ by modulating its radio frequency (RF) signal onto the laser signal and transmitting

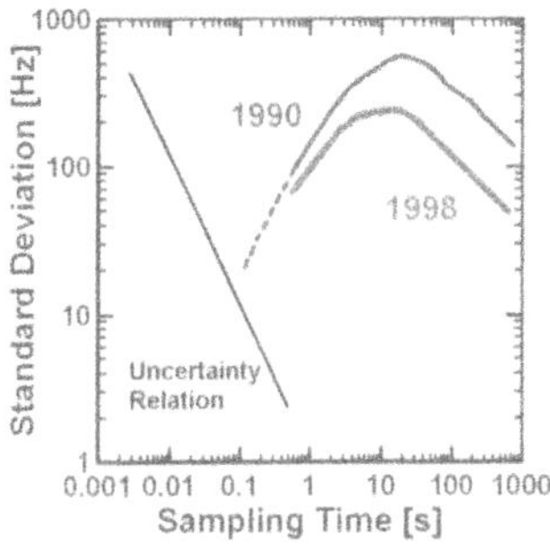

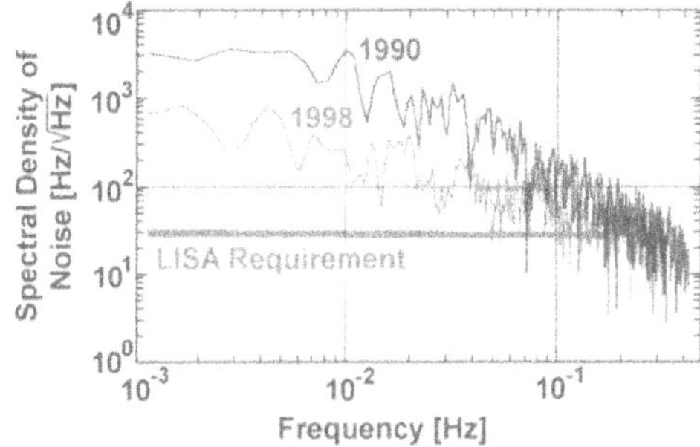

Fig. 6. Frequency stability of a dye laser in Theodor Hänsch's hydrogen laboratory in 1990 [15, 27] and of the monolithic space-qualified Nd:YAG lasers in 1998 [24]

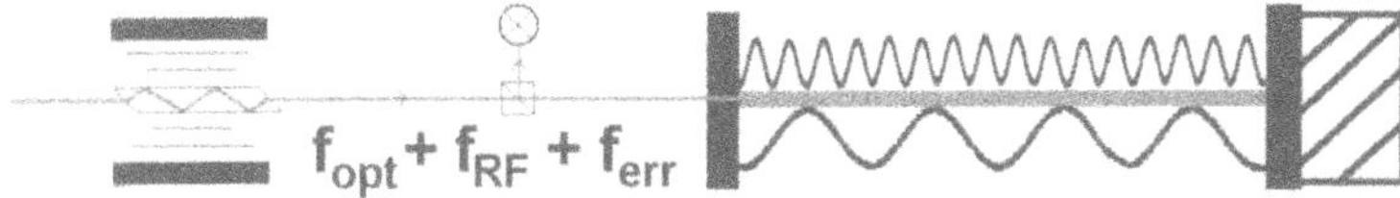

Fig. 7. Electro-optic sideband generation at 72 GHz [17] applied to the double frequency modulation technique [6]

it between the LISA spacecraft [5]. This measurement may be used to further stabilize the USO, or, at the time of launch, a more precise spacequalified USO may be available. In that case, the double frequency modulation method [6] may transfer the stability of the USO to the laser. In 1984, a reference cavity has been locked to RF standards within $\sim 10^{-7}$ of its optical linewidth. If a multiple of the free spectral range of a 0.5 m-sized cavity with a finesse of $\sim$ 100 000 (`http://www.newport.com`) were locked to an RF of $\sim$ 100 GHz, modulated onto the laser frequency [17], the cavity length could be stabilized within 3×10^{-15} (Fig. 7). This may push the LISA sensitivity to 10^{-25} if at the same time the noise due to the proof masses is reduced.

Nevertheless, nature may have decided $\sim$ 15 Gy ago to generate too little power of primordial gravitational waves to be detected by LISA. A future generation of missions may fly optical atomic clocks [2] or a set up that links microwaves with optical frequencies by a femtosecond laser comb [7].

Acknowledgements

The author has enjoyed three years with Theodor Hänsch and his group in the "hydrogen" laboratory at the Max-Planck Institut für Quantenoptik in Garching. Rudolf von Steiger has contributed to this text with discussions. Stein Haaland has calculated the Fourier transform of the signal in Fig. 3 of reference [27] used in Fig. 6.

References

1. W. Benz, R. Kallenbach, G.W. Lugmair (eds.), *From Dust to Terrestrial Planets* (Kluwer, Dordrecht 2000)
2. R.G. Brewer, R.G. DeVoe, R. Kallenbach, *Atomic Clock Employing Ion Trap of Mono- or Multi-planar Geometry.* US-Patent 5379000 (1992)
3. S. Burles, K.M. Nollett, M.S. Turner, Astrophys. J. **552**, L1 (2001)
4. R.R. Caldwell, B. Allen, Phys. Rev. D **45**, 3447 (1992)
5. S. Danzmann et al., 'LISA: Pre-Phase A Report'. MPQ Report **208** (1998)
6. R.G. DeVoe, R.G. Brewer, Phys. Rev. A **30**, 2827 (1984)
7. S.A. Diddams, D.J. Jones, J. Ye, S.T. Cundiff, J.L. Hall, J.K. Ranka, R.S. Windeler, R. Holzwarth, T. Udem, T.W. Hänsch, Phys. Rev. Lett. **84**, 5102 (2000)
8. P. Eberhardt, J. Geiss, N. Krögler. Earth Planet. Sci. Lett. **1**, 7 (1966)
9. R. Fabbri, M.D. Pollock, Phys. Lett. **125B**, 445 (1983)
10. J. Geiss, H. Reeves, Astron. Astrophys. **18**, 126 (1972)
11. J. Geiss, R. von Steiger, 'Production of Light Nuclei in the Early Universe'. ESA SP-420, 99 (1997)
12. J. Geiss, F. Bühler, H. Cerutti, P. Eberhardt, C. Filleux, 'Solar Wind Composition Experiment', Apollo 16 Prel. Sci. Rep., NASA SP-315, 14.1 (1972)
13. R.A. Hulse, J.H. Taylor, Astrophys. J. **345**, 434 (1989)
14. M. Jacob, 'The Early Universe as it is Seen in the Deep Structure of Matter'. ESA SP-420, 107 (1997)
15. R. Kallenbach, Very High Resolution Spectroscopy of the 1S–2S Two Photon Resonance of an Atomic Hydrogen Beam. PhD Thesis, University of München, MPQ Report **151**, München (1990)
16. R. Kallenbach, Isotopic Composition of the Solar Wind. Habilitationsschrift, University of Bern, Bern (2000)
17. R. Kallenbach, B. Scheumann, C. Zimmermann, D. Meschede, T.W. Hänsch, Appl. Phys. Lett. **54**, 1622 (1989)
18. R. Kallenbach et al., 'Fractionation of Si, Ne, and Mg Isotopes in the Solar Wind as Measured by SOHO/CELIAS/MTOF'. In: *Solar Composition and its Evolution - from Core to Corona*, ed. by C. Fröhlich, M.C.E. Huber, S.K. Solanki, R. von Steiger (Kluwer, Dordrecht 1998); Space Sci. Rev. **85**, 357 (1998)
19. R. Kallenbach, J. Geiss, W.K. Hartmann (eds.), *Chronology and Evolution of Mars* (Kluwer, Dordrecht 2001)
20. E.W. Kolb, M.S. Turner, *The Early Universe* (Addison Wesley, New York 1990)
21. A. Kosowsky, M.S. Turner, R. Watkins, Phys. Rev. D **45**, 4514 (1992)
22. M. Limongi, O. Straniero, A. Chieffi, Astrophys. J. Suppl. **129**, 625 (2000)
23. J.A. Peacock, *Cosmological Physics* (Cambridge Univ. Press, Cambridge 1999)
24. M. Peterseim, O.S. Brozek, K. Danzmann, I. Freitag, P. Rottengatter, A. Tünnermann, H. Welling, Proc. Second Intern. LISA Symposium, ed. by W.M. Folkner (AIP Press, New York 1998), pp. 148–155
25. G.A. Tammann, 'Observed Densities in the Universe'. In: *Primordial Nuclei and Their Galactic Evolution.* ed. by N. Prantzos, M. Tosi, R. von Steiger (Kluwer, Dordrecht 1998); Space Sci. Rev. **84**, 15 (1998)
26. T.X. Thuan, Y.I. Izotov, Space Sci. Rev. **84**, 83 (1998)
27. C. Zimmermann, R. Kallenbach, T.W. Hänsch, Phys. Rev. Lett. **65**, 571 (1990)

Measuring the Birefringence of the QED Vacuum

Siu Au Lee and William M. Fairbank, Jr.

In classical electrodynamics, the Maxwell equations are linear in the fields, and there is no interaction of light with light. In quantum electrodynamics, however, photons can couple with each other via the production of virtual electron–positron pairs (Fig. 1). Since the vacuum is now treated as a dielectric medium, it can become polarized by electric and magnetic fields. In this chapter, we discuss the quantum electrodynamics (QED) birefringence of the vacuum, the status of current efforts, and how an experiment utilizing femtosecond lasers may very well lead to the first direct observation of this effect.

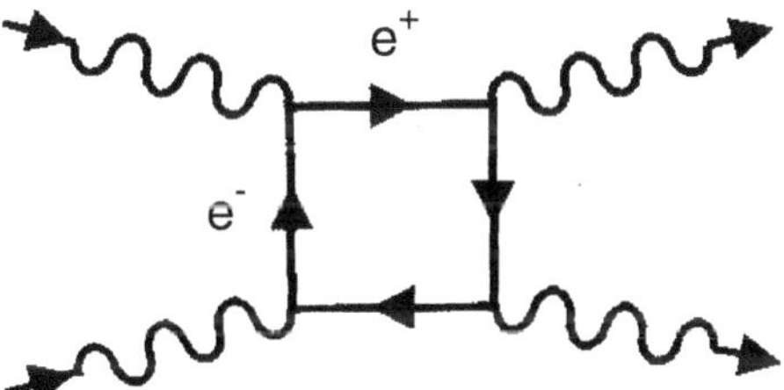

Fig. 1. Photon–photon scattering

1 Birefringence of the QED Vacuum

More than sixty years ago, Heisenberg and Euler derived an effective Lagrangian for slowly varying electric and magnetic fields in vacuum, including the corrections due to vacuum polarization [1]. To lowest order in the QED correction, the Lagrangian may be written as:

$$L = \frac{1}{8\pi}(E^2 - B^2) + \frac{1}{(4\pi)^2}\frac{\alpha^2\hbar^3}{90\ m_{\mathrm{e}}^4\ c^5}\left[4(E^2 - B^2)^2 + 7(E \cdot B)^2\right], \qquad (1)$$

where E and B are the electric field and magnetic field, respectively, α is the fine structure constant, and Gaussian units are used. The first term on the right hand side is the familiar Lagrangian for the linear Maxwell equations, and the second term represents the nonlinear interaction of light with light. In 1971, Adler [2] considered the effect of a strong magnetic field (occurring near pulsars) on the propagation of light. Using the Heisenberg–Euler Lagrangian, he showed that the magnetic field induced a change in the index of refraction of vacuum. For light linearly polarized at an angle θ to the external B field, the eigenstates of propagation are the two polarization states parallel ($\|$) and perpendicular ($\perp$) to B, and they have different indices of refraction. In other words, the vacuum becomes birefringent. The indices are

$$n_{\|} = 1 + \frac{7}{2}\,\xi\,\sin^2\,\theta, \qquad n_{\perp} = 1 + \frac{4}{2}\,\xi\,\sin^2\,\theta, \tag{2}$$

where

$$\xi = \frac{\alpha}{45\pi}\left(\frac{B}{B_{\mathrm{cr}}}\right)^2 \tag{3}$$

and

$$B_{\mathrm{cr}} = m_{\mathrm{e}}^2 c^3/e\hbar = 4.41 \times 10^{13}\,\mathrm{G} \tag{4}$$

is the critical field strength. For maximum effect, $\theta = 45°$, and the magnetic field induced birefringence of the vacuum is

$$\triangle n = n_{\|} - n_{\perp} = \frac{\alpha}{30\pi}\left(\frac{B}{B_{\mathrm{cr}}}\right)^2 . \tag{5}$$

The above equations are for light propagating in a constant magnetic field. However, the vacuum also becomes birefringent when the magnetic field is replaced by an electric field. (The scattering of photons off the electric field of a nucleus is known as Delbrück scattering.) Indeed, the critical E field, 4.41×10^{13} statV/cm, has the simple physical meaning that when a virtual electron–positron pair is separated by their Compton wavelength, the QED interaction energy and the mass energy are equal, i.e. $eE_{\mathrm{cr}}\lambda_{\mathrm{c}} = m_{\mathrm{e}}c^2$. Although the birefringence of the vacuum has been predicted for three decades, there has not been a direct observation of this effect. The reason is that the birefringence is exceedingly small for practical B and E fields. Even with a reasonably strong laboratory magnetic field of 10^5 G, the difference in index is only 4×10^{-22}.

Now consider a laser of wavelength λ, passing N times through a magnetic field region of length L. If the laser is initially linearly polarized at 45° to the magnetic field, it will become elliptically polarized. The ellipticity, defined as the ratio of the minor to major axes, is given by half the phase shift:

$$\Psi_{\mathrm{QED}} = N\,\frac{\triangle n L \pi}{\lambda} = \frac{NL}{\lambda}\left(\frac{\alpha}{30}\right)\left(\frac{B}{B_{\mathrm{cr}}}\right)^2 . \tag{6}$$

Thus, the general experimental strategy is to use as large a B field and as long a field region as possible, and to pass the laser beams many times through the B field region in order to enhance the ellipticity.

2 Experiments

2.1 First-Generation Experiment: BNL 840

The first major attempt to measure the magnetically induced QED birefringence of the vacuum was carried out at the Brookhaven National Laboratory (BNL 840) [3], making use of the prototype superconducting dipole magnets built for the Isabelle. This was a collaboration involving BNL, Fermilab, University of Rochester and the University of Trieste. The experiment was intended as a search for pseudoscalar particles such as the axion, but it applied to a QED birefringence measurement as well. In this experiment, the magnetic field was modulated between 2.6 to 3.9 T and the field length was 8.8 m. An argon ion laser at 514.5 nm and linearly polarized at 45° to the B field was passed through the magnetic field region many times in a Herriott optical delay-line cavity. The output laser beam was sent through a 1/4-wave plate to convert the ellipticity to a rotation, and next through a Faraday cell where its polarization was modulated. The laser beam was then detected through a crossed polarizer and its Fourier components at the sum and difference frequencies of the magnet and Faraday cell modulations were analyzed. This heterodyne scheme has the beauty that the detected signal was linear in the ellipticity of the light beam.

The major sources of noise were from the random motion in the light beam caused by mechanical vibrations of the optics and the pointing stability of the laser, and from the systematic fluctuations in the magnet modulation frequencies. The best results were obtained with 34 passes even though the system can accommodate more than 500 passes. The best ellipticity measurement was 2×10^{-9} rad, but it was nearly four orders of magnitude short of the predicted QED signal.

2.2 Second-Generation Experiments

The second-generation experiments are patterned after the BNL 840 but use high-finesse Fabry–Perot cavities to substantially increase the number of laser passes in the magnetic field. These experiments require highly stabilized cw lasers and state-of-the-art feedback controls to lock the laser frequency to the cavity.

PVLAS

The PVLAS (Polarizzazione del Vuoto con LASer) collaboration [4] consists of the Universities of Trieste, Pisa, Ferrara, Padova, and the Legnaro National Laboratory of INFN. The experiment uses a Nd:YAG laser travelling in a high-finesse cavity to enhance the QED signal. The Fabry–Perot cavity is 6.5 m in length and has a finesse of 100 000. The magnetic field is produced by a 6.5 T superconducting dipole magnet of 1 m length. The experimental setup is vertical in order to reduce the vibrations caused by ground motions. The whole magnet and cryostat assembly sits on a giant turntable and is rotated at 0.3 Hz to provide the magnetic modulation for heterodyne detection. The frequency of the linearly polarized laser is locked to the cavity, and the polarization state of the output laser beam is analyzed in the same way as the BNL 840 experiment. The predicted QED ellipticity for this experiment is 3×10^{-11} rad. To date the experiment has achieved a sensitivity of 2.5×10^{-7} rad/Hz$^{1/2}$. The dominant noise source is associated with the mechanical stress caused by the rotating magnet.

FNL 877

The Fermilab-877 collaboration consists of our group at Colorado State University, John Hall of JILA/National Institute of Standards and Technonology, and Frank Nezrick of Fermilab [5]. The experiment utilizes the prototype dipole magnets built for the Superconducting Super Collider (SSC). The SSC magnets provide an intense field (ramping from 1 T to 6 T in 5 mHz) and a long length (30 m). The output from a frequency-doubled diode-pumped Nd:YAG laser at 532 nm is sent into a 50 m long high-finesse ($F \sim 40\,000$) Fabry-Perot cavity. The predicted QED ellipticity is 6.3×10^{-10} rad, and the goal is to be able to measure the QED birefringence at the 1% level.

The long-term laser frequency is stabilized to about 100 Hz to an iodine hyperfine transition using saturated absorption. Short-term stabilization to the sub-Hz level is based on a fixed length (29 cm) zerodur cavity [6]. The high-finesse Fabry–Perot cavity presents many technological challenges. The most obvious problem is the random motion of the mirrors caused by ground vibrations. Typical ground motion at the Fermilab site is about 2 nm at 1 Hz. A motion of 7 pm would have detuned the cavity by one full-cavity linewidth. Therefore, it is necessary to isolate the cavity mirrors from seismic perturbations. The prototype seismic damping for the cavity mirrors was developed at Fermilab and consisted of a four-stage passive isolation and a digitally controlled active damping system [7]. A more subtle problem is the photo-desorption of hydrogen, arising from stray light scattered from the interferometer mirrors and striking the stainless steel wall of the vacuum tube inside the magnet bore. This process may produce a sufficient hydrogen density such that the Cotton–Mouton effect would be comparable to the QED signal. With judicious use of baffles and getters, the amount of scattered light hitting the walls and the hydrogen density can be reduced to a safe level. The birefringence of the cavity mirrors, induced by the high circulating laser power in the cavity, has yet to be solved. This experiment is currently on hold, as Fermilab has suspended the support for the operation of the SSC magnets.

3 Next Generation: Light by Light Scattering with Femtosecond Lasers

The experiments above are pushing the limit of what can be done with cw lasers and large-magnet technology. Suggestions for using an intense laser field to produce the vacuum birefringence started appearing in the mid-1980s [8], when large-scale picosecond lasers were coming on line, for example, at the Lawrence Livermore Laboratory. However, the idea of using a NOVA-type laser to do precision QED measurements was, and still is, very daunting.

The tremendous advances that Theodor Hänsch and colleagues have achieved in the field of femtosecond laser metrology have changed the whole landscape of precision experiments. We were inspired by the beautiful results obtained with femtosecond lasers, and began to wonder if these lasers could be used in the QED experiment. We find that a femtosecond laser experiment may very well lead to the first direct observation of the QED vacuum birefringence.

For photon energy $\ll m_e c^2$ and field strength $E \ll E_{\mathrm{cr}}$, the interaction of electromagnetic waves can be described again by the Heisenberg–Euler Lagrangian given in (1), with similar birefringence properties to those predicted for the static fields. For an experimental situation, we consider a pulsed laser with wavelength λ_1, an energy of ε per pulse and pulse width τ. This laser, called the pump laser, is focused to a waist radius w_0 and is used to generate the intense fields. The change in the index of refraction of the vacuum will be probed with a second, weaker laser with wavelength λ_2. The polarizations of the pump and probe lasers will be oriented at 45° to each other.

It should be noted that the QED birefringence is identically zero if the pump and the probe lasers propagate in the same direction. This is a simple consequence of the E dot B term in the Lagrangian. Specifically, if the two laser beams travel in the same direction, then $E_1 \cdot B_2 + E_2 \cdot B_1 = 0$, due to the right-handed coordinate system that the E and B fields obey. We will consider counter-propagating laser beams, where there is optimum overlap between the pump and the probe lasers. Since the probe laser will only feel an effect when the pump pulse is present, the interaction region is approximately $c\tau$ and is typically very small. On the other hand, it is possible to generate very high peak fields, $\sim 3 \times 10^9$ G. For a single path ($N = 1$), the induced ellipticity in the probe laser is, according to (6),

$$\Psi_{\mathrm{QED}} = \frac{2c\tau}{\lambda_2}\left(\frac{\alpha}{30}\right)\left(\frac{B_1}{B_{\mathrm{cr}}}\right)^2 , \tag{7}$$

where B_1 is the peak field of the pump laser, and the factor of 2 is introduced since both the E and B fields of the pump laser contribute. A tighter focus would give a larger intensity. However, diffraction effects limit the waist size to about λ. A tighter focus also leads to faster spreading of the beam diameter and reduces the field strength. It is reasonable to choose the waist radius such that the confocal parameter is approximately the pulse length.

As an example, we consider the pump laser to be an amplified Ti:sapphire laser operating at 800 nm, with 1 J per pulse, 50 fs pulse length and focused to a waist radius of 1 μm. The repetition rate of the pump laser is 10 Hz [9]. The B field at focus is 2.3×10^9 G. This is substantially larger than any magnetic fields generated by superconducting magnets. The probe beam is polarized at 45° to the pump beam. The QED-induced ellipticity of the probe beam is

$$\Psi_{\mathrm{QED}} = 2.5 \times 10^{-11}\,\mathrm{rad}. \tag{8}$$

This value of ellipticity in a single pass is already at the same level as that of the PVLAS experiment. Although this analysis has neglected the Gaussian nature of the laser beams, we do not expect the ellipticity to be much smaller.

The sensitivity of the experiment is set by the noise of the apparatus. For a first approximation, we assume that the probe laser is 1% of the pump laser. This results in a probe beam energy of 10 mJ per pulse. At 10 Hz repetition rate, we have 4×10^{17} photons/s. The statistical fluctuation is then $1/\sqrt{n} = 1.6 \times 10^{-9}$ in 1 s of integration time. This estimate roughly sets the limit of the noise contributions in the experiment. A detailed analysis of the experimental shot noise gives

$$\mathrm{Sensitivity} = 4 \times 10^{-9}/\sqrt{T_{\mathrm{int}}}\,\mathrm{rad\,s}^{-1/2} \tag{9}$$

at S/N = 1. This value is very similar to the value obtained above based purely on statistical arguments.

At first sight, an attractive way to improve the sensitivity would be to use a Fabry–Perot cavity to build up the laser intensity and to increase the number of passes. Chirped mirrors with $\approx 99.8\%$ reflectivity [10] can be used to maintain the pulse width while providing reasonable cavity finess. This method also allows for the possibility of detecting the ellipticity as a small frequency shift in the cavity resonances [5]. However, the low repetition rate (10 Hz) of the laser means that a synchronously pumped Fabry–Perot cavity will need a length of order 10^7 m. This is totally out of the question. A second scheme is to use a Herriott delay line to multipass the laser beams. This involves injecting the two beams through holes in opposite mirrors. Since the far-field diffraction angle of the laser beams is 0.25 rad, the crossing angle of the two beams in the interaction region must be at least this to achieve clean injection and subsequent reflections. The loss in signal and the increased experimental complexity may outweigh the gain obtained from multipassing.

Other systematic effects will no doubt limit the sensitivity. This is where using a femtosecond laser in a single pass experiment becomes attractive as compared to using magnets. The experiment does not require the large infrastructure and cost associated with operating the superconducting magnets. The optical interaction region is shortened from tens of meters to centimeters, and the experiment is much more robust against vibrations. Modulation of the intense field can be accomplished with a rotating waveplate rather than ramping a high-field magnet. Furthermore, we expect that the femtosecond laser technology will continue to progress, so that the experimental sensitivity will continue to improve.

For a first experiment, a reasonable goal is to expect the experiment to be performing at several times the shot-noise limit. This indicates that a femtosecond laser experiment can very well lead to the first direct observation of the birefringence of the QED vacuum.

4 Remarks

It was an exciting time for us to be graduate students at Stanford University, in the late 1960s and early 1970s. A young NATO postdoc, by the name of Theodor Hänsch, had joined the group of Prof. Arthur Schawlow and embarked on the first precision laser experiments of the hydrogen atom. Theodor had introduced us to the wonderful world of lasers and the beautiful physics investigations that would not otherwise be possible. Since then we have continued to benefit from his ingenuity and his deep physics insight.

Acknowledgements

We wish to thank Theodor Hänsch and John Hall for stimulating discussions concerning femtosecond laser metrology, and with Frank Nezrick on the use of high power lasers for the QED experiment. The work at Colorado State University was funded by a Precision Measurement Grant from the National Institute of Standards and Technology. The work at Fermilab was supported by the U.S. Department of Energy.

References

1. W. Heisenberg, H. Euler, Z. Phys. **38**, 314 (1936); Other derivations have been made by V. Weisskopf, Kgl. Danske Vid. Selskab., Math.-fys. Medd. XIV, 166 (1936) and by J. Schwinger, Phys. Rev. **82**, 664 (1951). See also W. Greiner, *Quantum Electrodynamics (Theoretical Physics Vol. 4)* (Springer, Berlin, Heidelberg 1996) pp. 277–285
2. S.L. Adler, Ann. Phys. **67**, 599 (1971)
3. R. Cameron, G. Cantatore, A.C. Melissinos, G. Ruoso, Y. Semertzidis, H.J. Halama, D.M. Lazarus, A.G. Prodell, F. Nezrick, C. Rizzo, E. Zavattini, Phys. Rev. D **47**, 3707 (1993)
4. The details of the experiment may be found at www.lnl.infn.it/~pvlas/home.htm
5. S.A. Lee, W.M. Fairbank Jr., W. Toki, J.L. Hall, P. Colestock, V. Cupps, H. Kautzky, M. Kuchnir, F. Nezrick, R. Noble, Fermilab P-877. See http://fnalpubs.fnal.gov/archive/proposals/P-877.html
6. Ch. Salomon, D. Hils, J.L. Hall, J. Opt. Soc. Am. B **5**, 1576 (1988)
7. F. Nezrick, in *Frontier Tests of QED and Physics of the Vacuum*,ed. by E. Zavattini, D. Bakalov, C. Rizzo (Heron Press, Sofia 1998) pp. 72–82

8. E.B. Aleksandrov, A.A. Ansel'm, A.N. Moskalev, Sov. Phys. JETP **62**, 680 (1985)
9. Lasers with similar specifications are already in existence in various universities and national laboratories around the world. One such system is under construction at Colorado State University, in the soft X-ray laser laboratory of Jorge Rocca.
10. F.X. Kärtner, N. Matuschek, T. Schibli, U. Keller, H.A. Haus, C. Heine, R. Morf, V. Scheuer, M. Tilsch, T. Tschudi, Opt. Lett. **22**, 831 (1997)

Observing Mechanical Dissipation in the Quantum Vacuum: An Experimental Challenge

Astrid Lambrecht

Vacuum field fluctuations exert radiation pressure on mirrors in the quantum vacuum. For a pair of mirrors this effect is well known as the Casimir force, that is an attractive force between two mirrors at rest in the vacuum. When a single mirror is moving in the vacuum, radiation pressure leads to a dissipative force which opposes the mirror's motion. Accordingly the electromagnetic field does not remain in the vacuum state but photons are emitted by the mirror into the vacuum. This motion-induced radiation and the associated radiation reaction force are dissipative effects related to motion in the quantum vacuum, although this motion has no further reference than the vacuum itself.

This article describes the photon emission of a high-finesse cavity oscillating globally in the quantum vacuum. Novel effects of quantum radiation like pulse shaping and frequency up-conversion are predicted, which could be used to experimentally demonstrate motion-induced dissipative effects.

1 Introduction

The quantum vacuum is the arena where fundamental physical processes take place, and is by no means a simple empty space where nothing ever happens or a pure abstract concept of quantum field theory. Many of these fundamental processes are nowadays well understood. However important questions remain unsolved, for example related to the infiniteness of the vacuum energy and to what is called the cosmological constant problem [1, 2]. When one empties out a space of all matter and lowers the temperature to absolute zero, one produces in a *Gedankenexperiment* the quantum vacuum state. This state corresponds to the fundamental state of the electromagnetic field and is characterized by vacuum fluctuations corresponding to a mean energy of $\frac{1}{2}\hbar\omega$ per field mode. Vacuum fluctuations have well-known observable effects. Consider for example an isolated atom in the vacuum. This atom interacts only with vacuum fluctuations and this interaction is responsible for the spontaneous emission processes during which the atom changes its internal state by falling from a higher energy level to a lower one. When fallen in the ground state, the state of lowest energy where it can no longer emit photons, the atom is still coupled to vacuum fluctuations and this coupling results in measurable effects like the Lamb shift of the atomic absorption frequencies. As this constitutes an important prediction of quantum field theory, the Lamb shift has been measured with very high precision repeatedly and the experimental verification of the Lamb shift as well as of other fundamental quantities has been and still is a very active area of research to which the name of Theodor Hänsch is unseparably associated [3, 4].

If one considers not a single atom but two atoms in the vacuum, the two atoms are coupled to the field fluctuations which produce an attractive force between them, the so-called van der Waals force. This force plays an important role in physical and chemical processes and its quantum theoretical interpretation has been studied since the first years of quantum theory. But vacuum fluctuations do not only influence microscopic objects like atoms. They also produce mechanical effects on macroscopic bodies. The most famous mechanical effect of the quantum vacuum was discovered in 1948 by Henrik Casimir who found that there exists a force between two mirrors placed in the quantum vacuum [5]. Vacuum fluctuations are modified by the cavity formed by the mirrors. As any classical field, they are enhanced or diminished depending on whether their frequency corresponds to a cavity mode or not. The spectral mode densities inside and outside the cavity are therefore different and the same is true for the vacuum radiation pressure. Hence the vacuum exerts a force which mutually attracts the mirrors to each other. This Casimir force, as it was called later on, depends only on the distance and on two fundamental constants, the speed of light c and the Planck constant $\hbar$. This is a remarkably universal feature in particular because the Casimir force is independent of the electronic charge in contrast to the van der Waals forces between atoms and molecules.

It has more recently been recognized that dynamical counterparts of this static force appear for moving scatterers. This is the so-called dynamical Casimir effect. Indeed, for some types of motion, the field does not remain in the vacuum state, but photons are produced through non-adiabatic processes [6]. Due to energy conservation, the scatterers' motion then has to be damped out and this damping is associated with dissipative radiation reaction forces. While the observation of the effects of vacuum fluctuations on a microscopic scale has given rise to a great number of successful experimental observations, the mechanical effects on macroscopic objects are much harder to observe. However, the Casimir force has now been measured repeatedly with increasing precision and the emission of quantum radiation by objects moving in the vacuum and the associated radiation reaction force raise intriguing questions with respect to the standard mechanical description of motion [2]. Indeed, they imply that dissipative effects are associated with the motion of mirrors in the vacuum, although this motion has no further reference than the vacuum itself. It thus seems that the quantum vacuum constitutes a sort of reference frame with respect to which motion takes place. It would therefore be very important to obtain experimental evidence of these dissipative processes associated with motion in the vacuum. However, vacuum radiation pressure scales with the Planck constant $\hbar$ and its mechanical effect is therefore very small on macroscopic objects.

During my postdoctoral stay in Theodor Hänsch's group, I have been working on experiments using cold Rb atoms for different purposes, and I have enjoyed the stimulating atmosphere in which various physical topics and ways to understand physical phenomena are debated. It is also during this stay that I started to think about the idea to study mechanical effects of vacuum fluctuations, especially the radiation emitted by moving mirrors in the vacuum which was to become my main research interest afterwards. In contrast to other approaches to study quantum vacuum radiation, where only perfectly reflecting mirrors were considered which do not exist in reality, my main concern was to find configurations where these effects might become observable in a laboratory. In the search of experimental configurations and difficulties, I have profited during many discussions with Theodor Hänsch from his great experience and intuition for the essential physics in a given problem.

In collaboration with Serge Reynaud and Marc-Thierry Jaekel from the Laboratoire Kastler Brossel and the Laboratoire de Physique Théorique in Paris, we have developed an approach which allowed us to consider more realistic configurations, that is mirrors and cavities with frequency-dependent reflection coefficients. In this article only the basic physics of the phenomenon, which is motion-induced radiation due to vacuum fluctuations, will be described. I will not present any calculations or detailed formulae (they can be found in the references) but rather concentrate on the main ideas concerning the possibility of experimental observation. In this respect the important questions are how to increase the orders of magnitude of the dissipative effects of the quantum vacuum and what are its precise signatures which could be searched for in an experiment.

2 An Interpretation in Terms of Parametric Processes

Motion-induced effects of vacuum radiation pressure do not require the presence of two mirrors but already exist for a single mirror moving in the vacuum. In this case, the radiation reaction force is known to arise for arbitrary motion [7]. It has to be stressed that naturally this force vanishes for a uniform velocity, as expected from the Lorentz invariance of the quantum vacuum. It also vanishes for a uniform acceleration. However, for example a mirror oscillating in the vacuum feels a dissipative force. As a consequence, its motion is damped to a motion with uniform acceleration where the dissipative force vanishes. The dissipated energy is emitted by the mirror in the form of photons into the free field vacuum.

This process can also be interpreted using analogies with optical parametric processes. Let us for example consider a single mirror oscillating in the quantum vacuum with an amplitude a and a frequency Ω as schematically

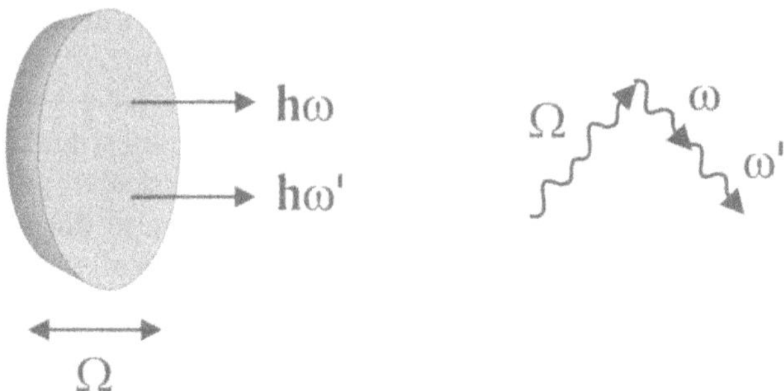

Fig. 1. Emission diagram for a parametric process. Through opto-mechanical coupling between vacuum fluctuations and the mirror's motion, a mechanical excitation Ω is transformed into two photons of frequency ω and ω' which are emitted into the vacuum

indicated in Fig. 1

$$q(t) = a \sin \Omega t \,. \tag{1}$$

Indeed, photon production by the oscillating mirror can be explained through opto-mechanical coupling between vacuum fluctuations and the mechanical motion. In a linear approximation, the "phonon" corresponding to the mechanical excitation frequency Ω is transformed into two photons of frequency ω and ω' emitted into the vacuum. Energy conservation of course imposes the condition that the sum of the photon frequencies must be equal to the oscillation frequency $\omega + \omega' = \Omega$. In the vacuum only spontaneous parametric processes can take place. The reason for this is that it is impossible to extract energy from the vacuum, which would be the case in a stimulated parametric process.

In view of experimental observation, it turns out that motion-induced radiation from a single oscillating mirror is too small to be detectable with current technology. This is why we started to consider a cavity instead of a single mirror in order to take advantage of the resonant enhancement of radiation inside the cavity to improve the orders of magnitude of motion-induced radiation. For a cavity oscillating in the vacuum, it is well known that the cavity field is parametrically excited when the mechanical cavity length is modulated periodically (breathing motion). If the cavity field is initially in the vacuum state, this excitation leads to a squeezed vacuum state [8] which differs from the pure vacuum state and in particular contains photons. This corresponds to a usual parametric amplifier or oscillator, which of course can be more easily and efficiently realized in an experiment by using a nonlinear crystal inside the cavity which produces a periodic change of the effective cavity length instead of the mechanical length. However these experiments can give no information about the relativity of motion in the quantum vacuum, which is my main concern here.

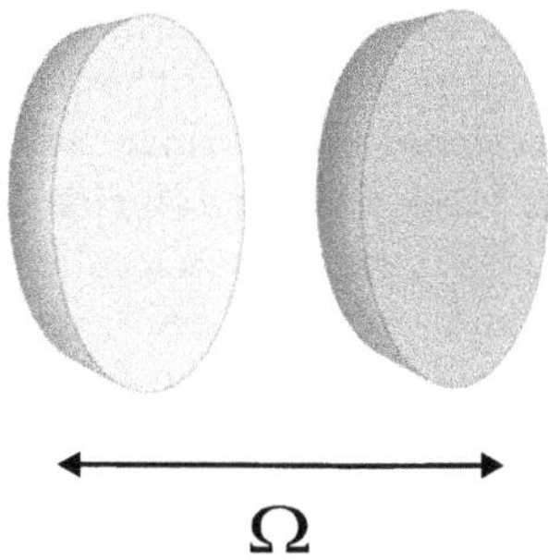

Fig. 2. Cavity oscillating globally in the vacuum, that is both mirrors oscillate with the same amplitude and frequency such that the mechanical cavity length remains constant

More strikingly, a resonant enhancement of radiation also exists when the cavity oscillates as a whole, with its mechanical length L kept constant as shown in Fig. 2. Here both mirrors oscillate with the same amplitude a and frequency Ω. While the breathing motion of the cavity takes place when the mechanical oscillation frequency equals an even multiple of the fundamental cavity resonance frequency $\frac{\pi c}{L}$, the global motion is realized for frequencies equal to odd multiples of the fundamental resonance frequency:

$$\begin{aligned} \Omega &= \frac{2\pi c}{L}, \frac{4\pi c}{L}, \cdots \qquad \text{breathing} \quad \text{motion} \\ \Omega &= \frac{3\pi c}{L}, \frac{5\pi c}{L}, \cdots \qquad \text{global} \quad \text{motion.} \end{aligned} \tag{2}$$

At first glance, it seems to be paradoxical that both the breathing and the global motion give rise to emission of radiation due to parametric excitation of the cavity field. The basic reason for this is that the optical length (which depends on retardation effects) as seen by the field varies in the same way for both kinds of motion although the mechanical cavity length is modulated in one case and constant in the other.

Motion-induced radiation emitted by a cavity oscillating globally is reminiscent of photon emission from a single oscillating mirror. However, the important point is that compared to the situation with a single oscillating mirror in the vacuum, motion-induced radiation from an oscillating cavity is resonantly enhanced by a factor of the order of the cavity finesse [9]. As the cavity finesse can be made large in an experiment, this fact allows us to improve considerably the order of magnitude of motion-induced radiation.

3 General Treatment

In the following discussion I will limit myself to one-dimensional space. As is well known from the analysis of squeezing experiments, the transverse structure of the cavity modes does not change appreciably the results obtained from this simplified model. Each transverse mode is correctly described by a two-dimensional model as soon as the size of the mirrors is larger than the spot size associated with the mode. The two-dimensional model thus corresponds to a conservative estimate where one transverse mode is efficiently coupled to the moving mirrors. A more precise evaluation for a realistic configuration should take diffraction into account and would probably lead to a result obtained by multiplying the one-dimensional result by the Fresnel number, i.e. the number of efficiently coupled transverse modes [10].

The interaction between the mirrors and vacuum fluctuations is described by scattering processes. The mirror is characterized by its reflection and transmission coefficients. At each reflection on the mirror's surface, the field undergoes a phase shift. If the mirror is at rest, the scattering process does not change the frequency of the field. However when the mirror is moving, the scattered field is dephased compared to the input field and this dephasing depends on the mirror's motion $q(t)$. Generally this dephasing changes the field frequency. For example if the mirror moves with uniform velocity the frequency of the reflected field is Doppler shifted compared to the frequency of the input field although no radiation is emitted by a mirrors moving with uniform velocity, in agreement with special relativity theory. For an oscillatory motion (1) the phase shift acquired by the field at a single reflection is given by the mirror's maximum velocity $v = a\Omega$ divided by the speed of light.

One can distinguish two regimes of emission of motion-induced radiation which correspond to a linear and a nonlinear operation mode of the cavity oscillating in the vacuum [11]. The crucial physical parameter to describe the operation mode of the system is the effective phase velocity given by the product of the mechanical velocity of the mirrors v divided by the speed of light c multiplied by the cavity finesse $\mathcal{F}$:

$$\eta = \frac{\mathcal{F}v}{c} . \tag{3}$$

It is reasonable to suppose that the mechanical velocity of a macroscopic mirror is limited by the speed of sound corresponding to the mirror's material. Thus naturally the ratio of the mirror's velocity over the speed of light is very small for any macroscopic realistic system. This is essentially the reason why motion-induced radiation from a single oscillating mirror cannot be observed. However, as illustrated by (3) for a cavity the simple mechanical velocity is multiplied by the cavity finesse which can be a large quantity therefore allowing us to gain many orders of magnitude. If the cavity finesse is small, η remains a small parameter, and the field undergoes only a few reflections inside the cavity. It accumulates a small phase shift due to the mirror's motion. In this case all processes taking place inside the cavity are linear ones. On the other hand, if the cavity finesse becomes high, such that the effective phase velocity approaches unity, the field is reflected a large number of times on each moving mirror and accumulates an important phase shift due to the mirror's motion. For this operation mode, nonlinear processes take place inside the cavity and new phenomena appear.

4 Signatures

If one is interested in the observation of motion-induced radiation, the signatures of the effect are extremely important in order to be able to identify the phenomenon. Motion induced radiation shows two very interesting signatures in the nonlinear regime, which we have studied in detail in view of the possible experimental realization [11]. The first signature is the energy density emitted by the cavity into free field vacuum and the energy density inside the cavity; the second feature is the radiation spectrum. Both of these signatures can be discussed in terms of the effective phase velocity η.

The variation of the energy density for different parameters η is presented in Fig. 3. In the linear regime where $\eta \ll 1$ the temporal variation of the emitted energy is sinusoidal which means that the mirror's mechanical motion (which is sinusoidal) is transformed linearly into the field. However, when the effective phase velocity increases, the emitted energy concentrates in pulses which are emitted by the cavity in regularly spaced time intervals. This pulse shaping becomes more pronounced, the width of the pulses is smaller, and the effective phase velocity becomes larger.

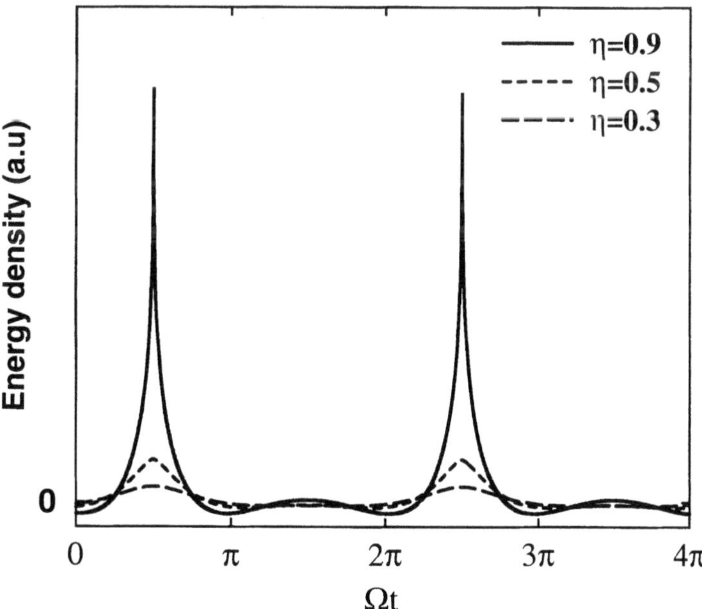

Fig. 3. Energy density emitted by the cavity as a function of time for different effective phase velocities η. With increasing values of η the energy starts to concentrate in pulses emitted periodically by the cavity

It is also interesting to notice that this system can exhibit self-sustained oscillations. Indeed, we find an oscillation threshold when the effective phase velocity equals unity. This can be seen in close analogy with a two-photon laser. Loosely speaking one might say that at each roundtrip the field acquires a "gain" characterized by v/c while it experiences a loss characterized by $1/\mathcal{F}$. If the gain equals the loss, self-sustained oscillations set in. In this sense, the device working above threshold might be considered as a sort of "mechanical laser", because the energy of the emitted photons does not come from an amplifying medium, but from the mechanical motion of the mirrors. Furthermore, the laser emission would be intrinsically mode-locked, resulting in emission of pulses, because of the phase relation existing between the different harmonics. However, if one wants to go beyond these qualitative statements, additional calculations are needed as our present calculations are restricted to the regime below threshold.

Figure 4 shows the radiation spectrum for an effective phase velocity $\eta = 0.9$. The spectrum shown here is plotted for a cavity oscillating globally at a frequency of $\Omega = 5\pi c/L$. This means that the cavity performs five

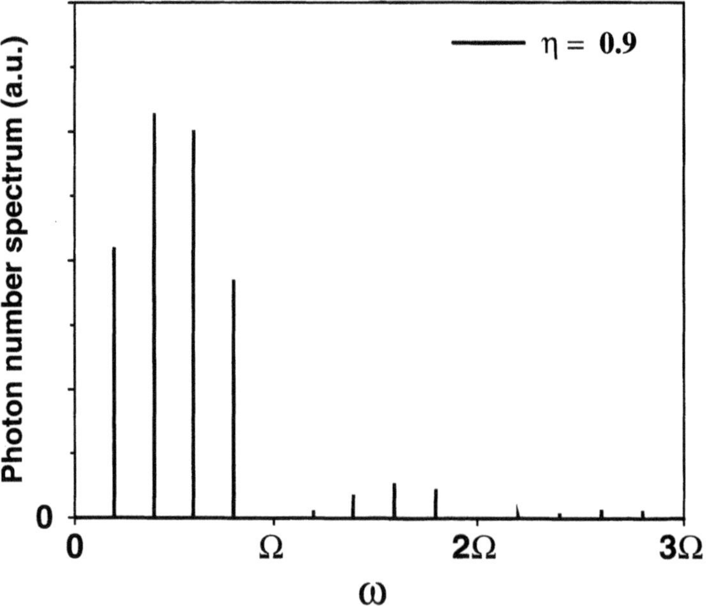

Fig. 4. Spectrum of the radiation emitted by the cavity for $\eta = 0.9$. The peaks correspond to cavity resonance frequencies. The spectrum is plotted for a cavity oscillating globally at a mechanical frequency $\Omega = 5\pi c/L$. Photons are created at frequencies higher than the mechanical oscillation frequency through frequency up-conversion in the opto-mechanical coupling between vacuum fluctuations and the mirror's motion. Furthermore the radiation spectrum vanishes for frequencies equal to a multiple integer of the mechanical excitation frequency

oscillations during one roundtrip of the field inside the cavity. Clearly photons can be created by higher-order harmonics of the motion as well as by the fundamental one as soon as the effective velocity becomes appreciable compared to the speed of light. As a striking consequence, photons are radiated at frequencies higher than the mechanical frequency Ω. A process of frequency up-conversion thus exists in the opto-mechanical coupling between vacuum fluctuations and the mechanical motion of the cavity. A second striking feature is that no photons are emitted at frequencies equal to multiple integers of the excitation frequency Ω.

The simple interpretation given in Sect. 2 of emission of motion-induced radiation in terms of a parametric process is therefore no longer valid in the nonlinear regime, where the field inside the cavity acquires a large phase shift due to numerous reflections on the two oscillating mirrors. The process of frequency up-conversion can be explained in the following way. While for a single reflection only spontaneous parametric processes are possible, the field inside the cavity undergoes a great number of roundtrips. At each reflection spontaneous parametric photons are emitted. However, before these photons leave the cavity through one of the mirrors, they can be re-absorbed in a higher-order parametric process as shown in Fig. 5 which produces a photon of a frequency larger than the mechanical oscillation frequency.

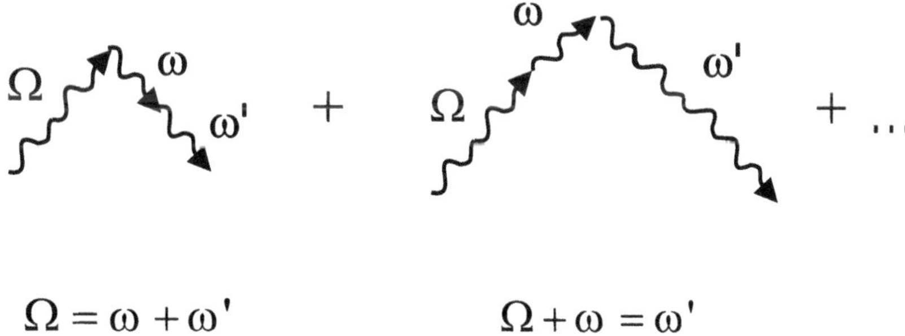

Fig. 5. Emission diagram for the nonlinear operation mode. Spontaneous parametric photons are emitted from a single reflection on the moving mirrors. Before these photons leave the cavity, they can be re-absorbed in a higher-order process giving rise to the emission of photons with frequencies higher than the mechanical oscillation frequency

5 Discussion

To evaluate the orders of magnitude of motion-induced radiation, let me first recall that if one wants to benefit from the resonant amplification of the radiation inside the cavity, the opto-mechanical resonance condition (2) has to be satisfied. The mechanical oscillation frequency and the photon frequency are then of the same order of magnitude. This restricts the frequency domain to the microwave region. It is possible to excite a mechanical oscillation at about 1 GHz. One could for example think of either a device using excitation via laser radiation pressure or piezo-electrical excitation.

In order to be sure that the observed radiation is due to the quantum vacuum and not to a thermal effect, the temperature of the device would have to be of the order of 10 mK [12]. In the microwave region, the finesse of a superconducting cavity can reach 10^9 [13]. A peak velocity $v \simeq 3$ cm/s, corresponding to an amplitude of about 10^{-11} m, would then be sufficient to obtain a radiated flux of 1 photons per second outside and a 10 photons per pulse inside the cavity. It is important to emphasize that this peak velocity is only a tiny fraction of the typical sound velocity in materials.

In this context we have also considered the phenomenon of sonoluminescence, where a collapsing air bubble in water produces intense light pulses. It turns out that motion-induced radiation emitted by the moving bubble is far too small to explain this phenomenon, because the air bubble in water is a very bad cavity with a quality factor of the order of 1 and cannot move with a speed exceedingly larger than the speed of sound [14].

Coming back to motion-induced radiation emitted by a high finesse cavity, the emitted photons may be detected outside the cavity by performing sensitive photon-counting detection of the radiated flux. Inside the cavity the state of the field could be probed with the help of Rydberg atoms [13]. However, to excite the motion, a huge force would have to be applied onto the mirrors which would create spurious signals against which one would have to distinguish experimentally the motion-induced photons. This is why the particular signatures of the effect are extremely important. The challenge of this experiment does indeed not come from one particular constraint, but from the fact that the conditions I just mentioned would have to be fulfilled simultaneously. Nevertheless, dissipative effects of vacuum fluctuations are a fundamental phenomenon related to important conceptual questions in physics and their observation would therefore be worth the effort.

References

1. For an introduction see for example R.J. Adler, B. Casey, O.C. Jacob: Am. J. Phys. **63**, 620 (1995) and references therein
2. S. Reynaud, A. Lambrecht, C. Genet, M.T. Jaekel: Contribution to a special issue Compte-Rendu des Académie des Sciences (2001); arXiv:quant-ph/0105053
3. W.E. Lamb, R.C. Retherford: Phys. Rev. **72**, 241 (1947)
4. G.W. Series: *The Spectrum of Atomic Hydrogen* (World Scientific, Singapore 1988) and references therein; F. Biraben, T.W. Hänsch, M. Fischer, M. Niering, R. Holzwarth, J. Reichert, Th. Udem, M. Weitz, B. de Beauvoir, C. Schwob, L. Jozefowski, L. Hilico, F. Nez, L. Julien, O. Acef, J.-J. Zondy, A. Clairon:

The Hydrogen Atom, ed. by S.G. Karshenboim, F.S. Pavone, G.F. Bassani, M. Inguscio, T.W. Hänsch, Lect. Notes Phys. (Springer, Berlin, Heidelberg 2001) and references therein
5. H.B.G. Casimir: Proc. K. Ned. Akad. Wet. **51**, 793 (1948)
6. G.T. Moore: J. Math. Phys. **11**, 2679 (1970)
7. S.A. Fulling, P.C.W. Davies: Proc. Soc. London A **348**, 393 (1976)
8. E. Giacobino, C. Fabre (eds.): Noise Reduction in Optical Systems, Spec. Issue, Appl. Phys. B **55**, 189 (1992)
9. A. Lambrecht, M.T. Jaekel, S. Reynaud: Phys. Rev. Lett. **77**, 615 (1996)
10. P.A. Maia Neto, S. Reynaud: Phys. Rev. A **47**, 1639 (1993)
11. A. Lambrecht, M.T. Jaekel, S. Reynaud: Eur. Phys. J. D **3**, 95 (1998)
12. A. Lambrecht, M.T. Jaekel, S. Reynaud: Europhys. Lett. **43**, 147 (1998)
13. G. Rempe, F. Schmidt-Kaler, H. Walther: Phys. Rev. Lett. **64**, 2783 (1990); M. Brune, F. Schmidt-Kaler, A. Maali, J. Dreyer, E. Hagley, J.-M. Raimond, S. Haroche: Phys. Rev. Lett. **76**, 1800 (1996)
14. A. Lambrecht, M.T. Jaekel, S. Reynaud: Phys. Rev. Lett. **78**, 2267 (1997)

Precision in Length

Gerd Leuchs

After a short prologue on "precision in time" the main part of the text addresses techniques in high precision length measurements, namely the use of non-classical light for the improvement of the sensitivity of interferometers. There now exist a number of schemes in quantum interferometry, which will be discussed from a common perspective.

1 Prologue

It is a great pleasure for me to write this article in honour of Theodor Hänsch, for whom I feel great appreciation. Long before I first met him, his name was familiar. When working as a student helper in Herbert Walther's laboratory in 1973 and 1974, then at Cologne, it was pointed out to me that the laboratory's nitrogen laser-pumped dye laser was built according to a design by Hänsch. Another project in a neighbouring laboratory had the goal of detecting the deflection of a sodium atomic beam by laser light. The director of my laboratory was Hartmut Figger. He informed me that Hänsch and Schawlow had just proposed that one could not only deflect an atomic beam but also cool it in the transverse dimension using laser radiation. I remember that I had taken apart a vacuum forepump and in the middle of cleaning I was wondering what kind of impact that proposal could possibly have. Of course, I had no idea what it would lead to. Now I know. At about that time a new graduate student, Kostadinos Siomos, now a professor at Athens, arrived from Heidelberg, where he had received his Diploma in physics. In Heidelberg he had overlapped with Theodor Hänsch before his move to Stanford and was full of admiration. Over the years more and more scientific results and innovations were published by Theodor Hänsch, mostly related to precision metrology. Several German Universities offered him prestigious positions but he declined them and remained at Stanford. Then Herbert Walther proposed a bold move. It was his initiative that led to offering Theodor Hänsch a joint appointment with the University of Munich and the Max-Planck-Institute for Quantum Optics. We, that is the younger scientists in quantum optics at Garching, were excited about the possibility of attracting Theodor Hänsch to Munich. Would he accept? We all thought that there was a very good chance, since the offer would be hard to beat. But time went by, one negotiation round was followed by another one and finally the University of Munich gave Theodor Hänsch a deadline for deciding. The deadline was March 31, 1986, as I vividly remember. The date passed and Theodor Hänsch had not accepted. On April 1 we sat there a bit sad and disillusioned. The silence was broken by ringing of the phone. The caller announced that Theodor Hänsch had finally agreed to take the job. We did not believe it and pointed out that we were perfectly aware of the date, that we were not in the mood for jokes, and that no one would pull our leg that day. But the voice on the phone did not give up and insisted that the telegram (This was before the internet.) had been sent off in California just before midnight local time, still on March 31! It takes a true precision metrologist like Theodor Hänsch to time things so closely. The latest proof of his expertise in time and frequency metrology is his ingenious invention for directly measuring the frequency of light, with which he has made the whole field of frequency chains obsolete.

My own work on precision measurement has been in the field of interferometry and length measurement. In the scientific part following below I will concentrate on precision in length measurement. As will be seen, the ideas which have been developed for quantum interferometry tie into the current concepts for quantum information technology.

2 The Beam Splitter – the Heart of Each Interferometer

A beam splitter has two input and two output ports. In the single-mode picture each of these ports is associated with a mode of the quantized electromagnetic field. The corresponding field operators are $\hat{a}_{\text{in1}}$, $\hat{a}_{\text{in2}}$, $\hat{a}_{\text{out1}}$, and $\hat{a}_{\text{out2}}$. They are related by the transmission and reflection coefficients $\underline{r}_1$, $\underline{r}_2$, $\underline{t}_1$, and $\underline{t}_2$ (see Fig. 1a):

$$\begin{aligned}\hat{a}_{\text{out1}} &= \underline{t}_1\hat{a}_{\text{in1}} + \underline{r}_1\hat{a}_{\text{in2}} \\ \hat{a}_{\text{out2}} &= \underline{r}_2\hat{a}_{\text{in1}} + \underline{t}_2\hat{a}_{\text{in2}}\,.\end{aligned} \tag{1}$$

The underlining of the coefficients indicates that they are complex valued to account for both amplitude and phase. The general relation between the corresponding coefficients for the electric fields has been derived by G.G. Stokes in a remarkable paper [1] which he based on the principle of reversion. He considered one incoming ray being split by the beam splitter. Then he argued that if the two rays exiting the beam splitter are time reversed, they will have to interfere upon rearrival at the beam splitter such as to generate the time-reversed incoming beam. In full generality, Stokes allowed for the reflection and transmission coefficients r_1, r_2, t_1, and t_2 to be all different in both amplitude and phase. Stokes' argument leads to the following formulae when using the complex notation (which he did not use). In this notation time reversal corresponds to taking the complex conjugate of the wave amplitude. Time reversal means that all the light goes back to

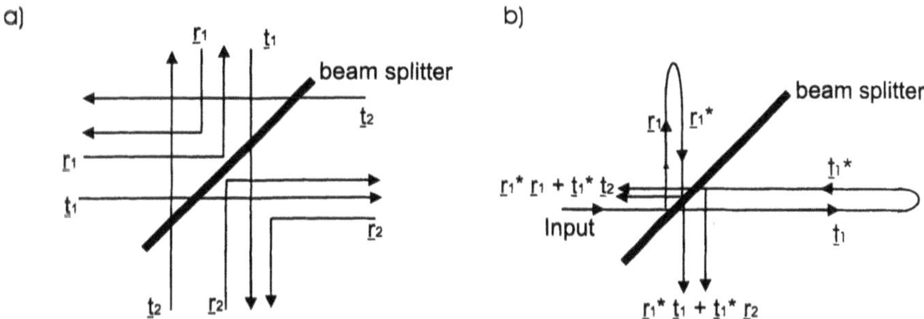

Fig. 1. Beam splitter showing the transmission and reflection coefficients in general (**a**) and the special scenario for time reversal (**b**)

input port 1 and no light out of input port 2. This leads to (see also Fig. 1b):

$$\begin{aligned} 1 &= \underline{r}_1^*\underline{r}_1 + \underline{t}_1^*\underline{t}_2 \\ 0 &= \underline{r}_1^*\underline{t}_1 + \underline{t}_1^*\underline{r}_2 \,. \end{aligned} \tag{2}$$

When writing $\underline{t}_k = t\,\exp(\mathrm{i}\tau_k)$ and $\underline{r}_k = r\,\exp(\mathrm{i}\rho_k)$, for $k = 1, 2$ one finds:

$$\begin{aligned} \tau_1 - \tau_2 &= 0, \text{ or an even multiple of } \pi \\ 2\tau_1 - \rho_1 - \rho_2 &= \pi, \text{ or an odd multiple of } \pi \,. \end{aligned} \tag{3}$$

One complication that may arise is that the medium on both sides of the beam splitter is not the same. Then the absolute squares of the transmission and reflection coefficients do not add up to one. The reason is that the velocity of light and the beam cross-section change and so does the Poynting vector [2]. If, however, the medium on both sides of the beam splitter is the same, then the beam cross-sections are likewise equal and the absolute squares of the transmission coefficient and the corresponding reflection coefficient do add up to one. We conclude that $t_1 = t_2$. In practice one often chooses the beam splitter to be either symmetric ($\tau_1 = \tau_2 = 0$ and $\rho_1 = \rho_2 = \pi/2$) or asymmetric with all coefficients being real ($\tau_1 = \tau_2 = 0$ and $\rho_1 = 0$, $\rho_2 = \pi$). In the following we will use the first version. For further references see Hamilton [2].

3 Pictorial Representation of the Field Variables Using Arrows

In quantum optics with intense light beams one may use the linearized approach [3] where the field operator of a mode is written as a classical mean field plus an operator describing the field uncertainty:

$$\hat{a} = \alpha + \delta\hat{a} \,,$$

with the expectation value of $\delta\hat{a}$ being zero. In a phase diagram where the imaginary part of the field is plotted versus the real part, $\delta\hat{a}$ leads to a region of uncertainty which is best described by the corresponding Wigner function [4]. Figure 2 shows a phase diagram where the contour of the Wigner function at half-maximum is indicated. For a field in a coherent state the uncertainties in amplitude and phase direction are the same and the contour is circular. The measured value of the real part of the field $a_{\cos} = (\hat{a}^{+} + \hat{a})/2$ and its imaginary part $a_{\sin} = \mathrm{i}(\hat{a}^{+} - \hat{a})/2$ are plotted on the axes in Fig. 2.

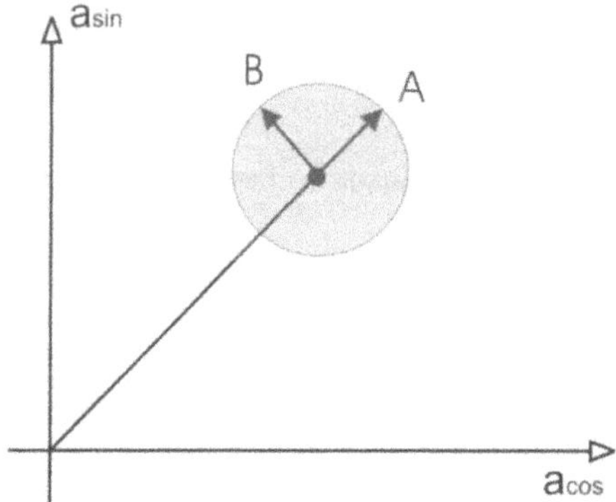

Fig. 2. Phase diagram representing a light field

Let us now introduce two orthogonal arrows which span the circular region of uncertainty of the field. In order to mimic the full uncertainty area one has to form a linear superposition of these two arrows with stochastic coefficients ranging between, say, -1 and $+1$ in a first approximation. For the coherent state e.g. there will be no correlation between the two stochastic variables. A better simulation would be to use stochastic coefficients obeying a probability distribution derived from the Wigner function of the state by taking the marginal distributions corresponding to the two arrow orientations. In the following we will keep track of the arrows only. The action of a beam splitter will be to transfer each arrow from an input port to both output ports with reduced amplitudes. In the model we have to properly take into account the Stokes relations. If the same stochastically varying input arrow contributes to two output ports, one may expect correlations between these two output fields and we will go through some examples. The geometrical visualization of the field variables can be helpful in describing an interferometer [5] and in understanding the generation of intense quantum entangled light beams [6]. For a general introduction see Leuchs [7].

To get used to the arrow description we consider the interference of two coherent states at a beam splitter. Coherent states have a circular region of uncertainty in phase space and as a result all four arrows describing the two coherent states have the same size. The situation is shown graphically

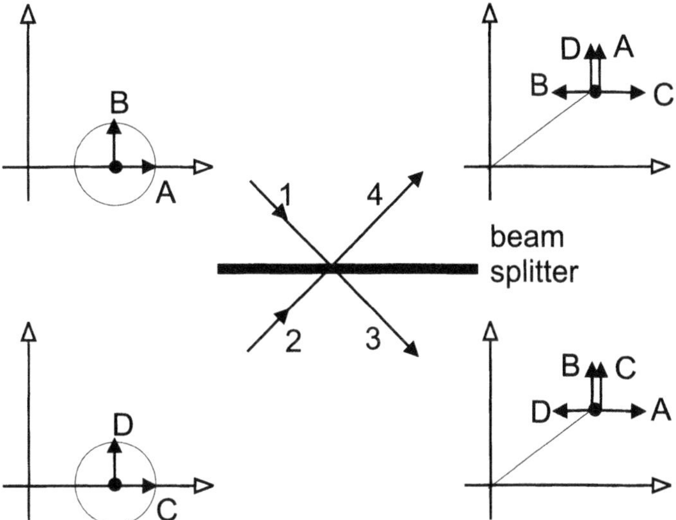

Fig. 3. Beam splitter with two coherent input states in the pictorial presentation

in Fig. 3. Note that the angles in phase space correspond to the classical optical phase of the light beams. The Stokes relations are obeyed by associating a 90° phase shift to each reflection, i.e. the factor "i", and 0° phase shift to each transmission. In practice this corresponds to a symmetric beam splitter such as a lamella or a beam-splitting cube. The amplitude reduction is not shown, for simplicity. The four input arrows A, B, C, and D determine the field uncertainties in the two output ports 3 and 4. The amplitude uncertainty in output 3 is determined by the projections of all arrows onto the amplitude direction: $A + B + C - D$. Each arrow would then still have to be multiplied with its individual stochastic coefficient. This and the additional amplitude reduction due to the projection is again not shown, as it will be the same factor for all arrows in the cases considered here. Likewise, the amplitude uncertainty at output 4 is determined by $A - B + C + D$.

Although the uncertainties in both output ports are governed by the same four arrows, they are not correlated. The reason for this lack of correlation can be traced back to the sum of two statistically independent stochastic variables and their difference being again statistically independent. Output 3 is governed by the sum of two statistically independent variables $(A + C) + (B - D)$ and output 4 is determined by the difference $(A + C) - (B - D)$. If, however, the two input states were amplitude-squeezed states with close to zero amplitude uncertainty $A, C, \approx 0$, and correspondingly larger uncertainty in the phase direction, the amplitude uncertainties of the two output ports would be anticorrelated "$(B - D)$" at port 3 and "$-(B - D)$" at port 4. The uncertainties in the phase directions are likewise correlated, as can be seen by going through similar arguments for the orthogonal projection. Actually, one squeezed input field is enough to obtain quantum correlations between the two output fields but the correlations will be less strong [8].

4 Interferometer Operating with a Coherent State and Squeezed Vacuum

The next step is to apply the pictorial approach to an interferometer. One scenario is that a coherent state enters through one input port and the other input mode is not excited and, therefore, is in a vacuum state (Fig. 4). After the first beam splitter the four arrows in each beam point in the same directions as in Fig. 3 . A difference being the missing mean amplitude at input 2, which has the consequence that inside the interferometer the arrows are now parallel or perpendicular to the mean field vector and not at $\pm 45°$ as in Fig. 3. Before going to the second beam splitter we now introduce a $-\pi/2$ phase shift (i.e. 90° clockwise). This is done, for example, by introducing a path-length difference between the two arms. It ensures that both output ports are at half-fringe height, i.e. that they are equally intense. At the outputs 3 and 4, again following the rules introduced above, we now have 8 arrows altogether. Two arrows marked with the same letter derive from one and the same stochastic input variable, so they can be added vectorially. As can be seen in Fig. 4, arrows A lead to correlated amplitude uncertainties in the two outputs 3 and 4, arrows B to correlated phase, arrows C to anti-correlated amplitude, and arrows D to anti-correlated phase uncertainties. Another way to say this is that the amplitudes in output ports 3 and 4 are given by $\mathrm{A}+\mathrm{C}$ and $\mathrm{A}-\mathrm{C}$, respectively. They are uncorrelated for the same argument as the one that was made when discussing Fig. 3. If, however, variable C is suppressed in input 2, i.e. if the field at input 2 is squeezed, then the amplitudes in the two output ports are correlated: both are proportional to A. When measuring the difference of the intensities in the two output ports, as is necessary if one wants to derive a feedback signal, one finds a quantum noise-suppressed signal and hence an improvement in the sensitivity for measuring arm length differences.

Without the $-\pi/2$ phase shift in one of the arms, output port 4 would be close to an interference minimum. Again, the pictorial argument can be made for the noise and again one finds a sensitivity improvement by quantum noise reduction. Again, C has to be suppressed at input port 2 at the expense of increasing D, which in turn does not affect the close to zero amplitude at port 4. The bright output beam at port 3 recovers the noise properties of input 1 and it can be recycled. The sensitivity improvement by using a squeezed vacuum was first proposed by Caves [9]. The effect of imperfections of the interferometer such as losses and non-unity fringe visibility were discussed by Gea-Banacloche et al. [10]. A similar scheme may be used to improve the sensitivity of Sagnac interferometers [11]. The compatibility of squeezing and recycling has been pointed out by Brillet et al. [12]. V. Chikarmane et al. [13]

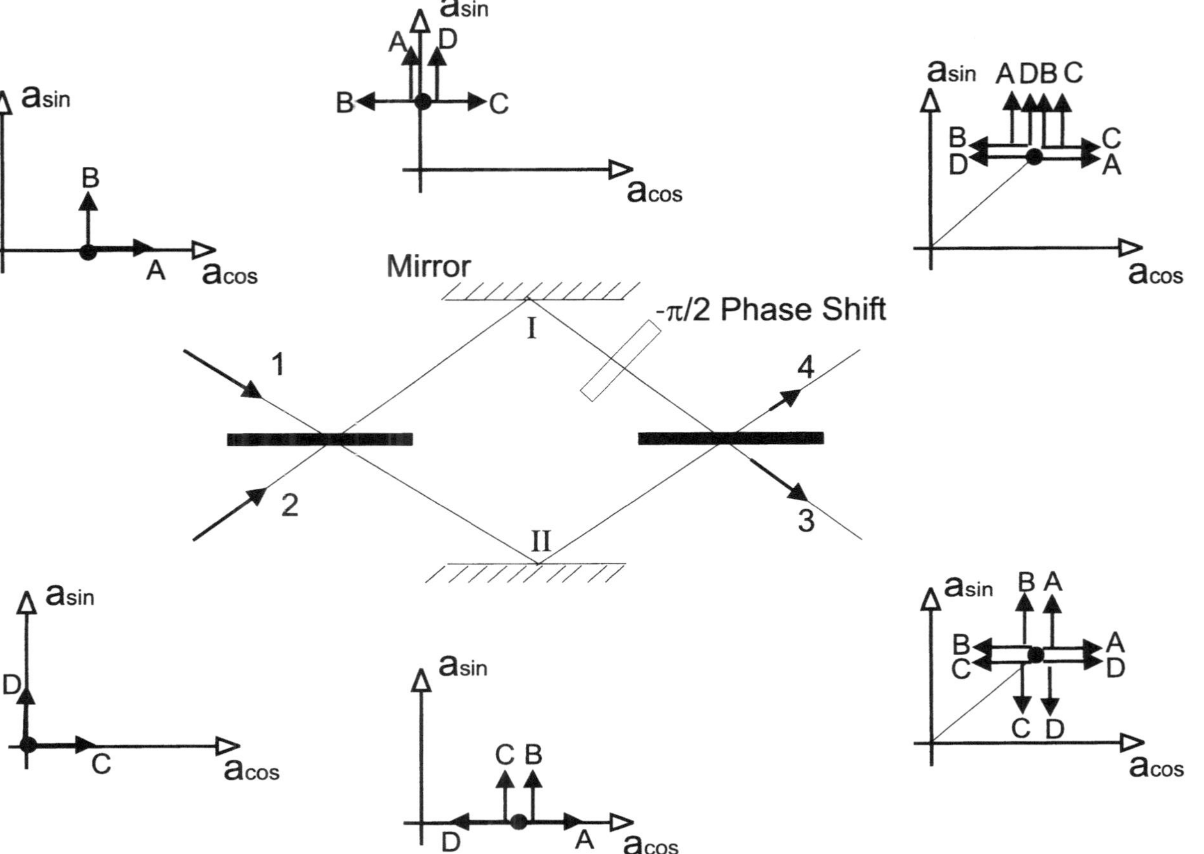

Fig. 4. An interferometer with a coherent and a vacuum input state

studied the effect of a squeezed-vacuum input port in the context of dual recycling.

When increasing the power at input 1 to further improve the interferometric sensitivity, one will eventually come into the regime where the intensity fluctuations in the two arms will wiggle the mirror appreciably through the light pressure force [14, 15]. It is important to see that only anti-correlated light pressures on the two mirrors will affect the output signal, because only this leads to changes of the arm length difference. The amplitude uncertainty in arm I (see Fig. 4) is proportional to $A + D$ and the one in arm II is proportional to $A - D$. In this regime an increase in arrow D due to squeezing of C seems counter-productive, because it will lead to larger light pressure-induced arm length difference fluctuations, which in turn introduce signal fluctuations at the output. There is an optimum called the Heisenberg limit. To reach this limit the input power at port 1 has to be chosen such that the light pressure-induced (arrow D) and the photon noise-induced signal noise (arrow C) are of the same magnitude [16]. When using squeezed light to reduce the light pressure or the photon counting uncertainty this will increase, respectively, the photon counting or the light pressure-induced uncertainty and there will be no gain in sensitivity if the optimum value for the light power is chosen. The effect of this type of squeezing is twofold: the Heisenberg limit is reached at a different light power and away from the optimum light power squeezing may increase the non-optimum sensitivity [31]. The name Heisenberg limit or standard quantum limit of an interferometer is used in this situation because this sensitivity limit may also be derived from the Heisenberg uncertainty relation applied to the position and momentum of the mirror masses [17]. In the derivation one assumes that the position and the momentum uncertainty are not correlated and that, therefore, the signal variance is the sum of the variances of the photon counting and the light pressure noise. This limitation is, however, not at all fundamental, as pointed out by Unruh [18]. Note that this statement also holds for standard quantum limits in general [19]. For the interferometer this point was studied later in more detail by Jaekel [20], Luis et al. [21], and Pace et al. [22]. For a treatment where the coherent input is replaced by a non-classical input see Shirasaki et al. [23]. The Heisenberg limit may be overcome by using correlated arrows C and D. Correlated arrows means that they can be added vectorially. This leads to one common arrow under 45° describing the field at input 2 (Fig. 4). A squeezed field at input port 2 will thus reduce the interferometric sensitivity below the Heisenberg limit provided that the squeezing ellipse is oriented under 45° with respect to the mean field at input 1. This holds in the case where the variances of the two types of noise, light pressure noise, and photon counting noise are about equal. Note that an orientation under $-45°$ would have the opposite effect and lead to a deterioration of the sensitivity. Using the pictorial approach it is also evident that this method of reducing the interferometric sensitivity below the Heisenberg limit using squeezed states is likewise compatible with recycling. The US-gravitational wave detection project expects to use these sort of techniques [30].

5 Interferometer Operating with Intense Amplitude-Squeezed Beams

In our next and final example we consider an interferometer operated with two intense beams at each input port (Fig. 5). Now the light pressure in arm I is proportional to A − B + C + D and in arm II is proportional to A + B + C − D. Hence the light pressure-induced arm length differences are proportional to the difference D − B. Here we want to focus on the situation where output port 3 is close to an interference minimum. Working at the minimum requires, of course, a modulation technique. To reach the minimum we introduce a phase shift of $-\pi/2 + \varphi$ in arm I, with $\varphi \ll 1$. The anti-correlated light pressure causes a positive phase shift φ. We follow the procedure described above to paint the arrows into the phase diagrams. With a closer look at output port 3 we find that the amplitude uncertainty of this close-to-dark beam is proportional to A − C.

A first conclusion is that if $A \approx C \approx 0$ then the interferometric sensitivity will be improved. It is interesting to note that amplitude-squeezed beams at the input (i.e. $A \approx C \approx 0$) will actually allow for sub-shot noise operation in a more general case for any value of the interference phase i.e. for any arm length difference [24] in contrast to the special case shown in Fig. 3. The light beams inside the interferometer will be entangled, and so the amplitudes in arms I and II, for example, are anti-correlated. On the other hand, squeezing the amplitudes A and C blows up the arrows B and D. This increases the light pressure-induced noise. At the input intensity where both types of noise, light pressure and photon counting, are balanced one may follow the same procedure as above and tilt the squeezing ellipse by 45° in the right direction. Here one wants to achieve A − C = −(D − B) so that both noises cancel as much as possible. Consequently the ellipses have to be tilted such that A = B and C = D. The phase diagram in Fig. 6 shows the required tilt of the squeezing ellipse. Such a tilted ellipse is generated if an initially coherent field propagates through a Kerr medium. Instead of sending squeezed beams to the input ports one may place Kerr media at the input ports to generate the squeezing. Bondurant and Shapiro [25] showed that a Kerr medium inside the interferometer would likewise lead to an improvement of interferometric sensitivity, owing to the similarity of the Hamilton operators describing the Kerr effect and the radiation pressure interaction. Vyatchanin et al. [26] carried this further and showed that the radiation pressure alone can be exploited for a sensitivity improvement provided that the right quadrature is measured at the interferometer output. Another proposal for quantum interferometry was made by Holland and Burnett [27]. They, however, did not consider light pressure effects and suggested operating the interferometer with correlated

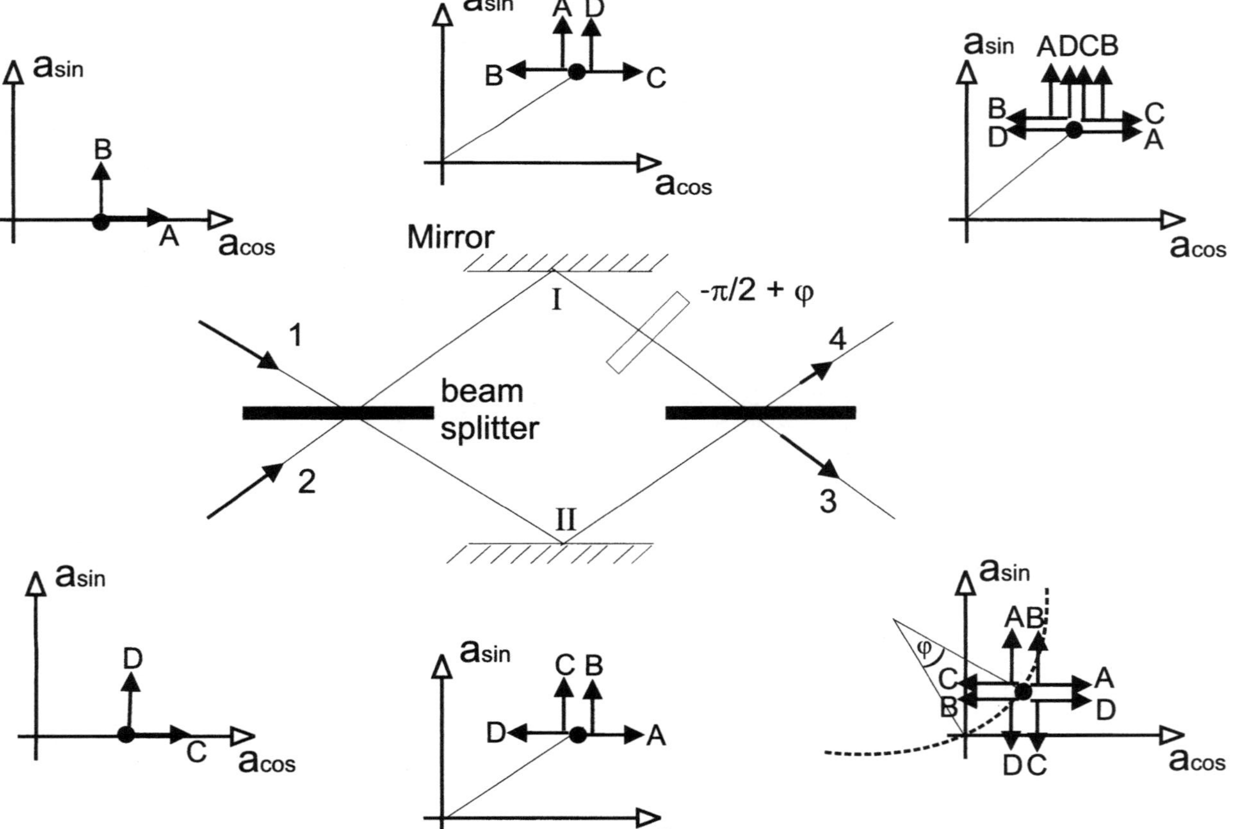

Fig. 5. An interferometer with two intense input beams with which the effect of two intense amplitude squeezed input beams can be demonstrated

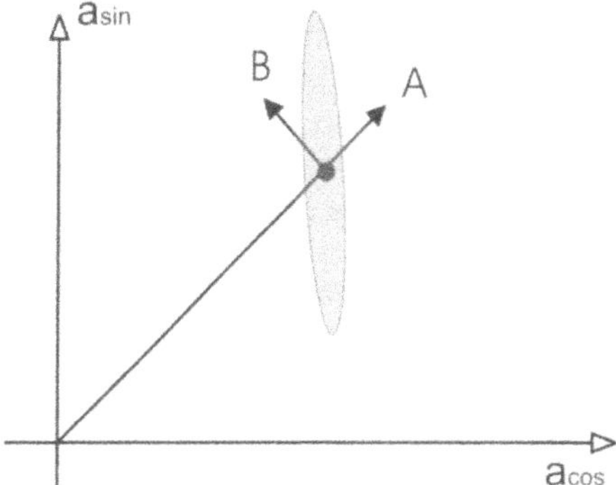

Fig. 6. Phase diagram showing correlated uncertainties

Fock states. Experiments were performed by Trifonov et al. [28]. They also refer to a Heisenberg limit. This is, however, different from the one mentioned above and the two should not be confused. A scheme at the many photon level using a non-degenerate parametric amplifier for the generation of intense, entangled beams was demonstrated by Zhang et al. [32]. It is related to the scheme shown in Fig. 5.

Quantum interferometry provides a rich spectrum of possible experiments. The interferometric sensitivity can be enhanced beyond limits, which were once believed to be fundamental. As shown, a wide variety of non-classical light beams at the interferometer input ports will lead to sensitivity-enhanced quantum interferometry. It was found that squeezed beams can be turned into entangled beams. In general, quantum interferometry provides basic building blocks necessary for quantum information processing. Recently it has been proposed that such quantum entangled light beams may be used to improve the synchronization of clocks [29], a subject which is still under discussion. This closes the loop back to precision in time.

Acknowledgements

I greatfully appreciate most valuable discussions with Hans A. Bachor, Julio Gea-Banacloche, Roy J. Glauber, Oliver Glöckl, Murray W. Hamilton, James Hough, Natalia Korolkova, Ping-Koy Lam, Rodney Loudon, Christoph Marquardt, Timothy C. Ralph, Marlan. O. Scully, and Christine Silberhorn. I thank Sebastian Schrenk for his help with the drawings.

References

1. G.G. Stokes, Cambridge and Dublin Math. J. IV, 1 (1849)
2. M.W. Hamilton, Am. J. Phys. **68**, 186 (2000)

3. S. Reynaud, A. Heidmann, E. Giacobino, C. Fabre, Progress in Optics XXX, ed. by E. Wolf (Elsevier, Amsterdam 1992) p. 1
4. W. Schleich, *Quantum Optics in Phase Space* (Wiley-VCH, Heidelberg, New-York 2000)
5. W. Winkler, G. Wagner, G. Leuchs, 'Interferometric detection of gravitational radiation and non-classical light', in *Fundamentals of Quantum Optics II*, ed. by F. Ehlotzky, (Springer, Berlin, Heidelberg 1987) pp. 92–108
6. G. Leuchs, T.C. Ralph, C. Silberhorn, N. Korolkova, J. Mod. Op. **46**, 1927 (1999)
7. G. Leuchs, 'Photon statistics, anti-bunching, and squeezed states', in *Frontiers of non-equilibrium statistical physics*, ed. by G.T. Moore, M.O. Scully (Plenum Press, New York, London 1986) pp. 329–360, Contemp. Phys. **29**, 299 (1988)
8. U. Leonhard, 'Measuring the quantum state of light' (Cambridge University Press, Cambridge, New York 1997)
9. C.M. Caves, Phys. Rev. D **23**, 1693 (1981)
10. J. Gea-Banacloche, G. Leuchs, J. Mod. Optics **34**, 793 (1987); J. Opt. Soc. Am. B **4**, 1667 (1987); J. Mod. Optics **36**, 1277 (1989)
11. G. Leuchs, M.W. Hamilton, 'Possible Applications of squeezed states in interferometric tests of general relativity', in *Frontiers in Quantum Optics*, ed. by E.R. Pike, S. Sakar (Hilger, Bristol 1986) pp. 106–119
12. A. Brillet, J. Gea-Banacloche, G. Leuchs, C.N. Man, J.Y. Vinet, 'Advanced techniques: recycling and squeezing,' in *Detection of Gravitational Waves*, ed. by D. Blair (Cambridge University Press, Cambridge, New York 1991) pp. 369–405
13. V. Chickarmane, S.V. Dhurandhar, Phys. Rev. A **54**, 786 (1996); V. Chickarmane, S.V. Dhurandhar, T.C. Ralph, M. Gray, H-A. Bachor, D.E. McClelland, Phys. Rev. A **57**, 3898 (1998)
14. P. Meystre, E.M. Wright, J.D. McCullen, E. Vignes, J. Opt. Soc. Am. B **2**, 1830 (1985); A. Dorsel, J.D. McCullen, P. Meystre, E. Vignes, H. Walther, Phys. Rev. Lett. **51**, 1550 (1983)
15. P.F. Cohadon, A. Heidmann, M. Pinard, Phys. Rev. Lett. **83**, 3174 (1999)
16. C.M. Caves, Phys. Rev. Lett. **45**, 75 (1980); R. Loudon, Phys. Rev. Lett. **47**, 815 (1981)
17. W.A. Edelstein, J. Hough, J.R. Pugh, W. Martin, J. Phys. E: Scientific Instruments **11**, 710 (1978); C.M. Caves, K.S. Thorne, R.W.P. Drever, V.D. Sandberg, M. Zimmerman, Rev. Mod. Phys. **52**, 341 (1980)
18. W.G. Unruh, 'Quantum Noise in the Interferometer Detector', in *Quantum Optics, Experimental Gravitation, and Measurement Theory*, ed. by P. Meystre, M.O. Scully, (Plenum Press, New York, London 1983) pp. 647–660
19. H.P. Yuen, Phys. Rev. Lett. **51**, 719 (1983)
20. M.T. Jaeckel, S. Reynaud, Europhys. Lett. **13**, 301 (1990)
21. A. Luis, L.L. Sánchez-Soto, Phys. Rev. A **45**, 8228 (1992); A. Luis, L.L. Sánchez-Soto, Opt. Commun. **89**, 140 (1992)
22. A.F. Pace, M.J. Collett, D.F. Walls, Phys. Rev. A **47**, 3173 (1993)
23. M. Shirasaki, I. Lyubomirsky, H.A. Haus, J. Opt. Soc. Am. B **11**, 857 (1994)
24. Ch. Silberhorn, P.K. Lam, O. Weiß, F. König, N. Korolkova, G. Leuchs, Phys. Rev. Lett. **86**, 4267 (2001)
25. R.S. Bondurant, Phys. Rev. A **34**, 3927 (1986)
26. S.P. Vyatchanin, E.A. Zubova, Opt. Commun. **111**, 303 (1994); Opt. Commun. **109**, 492 (1994)

27. M.J. Holland, K. Burnett, Phys. Rev. Lett. **71**, 1355 (1993); T. Kim, O. Pfister, M.J. Holland, J. Noh, J.L. Hall, Phys. Rev. A **57**, 4004 (1998) and references therein
28. A. Trifonov, T. Tsegaye, G. Björk, J. Söderholm, E. Goobar, M. Atatüre, A.V. Sergienko, J. Opt. B **2**, 105 (2000); A. Trifonov, G. Björk, J. Söderholm, Phys. Rev. Lett. **86**, 4423 (2001)
29. R. Josza, D.S. Abrams, J.P. Dowling, C.P. Williams, Phys. Rev. Lett. **85**, 2010 (2001); I.L. Chuang, Phys. Rev. Lett. **85**, 2006 (2001); V. Giovanetti, S. Lloyd, L. Maccone, F.N.C. Wong, Phys. Rev. Lett. **87**, 117902 (2001)
30. H.J. Kimble, Y. Levin, A.B. Matsko, K.S. Thorne, S.P. Vyatchinin, arXiv: gr-qc/000826
31. M. Xiao, L.-A. Wu, H.J. Kimble, Phys. Rev. Lett. **59**, 278 (1987); P. Grangier, R.E. Slusher, B. Yurke, A. LaPorta, Phys. Rev. Lett. **59**, 2153 (1987)
32. Y. Zhang, H. Wang, X. Li, J. Jing, C. Xie, K. Peng, Phys. Rev. A **62**, 023813 (2000)

Part IV

Cold Atoms and Ions

The photo shows the directors of the Max-Planck-Institut für Quantenoptik in 1998 in the foyer of the MPQ in Garching. *From left to right:* Profs. K. Kompa, H. Walther, and T.W. Hänsch

Probing an Optical Field with Atomic Resolution

Günter R. Guthöhrlein, Matthias Keller, Wulfhard Lange, and Herbert Walther

This paper describes the investigation of an optical field with atomic resolution. The probe is a single trapped ion in a linear Paul trap. The trap allows the longitudinal direction to be scanned by displacing the ion along the trap axis. In the two transversal directions the ion is confined by the trap potential and therefore the optical field is scanned by displacement of the sample using a piezo drive. The method is demonstrated by measuring particular modes of an optical cavity.

1 Prologue

The main part of Theodor Hänsch's scientific work is concerned with high-resolution spectroscopy and the steady improvement of related techniques, including the measurement of optical frequencies. An impressive example of his outstanding work is his study of the hydrogen spectrum and the enormous improvement of the resolution achieved over the years. However, being the type of person who is always interested in new experimental gadgets and tricks, he has also examined other areas of physics. One of these is the technique of near-field microscopy, where he has achieved quite an improvement in optical resolution using the modified total reflection of plasmons. We therefore believe that our present paper may be of interest to him since it deals with another improvement of near-field techniques using a laser-cooled ion as a probe. His interest may be further enhanced by the fact that the cooling technique used for the ion also results from his earlier work. The work presented here evolved as a sideline of our work on cavity quantum electrodynamics of trapped ions.

2 Introduction

The resolution achieved in near-field microscopy is usually determined by the size of the near-field probe. While initial experiments used optical fibres with sharply pointed tips, better resolution was obtained by employing single molecules [1]. The intensity distribution of the optical near field is mapped by detecting the fluorescent light emitted by the molecule as a function of using its position. Using this technique, a resolution below 100 nm is achieved [2–5]. An inherent drawback of the method is the need for a host-crystal matrix or a substrate to accommodate and immobilise the probe molecule. Therefore this type of measurement causes perturbations of the optical field.

With the technique presented here, the perturbations through the host crystal are avoided by using a single ion in a radio-frequency trap as a near-field probe. In many ways, a trapped ion is the perfect nanoscopic detector of the radiation field. The absence of a medium from the interaction region leaves the optical field unaffected by the probe. The resolution may be adjusted by varying the strength of the confining radio-frequency field, and is then limited only by the size of the wave function of the ion in its motional ground state. In addition to the high spatial resolution, the spectral resolution of a single ion may be an additional advantage, because some of the narrowest spectral lines investigated so far were measured with trapped ions. (For details of the setup see also [6]).

3 Experiment

For the experiments about to be described we used a ^{40}Ca ion, which has a resonance line $4^2S_{1/2} - 4^2P_{1/2}$ at a wavelength of $\lambda = 397$ nm. The fluorescent light emitted by the ion is collected with a lens (numerical aperture = 0.17) and detected with a photomultiplier tube (overall detection efficiency $\eta \approx 10^{-4}$). The observed fluorescence rate R is proportional to the local intensity of the optical field at the position $\boldsymbol{r}$ of the ion, i.e. $R \propto I(\boldsymbol{r})$, provided there is no saturation of the atomic transition. By scanning the position of the ion in the field and detecting the fluorescence rate at each point, a high-resolution map of the optical intensity distribution is obtained. It should be noted that a single ion can also probe the amplitude distribution $\boldsymbol{E}(\boldsymbol{r})$ of the light field and hence measure its phase. To this end, heterodyne detection of the fluorescent light must be used, with the exciting laser as a local oscillator [7, 8].

The optical field we used for this test experiment was the eigenmodes of a Fabry–Perot resonator formed by two mirrors (radius of curvature = 10 mm) at a distance of $L = 6\,\text{mm}$ (Fig. 1). The transverse mode pattern is described by Hermite–Gauss functions with a beam waist $w_0 \approx 24\,\mu\text{m}$, while in the direction of the cavity axis a standing wave builds up. In the experiment, a particular cavity mode is excited by a laser beam with a power of a few hundred nanowatts at 397 nm. The length of the cavity is actively stabilised to this mode.

An ion is loaded in the trap after electron-impact ionisation of calcium atoms. Since the electron beam and the calcium beam would degrade the optical mirrors and make stable trapping difficult, we use a linear trap and load it in a region spatially separated from the observation zone, as shown in Fig. 1. Subsequently, DC electrodes along the axis are employed to shuttle the

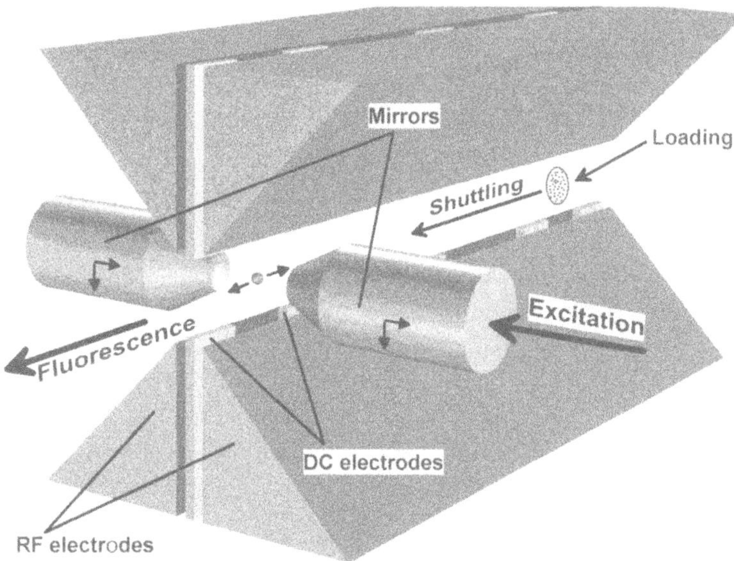

Fig. 1. Experimental arrangement of trap electrodes and cavity mirrors. The ion is loaded at the rear end of the trap and shuttled to the mirror region. Fluorescence is observed from the side of the cavity. For scans in the direction of the trap axis, the ion is moved with DC electrodes. In all other directions, the cavity is translated relative to the ion's position, as indicated by arrows

ion over a distance of 25 mm to the uncontaminated end of the trap, where the cavity is located, oriented at right angles to the trap axis. Residual DC fields in the radial direction must be carefully compensated using correctional DC voltages to place the ion precisely on the nodal line of the RF field (coinciding with the trap axis), to prevent the trapping field from exciting the micromotion of the ion.

In the direction of the trap axis, the ion is confined in a DC potential well which is approximately harmonic with an oscillation frequency of $\omega_z/2\pi \approx$ 300 kHz. By applying asymmetric voltages, the minimum of the potential well and hence the equilibrium position of the ion are moved along the trap axis. By simultaneously monitoring the fluorescence we sampled one-dimensional cross-sections of the cavity mode. The width of the ion's wave function in the axial potential well is a few hundred nanometres, which provides sufficient resolution to map the transverse mode pattern with an intensity distribution varying on a scale given by the cavity waist w_0.

Figure 2 shows scans of the first four TEM_{0n} modes of the cavity obtained in this way. The fluorescence data are not entirely symmetric because of a small displacement and rotation of the cavity eigenmodes with respect to the trap axis. In each plot, an inset indicates the path along which the ion is scanned. The solid curves in Fig. 2 are obtained from a fit using Hermite–

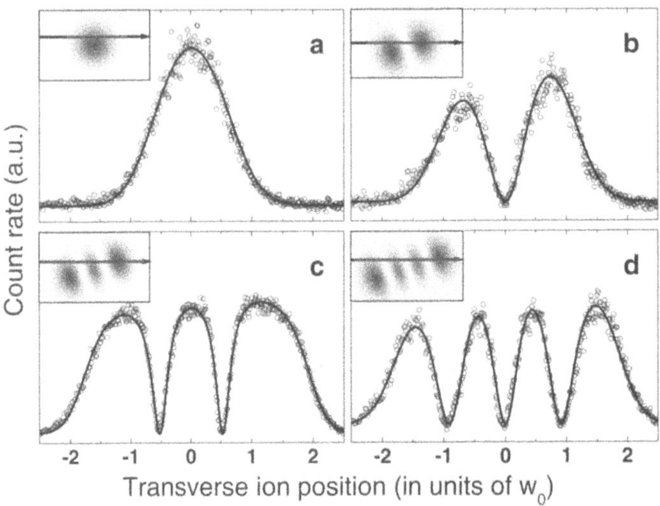

Fig. 2a–d. Transverse profiles of the Hermite–Gauss modes of the cavity, obtained by monitoring the ion's fluorescence while scanning over a range of 120 μm. The *solid line* is a fit including saturation of the transition. The *inset* shows the calculated intensity distribution of the mode and indicates the scan path. The modes are: (**a**) TEM_{00}; (**b**) TEM_{01}; (**c**) TEM_{02}; and (**d**) TEM_{03}

Gauss functions and take into account saturation of the ion's transition. The influence of saturation is apparent in Fig. 2c, where a slightly higher intensity was injected into the cavity. In all cases, the correspondence with the measured fluorescence is excellent. Higher-order Gauss–Laguerre modes were also studied. One of the results is shown in Fig. 3.

As mentioned above, the ion trap confines the ion's motion to the trap axis. It cannot be removed from there since this would lead to micromotion. To scan other dimensions of the field, the sample must therefore be moved. In our experiment this is done by piezoelectrically translating the entire cavity assembly perpendicularly to the trap axis.

RF confinement of the ion perpendicular to the trap axis is also harmonic, but the corresponding oscillation frequency $\omega_r/2\pi \approx 1.1$ MHz is larger than the axial frequency so that field structures in the radial direction of the trap are better resolved. The resolution achieved with our method may be determined most accurately by probing the standing-wave field created between the cavity mirrors, which varies on a scale of $\lambda/2$. To this end the cavity was moved parallel to its axis while keeping the ion stationary and monitoring its fluorescence. Figure 4 shows the mapping of the cavity field obtained in this way. A pronounced standing-wave pattern is observed with a visibility of 40%.

At present, the visibility in our experiment is given by the thermal wave packet spread of the calcium ion under conditions of Doppler cooling. In

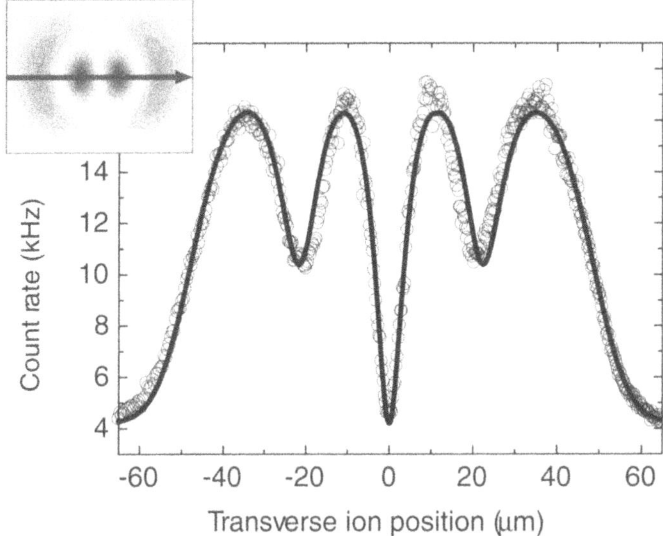

Fig. 3. Scan of the TEM_{11} Gauss–Laguerre mode. The theoretical fit to the fluorescence intensity includes saturation and the residual motion of the Doppler-cooled ion perpendicular to the scan direction

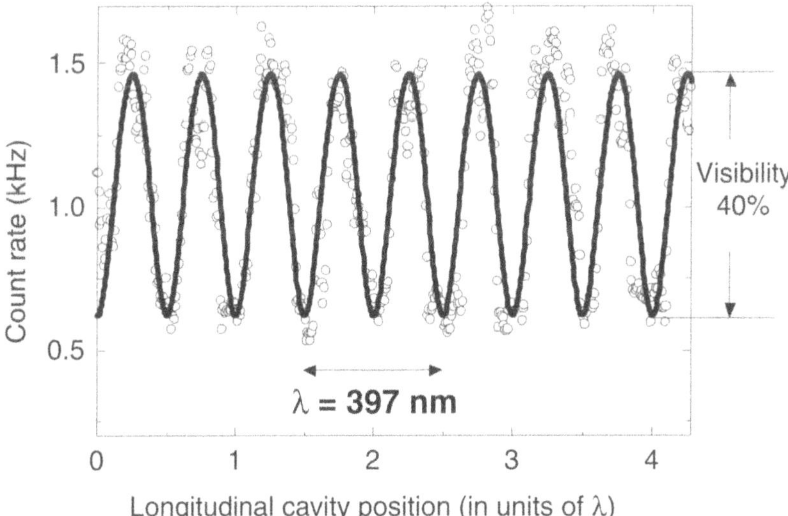

Fig. 4. Single-ion mapping of the longitudinal structure of the cavity field. The visibility is determined by the residual thermal motion of the Doppler-cooled ion. It corresponds to a resolution of 60 nm. The localisation of the ion's wave packet in this measurement is 16 nm

the direction of the cavity axis, cooling is provided by the cavity field itself, which, for this purpose, is red-detuned with respect to the ion's transition. The limiting temperature is then determined by the linewidth Γ of the atomic transition and is given by $\hbar\Gamma/2k_{\mathrm{B}} \approx 500\,\mu\mathrm{K}$. The transverse degrees of freedom of the ion are cooled with an additional laser beam injected from the side of the cavity, which is switched off during the measurement. Assuming thermal equilibrium, there is a direct connection between the spread, b, of the ion's wave function, which is identical to the resolution, and the visibility V:

$$V = \exp(-2\pi b/\lambda^2)\ . \tag{1}$$

The observed visibility corresponds to a wave packet 60 nm wide, indicating a temperature of the ion slightly above the Doppler temperature. The resolution of single-ion mode mapping in this case is almost one order of magnitude smaller than the wavelength of the transition.

Even higher resolution is within reach of standard ion-trapping technology. By using larger driving frequencies and higher voltages, the trap is made steeper, implying a wave packet size below 10 nm. Another factor of two is gained by cooling the ion to the ground state of motion, which has been achieved in several experiments [9–11]. Finally, the technique of motional squeezing [12] of the ion's wave function may be employed to create a "breathing" wave packet which periodically reaches a position uncertainty below 1 nm. In this case, stroboscopic detection must be used. A single ion as a probe of the optical field thus has the potential for atomic-scale resolution, and it does not disturb the field to be measured. See Fig. 5 for a survey of the improvements discussed.

An important issue, especially with regard to applications in cavity quantum electrodynamics, is the question of how well the ion is localised in the optical field, i.e. how accurately the centroid of the ion's wave function may be determined. With the spatial variation of the local fluorescence rate as the only parameter available for inferring a change in position of the ion, the largest position sensitivity in our setup is obtained at the maximum slope of the standing wave. At this point, the localisation Δx is completely determined by the signal-to-noise ratio SNR and the visibility V:

$$\Delta x \approx \frac{\lambda}{4\pi\ V\ \mathrm{SNR}}\ . \tag{2}$$

With a signal-to-noise ratio of 5:1, reached with an integration time of 200 ms at each point, a localisation of approximately 16 nm or $\lambda/25$ was measured. Even in the presence of noise, our measurement with a scanned single particle thus constitutes the most precise measurement of the spatial structure of an optical field to date [2–5]. Formally, Δx may be further reduced if the signal-to-noise ratio is improved by using a longer averaging time.

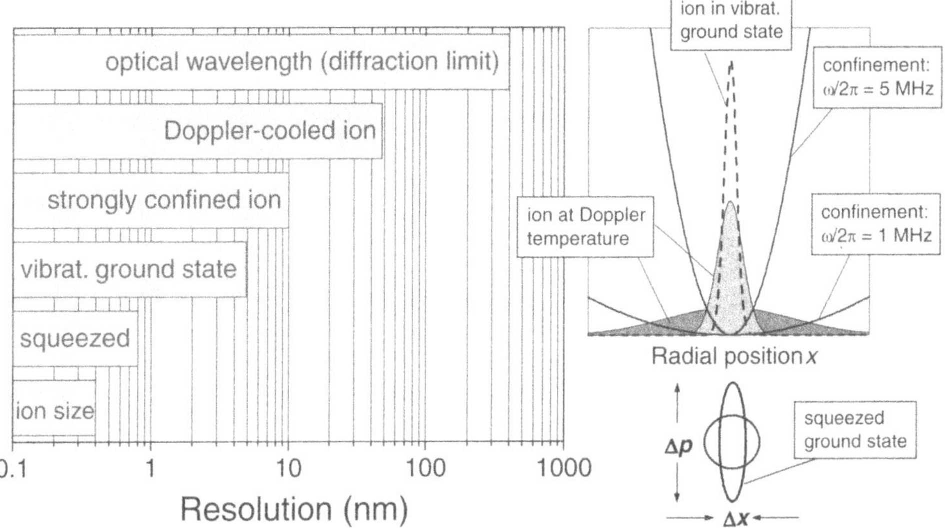

Fig. 5. Resolution of single-ion mapping achievable using different techniques (for details see text)

4 Conclusion

The setup described here has also other interesting applications, especially in cavity quantum electrodynamics. We are now implementing two experiments exploiting the localisation of an ion in a cavity. By pulsed excitation of a maximally coupled ion, single-photon wave packets may be emitted from the cavity on demand [13–15] (single-photon gun). Under conditions of strong coupling, a single calcium ion in the cavity provides sufficient gain to build up a laser field [16]. Like a single ion in free space, which was previously shown to be an excellent source of antibunched light [17, 8], radiation from a single-ion laser has nonclassical photon statistics and correlations.

An equally attractive goal in the area of cavity quantum electrodynamics is the simultaneous interaction of two or more ions with a single cavity mode. The linear geometry of our trap allows several ions to be stored within the mode volume. As a first test, we placed an array of two ions in the cavity field and observed the total fluorescence. We succeeded in matching the ion crystal to the two maxima of the TEM_{01} mode of the cavity. In such a configuration the cavity field may be used to entangle the two ions [18, 19]. This is a promising alternative to schemes involving the ions' motional degrees of freedom, since there is no need to cool the vibrational modes of the string below the Doppler temperature. Using a cavity to perform quantum operations on adjacent pairs of ions in a long string affords a viable route to a scalable quantum computer.

References

1. E. Betzig, R.J. Chichester, Science **262**, 1422 (1993)
2. J. Michaelis, C. Hettich, A. Zayats, B. Eiermann, J. Mlynek, V. Sandoghdar, Opt. Lett. **24**, 581 (1999)
3. B. Sick, B. Hecht, U.P. Wild, L. Novotny, J. Microsc. **202**, 365 (2001)
4. S. Gotzinger, S. Demmerer, O. Benson, V. Sandoghdar, J. Mircosc. **202**, 117 (2001)
5. J.A. Veerman, M.F. Garcia-Parajo, L. Kuipers, N.F. van Hulst, J. Microsc. **194**, 477 (1999)
6. G.R. Guthöhrlein, M. Keller, K. Hayasaka, W. Lange, H. Walther, Nature, in press
7. J.T. Höffges, H.W. Baldauf, W. Lange, H. Walther, J. Mod. Opt. **44**, 1999 (1997)
8. J.T. Höffges, H.W. Baldauf, T. Eichler, S.R. Helmfrid, H. Walther, Opt. Commun. **133**, 170 (1997)
9. C. Monroe et al., Phys. Rev. Lett. **75**, 4011 (1995)
10. E. Peik, J. Abel, T. Becker, J. von Zanthier, H. Walther, Phys. Rev. A **60**, 439 (1999)
11. C.F. Roos et al., Phys. Rev. Lett. **85**, 5547 (2000)
12. D.M. Meekhof, C. Monroe, B.E. King, W.M. Itano, D.J. Wineland, Phys. Rev. Lett. **76**, 1796 (1996)
13. A. Kuhn, M. Hennrich, T. Bondo, G. Rempe, Appl. Phys. B **69**, 373 (1999); P.W.H. Pinkse, T. Fischer, P. Maunz, G. Rempe, Nature (London) **404**, 365 (2000); J. Ye, D.W. Vernooy, H.J. Kimble, Phys. Rev. Lett. **83**, 4987 (2000); C.J. Hood, T.W. Lynn, A.C. Doherty, A.S. Parkins, H.J. Kimble, Science **287**, 1447 (2000)
14. C.K. Law, J.H. Eberly, Phys. Rev. Lett. **76**, 1055 (1996); C.K. Law, H.J. Kimble, J. Mod. Opt. **44**, 2067 (1997); P. Domokos, M. Brune, J.M. Raimond, S. Haroche, Eur. Phys. J. D **1**, 1 (1998)
15. M. Hennrich, T. Legero, A. Kuhn, G. Rempe, Phys. Rev. Lett. **85**, 4872 (2000)
16. G.M. Meyer, H.-J. Briegel, H. Walther, Europhys. Lett. **37**, 317 (1997)
17. F. Diedrich, H. Walther, Phys. Rev. Lett. **58**, 203 (1987)
18. T. Pellizzari, S.A. Gardiner, J.I. Cirac, P. Zoller, Phys. Rev. Lett. **75**, 3788 (1995)
19. S.B. Zheng, G.C. Guo, Phys. Rev. Lett. **85**, 2392 (2000)

From Spectral Relaxation to Quantified Decoherence

Christoph Balzer, Thilo Hannemann, Dirk Reiß, Werner Neuhauser, Peter E. Toschek, and Christof Wunderlich

Quantum information processing (QIP) requires thorough assessment of decoherence. Atoms or ions prepared for QIP often become addressed by radiation within schemes of alternating microwave-optical double resonance. A well-defined amount of decoherence may be applied to the system when spurious resonance light is admitted simultaneously with the driving radiation. This decoherence is quantified in terms of longitudinal and transverse relaxation. It may serve for calibrating observed decoherence as well as for testing error-correcting quantum codes.

The spectroscopic determination of atomic relaxation is among the earliest applications of laser spectroscopy [1]. Over the decades that have passed since then, the corresponding techniques have been dramatically refined, and the demand for reliable data has expanded. An important aspect is the availability of respective data derived from observations on individual trapped ions [2].

Laser-cooled ions prepared in an electrodynamic or electromagnetic trap and individually addressed are convenient building blocks for the storage and processing of quantum information [3]. In fact, they represent scalable systems, in contrast with certain competitive approaches, e.g. spin resonance spectroscopy [4]. The more important is the realization and full understanding of the radiative interaction of such a system: The implementation of the *coherent* dynamics of an individual atomic particle being radiatively driven is prerequisite for the demonstration of any quantum-logical gate, in particular of the basic Hadamard transformation [5]. The demonstration of the coherently controlled dynamics of trapped ions has been reported with the vibrational sidebands (or "one-phonon lines") and carrier line ("zero-phonon") of a laser-driven Raman transition between ground-state hyperfine levels [6], with a microwave-driven hyperfine carrier line [7], with the vibrational sidebands of a dipole-forbidden optical line [8], and with a corresponding optical carrier line [9]. Some of these schemes are troubled by imperfectly identified kinds of decoherence that would thwart the performance of a substantial number of successive steps of coherent interaction. This decoherence may range from the residual decay of the involved metastable states to spurious light scattering, and to parasitic phase fluctuations of the applied radiation.

For instance, the decoherence on the vibrational dynamics of ions localized in a Paul trap [5] has been suggested recently to result from parasitic spontaneous scattering of the two off-resonant light fields that serve as the pump and Stokes waves for Raman excitation of the hyperfine resonance [10]. Also, various types of reservoirs have been implemented that make the specific motional states decay in characteristic ways [11].

In a microwave-optical double-resonance experiment, we have found decoherence to appear during the coherent evolution of an individual trapped and cooled $^{171}Yb^{+}$ ion that was microwave-driven on the ground-state hyperfine transition. The evolution was monitored via the corresponding spin nutation. For this purpose, the ion was *alternatingly* driven and probed by highly monochromatic microwave radiation and by resonant scattering of laser light, respectively. The observed decoherence was brought about by a minute amount of residual laser light that impinged upon the ion while the microwave drive was applied. Blocking out this spurious light eliminated the decoherence. We shall show that well-defined quantities of longitudinal and/or transversal relaxation may be admitted to the ion by controlling the strength and polarization of the light and the level and direction of the ambient magnetic field. Such a "designed" decoherence seems useful for quantitative comparison with observed decoherence and as a testing ground for the sensibility of quantum algorithms to the action of decoherence, as well as for the function of error-correcting codes.

The double-resonance experiment was performed on a single $^{171}Yb^{+}$ ion in a 2 mm-sized electrodynamic trap. The $F = 0 \rightarrow 1$ hyperfine resonance of the ion's $S_{1/2}$ ground state was driven, for $\delta t = 100$ or $400\,\mu s$, by 12.6 GHz microwave radiation. Probing the $F = 1$ state immediately followed, when the ion was illuminated by a 5 ms pulse of 369 nm laser light. The excitation and observation of resonance scattering on the $S_{1/2}(F = 1) \rightarrow P_{1/2}(F = 0)$ line proves the preceding attempt of microwave excitation to have succeeded; the absence of scattering proves this attempt to have failed. A frequency-doubled Ti:sapphire laser generated 369 nm light of about 100 kHz bandwidth. This light was scattered on the ion's $S_{1/2}$–$P_{1/2}$ resonance line, such that the ion was cooled deep into the Lamb–Dicke regime. Occasional optical pumping of the ion into its metastable $D_{3/2}$ level was counteracted by using 935 nm light from a diode laser to repump the ion into the ground state. Trajectories of measurements have been recorded, each of which includes an initial preparatory light pulse that pumps the ion into state $F = 0$, a microwave driving pulse of length $\tau = N\delta t$, a probe-light pulse, and simultaneous recording of the presumptive fluorescence scattered off the ion (Fig. 1). Within a trajectory, N varied from 1, indicating the initial measurement, through 300, the final one. The recorded signals form quasi-random sequences of "on" and "off" results; however, the superposition of many accumulated trajectories yields the probability $P_1(\theta)$ for occupation of the $F = 1$ ground level by the ion, as a function of the driving time τ. The data of 50 trajectories have been superimposed, and

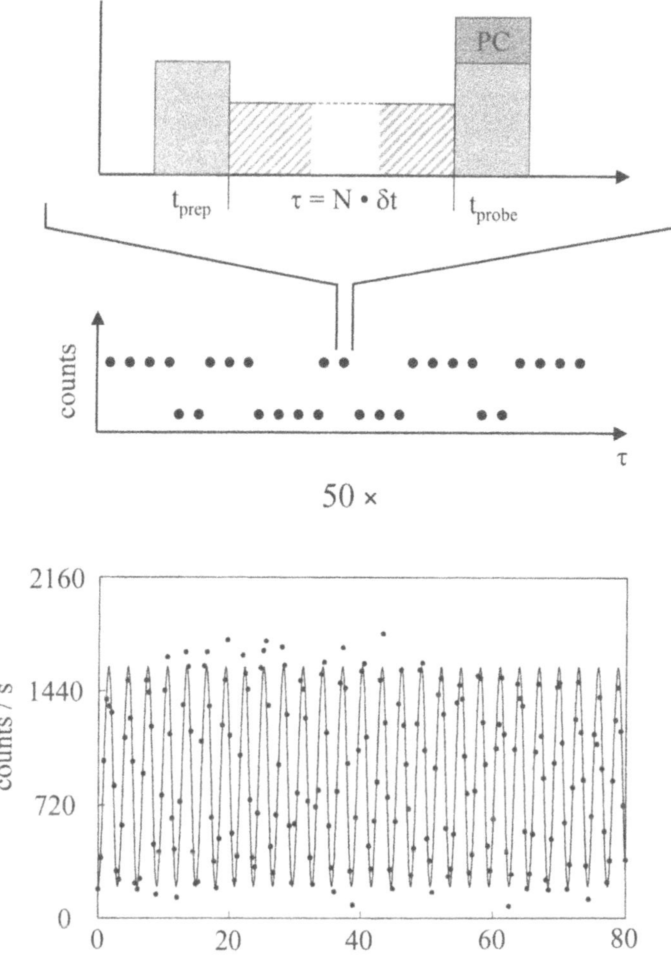

Fig. 1. An individual measurement includes a pulse of resonant light that prepares the ion in the $F = 0$ state of the ground-state hyperfine doublet, a microwave driving pulse that is N times a unit length δt, and a probe light pulse. The scattered light is detected by a photon counting (PC) system (*top*). With each subsequent measurement, N is incremented by a unit, up to $N_{\max}$, and yields a sequence of $N_{\max}$ data, "on" or "off" (*center*). Accumulation of 50 sequences reveals rf nutation on the ground-state hyperfine transition (*bottom*)

modulation of P_1 emerges that is ascribed to Rabi nutation of the ion driven by the microwave pulses of area $\theta = \Omega\tau$, whose duration τ was stepwise extended (Fig. 1, bottom).

The nutational oscillation shows high contrast. However, when a spurious level of probe light is admitted to the ion *simultaneously* with the driving pulse, the evolution of the probability of light scattering is dramatically modified (Fig. 2a): (i) the Rabi nutation becomes damped with its time constant being reduced upon increased intensity of the background light, and (ii) at long times τ, the probability $P_1(\theta)$ saturates to a level somewhere between 1/2 and unity, and not neccessarily to level 1/2.

In order to account for these observations, the ion, the light fields, and their interaction have been modelled by four-level Bloch equations (Fig. 2b). The corresponding level scheme is shown in Fig. 3. The Zeeman sublevels $m_F = +1$ and -1 have been combined and make up for the effective level 2 not affected by the driving microwave radiation on the hyperfine transition $0-1$, with $\Delta m_F = 0$.

The minute light intensity applied to the ion during the driving intervals is far below saturation of the resonance line, $I = \Omega_1^2/\gamma_1\Gamma_3 \ll 1$, where Ω_1 is the Rabi frequency of the ion generated by the laser light, Γ_3 and $\gamma_1 = \Gamma_3/2+\gamma_{1\text{ph}}$

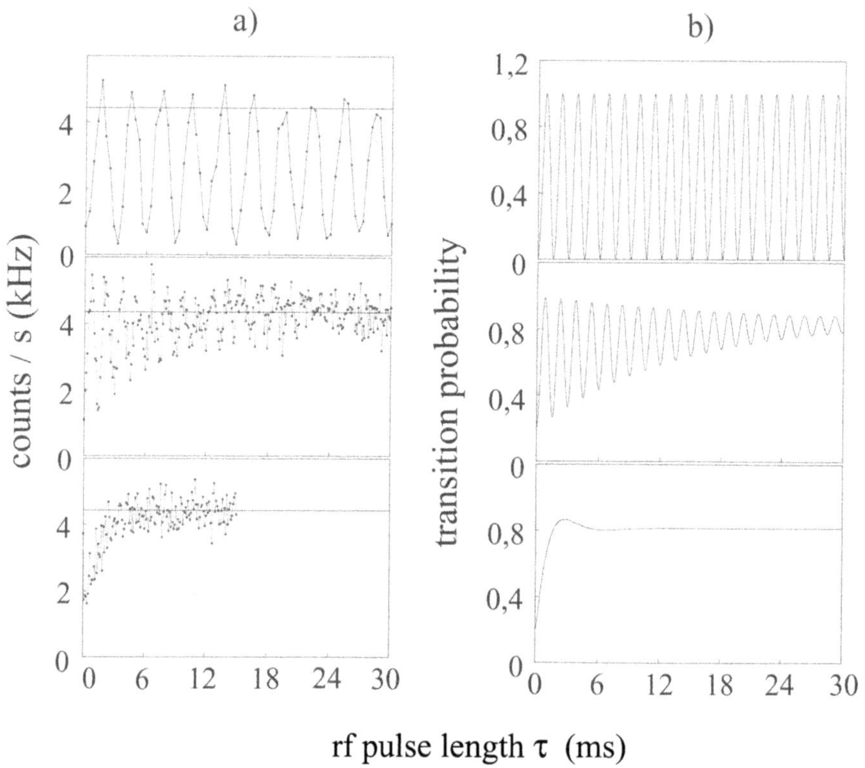

Fig. 2a,b. Scattered-light response vs. length of microwave driving pulse. (**a**) Observed, with level of light applied to the ion *during* the drive: no light (*top*), 2 nW (*center*), 20 nW (*bottom*). (**b**) Simulated: Rabi frequency $\Omega_1 = 0$ (*top*), 50 kHz (*center*), 500 kHz (*bottom*)

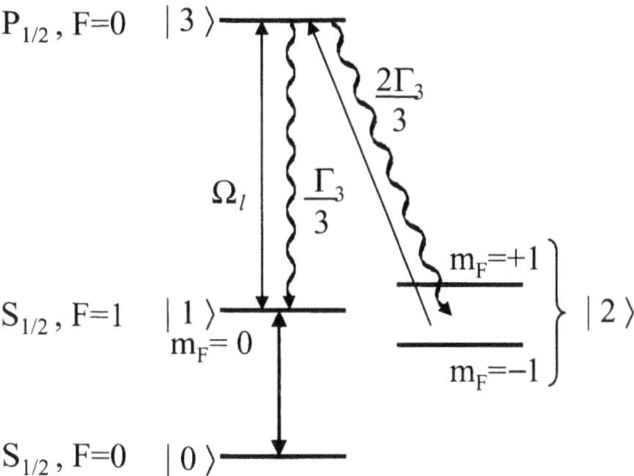

Fig. 3. Simplified level scheme of the $^{171}Yb^{+}$ ion used in the microwave-optical double resonance experiment

are the constants of energy relaxation of the resonance level ($P_{1/2}$) and of the phase relaxation of the laser-excited (electric) dipole, respectively. In the present experiment, the extra rate is small, $\gamma_{\mathrm{lph}} \ll \Gamma_3$, and it is neglected. In the above limit it is appropriate, for the interpretation of the ionic evolution, to restrict the modelling of the *optical* part of the dynamics – i.e. the part related to the three levels $1, 2, 3$ – to optical pumping and decay, in terms of rate equations. This rate evolution is coupled to the microwave-driven coherent evolution, on the two-level system $0-1$, via level 1 that represents state $S_{1/2}(F=1, m_F=0)$. Now, the latter coherent dynamics is affected by light-generated decoherence of two kinds: (i) a net loss of population, from the driven two-level system $0-1$, by optically pumping the ion into state 2, i.e. the $F=1$ Zeeman sublevels $m_F = \pm 1$, with subsequent repumping, and (ii) additional loss of phase coherence of the driven spin dynamics by Rayleigh scattering into the eigenstate 1, i.e. $F=1$, $m_F=0$.

The loss of some population from state 1 to state 2 makes the probability P_1 of finding the ion in the entire probed hyperfine state $F=1$ saturate *above* 1/2, since the Zeeman sublevels ± 1 retain part of the population from the coherent transfer to state $F=0$. This fractional population, however, may become re-excited by the linearly polarized laser light during the probing interval, since the width of the resonance line far exceeds the small Zeeman splitting of the $S_{1/2}(F=1)$ state. Complete pumping to level 2 ($m_F = \pm 1$) makes P_1 saturate at unity. The scattering rates are $\beta_{\mathrm{f}} n_{\mathrm{i}} r_{\mathrm{i}}$, where n_{i} is the population in the initial state, β_{f} is the branching ratio of the decay of state 3 into the final state ($\beta_1 = \frac{1}{3}$, $\beta_2 = \frac{2}{3}$), and the scattering rate per atom is the average population $\langle P_3 \rangle$ of state 3 times the decay rate Γ_3, such that

$$r_1 = \langle P_3(0) \rangle \Gamma_3 \tag{1}$$

$$r_2 = [\langle P_3(+1)\rangle \ + \ \langle P_3(-1)\rangle] \ \Gamma_3 \ . \tag{2}$$

Here ($m \equiv m_F$),

$$\langle P_3(m)\rangle \ = \frac{1}{2} \frac{I(m)\mathfrak{L}(B,m)}{1 + I(m)\mathfrak{L}(B,m)} \ , \tag{3}$$

where $I(\pm 1) \ = \ I_0 \sin^2 \alpha$, $I(0) \ = \ I_0 \cos^2 \alpha$, I_0 is the densitiy of light flux at the ion's location, α is the angle subtended by the direction of the light polarization with respect to the magnetic field B. $\mathfrak{L}(B,m)$ is defined by

$$\mathfrak{L}(B,m) = \frac{(\Gamma_3/2)^2}{(\Gamma_3/2)^2 + (\omega_0 - \omega + m\delta)^2} \ ,$$

where ω and ω_0 are the light and resonance frequencies, respectively, $\delta = g_F \mu_B B/\hbar$, $g_F \simeq 1$ is the Landé factor of state 2, and μ_B is the Bohr magneton [12]. Now, let us further restrict the model to the two-level system made up by states 0 and 1, and attribute to it conventional rates per atom of phase and energy relaxation, γ and Γ, respectively. These rates will become *identified* with quantities taken from the above three-level model: the spin nutation gives rise to oscillation of the population difference in the two-level system that is supposed to damp out exponentially with an effective constant γ of phase relaxation [12] such that

$$\gamma = \frac{\Gamma}{2} + \gamma_{\mathrm{ph}} \mathrel{\widehat{=}} r_1 \ , \tag{4}$$

where γ_{ph} is the contribution of extra phase perturbation not related to intrinsic relaxation. Moreover, the flow equilibrium established by the scattering as well as the quasi-steady state on the microwave-driven line after dephasing require

$$n_0 = n_1 \ = \frac{1 - n_2}{2} \tag{5}$$

and

$$\frac{1}{3} n_2 \, r_2 \ = \frac{2}{3} n_1 \, r_1 \ . \tag{6}$$

Thus, $n_2 = \frac{r_1}{r_1 + r_2}$, and the normalized probe signal, i.e. the probability of finding the system in the upper state 1 of a relaxing two-level system is

$$\begin{aligned} P_1^{(3)} &= n_1 + n_2 \ = \frac{1 + n_2}{2} \\ &= 1 - \frac{1}{2} \frac{r_2/r_1}{1 + r_2/r_1} \\ &= 1 - P_0^{(3)} \ . \end{aligned} \tag{7}$$

One may identify $P_0^{(3)}$ with an effective two-level excitation probability $P_1^{(2)} = \frac{1}{2}\frac{I}{1+I}$, where $I = \Omega^2/\Gamma\gamma$, and Ω is the Rabi frequency of the microwave. This interpretation (i) yields the effective rate constant of energy relaxation,

$$\Gamma \mathrel{\hat{=}} \frac{\Omega}{r_2}\,, \tag{8}$$

and (ii) shows that this effective relaxation makes the ion decay into the *excited* state 1. Note that the two effective rate constants γ and Γ of (4) and (8) may be set *separately* by suitable selection of the light polarization and/or of the ambient magnetic field. Fig. 4 shows trajectories calculated

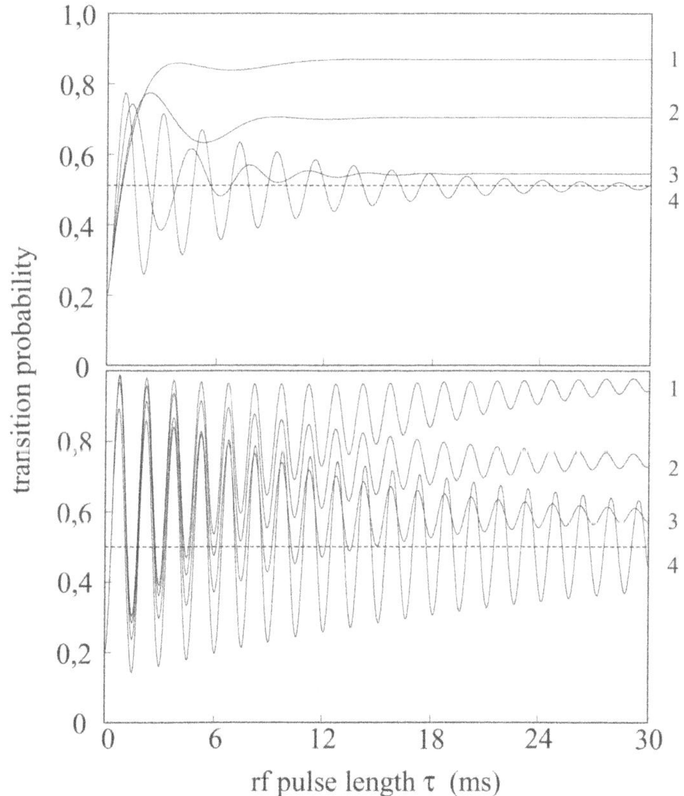

Fig. 4. Probability $P_1(\theta)$ of the ion to be found in the upper state 1 after having been driven by a microwave pulse of area $\theta = \Omega\tau$. Weak resonance scattering excited by laser light of intensity $I = \Omega_1^2/\Gamma_3\gamma_1$ mimics phase relaxation and energy decay into state 1. Initial population in 0: 0.8, in 1: 0.2. Detuning $\delta = -0,28$, $\delta_1 = -3 \times 10^3$. $\Gamma_3 = 18 \times 10^3$, $\Omega = 4,2$. *Top:* $\sqrt{2r_1\gamma_1} = 700$, $\sqrt{r_2\gamma_1} = 70(1)$, 140(2), 350(3), 700(4). *Bottom:* $\sqrt{2r_1\gamma_1} = 70$, $\sqrt{r_2\gamma_1} = 7(1)$, 70(2), 140(3), 350(4). All frequencies in $2\pi \times$ kHz

with selected values of $T_0 = \Gamma^{-1}$ and $T_2 = \gamma^{-1}$. They demonstrate various degrees of damping as well as different levels of saturation of $P_1(\tau)$.

With a single atom, competing spontaneous decay into non-degenerate levels may preserve coherence in the atom [13]. Therefore, complete modelling of the rate of optical pumping represented by (2) would require one to include an interference term of the transition amplitudes of the back-and-forth optical pumping, via the states $F = 1$, $m_F = +1$ and -1, that give rise to indiscernible pathways. This term displays a resonance in zero magnetic field and represents what is called "zero-field level crossing" [14], a ground-state Hanle effect. The width of the crossing resonance is determined by the lifetime of the intermediate interfering states. The $F = 1$ levels show an effective lifetime on the order of milliseconds, or longer. Thus, the interference term in rate r_2 would make this rate vary across a spectral tuning range of the laser less than 1 kHz wide. This spectral feature is not resolved by the emission bandwidth of the laser.

Historically it is interesting, that the Hanle effect when observed on Ne atoms irradiated by the light field of a HeNe laser was one of the earliest laser-spectroscopic topics; it was dealt with in the diploma thesis of Theodor Hänsch [15, 16].

The availability of easily quantifiable longitudinal and transversal relaxation that is *light-induced* upon individual atomic systems displays important advantages when it comes to the application of such a system to manipulations in QIP impaired by loss of coherence. In particular, codes of information processing may be tested for their applicability under this challenging but commonplace condition. On the other hand, codes for error correction may be made to demonstrate their capacity upon increasing levels of decoherence fed into the system. The light-induced decoherence, as demonstrated in this chapter, is readily applicable to individually addressed quantum systems, it may be switched on and off immediately, and it is reproducible.

In summary, a simple method has been outlined that adds a predetermined degree of decoherence on the coherent radiative interaction of an individual atom or ion that is to be used for information processing. Although, in the present experiment, the coherent drive was microwave radiation resonant with a ground-state hyperfine transition, the same principle seems to apply to a system where a dipole-forbidden optical transition is driven by laser light.

Acknowledgements

This work was supported by the Hamburgische Wissenschaftliche Stiftung and by the Deutsche Forschungsgemeinschaft.

References

1. Th. Hänsch, P. Toschek, Phys. Lett. **20**, 273 (1966)
2. P.E. Toschek, 'Atomic particles in traps', Les Houches Session 1982, in *New Trends in Atomic Physics*, ed. by G. Grynberg, R. Stora, eds. (North-Holland, Amsterdam 1982) p. 383
3. e.g. see A. Steane, Appl. Phys. B **64**, 623 (1997)
4. e.g. see J.A. Jones, Fortschr. Phys. **48**, 909 (2000)
5. D.M. Meekhof, C. Monroe, B.E. King, W.M. Itano, D.J. Wineland, Phys. Rev. Lett. **76**, 1796 (1996)
6. C. Monroe, D.M. Meekhof, B.E. King, W.M. Itano, D.J. Wineland, Phys. Rev. Lett. **75**, 4714 (1995)
7. R. Huesmann, Chr. Balzer, Ph. Courteille, W. Neuhauser, P.E. Toschek, Phys. Rev. Lett. **82**, 1611 (1999)
8. B. Appasamy, Y. Stalgies, P.E. Toschek, Phys. Rev. Lett. **80**, 2805 (1998); Ch. Roos, Th. Zeiger, H. Rohde, H.C. Nägerl, J. Eschner, D. Leibfried, F. Schmidt-Kaler, R. Blatt, Phys. Rev. Lett. **83**, 4713 (1999)
9. Chr. Balzer, R. Huesmann, W. Neuhauser, P.E. Toschek, Opt. Commun. **180**, 115 (2000)
10. C. Di Fidio, W. Vogel, Phys. Rev. A **62**, 031802-1 (2000)
11. Q.A. Turchette, C.J. Myatt, B.E. King, C.A. Sackett, D. Kielpinski, W.M. Itano, C. Monroe, D.J. Wineland, Phys. Rev. A. **62**, 053807 (2000)
12. P. Meystre, M. Sargent III, *Elements of Quantum Optics* (Springer, Berlin 1990) Sect. 15
13. M. Schubert, I. Siemens, R. Blatt, W. Neuhauser, P.E. Toschek, Phys. Rev. A **52**, 2994 (1995)
14. e.g. see H. Haken, H.C. Wolf, *Atom- und Quantenphysik* (Springer, Berlin, Heidelberg 1990)
15. Th. Hänsch, Diplomarbeit, Heidelberg 1966, unpublished
16. Th. Hänsch, P. Toschek, Phys. Lett. **22**, 150 (1966)

Laser Cooling of Trapped Ions

Ferdinand Schmidt-Kaler, Jürgen Eschner, Rainer Blatt, Dietrich Leibfried, Christian Roos, and Giovanna Morigi

Laser cooling was first proposed in 1975 by Hänsch and Schawlow, and simultaneously by Wineland and Dehmelt. After some general remarks on laser cooling in traps we report on experiments featuring two special laser cooling techniques for ions which are stored in Paul traps. With both techniques we demonstrate ground state cooling of a single trapped ion. Ground state cooling of one or a string of ions might help to improve ion-based frequency standards, and is a prerequisite for an ion-based quantum processor.

The first method is based on electromagnetic induced transparency (EIT) and utilizes the absorption profile of a three-level system which is tailored by quantum interference. A single Ca^+ ion was recently cooled to its vibrational ground state with this method. It will be particularly useful for the cooling of larger ion strings, since it works for a several MHz broad band of vibrational frequencies simultaneously. Thus, the cooling method could be of great practical importance for initializing a quantum processor based on trapped ions.

We also report on ground state cooling of a single Ca^+ ion with the resolved sideband cooling technique on a narrow quadrupole transition, leading to 99.9% ground state population. Starting from this Fock state $|n = 0\rangle$ we have demonstrated coherent quantum state manipulation on an optical transition.

1 Introduction

Most people agree that the era of laser cooling started in 1975 when Hänsch and Schawlow [1] and almost simultaneously Wineland and Dehmelt [2] proposed methods that we would now call Doppler cooling. While Hänsch and Schawlow's arguments were based on free atoms with a continuous Doppler spectrum, Wineland and Dehmelt's proposal was based on trapped atoms with an absorption spectrum characterized by motional sidebands. Still the basic idea and underlying physics are very similar.

Laser cooling of free and trapped atoms has blossomed in the meantime and it is probably not unfair to say that it was one of the driving forces in the explosion of work done in atomic physics in the last three decades.

The more subtle points of laser cooling of trapped atoms can be especially well studied with trapped ions where relatively high trap frequencies are easily achieved. This facilitates the resolution of motional sidebands, and has led to the first cooling to the motional ground state in 1989 [3].

Recently work in this field has gained momentum since a string of ions in a linear Paul trap is considered to be a promising candidate for a scalable implementation of quantum computation [4, 5] (see Fig. 1 for the fluorescence image of a linear $^{40}Ca^{+}$ ion crystal taken with a CCD camera). The cooling experiments we report on here were motivated by this goal. Long lived internal states of the ions serve to hold the quantum information (qubits), and are manipulated coherently by laser beams which are focused to individual ions in the string. This way, single-bit logic gates are implemented. The excitation of the ion's common vibrational motion (gate mode) provides the coupling between qubits which is necessary for two-bit quantum gates. Any quantum algorithm can be decomposed into single- and two-bit logic gates.

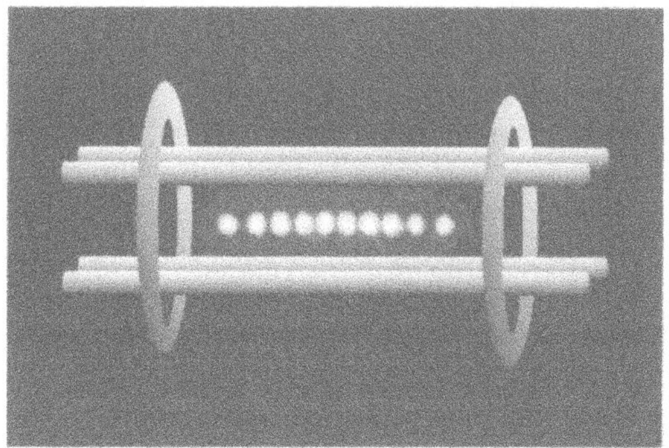

Fig. 1. Scheme of a linear Paul trap made out of four rods (radial confinement) and two rings (axial confinement), with an ion string (not to scale)

Single and two-qubit logic gates demand individual addressing [6] and individual readout of the qubits [7], so inter-ion distances should be large enough to optically resolve every single ion in the string. The minimum distance δx decreases with the axial trap frequency ν_{ax} and the number of ions N like $\delta x \sim \nu_{\mathrm{ax}}^{-2/3}\ N^{-0.559}$ [8], so the upper limit for the axial trap frequency set by this demand is typically in the range of a MHz [9].

The Cirac–Zoller proposal for a two-bit quantum gate [4], requires that initially the gate mode is optically cooled to the ground state. In 1999, Mølmer and Sørensen proposed a two-bit gate which relaxes this condition and only requires cooling into the Lamb–Dicke regime, typically with a mean vibrational quantum number $\bar{n}\ \leq 1$ [10, 11]. In general, for a string of N trapped ions in a linear Paul trap, we describe the system by $3N$ modes of vibration. One of them is chosen for the gate mode. All other vibrational modes (spectator modes) affect the quantum gate fidelity if they are not cooled deeply into the Lamb–Dicke regime [12]. In any case, and whatever type of gate is chosen for the quantum processor, one has to cool the ions to the Lamb–Dicke regime where the extension of the ion wave function is small compared to the excitation wavelength on the qubit transition. In typical experimental situations this requires cooling techniques which lead to lower mean vibrational state than simple Doppler cooling. Resolved sideband cooling on narrow quadrupole or Raman transitions has yielded sub-Doppler temperatures [3, 13–15] but proves increasingly inefficient for cooling many modes. The recently proposed method of EIT cooling [16] promises a higher bandwidth of mode frequencies that can be cooled in parallel typically with reduced experimental effort as compared to sideband cooling. In this chapter experiments on EIT cooling and resolved sideband cooling of Ca^+ ions are reviewed and the perspective of these techniques in the context of quantum information with trapped ions is discussed.

The first part (Sect. 2) of this chapter describes EIT cooling and how to reach $\bar{n} \leq 1$ for a wide band of vibrational modes. In particular, we discuss the cooling of a linear string of N ions. At this point, the conditions for a Mølmer–Sørensen gate are already fulfilled. A short pulse of ground state sideband cooling for the gate mode, as described in the second part (Sect. 6), leads then to the situation which is required for a gate relying on the motional ground state such as the Cirac–Zoller gate. The third part (Sect. 7) contains two examples of state engineering and coherent dynamics starting from the motional ground state.

2 The Principle of EIT Cooling

The use of electromagnetic induced Transparency for cooling was proposed by Morigi, Eschner, and Keitel recently. We aim here to sketch the principle briefly, for a more detailed discussion we refer to [16].

The interaction of ions with light fields can lead to an exchange of energy and momentum. In the specific case of a periodically oscillating trapped particle, the atomic spectrum acquires sidebands spaced by the oscillation frequency ν_{trap}. The most important scaling factor for the interaction of a trapped particle with light fields is the Lamb–Dicke parameter, $\eta = \sqrt{E_{\mathrm{rec}}/(h\nu_{\mathrm{trap}})}$ that compares E_{rec}, the recoil energy of the scattered photons, to the level spacing of the harmonically bound particle.

We assume Doppler precooling which will leave the ion close to the Lamb–Dicke regime with $\eta^2 \ll 1$, but $\eta^2\bar{n} \approx 1$, where $\bar{n}$ denotes the mean vibration quantum number. Then only the carrier and the first-order sidebands have nonnegligible strength, and the spectrum consists of three resonances: the carrier (at zero detuning from the electronic resonance of the atom at rest) and the red and the blue sideband with detunings of $\pm\,\nu_{\mathrm{trap}}$ and oscillator strength $\eta^2\bar{n}$ (red sideband) and $\eta^2(\bar{n}+1)$ (blue sideband) relative to the carrier. If now the ion absorbs a laser photon on its red sideband and decays under the emission of a photon at the carrier frequency, this will lead to cooling. Similar, a blue sideband absorption leads to heating. Carrier absorption leads to no change if the atom also decays on the carrier transition and to diffusion if the decay is on the sidebands. The balance between cooling processes and the counteracting heating and diffusion determines the final mean phonon number $\bar{n}$. The key point is that the likelihood of a transition leading to cooling scales with $\bar{n}$ and therefore vanishes when $\bar{n}$ goes to zero while the likelihood for heating and diffusion processes remains at a nonvanishing value.

Therefore cooling to $\bar{n} \approx 0$ requires a strong imbalance between cooling and heating transitions, which can be achieved with well-resolved sidebands (see Sect. 6) or, in the EIT scheme, by utilizing a dark resonance (see Fig. 2, from [16]). A strong dressing laser r is blue detuned from the $|r\rangle$ to $|e\rangle$ ion transition. A weaker probe laser field g tests the absorption on the $|g\rangle$ to $|e\rangle$ transition. For zero relative detuning with $\Delta_g = \Delta_r$ a dark resonance occurs, a coherent superposition of both ground states $|g\rangle$ and $|r\rangle$ with zero absorption. This phenomenon is known as electromagnetically induced transparency (EIT). The inset of Fig. 2 sketches the full absorption profile for the probe laser, which exhibits a broad Lorentzian near $\Delta_g = 0$ and a bright resonance slightly above the dark resonance by $\delta = +\Omega_r^2/4\Delta_r$ (due to the light-shift of the dressing laser field).

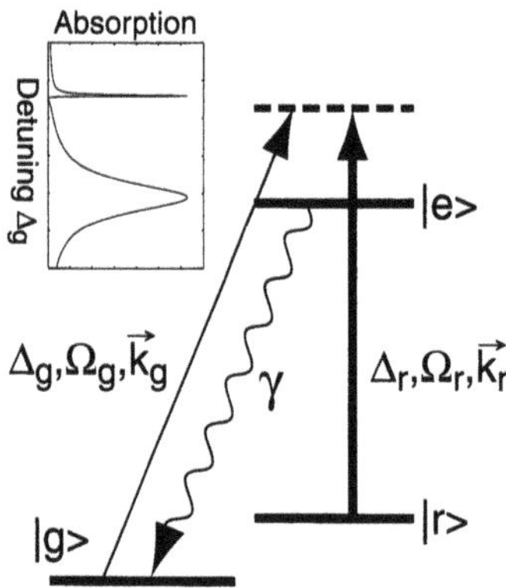

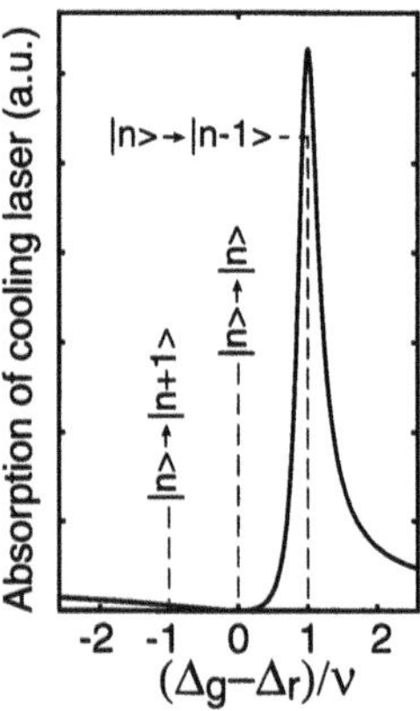

Fig. 2. *Left*: Three-level system and transitions used the cooling scheme. The strong dressing laser with Rabi frequency Ω_r generates two dressed states out of the original $|e\rangle$ state, which are seen in the absorption of a weak probe laser field (Ω_g). The inset shows the absorption rate on $|g\rangle \rightarrow |e\rangle$. *Right*: Detail of the absorption for the probe laser. For an ion at rest the absorption probability vanishes for $\Delta_g = \Delta_r$ ($|n\rangle \rightarrow |n\rangle$ transition). Due to the asymmetric shape of the absorption profile, the probability for a cooling transition from $|n\rangle$ to $|n-1\rangle$ is much higher than for the transition to $|n+1\rangle$

If the laser frequencies are set to the EIT condition $\Delta_g = \Delta_r$ no carrier absorption happens. On the other hand the ion's vibrational sidebands do not coincide with the dark resonance, so probe beam absorption on the sidebands is possible. By choosing the AC-Stark shift δ to be equal to ν_{trap} the bright resonance will exactly coincide with the red sideband. The situation is depicted on the left side of Fig. 2. Due to the bright resonance the absorption on the red sideband is much more likely to occur than the absorption on the blue sideband. As a result, the phonon number n is reduced by one in most absorption–emission cycles. The carrier absorption which is leading to diffusion is completely suppressed by the dark resonance. This mechanism can lead to very efficient cooling with final thermal distributions close to the ground state.

3 Levels and Transitions in Ca^+

When the proposal on EIT-cooling came out, we instantaneously realized that we could use the existing experimental setup of a single ion Paul trap to demonstrate the method. In the following sections, 2 to 5.4, we discuss the measurements which have been published by Roose et al. [17] only a few months after the proposal.

The theoretical background of the method as described above has to be modified only slightly to apply it to the four-level system present in our experiment with a single trapped Ca^+ ion. We implemented the EIT scheme on the $S_{1/2} \rightarrow P_{1/2}$ transition, whose Zeeman sublevels $m = +1/2$ and

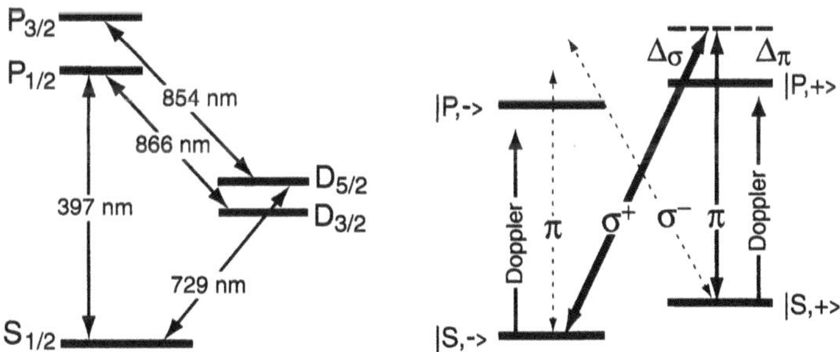

Fig. 3. Levels and transitions in $^{40}Ca^+$ used in the experiment (*left side*). The $S_{1/2}$ and $P_{1/2}$ transition is used for Doppler cooling, and EIT cooling, and the scattered photons are observed to detect the ion's quantum state. The narrow $S_{1/2}$ and $D_{5/2}$ transition serves to investigate the vibrational state (see Sect. 5). The Zeeman sublevels of the $S_{1/2}$ and $P_{1/2}$ states (denoted by $|S, \pm\rangle$ and $|P, \pm\rangle$) and laser frequencies which are relevant for the cooling are shown at the *right side*. *Solid lines* indicate here transitions which are necessary for the EIT cooling, π and σ^+, with $\Delta_\pi = \Delta_\sigma$, and the Doppler precooling transition. *Dashed lines* indicate further transitions which involve the level $|P, -\rangle$

$m = -1/2$ constitute a four-level system [17]. We denote these levels by $|S, \pm\rangle$ and $|P, \pm\rangle$, see Fig. 3. Three of the levels, $|S, \pm\rangle$ and $|P, +\rangle$, coupled by σ^+- and π-polarized laser beams, form an effective three-level system of the kind considered above [16], with only slight modifications due to the fourth level $|P, -\rangle$.

4 Experimental Setup for EIT Cooling

For the demonstration of the EIT cooling method, we have chosen a single trapped Ca^+ ion. This system reveals a negligible heating rate and the cooling results can be revealed with high precision on the narrow $S_{1/2}$ to $D_{5/2}$ quadrupole transition [18]. We will briefly discuss the experimental setup; for details we refer to [18, 7, 17]. First we describe all light sources which are necessary for the Ca^+ experiment.

For the excitation of the $S_{1/2} \rightarrow P_{1/2}$ dipole transition at 397 nm, we use light from a Ti:sapphire laser near 793 nm which is frequency doubled in a LBO crystal inside a build-up cavity. Up to 5 mW of light near 397 nm are sent through a single mode UV fiber to the trap setup. As displayed in Fig. 4, we use the beam deflected into $+1^{st}$ order from an acousto-optical modulator (AO 1) for Doppler cooling ($\Delta = 20\,\mathrm{MHz}$).

For the demonstration of EIT cooling, two beams for EIT cooling are derived from the UV-fiber output as the $+2^{nd}$ order Bragg reflexes of two acousto-opical modulators (AO 2, AO 3) driven at 86 MHz and 92 MHz, respectively. We have chosen a detuning of $\Delta_\sigma = \Delta_\pi \simeq 75$ MHz (the natural linewidth of the transition is 20 MHz). The difference in driving frequencies compensates for the 12 MHz Zeeman splitting of the $|S, \pm\rangle$ states in the quantization B-field of 4.4 gauss. The three output beams are labeled 397 nm-Doppler, 397 nm-σ, and 397 nm-π in Fig. 4. They are focused onto the single ion in the Paul trap. For reproducibility and fine tuning, the light power (typically a few 10 μW in a waist of $\simeq$ 60 μm) is adjusted with the power of the

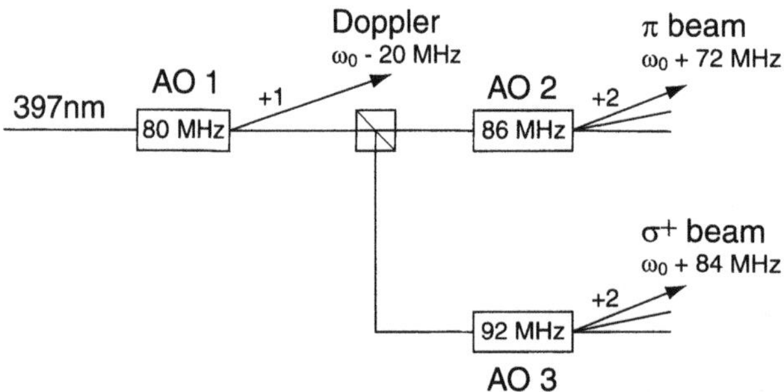

Fig. 4. Optical setup for light beams at 397 nm. Generation of optical frequencies as used for Doppler- and for EIT cooling. Both EIT beams are blue detuned. The Doppler cooling beam is red detuned from the Ca^+ resonance

AO rf-drive which can be varied for all AO's. Also, the rf-power, and therefore the light beams, can be switched on and off within a few μs, controlled by a computer program.

Light from a second Ti:sapphire laser near 729 nm is used to test the excitation from the $S_{1/2}$ ground state on the quadrupole transition to the metastable $D_{5/2}$ state (spont. lifetime 1 s). The laser frequency is stabilized to a bandwidth of $\delta\nu \leq 100$ Hz, using a Pound–Drewer–Hall scheme and a stable high finesse optical reference cavity. The laser frequency can be tuned by an acousto-optical modulator, and the laser light intensity is controlled by the rf-drive power. With this laser we can selectively excite the sidebands of vibration of the single ion in the harmonic trapping potential, a few MHz detuned from the $S_{1/2} - D_{5/2}$ carrier transition. The relative excitation strength of the first red and blue sideband reveals information about the ion's vibrational quantum state (Sect. 5).

The light of grating stabilized laser diodes near 866 nm and 854 nm serves to pump the ion out of both metastable D levels. While the light of the laser at 866 nm is tuned continuously throughout the whole experiment, the laser at 854 nm is also switched on and off by an acousto-optical modulator which is computer controlled.

Our ion trap is a 3D quadrupole Paul trap: A ring with inner diameter of 1.4 mm is formed from a Mb wire of 0.2 mm diameter and two endcaps are made out of sharpened tips of the same material at 1.2 mm distance [18]. The ring electrode is driven by $\simeq 600$ V radio frequency at 20.9 MHz, while the tips are held near ground potential. Additional compensation electrodes, together with a differential voltage applied to the tips, can be used to shift the ion into the center of the rf-potential to minimize micromotion. For the experiments described below, a single $^{40}\mathrm{Ca}^+$ ion is generated by electron bombardment of a weak thermal Calcium beam. We observe oscillation frequencies (ν_x, ν_y, ν_z) of (1.69, 1.62, 3.32) MHz.

5 Measuring the Vibrational Quantum State of an Ion

The excitation probabilities P_{red} and P_{blue} on the $|S,n\rangle$ to $|D,n-1\rangle$ transition (red sideband) and the $|S,n\rangle$ to $|D,n+1\rangle$ transition (blue sideband) are proportional to n and $n+1$, respectively. P_{red} drops to zero for $|n=0\rangle$, see Fig. 5. For a thermal phonon distribution with mean phonon number $\bar{n}$, the probability for finding n phonons is $p_n(\bar{n}) = \bar{n}^n/(\bar{n}+1)^{n+1}$. Thus $\bar{n}$ is directly related to the measured ratio $P_{\mathrm{red}}/P_{\mathrm{blue}} = \bar{n}/(\bar{n}+1)$. It can be shown that this result holds for incoherent *and* coherent excitation if the phonon distribution $p_n(\bar{n})$ is thermal. Our cooling results will be specified either as $\bar{n}$ or by the ground state occupation p_0 for the three trap vibrational modes.

One way to determine the cooling result is to first excite a vibrational sideband at 729 nm and then probe the $\mathrm{S}_{1/2}$–$\mathrm{P}_{1/2}$ transition at 397 nm (shelving method) [3]. If sideband excitation to the $\mathrm{D}_{5/2}$ state was successful,

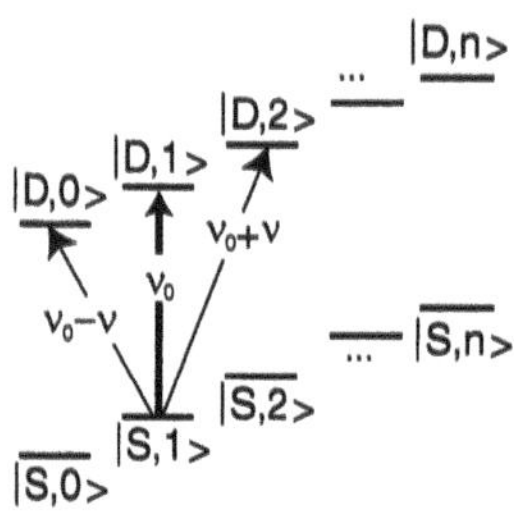

Fig. 5. Principle of temperature measurement. Transitions from $|\mathrm{S}, n\rangle$ to $|\mathrm{D}, n\pm1\rangle$ are selectively laser-excited and the relative excitation probability is measured. ν_0 is the bare atomic transition frequency

no fluorescence is emitted on the $\mathrm{S}_{1/2}-\mathrm{D}_{5/2}$ transition, whereas in the other case we observe many scattered photons. This procedure is repeated 100 times on both the red and the blue sideband, and the difference in their excitation probabilities yields $\bar{n}$ and p_0.

The second method we used is driving coherent Rabi oscillations on the blue sideband [18, 19]. We record the excitation probability $P_{\text{blue}}(t)$ as a function of the 729 nm excitation time t. In the Lamb–Dicke regime $P_{\text{blue}}(t) = \sum p_n(\bar{n}) \sin(\eta \Omega_{\text{Rabi}} \sqrt{n+1}\ t)$, where Ω_{Rabi} denotes the $\mathrm{S}_{1/2}-\mathrm{D}_{5/2}$ carrier Rabi frequency and η the Lamb–Dicke factor [20]. The analysis of the time evolution $P_{\text{blue}}(t)$ reveals the mean phonon number $\bar{n}$ or can even be used to measure all coefficients of the phonon distribution $p_n(\bar{n})$.

5.1 Setting the Power Level for the EIT Beams

Applying the EIT cooling scheme to our system, we find that the intensity of the EIT σ^+ beam should be such that the AC Stark shift equals the frequency of the vibrational mode to be cooled (the σ^+ beam is used as the dressing laser, see Fig. 2). If the waist sizes of the beams are known, their intensities can be calculated. There are also a few possible methods to experimentally determine the AC Stark shift or the corresponding Rabi frequencies:

(a) A rough estimate may be gained from the fluorescence rate as a function of light power. Finding first the approximate saturation power level, one can then deduce the necessary laser power for both EIT beams.
(b) In the $\{\mathrm{S}_{1/2}, \mathrm{P}_{1/2}, \mathrm{D}_{3/2}\}$ three-level system excited by the 866 nm and 397 nm lasers, one can measure the fluorescence rate if one of the laser frequencies is tuned over the resonance and the other is kept fix. Optical Bloch equations are then used to fit the excitation spectrum and determine the relevant Rabi frequencies [21].
(c) The time constant of optical pumping between the $|\mathrm{S}, \pm\rangle$ states may be used. To probe their population, one of the two states must be selectively excited to the $\mathrm{D}_{5/2}$ level.
(d) One can measure the broadening and AC-Stark shift of the narrow $\mathrm{S}_{1/2}$ to $\mathrm{D}_{5/2}$ transition when the ion is simultaneously illuminated with a pulse of 729 nm and one of the EIT beams.

We used the last method to determine the AC Stark shift δ as a function of power in the EIT beams. From this, we could extrapolate to the power level where $\delta = \nu_{\text{trap}}$. In more detail, the ion was excited on the $S_{1/2}(m = 1/2) \rightarrow D_{5/2}(m = 5/2)$ transition with a 3 µs pulse at 729 nm and *simultaneously* with a EIT σ^+ pulse. Although the method suffered from asymetric excitation line shapes and from rapid optical pumping out of the $S_{1/2}, m = 1/2$ level, it led to a good estimate for the right power level. For a final optimization of the EIT σ^+ beam power we used the cooling results for a specific vibrational mode [17].

5.2 EIT Cooling Experimental Procedure

We investigated EIT cooling of the y and the z oscillation at 1.62 and 3.32 MHz, respectively. The experiments proceeded in three steps: Doppler precooling, EIT cooling and finally the determination of the vibrational quantum state, see Fig. 6.

(i) We first Doppler precooled the ion on the $S_{1/2}$ to $P_{1/2}$ transition at 397 nm (natural linewidth $\Gamma \simeq 20$ MHz). A detuning of approximately -20 MHz with respect to the $S_{1/2}$–$P_{1/2}$ transition line was chosen for optimum Doppler cooling results. To avoid optical pumping into the $D_{3/2}$ states, we used the repumping beam near 866 nm [7]. The Doppler cooling limit on this transition of 0.5 mK corresponds to mean vibrational quantum numbers of $\bar{n}_z \approx 3$ and $\bar{n}_x \approx \bar{n}_y \approx 6$. The cooling limits reached in our experiment are higher, due to the fact that the simple assumption of a two-level system in the determination of the Doppler limit does not hold in our case. We experimentally determined the mean excitation numbers after Doppler cooling to

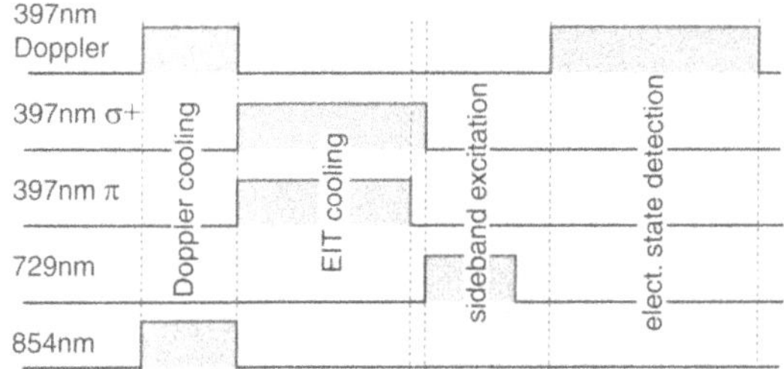

Fig. 6. The pulse sequence for measuring the vibrational state after EIT cooling. After a 1.5 ms period of Doppler cooling there follows a period of EIT cooling (varied between 0 and 7.9 ms). The EIT σ beam is left on for 50 µs longer than the EIT π beam to optically pump the ion into the $|S, +\rangle$. The analysis of the vibrational state uses a pulse (of length t) at 729 nm for the sideband excitation, followed by a pulse of 397 nm Doppler to deduce the quantum shelving signal. The laser at 866 nm continuously empties the $D_{3/2}$ level. The whole sequence of 20 ms duration is repeated 100 times and the average excitation rate is calculated. To synchronize to 50 Hz magnetic field fluctuations the sequence is triggered by the electric power line phase

be $\bar{n}_z = 6.5(1.0)$ and $\bar{n}_y = 16(2)$. A diode laser at 854 nm serves to repump the ion from the $D_{5/2}$ to the $S_{1/2}$ level. The Doppler cooling is applied 1.5 ms.

(ii) After Doppler cooling we apply both EIT light beams, 397 nm-σ and 397 nm-π. A 1.8 ms pulse duration was sufficient and longer cooling time did not lower the mean phonon number. The power level was set as discussed above, see Sect. 5.1. For our setup with laser beam waists of $\simeq 50$ µm, a laser power of ≤ 50 µW was used for the EIT σ beam.

The k-vectors of the two EIT cooling beams enclose an angle of 125° and illuminate the ion in such a way that their difference $\Delta \boldsymbol{k}$ has a component along all trap axes ($(\phi_x, \phi_y, \phi_z) = (66°, 71°, 31°)$, where ϕ_i denotes the angle between $\Delta \boldsymbol{k}$ and the respective trap axis). As a result, *all* vibrational directions were cooled. Since our current vacuum recipient does not allow for two beams which are under 90°, we cannot realize the ideal constellation shown in Fig. 3, but the EIT-π beam has some σ^+ component since it is not at right angles with the quantization magnetic field. The transitions due to this nonideal constellation are indicated with dashed lines in Fig. 3 and result in a slightly higher final mean vibration quantum number. A more detailed discussion is found in [17].

(iii) Finally we analyze the vibrational state after EIT cooling by spectroscopy on the $S_{1/2} \to D_{5/2}$ quadrupole transition at 729 nm.

5.3 Cooling Results for a Single Mode of Vibration

With the AC Stark shift δ set to the frequency of the radial y-mode, we monitored the vibrational state after EIT cooling by exciting the blue sideband of the $|S,+\rangle \to D_{5/2}(m = +5/2)$ transition with a 729 nm pulse and then measuring the $|S,+\rangle$ level occupation as a function of the pulse length t [17, 18]. The observed Rabi-oscillations were subsequently fitted to determine the mean vibrational occupation number $\bar{n}_y$ [19], see Fig. 7. The

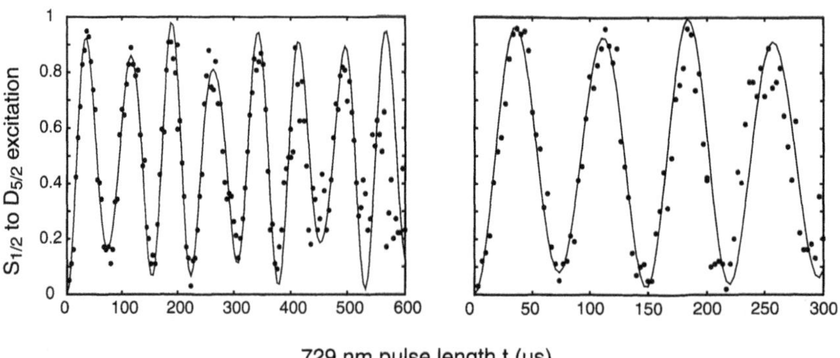

Fig. 7. Rabi oscillations on the blue sideband of the radial y-mode (*left side*) and the axial z-vibrational modes of a single ion (*right side*). From the theoretical curves for $P_{\text{blue}}(t)$ (see Sect. 5), mean phonon numbers of 0.18 and 0.10 are obtained

lowest mean vibrational number $\bar{n}_y = 0.18$ observed corresponds to a 84% ground state probability. We repeated this experiment on the z-mode at $\nu_z =$ 3.3 MHz after having increased the intensity of the σ^+ beam to adjust δ. For this mode, a minimum mean vibrational number of $\bar{n}_z = 0.1$ was obtained, corresponding to a 90% ground state probability.

We found the cooling results largely independent of the intensity of the π beam as long as it is much smaller than the σ^+ intensity. In our experiment the intensity ratio was $I_\sigma/I_\pi \simeq 100$ and we varied the intensity of the π-beam by a factor of 4, with no observable effect on the final $\bar{n}$.

By determining the dependence of the mean vibrational quantum number on the EIT pulse length τ, we observed the mean quantum number $\bar{n}_y$ to initially drop in an exponential fashion with a time constant 250 µs (dashed line in Fig. 8).

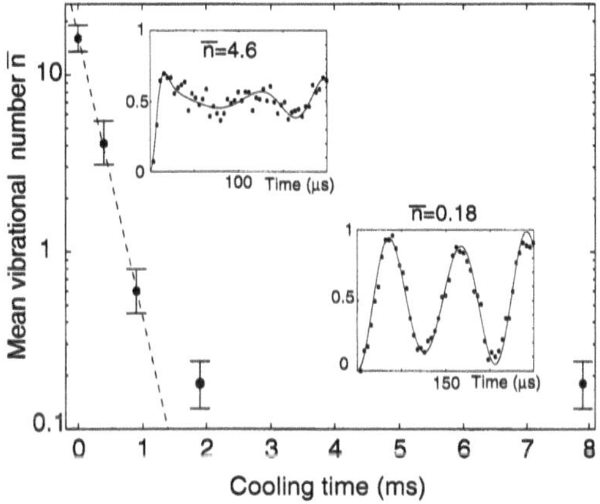

Fig. 8. Mean vibrational quantum number $\bar{n}_y$ versus EIT cooling pulse length τ. The insets show Rabi oscillations excited on the upper motional sideband of the $|\mathrm{S},+\rangle \rightarrow \mathrm{D}_{5/2}(m = +5/2)$ transition, after 0.4 ms (*left*) and 7.9 ms (*right*) of EIT cooling. A thermal distribution is fitted to the data to determine $\bar{n}_y$ in both cases

5.4 Cooling of Two Modes

To show that the EIT method is suitable to simultaneously cool several vibrational modes with vastly different frequencies of oscillation, we chose the axial z-mode at 3.3 MHz, and the radial y-mode at 1.62 MHz, which have a frequency difference of 1.7 MHz. The intensity of the σ^+-beam was set such that the AC Stark shift was roughly halfway between the two mode frequencies. Again we applied the EIT cooling beams for 7.9 ms after Doppler cooling. This time we determined the final $\bar{n}$ by comparing the excitation probability on the red and the blue sideband of the $\mathrm{S}_{1/2}(m = 1/2) \rightarrow \mathrm{D}_{5/2}(m = 5/2)$

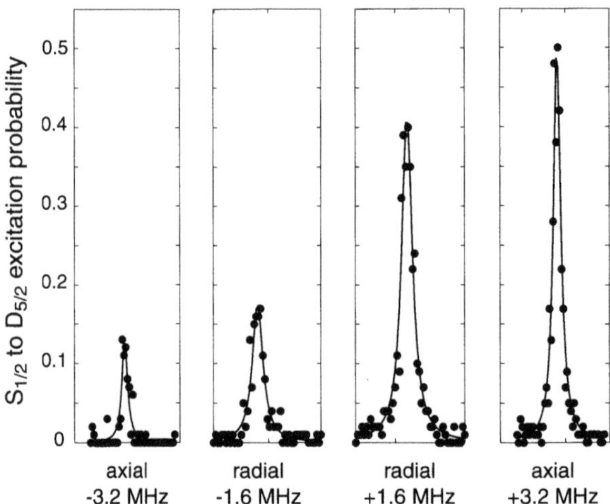

Fig. 9. Cooling of two modes at 1.6 MHz and 3.2 MHz simultaneously. From the sideband excitation rate we deduce a ground state occupation number of 74% for the axial mode (3.2 MHz) and 58% for the radial mode (1.6 MHz)

transition [3]. We find both modes cooled deeply inside the Lamb–Dicke regime ($\eta_i\sqrt{\bar{n}_i} \sim 0.02 \ll 1$, $i = y, z$), with $(p_0)_y = 58\%$ and $(p_0)_z = 74\%$ ground state probability, see Fig. 9.

5.5 EIT Cooling of Linear Ion Strings

Since EIT cooling allows simultaneous cooling of several modes at different frequencies, it seems to be particularly suited for ion strings in linear traps. The frequencies of the axial vibrational eigenmodes of a linear string have been calculated [22, 8] and measured [23]. For a 10-ion string trapped in a linear trap with a center-of-mass axial frequency of 0.7 MHz, the closest inter-ion spacing is found to be 3.0 µm (4.5 µm for $N = 5$). The axial vibration frequencies are 0.7 MHz, 1.22 MHz, ... 4.6 MHz. The radial trap frequency in a linear trap must be made sufficiently high in order to prevent a transition from the linear configuration of N ions to a zig-zag configuration. It was estimated that this transition occurs at $(\nu_{\mathrm{ax}}/\nu_{\mathrm{rad}})^2 = 2.94N^{-1.8}$, thus the radial trap frequency must exceed 3.25 MHz to keep 10 ions in a linear string (1.75 MHz for $N = 5$) [24]. Typically, the radial frequency is chosen higher. The linear ion trap experiment at Innsbruck [25] uses $\nu_{\mathrm{rad}} \sim 4$ MHz, the Be^+ experiments at NIST [26, 27] have $\nu_{\mathrm{rad}} \geq 20$ MHz. Apart from the purely axial modes there exist $2N$ radial (bending) modes. Their frequencies are in a band between the axial center-of-mass frequency ν_{ax} and ν_{rad}.

We now estimate the performance of EIT cooling for a 10-ion string. The result is displayed in Fig. 10. For a 10-ion string indeed *all* $3N$ vibrational

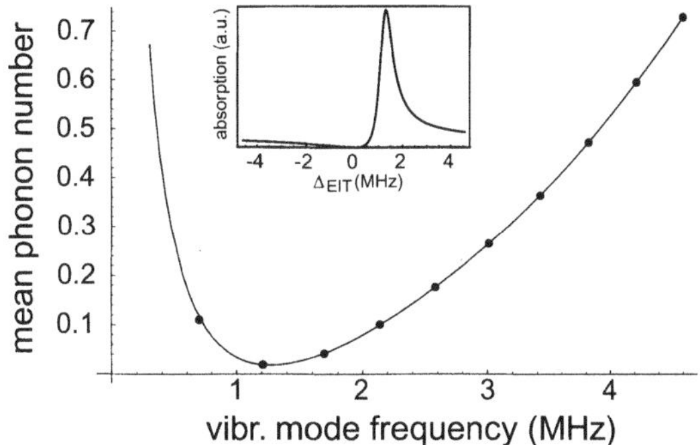

Fig. 10. EIT cooling of the axial modes of a linear string based on the $S_{1/2}$–$P_{1/2}$ transition with Ω_σ= 20 MHz, Ω_π= 0.5 MHz, Γ_P= 20 MHz, and a detuning of $\Delta_\sigma = \Delta_\pi$= 75 MHz. The axial trap frequency is 0.7 MHz. For the calculation we have chosen the light intensity such that the bright state is AC-Stark shifted by $\sim$ 1.3 MHz (see *inset*). The bright resonance with a width of $\sim$ 0.5 MHz leads to cooling for all axial modes. The mean phonon numbers (*black dots*) of all axial modes are plotted versus the mode frequencies

modes are cooled to a mean phonon number $\bar{n}$ below one. This is promising for the application of cold ion strings for quantum information processing [4, 10, 11, 28]. As discussed in the introduction, it is required that all modes which couple to the laser light (spectator modes) must be cooled well into the Lamb–Dicke regime. The reason for that is that thermally excited spectator modes cause a variation of the carrier Rabi frequency Ω which reduces the precision of quantum logic operations. For the case of $3N-1$ spectator modes i with n_i, and Lamb–Dicke factors η_i, the initial state is a mixed state. Each time the experiment will happen with slightly different initial conditions and the Rabi frequency for carrier or sideband transitions will differ. The relative variation scales as $\Delta\Omega/\Omega = \sqrt{\sum \eta_i^4 \; \bar{n}_i \; (\bar{n}_i + 1)/(3N-1)}$ where i runs over all spectator modes (see eqn. 126 in ref. [12]). For our specific example we find $\Delta\Omega/\Omega \simeq 3 \cdot 10^{-4}$, certainly not limiting the fidelity of gates in the near future.

6 Sideband Cooling

6.1 Principle of Sideband Cooling

As already mentioned in the introduction, ground state cooling is required for a succesful two bit gate operation according to the proposal of Cirac and Zoller [4]. This selection presents a discussin of the cooling results which have been published in Roos et al. [18] for a single Ca^+ ion in a spherical trap and which have be shortly later later extended to the case of two ion crystals, published in Schmidt-Kaler et al. [29] and in Rhode et al. [15].

In our resolved sideband cooling experiments we used a two-stage cooling process. First the ion was cooled to the Doppler limit by driving the $S_{1/2}$ to $P_{1/2}$ dipole transition. In the second stage we drove the red sideband of the narrow $S_{1/2}$ to $D_{5/2}$ quadrupole transition, thus removing one phonon with each electronic excitation. The cooling cycle was closed by a spontaneous decay to the ground state which conserves the phonon number with probability $1 - \eta^2\bar{n}$, very close to 1 inside the Lamb–Dicke regime. Once the vibrational ground state is reached the ion decouples from the laser. The long lifetime of the upper state on a bare quadrupole transition would lead to slow cooling cycles and therefore long cooling times. However, the cooling rate can be greatly enhanced by (i) strongly saturating the quadrupole transition and (ii) shortening the lifetime of the excited state by coupling it to a quickly decaying state on a dipole-allowed transition [3, 18].

6.2 Experimental Procedure for Sideband Cooling

As in the EIT cooling procedure (Sect. 5.2) we use a pulsed technique of consecutive steps, where only step (ii) is replaced by the sideband cooling step, see Fig. 11. The $S_{1/2}(m = 1/2) \leftrightarrow D_{5/2}(m = 5/2)$ transition is excited on one of the red sidebands at approximately 1 mW laser power focused to a waist size of 30 μm. The laser at 854 nm is switched on to couple the $D_{5/2}$ level to the $P_{3/2}$ level with a strength set for optimum cooling. Optical pumping to the $S_{1/2}(m = -1/2)$ level is prevented by interspersing short laser pulses of σ^+-polarized light at 397 nm (see dashed line in Fig. 11). The duration of those pulses is kept at a minimum to prevent excess heating.

The sequence is repeated typically several 100 times to determine the excitation probability P_D on the red and blue sidebands for a quantitative determination of the vibrational ground state occupation probability p_0, as explained in Sect. 5.

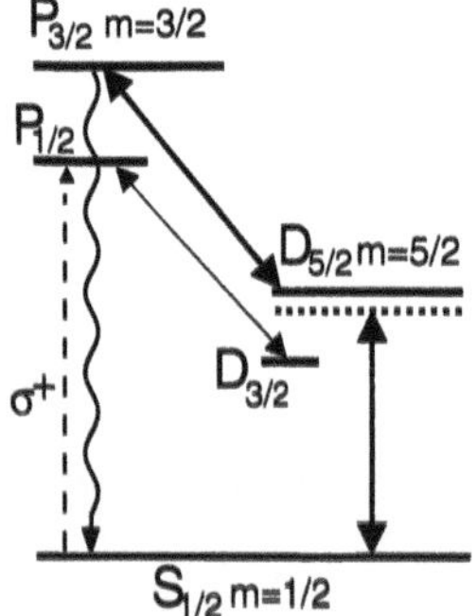

Fig. 11. Sideband cooling of Ca^+ ions on the $S_{1/2}$–$D_{5/2}$ transition; see text for details

6.3 Sideband Cooling Results

The ground state occupation after sideband cooling is determined by probing sideband absorption immediately after the cooling pulse. Figure 12 shows $P_D(\nu)$ for frequencies centered around the red and blue ν_z sideband. Comparison of the sideband heights yields a 99.9% ground state occupation for the axial mode when $\nu_z = 4.51$ MHz. By cooling the radial mode with $\nu_y = 2$ MHz, we transfer 95% of the population to the vibrational ground state. The x-direction is left uncooled because it is nearly perpendicular to the cooling beam. We also succeeded in simultaneously cooling all three vibrational modes by using a second cooling beam and alternating the tuning of the cooling beams between the different red sidebands repeatedly.

The best cooling results of 99.9% ground state occupation were achieved with a cooling pulse duration of $\tau_{\text{cool}} = 6.4$ ms. The power of the cooling laser was set to about 1 mW which yielded the lowest value of $\bar{n}$ in the experiments. To study the cooling dynamics from the Doppler limit into the final state we varied τ_{cool} between zero and the maximum of 6.4 ms and determined the resulting ground state occupation. We find that initially $\bar{n}$ decreases rapidly, then it tends to its finite final value. The decay constant, or cooling rate, determined from the data is 5 ms^{-1}. Both this value and the exponential behavior are consistent with the expected three-level dynamics during the sideband cooling process, taking into account our experimental parameters. The finite cooling limit is determined mainly by nonresonant excitation of the ion out of the ground state and heating in the subsequent spontaneous emission.

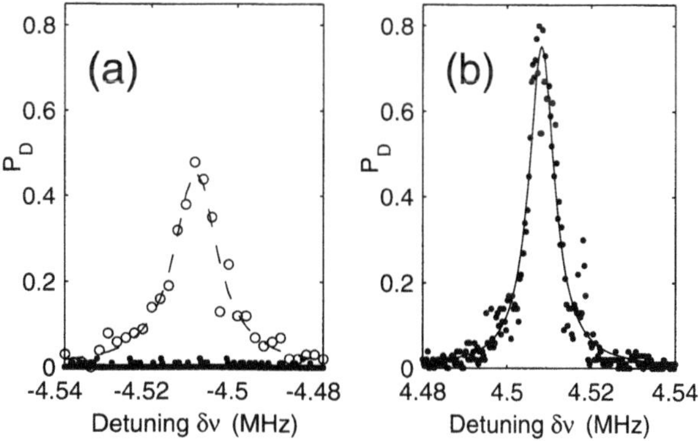

Fig. 12a,b. Sideband absorption spectrum on the $S_{1/2}(m = +1/2) \leftrightarrow D_{5/2}(m = +5/2)$ transition after sideband cooling (*full circles*). The frequency is centered around the (**a**) red and (**b**) blue sideband at $\nu_z = 4.51$ MHz. *Open circles* in (**a**) show the red sideband after Doppler cooling. Each data point represents 400 individual measurements

7 State Engineering and Rabi Oscillations

Starting from the vibrational ground state, arbitrary quantum states can be created. To demonstrate coherent state engineering and investigate decoherence we excited Rabi oscillations with the ion initially prepared in Fock states of its motion. Radiation at 729 nm was applied on the blue sideband transition $|S, n_z\rangle \leftrightarrow |D, n_z{+}1\rangle$ for a given interaction time t and the excitation probability P_{blue} was measured as a function of t. The Rabi flopping behavior allows us to analyze the purity of the initial state and its decoherence [30, 19]. Figure 13a shows $P_{\text{blue}}(t)$ for the $|n = 0\rangle$ state prepared by sideband cooling. Rabi oscillations at $\Omega_{01}/(2\pi) = 21$ kHz are observed with high contrast indicating that coherence is maintained for times well above 1 ms. For the preparation of the Fock state $|n = 1\rangle$, we start from $|S, n = 0\rangle$, apply a π-pulse on the blue sideband and an optical pumping pulse at 854 nm to transfer the population from $|D, n = 1\rangle$ to $|S, n = 1\rangle$. The fidelity of this process was limited to about 0.9 by the nonideal population transfer during the initial π-pulse (cf. the first oscillation in the upper curve of Fig. 13) and the recoil heating of the optical pumping. As shown in the lower curve of Fig. 13 for the $|n = 1\rangle$ initial state, we also observe high-contrast Rabi oscillations, now at $\Omega_{12}/(2\pi) = 30$ kHz. The ratio of the Rabi frequencies ($\Omega_{n,n+1} \propto \sqrt{n+1}$ [20]) agrees with $\Omega_{01}/\Omega_{12} = 1/\sqrt{2}$ to within 1%. The Fourier transform of the flopping signals yields directly the occupation probabilities for the contributing Fock states $|n = 0, 1, 2, 3, ...\rangle$ [30, 19] and allows one to calculate the purity of the prepared and manipulated states. For the "vacuum" state $|n = 0\rangle$, we obtain $p_0 = 0.89(1)$ with impurites of $p_1 = 0.09(1)$ and $p_{n\geq 2} \leq 0.02(1)$. For the Fock state $|n = 1\rangle$ the populations are $p_0 = 0.03(1)$, $p_1 = 0.87(1)$, $p_2 = 0.08(2)$, and $p_{n\geq 3} \leq 0.02(1)$. The measured transfer fidelity of about 0.9 agrees well with our expectation. Note that the Rabi flopping data were taken with less efficient cooling compared to the data shown in Fig. 12 (lower trap frequency), and the number state occupation from the Fourier analysis is consistent with the temperature we determined by sideband measurements.

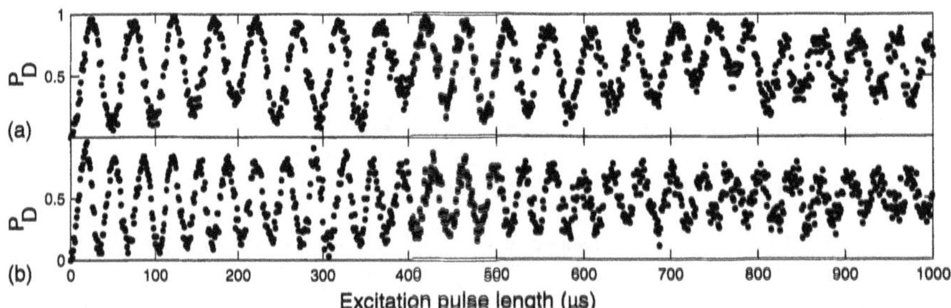

Fig. 13. (**a**) Rabi oscillations on the blue sideband for the initial state $|n = 0\rangle$. Coherence is maintained for up to ~ 1 ms. (**b**) Rabi oscillations as in (**a**) but for an initial vibrational Fock state $|n = 1\rangle$

8 Conclusion

Novel methods of optical cooling are developed and methods are still optimized, even 25 years after the first proposals and after thousands of experiments around the world on laser cooling of atoms and ions. Probably not even Theodor Hänsch had imagined the enormous relevance and rich applications when he started thinking about optical cooling of atoms.

Acknowledgements

We express our thanks to the PhD's and Diploma students working at the Calcium traps, H. Rohde, S. Gulde, A. Kreuter, and A. Mundt, and to P. Barton who was with us for two years as a PostDoc. As well we thank the Members of the SFB in Innsbruck and Vienna for discussions. This work is supported by the Austrian "Fonds zur Förderung der wissenschaftlichen Forschung" (SFB15 and START-grant Y147-PHY), by the European Commission (TMR networks "Quantum Information" (ERB-FRMX-CT96-0087) and "Quantum Structures" (ERB-FMRX-CT96-0077)), and by the "Institut für Quanteninformation GmbH".

References

1. T.W. Hänsch, A.L. Schawlow, Opt. Commun. **13**, 68 (1975)
2. D.J. Wineland, H.G. Dehmelt, Bull. Am. Phys. Soc. **20**, 637 (1975)
3. F. Diedrich, J.C. Bergquist, W.M. Itano, D.J. Wineland, Phys. Rev. Lett. **62**, 403 (1989)
4. J.I. Cirac, P. Zoller, Phys. Rev. Lett. **74**, 4091 (1995)
5. *The Physics of Quantum Information* Springer, Berlin. ed. D. Bouwmeester, A. Ekert, A. Zeilinger (2000)
6. H.C. Nägerl, Ch. Roos, H. Rohde, D. Leibfried, J. Eschner, F. Schmidt-Kaler, R. Blatt, Phys. Rev. **A 60**, 145 (1999)
7. H.C. Nägerl, W. Bechter, J. Eschner, F. Schmidt-Kaler, R. Blatt, Appl. Phys. B **66**, 603 (1998)
8. D.V.F. James, Appl. Phys. B **66**, 181 (1998)
9. Instead of using well-focused laser beams, there might be alternative methods for individual addressing which either use a large magnetic field gradient allowing us to resolve the qubits due to their different Zeeman shift, see F. Mintert, Ch. Wunderlich, quant-ph/0104041, or to use the micromotion of ions in the trap, see D. Leibfried, Phys. Rev. A **60**, 3335 (1999)
10. K. Mølmer, A. Sørensen, Phys. Rev. Lett. **82**, 1835–1838 (1999)
11. K. Mølmer, A. Sørensen, Phys. Rev. **A 62**, 022311 (2000)
12. D.J. Wineland, C. Monroe, W.M. Itano, D. Leibfried, B. King, D.M. Moekhof, J. Res. Natl. Inst. Stand. Technol. **103**, 259 (1998)
13. C. Monroe, D.M. Moekhof, B.E. King, W.M. Itano, D.J. Wineland, Phys. Rev. Lett. **75**, 4011 (1995)

14. B.E. King, C.J. Myatt, Q.A. Turchette, D. Leibfried, W.M. Itano, C. Monroe, D.J. Wineland, Phys. Rev. Lett. **81**, 1525 (1998)
15. H. Rohde, S.T. Gulde, C.F. Roos, P.A. Barton, D. Leibfried, J. Eschner, F. Schmidt-Kaler, R. Blatt, J. Opt. B: Quantum Semiclass. Opt. **3**, 34 (2001)
16. G. Morigi, J. Eschner, Ch. Keitel, Phys. Rev. Lett. **85**, 4458 (2000)
17. Ch. Roos, D. Leibfried, A. Mundt, F. Schmidt-Kaler, J. Eschner, R. Blatt, Phys. Rev. Lett. **85**, 5547 (2000)
18. Ch. Roos, T. Zeiger, H. Rohde, H.C. Nägerl, J. Eschner, D. Leibfried, F. Schmidt-Kaler, R. Blatt, Phys. Rev. Lett. **83**, 4713 (1999)
19. D.M. Meekhof, G. Monroe, B.E. King, W.M. Itano, D.J. Wineland, Phys. Rev. Lett. **76**, 1796 (1996)
20. C.A. Blockley, D.F. Walls, H. Risken, Europhys. Lett. **17**, 509 (1992); J.I. Cirac, R. Blatt, S. Parkins, P. Zoller, Phys. Rev. A **49**, 1202 (1994)
21. G. Janik, W. Nagourney, H. Dehmelt, J. Opt. Soc. Am. B **2**, 1251 (1985); M. Schubert, I. Siemers, R. Blatt, W. Neubauser, P.E. Toschek, Phys. Rev. Lett. **68**, 3016 (1992); Phys. Rev. A **52**, 2994 (1995)
22. A. Steane, Appl. Phys. B **64**, 632 (1997)
23. H.C. Nägerl, D. Leibfried, F. Schmidt-Kaler, J. Eschner, R. Blatt, Opt. Expr. **3**, 89 (1998)
24. D.G. Enzer, M.M. Schauer, J.J. Gomez, M.S. Gulley, M.H. Holzscheiter, P.G. Kwiat, S.K. Lamoreaux, G.G. Peterson, V.D. Sandberg, D. Tupa, A.G. White, R.J. Huges, Phys. Rev. Lett. **85**, 2466 (2000)
25. http://heart-c704.uibk.ac.at/
26. Q. Turchette, D. Kielpinski, B.E. King, D. Leibfried, D.M. Meekhof, C.J. Myatt, M.A. Rowe, C.A. Sackett, C.S. Wood, W.M. Itano, C. Monroe, D.J. Wineland, Phys. Rev. **A 61**, 163418 (2000)
27. http://www.bldrdoc.gov/timefreq/ion/index.htm
28. D. Jonathan, M.B. Plenio, P.L. Knight, Phys. Rev. **A 62**, 042307 (2000)
29. F. Schmidt-Kaler, C. Roos, H.C. Nägerl, H. Rohde, S. Gulde, A. Mundt, M. Lederbauer, G. Thalhammer, T. Zeiger, P. Barton, L. Hornekaer, G. Reymond, D. Leibfied, J. Eschner, R. Blatt, J. Mod. Opt. **47**, 2573 (2000)
30. M. Brune, F. Schmidt-Kaler, A. Maali, J. Dreyer, E. Hagley, J.M. Raimond, S. Haroche, Phys. Rev. Lett. **76**, 1800 (1996)

Conditional Spin Resonance with Trapped Ions

Christof Wunderlich

1 Motivation

Digital information processing builds upon elementary physical elements ("bits") that may occupy either one of two possible states labeled 0 and 1, respectively. If a quantum system, for example, an individual atom having discrete energy eigenstates is chosen as an elementary switch ("qubit"), then the general state of this system will be a superposition of the two computational basis states, i.e. the states chosen to represent the logic 0 and 1. When applying the superposition principle to a register comprising N qubits, one immediately sees that such a register can exist in a superposition of 2^N states thus representing 2^N binary encoded numbers simultaneously. Any operation on this register will act on all states at once, effecting parallel processing on an exponentially growing (with N) number of states. The outcome of a measurement on this register after such an operation will, of course, yield just one out of 2^N possible results with a certain probability.

In order to take advantage of quantum parallelism for efficient computing, a second ingredient is necessary: interference. A useful quantum algorithm has to exploit this parallelism, and, at the same time, make different computational paths interfere such that only the correct result survives after the last computational step [1]. An important example is Shor's algorithm for the factorization of large numbers [2]. Once created, coherent superpositions have to remain intact while a quantum algorithm is carried out, i.e. qubits must not interact in an uncontrollable way with their environment. This would lead to decoherence, an important issue, not only in the realm of quantum information processing (QIP), but also related to the notion of measurement in quantum mechanics [3, 4].

A quantum computer is ideally suited for the simulation of quantum mechanical systems [5, 6], for example, to determine eigenvalues and eigenvectors of many-body systems [7]. Calculating the dynamics of chaotic systems is another useful line of action for a quantum computer, even for one that consists of only a few qubits [8]. Beneficial both for fundamental research and applications is the ability of a quantum computer – comprising a modest number of qubits and working with limited accuracy – to simulate the dynamics of a macroscopic ensemble of classical particles, a task not suitable even for modern supercomputers [9].

In the course of a quantum computation entangled states of qubits are created exhibiting correlations between individual qubits that possess no classical analog. Fundamental questions concerning the role of entanglement, not only in QIP, but also in the framework of general physics [10] add more motivation to exploring the field of QIP. In 1935 Einstein, Podolsky and Rosen [11] scrutinized quantum mechanical predictions for two entangled particles and found nonlocal correlations between these particles. This, what Einstein called, "spooky action at a distance" prompted him to call into question quantum theory. During the last decade various experiments succeeded in preparing and analyzing entangled states of different physical systems [12, 13], which marked the beginning of controlled manipulation of entanglement of massive particles. On the theoretical side, too, the search for better understanding, quantification, and use of entanglement as a resource for QIP is a very active field [14].

2 Trapped Ions and QIP

QIP is an interdisciplinary field of research, whose results will have significant impact both on basic research and applied sciences. Theory in this field is still well ahead of experimental progress and manageable experimental systems are needed. Essential characteristics of a device designed for quantum computing include [15] the scalability of the system, the ability to reset the qubits' states to a known one, and to make qubit-specific measurements. Furthermore, decoherence times have to be much longer than the typical gate operation time. Finally, a set of quantum gates is needed to construct any desired unitary transformation of N qubits. A sequence of unitary transformations that make up a quantum algorithm can be broken down into two operational elements sufficient for the synthesis of any quantum algorithm [16]: (i) the preparation of individual qubits in arbitrary superposition states, and (ii) the execution of conditional dynamics on different qubits, which is at the heart of quantum computing. It is this last requirement we will be mainly concerned with in this contribution.

A promising system for QIP are electrodynamically trapped ions where two internal states of each ion, labelled $|0\rangle$ and $|1\rangle$ in the remainder of this chapter, are chosen as one qubit [17]. Conditional dynamics with N trapped ions require coupling of their internal and external degrees of freedom. Following the first preparation and detection of a single atom reported in [18] – a prerequisite for many important studies with trapped ions – the principal elements of ion trap quantum computing have been realized experimentally (for instance, [19–24]).

Theodor Hänsch once illustrated the principles of electrodynamic trapping using macroscopic charged particles [25]. After testing various kinds of electrode configurations he finally arrived at the ultimate simplification: a conventional paper clip, connected to a regular power socket sufficed to stably trap charged lycopodium seeds. He documented his efforts with a humorous video that, for example, shows the periodic motion of particles in step with ballet music. This may shed a little light on Theodor Hänsch's imaginative, playful approach to physics that enabled him to make so many outstanding contributions.

The vibrational motion of a collection of ions (the "bus-qubit") is used as a means of communication between individual qubits to implement conditional quantum dynamics in ion traps [17]. An example may serve to illustrate how qubit A in a trap is manipulated conditioned on the state of qubit B. Initially the ion string is cooled to the ground state of the vibrational mode to be used as the bus-qubit. A pulse of electromagnetic radiation is applied first to qubit A, then to qubit B. Each of these pulses switches the respective qubit's state between $|0\rangle$ and $|1\rangle$ (π-pulse). If qubit A is in state $|0\rangle$ initially, and is driven by radiation detuned below its resonance by the frequency of a vibrational mode of the ion string (the so-called red sideband), then the internal excitation to state $|1\rangle$ cannot take place because of energy conservation. However, if qubit A is initially in $|1\rangle$, then its deexcitation will be succesful and accompanied by the creation of one vibrational quantum. Analogously, qubit B – initially in its ground state – can only be excited to $|1\rangle$ by the second radiation pulse (red sideband), if the vibrational motion has previously been excited, i.e. if qubit A was in $|1\rangle$.

This example shows that cooling of the ions' motional degrees of freedom is indispensable for QIP. Optical cooling of atoms, suggested by Hänsch and Schawlow [26] and for trapped atoms by Wineland and Dehmelt [27], has for the first time been observed on a collection of trapped ions [28].

2.1 Why is Optical Radiation Used?

Common to all experiments – related either to QIP or other research fields – that require some kind of coupling between internal and external degrees of freedom of atoms is the use of optical radiation for this purpose. The parameter determining the coupling strength between internal and motional dynamics is the so-called Lamb–Dicke parameter

$$\eta \equiv \sqrt{\frac{(\hbar k)^2}{2m}/\hbar\nu_1} = \Delta z_1 \, k \tag{1}$$

the square of which gives the ratio between the change in kinetic energy of the atom due to the absorption or emission of a photon and the quantized energy spacing of the harmonic trapping potential characterized by the angular frequency ν_1 (k is the wavevector of the light field, m the mass of the atom, and $\Delta z_1 = \sqrt{\hbar/2m\nu_1}$ signifies the spatial extent of the vibrational ground state wavefunction of the atom). Only if η is nonvanishing will the absorption or emission of photons be possibly accompanied by a change of the motional state of the atom. This is apparent when the Hamiltonian describing the coupling between an applied electromagnetic field of angular frequency ω and a harmonically trapped 2-state atom is considered:

$$H_I = \frac{1}{2}\hbar\Omega_{\rm R}(\sigma_+ + \sigma_-)\left[\exp\left[{\rm i}\left(\eta(a^\dagger + a) - \omega t + \phi'\right)\right] + {\rm h.c.}\right] , \qquad (2)$$

where $\Omega_{\rm R} = \boldsymbol{d}\cdot\boldsymbol{F}/\hbar$ is the Rabi frequency with $\boldsymbol{d}\cdot\boldsymbol{F}$ signifying either magnetic or electric coupling between the atomic dipole and the respective field component. $\sigma_{+,-} = 1/2\,(\sigma_x \pm \sigma_y)$ are the atomic raising and lowering operators, respectively, $\Delta z_1(a^\dagger + a)$ is the position operator, and ϕ' is the initial phase of the driving field. Trapping a $^{171}Yb^+$ ion, for example, with $\nu_1 = 2\pi\,100\,\text{kHz}$ gives $\Delta z_1 \approx 17\,\text{nm}$ and (1) shows that driving radiation in the optical regime is necessary to couple internal and external dynamics of trapped atoms.

Here, and in the remainder of this article, we consider a Paul trap [29] in a linear configuration where a time-dependent two-dimensional quadrupole field strongly confines the ions in the radial direction yielding an average effective harmonic potential [30]. An additional static electric field is chosen such that the ions are harmonically confined also in the axial direction [31]. If the confinement of N ions is much stronger in the radial than in the axial direction, the ions will form a linear chain [32] with typical inter-ion distance $\delta z = \zeta\, 2N^{-0.57}$ where $\zeta \equiv (e^2/4\pi\epsilon_0 m\nu_1^2)^{1/3}$ [33]. The distance between neighboring ions δz is determined by mutual Coulomb repulsion of the ions and trap frequency ν_1 in the axial direction. Manipulation of individual ions is achieved by focusing laser light to a spot size smaller than δz. Typically, δz is of the order of a few µm; for example, $\delta z \approx 7\,\mu\text{m}$ for $N = 10$ $^{171}Yb^+$ ions with $\nu_1 = 2\pi\,100\,\text{kHz}$. Again, only optical radiation is useful for this purpose.

2.2 Spin Resonance

Many phenomena that were only recently studied in the optical domain form the basis for techniques belonging to the standard repertoire of coherent manipulation of nuclear and electronic magnetic moments associated with their spins. One reason for the tremendous and fast success of nuclear magnetic resonance (NMR) experiments in the field of QIP is the high level of sophistication that experimental techniques in this field have reached over decades. This is an impressive example for a successful technology whose basis was developed by physicists [34] and that has overcome the boundaries between disciplines of science. For many years researchers, for example, in chemistry and in the life sciences, have routinely used commercial NMR equipment. The technological basis for NMR – apart from the preparation of the samples to be investigated – is the generation and coherent manipulation of electromagnetic radiation in the radiofrequency (rf) and microwave (mw) regime. This treasure of knowledge and technology could immediately be exploited, again for fundamental research, in the emerging field of QIP, where even complete algorithms based on quantum logic have been demonstrated [35, 36].

There are also drawbacks associated with NMR quantum computing: for example, considerable effort has to be devoted to the preparation of pseudo-pure states of a macroscopic ensemble of spins with an initial thermal population distribution. This preparation leads to an exponentially growing cost (with the number N of qubits) either in signal strength or the number of experiments involved [37], since the fraction of spins in their ground state is proportional to $N/2^N$. Extending NMR quantum computing to larger numbers of qubits than in present experiments will also require molecules with more nuclear spins distinct in their resonance frequencies and, at the same time, with appreciable coupling constants.

Trapped ions, on the other hand, provide individual qubits – for example hyperfine states – well isolated from their environment. However, the application of mw radiation for quantum logic operations with a string of ions is not possible, since (i) this long wavelength radiation does not couple internal and external degrees of freedom of the ions, and (ii) focusing down to the required small spot sizes for access to individual qubits is not possible. It would be desirable to combine the advantages of trapped ions and NMR techniques in future experiments.

3 A Modified Ion Trap

An axial magnetic field gradient applied to an electrodynamic trap indeed has the desired effect of coupling internal state dynamics and motion of the ions when mw driving radiation is applied [38]. In addition, the field gradient serves to separate qubit resonances of individual ions making them distinguishable in frequency space. Thus microwave radiation can be used to coherently manipulate hyperfine states of individual ions and condition their internal dynamics on the states of other qubits. The treatment put forward in [38] is generalized in what follows and it is shown that mutual spin–spin coupling between qubits arises in such a modified ion trap analogous to the coupling Hamiltonian in molecules used for NMR. The size of this NMR-type coupling is proportional to the square of the ratio between the magnetic field gradient $\partial_z B$ and ν_1.

The nonrelativistic Hamiltonian describing the internal dynamics of a diatomic molecule may be written as [39]

$$H_{\mathrm{M}} = T_{\mathrm{N}} + T_{\mathrm{el}} + V(\boldsymbol{r}, R) \tag{3}$$

where T_{N} and T_{el} represent the kinetic energy operator of nuclear and electronic motion, respectively. All electrostatic potential energy terms are contained in $V(\boldsymbol{r}, R)$, with $\boldsymbol{r}$ denoting the collection of electronic coordinates and R the internuclear distance. Neglecting initially the nuclear kinetic energy yields the Schrödinger equation for the electronic wavefunctions

$$\left[T_{\mathrm{el}} + V(\boldsymbol{r}, R)\right] \Phi_{\mathrm{a}}(\boldsymbol{r}, R) \equiv H_{\mathrm{el}} \Phi_{\mathrm{a}}(\boldsymbol{r}, R) = E_{\mathrm{el,a}}(R) \Phi_{\mathrm{a}}(\boldsymbol{r}, R) \; . \tag{4}$$

These Born–Oppenheimer (BO) wavefunctions depend on R as a parameter. With $\langle \Phi_\mathrm{a}|T_\mathrm{N}|\Phi_\mathrm{a}\rangle\chi \approx T_\mathrm{N}\langle \Phi_\mathrm{a}|\Phi_\mathrm{a}\rangle\chi = T_\mathrm{N}\chi$ the Schrödinger equation for the nuclear motional wavefunction χ

$$(T_\mathrm{N} + E_\mathrm{el,a})\chi = E_\mathrm{T}\chi \tag{5}$$

determines the dynamics of the nuclei on the BO potential energy curves $E_\mathrm{el,a}$.

We now turn to the description of a linear chain of N harmonically trapped, singly ionized two-level ions in an analogous way. The electronic part of the total Hamiltonian can be solved independently for each ion, since the distance δz between different ions is much larger than the extent of individual spatial wavefunctions. Two Zeeman states, $|E_{0n}\rangle$ and $|E_{1n}\rangle$ of each ion serve as one qubit ($n = 1, \dots, N$). The overall electronic state of the ions obeys $H_\mathrm{el}\Phi_\mathrm{a}(\boldsymbol{z}) = E_\mathrm{el,a}(\boldsymbol{z})\Phi_\mathrm{a}(\boldsymbol{z})$ with

$$H_\mathrm{el} = \frac{1}{2}\hbar \sum_{n=1}^{N} \omega_n(z_n)\sigma_{z,n} \tag{6}$$

and $\Phi_\mathrm{a}(\boldsymbol{z}) = \prod_{n=1}^{N} |E_{cn}(z_n)\rangle$, where $a = 1 \dots 2^N, c = 0, 1$; z_n denotes the axial coordinate of ion n, and σ_z is the usual Pauli matrix. The qubit transition frequency is $\omega_n = (E_{1n} - E_{0n})/\hbar$. A magnetic field applied to the linear arrangement of ions shifts the qubit states $|E_{cn}\rangle$ depending on the location z_n of the n-th ion (here, $\boldsymbol{B} = bz \cdot \hat{z} + B_0$ is assumed for clarity, with $\hat{z}$ being the unit vector in the axial direction). The complete Hamiltonian for the ion chain is given by

$$\begin{aligned} H &= H_\mathrm{el}(\boldsymbol{z}) + T_\mathrm{A}(\boldsymbol{z}) + V_\mathrm{A}(\boldsymbol{z}) \\ &= H_\mathrm{el}(\boldsymbol{z}) + \frac{1}{2m}\sum_{n=1}^{N} p_{z,n}^2 + \frac{m}{2}\sum_{n=1}^{N} \nu_1^2 z + \frac{e^2}{8\pi\epsilon_0}\sum_{n\neq l}^{N} \frac{1}{|z_n - z_l|} \, . \end{aligned} \tag{7}$$

The potential energy relevant for the motion of the ions is obtained from $\langle \Phi_\mathrm{a}|(H_\mathrm{el}+V_\mathrm{A}(\boldsymbol{z}))|\Phi_\mathrm{a}\rangle = E_\mathrm{el,a}+V_\mathrm{A}(\boldsymbol{z})$. When there is no field gradient present, i.e. $b = 0$, the electronic energy is independent of z and simply gives an additive constant. Therefore, only T_A and V_A have to be considered in this case. Expanding V_A around the equilibrium positions $z_{0,n}$ of the ions in terms of $q_n \equiv z_n - z_{0,n}$ up to second order yields the dynamical matrix $\hat{A}$ with $A_\mathrm{ln} \equiv \partial_{z_l}\partial_{z_n} V_A$ and the Hamiltonian of a harmonic oscillator is obtained

$$T_\mathrm{A} + V_\mathrm{A} = \frac{1}{2m}\sum_{n=1}^{N} P_{Q,n}^2 + \frac{m}{2}\sum_{n=1}^{N} \nu_n^2 Q_n \tag{8}$$

with N uncoupled vibrational modes [40]. The normal coordinates $\boldsymbol{Q}$ and local coordinates $\boldsymbol{q}$ are connected via $\boldsymbol{q} = \hat{S}\boldsymbol{Q}$ where $\hat{S}$ is the unitary transformation matrix that diagonalizes $\hat{A}$. Further, $P_{Q,n} = m\dot{Q}_n$.

Taking into consideration the field gradient, a new term in the potential energy arises for ion j:

$$\langle \Phi_{\mathrm{a}} | H_{\mathrm{el},j}(\boldsymbol{z}) | \Phi_{\mathrm{a}} \rangle = E_{cj}(z_{0,j}) + \underbrace{\frac{\hbar}{2} \frac{\partial \omega_j}{\partial z_j}\bigg|_{z_{0,j}} q_j (-1)^{c+1}}_{V_{\mathrm{B}}} . \tag{9}$$

An order of magnitude estimate of the size of the additional potential energy term V_{B} experienced by ion j is obtained upon substitution of $q_j \approx \Delta z_1$ into (9). The new term V_{B} has to be compared to $\hbar\nu_1$, the ground state energy of the unperturbed lowest oscillator mode:

$$\varepsilon \equiv \frac{|V_{\mathrm{B}}|}{\hbar \nu_1} = \frac{|\partial_z \omega_j| \, \Delta z_1}{\nu_1} . \tag{10}$$

As long as ε is much smaller than unity, the eigenfrequencies of the oscillator modes only negligibly depend on the additional potential term introduced by the Zeeman shift of the ionic qubit states. Therefore, the part of the Hamiltonian that describes the motional state of the ion string is well approximated by the unperturbed harmonic oscillator, and the complete Hamiltonian reads

$$\begin{aligned} H = {} & \frac{\hbar}{2} \sum_{n=1}^{N} \omega_n(z_{0,n}) \sigma_{z,n} + \frac{1}{2m} \sum_{n=1}^{N} P_{Q,n}^2 + \frac{m}{2} \sum_{n=1}^{N} \nu_n^2 Q_n^2 \\ & + \frac{\hbar}{2} \sum_{n=1}^{N} \left[\frac{\partial \omega_n}{\partial z_n}\bigg|_{z_{0,n}} \sigma_{z,n} \sum_{l=1}^{N} S_{ln} Q_l \right] \\ = {} & \frac{\hbar}{2} \sum_{n=1}^{N} \omega_n(z_{0,n}) \sigma_{z,n} + \frac{1}{2m} \sum_{n=1}^{N} P_{Q,n}^2 \\ & + \frac{m}{2} \sum_{l=1}^{N} \nu_l^2 \left[Q_l + \frac{\hbar}{2m\nu_l^2} \sum_n \frac{\partial \omega_n}{\partial z_n}\bigg|_{z_{0,n}} \sigma_{z,n} S_{ln} \right]^2 \\ & \underbrace{- \frac{\hbar}{4m} \sum_{l=1}^{N} \frac{1}{\nu_l^2} \left[\sum_n \frac{\partial \omega_n}{\partial z_n}\bigg|_{z_{0,n}} \sigma_{z,n} S_{ln} \right]^2}_{H_{\mathrm{SS}}} \end{aligned} \tag{11}$$

with the electronic energy expanded up to first order in q_n. The unitary transformation $\tilde{H} = U^\dagger H U$ with

$$U = \exp \left[-\mathrm{i} \sum_l \left(\frac{1}{2m\nu_l^2} \sum_n \frac{\partial \omega_n}{\partial z_n}\bigg|_{z_{0,n}} \sigma_{z,n} S_{ln} \right) P_{Q,l} \right] \tag{12}$$

yields

$$\tilde{H} = \frac{\hbar}{2} \sum_{n=1}^{N} \omega_n(z_{0,n}) \sigma_{z,n} + \sum_{n=1}^{N} \frac{P_{Q,n}^2}{2m} + \frac{m}{2} \nu_n^2 Q_n^2 - H_{\mathrm{SS}} . \tag{13}$$

Expressing the harmonic oscillator in (13) in terms of creation and annihilation operators $a_n^\dagger$ and a_n, respectively, using the definitions

$$\varepsilon_{nl} \equiv S_{nl} \frac{\partial_z \omega_l \Delta z_n}{\nu_n} , \tag{14}$$

$$J_{nl} \equiv \sum_{j=1}^{N} \nu_j \varepsilon_{jn} \varepsilon_{jl} , \tag{15}$$

and after dropping constant terms, (13) reads

$$\tilde{H} = \frac{\hbar}{2} \sum_{n=1}^{N} \omega_n(z_{0,n}) \sigma_{z,n} + \sum_{n=1}^{N} \hbar \nu_n (a_n^\dagger a_n) - \frac{\hbar}{2} \sum_{n<l}^{N} J_{nl} \sigma_{z,n} \sigma_{z,l} . \tag{16}$$

$\tilde{H}$ describes a linear string of ions with each ion representing an individually accessible qubit with characteristic resonance frequency. The last term in this Hamiltonian expresses a pairwise coupling between qubits, analogous to the well-known spin–spin coupling in molecules used for NMR experiments. The collection of trapped ions can be viewed as an N-qubit molecule with adjustable coupling constants (compare Sect. 3.2).

3.1 Adding a Driving Field

The additional term in the Hamiltonian governing the dynamics of qubit j when irradiated with electromagnetic radiation at frequency ω close to its resonance is given by

$$\begin{aligned} H_{\mathrm{M}} &= \frac{\hbar}{2} \Omega_{\mathrm{R}} (\sigma_j^+ + \sigma_j^-) \left[\exp[\mathrm{i}(k z_j - \omega t + \phi')] + \exp[-\mathrm{i}(k z_j - \omega t + \phi')] \right] \\ &= \frac{\hbar}{2} \Omega_{\mathrm{R}} (\sigma_j^+ + \sigma_j^-) \left[\exp\left[\sum_n \mathrm{i} S_{nj} \eta_n (a_n^\dagger + a_n) - \mathrm{i}\omega t + \mathrm{i}\phi \right] + \mathrm{h.c.} \right] . \end{aligned} \tag{17}$$

First performing the unitary transformation $\tilde{H}_{\mathrm{M}} = U^\dagger H_{\mathrm{M}} U$ where it is convenient to express U given in (12) as

$$U = \exp\left[\frac{1}{2} \sum_{n=1}^{N} \sum_{l=1}^{N} \varepsilon_{nl} (a_n^\dagger - a_n) \sigma_{z,l} \right] , \tag{18}$$

then transforming $\tilde{H}_{\mathrm{M}}$ into the interaction picture defined by $\tilde{H}_{\mathrm{M}}^I = \exp(\frac{\mathrm{i}}{\hbar} \tilde{H} t) \tilde{H}_{\mathrm{M}} \exp(-\frac{\mathrm{i}}{\hbar} \tilde{H} t)$, and finally omitting terms with time dependent factors that contain the sum of ω and ω_j (rotating wave approximation) gives

$$\tilde{H}_{\mathrm{M}}^{I} = \frac{\hbar}{2}\Omega_{\mathrm{R}}\Big[\exp\left[\mathrm{i}\left(\omega_j - \omega - \frac{1}{2}\sum_n \nu_n \varepsilon_{nj}\right)t + \mathrm{i}\phi\right]\sigma_j^{+}$$
$$\exp\left[\mathrm{i}\left(\sum_n (\eta_n S_{nj} + i\varepsilon_{nj})a_n^{\dagger}(t) + (\eta_n S_{nj} - \mathrm{i}\varepsilon_{nj})a_n(t)\right.\right.$$
$$\left.\left. + \ \mathrm{i}\eta_n S_{nj}\sum_l \varepsilon_{nl}\sigma_{z,l}^{(1-\delta_{lj})}\right)\right] + \mathrm{h.c.}\Big] \tag{19}$$

with $a_n(t) = a_n \exp(-\mathrm{i}\nu_n t)$ and $a_n^{\dagger}(t) = a_n^{\dagger}\exp(\mathrm{i}\nu_n t)$. If the driving radiation ω pertains to the rf or mw regime, then η_n is close to zero and the last term in the exponent in (19) can be neglected ($\eta_1 \approx 10^{-6}$ for 10 Yb^{+} ions with transition frequency $\omega_0 = 2\pi\,12.6\,\mathrm{GHz}$ and $\nu_1 = 2\pi\,100\,\mathrm{kHz}$). With the definitions

$$\eta'_{nj}\exp(\mathrm{i}\phi_j) \equiv \eta_n S_{nj} + \mathrm{i}\varepsilon_{nj}\ ,$$
$$\phi_j \equiv \frac{\pi}{2} - \tan\frac{\eta_n S_{nj}}{\varepsilon_{nj}} \approx \frac{\pi}{2}\ ,$$

and

$$\Delta_j \equiv \frac{1}{2}\sum_n \nu_n \varepsilon_{nj} \tag{20}$$

the Hamiltonian in (19) can be rewritten as

$$\tilde{H}_{\mathrm{M}}^{I} = \frac{\hbar}{2}\Omega_{\mathrm{R}}\Big[\exp\left[\mathrm{i}(\omega_j + \Delta_j - \omega)t + \mathrm{i}\phi\right]\sigma_j^{+}$$
$$\exp\left[\mathrm{i}\sum_n \eta'_{nj}\left(a_n^{\dagger}(t)e^{\mathrm{i}\phi_j} + a_n(t)e^{-\mathrm{i}\phi_j}\right)\right] + \mathrm{h.c.}\Big]\ . \tag{21}$$

The exact value of ν_n depends on the internal state configuration of the ion chain. However, after summing over all vibrational modes, Δ_j in (21) is nearly independent of the ions' internal states and reflects a constant shift in the qubit's resonance frequency. The Hamiltonian (21) is formally the same as the one valid for the interaction between trapped ions and optical radiation, except that the parameter combination $\eta_n S_{nj}$ determining the coupling strength between external and internal dynamics has now been replaced by the effective Lamb–Dicke parameter $\eta'_{nj} \approx \varepsilon_{nj}$. Any operation that requires coupling between motion and internal dynamics and thus usually requires optical radiation can be carried out using radiation in the rf or mw regime. For example, conditional quantum dynamics on a collection of qubits may be implemented according to the schemes proposed in [17, 41, 42].

Sideband cooling is achieved when combining excitation on the so-called red sideband resonance of an internal ionic transition with a suitable dissipative process, similar to sideband cooling in the optical regime. Optical sideband cooling has proven efficient for the preparation of trapped ions close to their motional ground state [19–21].

3.2 Spin Resonance with Trapped Ions

The additional spin–spin coupling term in (16) is considered to be a disturbance when schemes for quantum logic are applied – specifically designed for trapped ions – that in one way or the other rely on the existence of motional sidebands accompanying qubit transitions. The error introduced by this term is negligible compared to other technological limitations, and does not impose a new limit on the precision of ion trap quantum logic operations [38].

Instead of employing usual ion trap schemes, this spin–spin coupling term may be directly used to implement conditional dynamics using NMR methods. To obtain an order of magnitude estimate of the coupling constant J in (15) we take $\partial_z \omega_j = (\mu_B/\hbar)\partial_z B \; \forall \; j$. Here, state $|1\rangle$ experiences a linear Zeeman shift and there is no shift for $|0\rangle$. This is the case, for example, with the ground state of $^{171}\mathrm{Yb}^+$when $|0\rangle$ and $|1\rangle$ are identified with $|S_{1/2}, F=0\rangle$ and $|S_{1/2}, F=1, m_F=1\rangle$, respectively. With $S_{jn} \approx N^{-1/2} \approx S_{jl}$ we obtain

$$J \approx \frac{1}{4Nm\hbar}\left(\mu_B \frac{\partial B}{\partial z}\right)^2 \frac{1}{\nu_1^2} \sum_{j=1}^{N} \frac{1}{\lambda_j^2} , \tag{22}$$

where λ_j^2 denotes the j-th eigenvalue of the dynamical matrix $\hat{A}$. For 10 $^{171}\mathrm{Yb}^+$ions, $\nu_1 = 2\pi \; 100\,\mathrm{kHz}$, and $\partial_z B = 10\,\mathrm{T/m}$, $J/2\pi \approx 40\,\mathrm{Hz}$. The magnitude of J is comparable to values that occur in NMR experiments where it depends on the type of molecule and nuclei used. For example, in [35] $J/2\pi = 7.2\,\mathrm{Hz}$ with protons is quoted; protons and carbon nuclei coupled by $J_{\mathrm{HC}}/2\pi = 103\,\mathrm{Hz}$ and $J_{\mathrm{CC}}/2\pi = 201\,\mathrm{Hz}$ are described in [43]; values of $J/2\pi$ ranging from 0.9 Hz to 163 Hz with the same nuclei in a different molecule are reported in [44]), and protons, nitrogen, carbon, and fluorine nuclei with $J/2\pi$ between 2.7 Hz and 366 Hz are described in [45]. Here, J can be given a desired value by variation of ν_1 that characterizes the trapping potential, and of the field gradient $\partial_z B$. If a gradient is applied that changes with z, then the coupling constants J_{nl} can assume different values for different pairs of spins.

The variation of the field gradient along the z-axis is also useful to simultaneously cool all vibrational modes of the ion string as will be shown elsewhere [46].

4 Concluding Remarks

Hitherto it was accepted that electromagnetic radiation does not couple internal and motional degrees of freedom of trapped atoms when long-wavelength radiation is used, since the Lamb–Dicke parameter is negligibly small. Here, physical conditions are described under which this coupling does occur for electrodynamically trapped ions and can be used for QIP and other experiments that require coherent conditional dynamics. It has been shown that individual qubits can be distinguished by frequency using microwave radiation.

To date, experiments using spin resonance on the one hand and trapped ions on the other undoubtedly have been most successful in the implementation of quantum computing. This proposal combines the respective advantages of these two types of experimental techniques: qubits in ion traps can be individually addressed, they are well isolated from the environment, and their number and mutual coupling is variable over a wide range. On the other hand, microwave and radiofrequency technology for NMR experiments has been developed over decades. Thus a new avenue for QIP research is opened up that may lead to simpler and more precise experimental procedures.

Acknowledgements

Fruitful discussions with D. Reiß and helpful comments on the manuscript by W. Neuhauser are gratefully acknowledged. This work was supported by the Deutsche Forschungsgemeinschaft.

References

1. R. Cleve, A. Ekert, C. Macchiavello, M. Mosca, Proc. R. Soc. Lond. A **454**, 339 (1998)
2. P.W. Shor, 'Algorithms for quantum computation'. In: *Proceedings of the 35th Annual Symposium of Foundations of Computer Science, Nov. 1994*, pp. 124–134
3. M. Brune, E. Hagley, J. Dreyer, X. Maitre, A. Maali, C. Wunderlich, J.M. Raimond, S. Haroche, Phys. Rev. Lett. **77**, 4887 (1996)
4. P.E. Toschek, Ch. Wunderlich, Eur. Phys. J. D **14**, 387 (2001)
5. R. Feynman, Int. J. Theoret. Phys. **21**, 467 (1982)
6. K. Mølmer, A. Sørensen, quant-ph/0004014 (2000)
7. D.S. Abrams, S. Lloyd, Phys. Rev. Lett. **83**, 5162 (1999)
8. B. Georgeot, D.L. Shepelyansky, Phys. Rev. Lett. **86**, 5393 (1999)
9. B. Georgeot, D.L. Shepelyansky, quant-ph/0105149 (2001)
10. C. Brukner, M. Cukowski, A. Zeilinger, quant-ph/0106119
11. A. Einstein, B. Podolsky, N. Rosen, Phys. Rev. **47**, 777 (1935)
12. E. Hagley, X. Maitre, G. Nogues, C. Wunderlich, M. Brune, J.M. Raimond, S. Haroche, Phys. Rev. Lett. **79**, 1 (1997)
13. Q.A. Turchette, C.S. Wood, B.E. King, C.J. Myatt D. Leibfried, W.M. Itano, C. Monroe, D.J. Wineland, Phys. Rev. Lett. **81**, 3631 (1998)
14. M. Lewenstein, D. Bruß, J.I. Cirac, B. Kraus, M. Kuś, J. Samsonowicz, A. Sanpera, R. Tarrach, J. Mod. Optics **47**, 2481 (2000)

15. D.P. DiVincenzo, Fortschr. Phys. **48**, 771 (2000)
16. D.P. DiVincenzo, Phys. Rev. A **51**, 1015 (1995); A. Barenco, C.H. Bennett, R. Cleve, D.P. DiVincenzo, N. Margolus, P. Shor, T. Sleator, J.A. Smolin, H. Weinfurter, Phys. Rev. A **52**, 3457 (1995)
17. J.I. Cirac, P. Zoller, Phys. Rev. Lett. **74**, 4091 (1995)
18. W. Neuhauser, M. Hohenstatt, P.E. Toschek, H.G. Dehmelt, Phys. Rev. A **22**, 1137 (1980)
19. B.E. King, C.S. Wood, C.J. Myatt, Q.A. Turchette, D. Leibfried, W.M. Itano, C. Monroe, D.J. Wineland, Phys. Rev. Lett. **81**, 1525 (1998)
20. E. Peik, J. Abel, Th. Becker, J. von Zanthier, H. Walther, Phys. Rev. A **60**, 439 (1999)
21. Ch. Roos, Th. Zeiger, H. Rohde, H.C. Nägerl, J. Eschner, D. Leibfried, F. Schmidt-Kaler, R. Blatt, Phys. Rev. Lett. **83**, 4713 (1999)
22. R. Huesmann, Ch. Balzer, Ph. Courteille, W. Neuhauser, P.E. Toschek, Phys. Rev. Lett. **82**, 1611 (1999)
23. C.A. Sackett, D. Kielpinski, B.E. King, C. Langer, V. Meyer, C.J. Myatt, M. Rowe, Q.A. Turchette, W.M. Itano, D.J. Wineland, C. Monroe, Nature **404**, 256 (2000)
24. D. Reiß, K. Abich, W. Neuhauser, Ch. Wunderlich, P.E. Toschek, quant-ph/0104002 (2001).
25. Trajectories of electrodynamically trapped, macroscopic charged particles were also demonstrated by R.F. Wuerker, H. Shelton, R.V. Langmuir, J. Appl. Phys. **30**, 342 (1959)
26. T.W. Hänsch, A.L. Schawlow, Opt. Commun. **13**, 68 (1975)
27. D.J. Wineland, H.G. Dehmelt, Bull. Am. Phys. Soc. **20**, 637 (1975)
28. W. Neuhauser, M. Hohenstatt, P.E. Toschek, H.G. Dehmelt, Phys. Rev. Lett. **41**, 233 (1978); D.J. Wineland, R.E. Drullinger, F.L. Walls, Phys. Rev. Lett. **40**, 1639 (1978)
29. W. Paul, O. Osberghaus, E. Fischer, Forschungsberichte des Wirtschafts- und Verkehrsministeriums Nordrhein-Westfalen Vol. **415** (Westdeutscher Verlag, Cologne 1958)
30. P.K. Ghosh, *Ion Traps* (Clarendon Press, Oxford 1995) Chap. 2
31. J.D. Prestage, G.J. Dick, L. Maleki, J. Appl. Phys. **66**, 1013 (1989); M.G. Raizen, J.M. Gilligan, J.C. Bergquist, W.M. Itano, D.J. Wineland, Phys. Rev. A **45**, 6493 (1992)
32. J.P. Schiffer, Phys. Rev. Lett. **70**, 818 (1993); D.H.E. Dubin, Phys. Rev. Lett. **71**, 2753 (1993)
33. A. Steane, Appl. Phys. B **64**, 623 (1997)
34. I.I. Rabi, S. Millman, P. Kusch, J.R. Zacharias, Phys. Rev. **55**, 526 (1939); E.M. Purcell, H.C. Torrey, R.V. Pound, Phys. Rev. **69**, 37 (1946); F. Bloch, W.W. Hansen, M. Packard, Phys. Rev. **70**, 474 (1946)
35. J.A. Jones, M. Mosca, J. Chem. Phys. **109**, 1648 (1998)
36. I.L. Chuang, L.M.K. Vandersypen, X. Zhou, D.W. Leung, S. Lloyd, Nature, **393**, 143 (1998)
37. L.M.K. Vandersypen, C.S. Yannoni, I.L. Chuang, quant-ph/0012108
38. F. Mintert, Ch. Wunderlich, quant-ph/0104041
39. H. Lefebvre-Brion, R.W. Field *Perturbations in the Spectra of Diatomic Molecules* (Academic Press, Orlando 1986) Chap. 2
40. D.F.V. James, Appl. Phys. B **66**, 181 (1998)

41. A. Sørensen, K. Mølmer, Phys. Rev. Lett. **82**, 1971 (1999)
42. D. Jonathan, M.B. Plenio, P.L. Knight, Phys. Rev. A **62**, 042307 (2000)
43. M.A. Nielsen, E. Knill, R. Laflamme, Nature **396**, 52 (1998)
44. E. Knill, R. Laflamme, R. Martinez, and C.-H. Tseng, Nature **404**, 368 (2000)
45. R. Marx, A.F. Fahmy, J.M. Myers, W. Bermel, S.J. Glaser, Phys. Rev. A **62**, 012310 (2000)
46. Ch. Wunderlich, D. Reiß, unpublished. Even if $\partial^2 B/\partial z^2$ assumes the required nonzero value, the expansion in (9) up to first order in q remains a very good approximation as long as $\zeta/\Delta z_1 \ll 1$

From Diode Laser to Atom Laser

Tilman Esslinger and Immanuel Bloch

We recollect the development of a particularly simple design for grating-stabilized diode lasers in the group of Theodor Hänsch. It is compact, reliable, easy to build and allows simultaneous operation of more than ten systems. The laser design is used in numerous experiments, amongst them the Munich atom laser experiment which is briefly described.

The question was puzzling and I could not answer it immediately: "Which is the most important part of your experiment?" The distinguished guest scientist asked this question after we had left the laboratory and were heading towards one of the nearby cafes in Schwabing. After a period of consideration I was convinced that I had found the correct answer. I recollected the history of this vital part of the experiment before answering: "The grating-stabilized diode lasers are the most important part of the experiment."

Since their first use in atomic physics in the early 1980s, diode lasers have become an essential part in many modern experiments [1]. This is primarily due to the high reliability and the low cost of these devices, which facilitates experiments involving a number of lasers operating at the same time. The imperfect spectral features of free-running laser diodes, such as mode hopping and the broad linewidth, can be dramatically improved by exploiting their sensitivity to optical feedback [2]. Two types of stabilization arrangements have been developed, either using coupling to an external high-Q cavity [3] or to a diffraction grating [4–6]. Stabilization of laser diodes achieved by means of feedback from a diffraction grating is simpler to realize and still provides a linewidth reduction to the 100 kHz level, which is sufficient for most experiments. In this arrangement the light diffracted from a grating is coupled back into the diode, so that the grating and the diode's rear facet form a resonator. The diode chip with its reflecting facets acts like an intracavity etalon. The external diffraction grating selects a single mode of the chip. The use of a grating results in a linewidth reduction of two orders of magnitude and allows the selection of any desired wavelength within a wide range. All this is achieved without further electronic feedback. The simplicity of this stabilization arrangement has led a number of research groups to design laser systems utilizing the technique.

The first tunable grating stabilized diode laser in our group was developed, machined and assembled by Theodor Hänsch himself in November 1990.

A photograph of this first diode laser is shown in Fig. 1. The setup consists of four elements: the laser diode, the collimating lens, a cylindrical weakly concave lens and the reflection grating. All components, with exception of the cylindrical lens, are held by mounts which are glued to an aluminum base plate. The cylindrical lens can be positioned in the plane perpendicular to the laser beam with a small x-y translational stage. By moving the cylindrical lens the angle between the laser beam and the reflection grating can be adjusted, facilitating wavelength tuning over a large range. The holographic diffraction grating itself is held by a modified piezo tube, which allows tilting of the grating over a small angle and thereby scanning of the laser frequency. The base plate is mounted on top of a small Peltier element which is used to actively stabilize the temperature of the laser cavity. A simple housing protects the laser resonator from air turbulences.

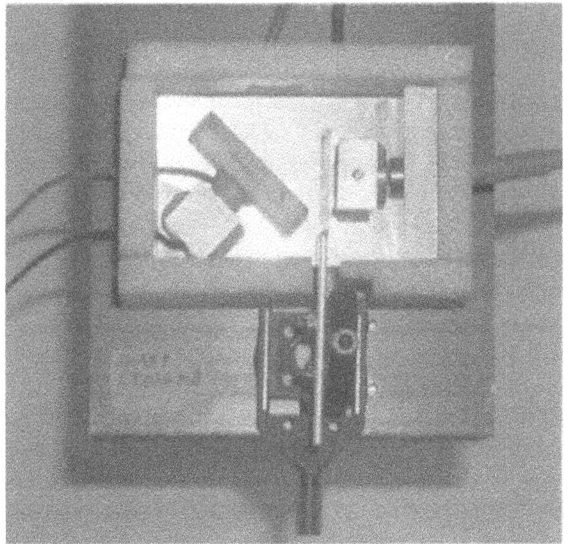

Fig. 1. Theodor Hänsch's original grating-stabilized diode laser setup

Theodor Hänsch proudly presented his experimental results with the diode laser in our group seminar. The data showed saturation spectra of both the D_1 and D_2 line of rubidium, demonstrating the still unmatched tunability range from 780 nm to 795 nm (using an off-the-shelf diode laser!). At the same time the laser linewidth was measured to be well below 1 MHz. The design of the original prototype was then continuously modified over the next one or two years until the design emerged which is now widely in use, not only in Theodor Hänsch's group. The commercially available diode laser DL100 by TOPTICA is also a direct derivative of this developement. Theodor Hänsch's diode laser design really kicked off the now numerous rubidium experiments in his group. One of those experiments is the Bose–Einstein condensation (BEC) and atom laser experiment which will be described below.

The quest for low temperatures and high densities of trapped gases of neutral atoms brought about a spectacular breakthrough: in 1995 Bose–Einstein condensation was achieved in a dilute gas of atoms [7–9]. A novel form of matter had been created in which an almost macroscopic number of atoms share the same quantum state. This startling experimental achievement rapidly led to a broader understanding of degenerate quantum gases and continues to provide fertile ground for studying fundamental concepts of quantum and condensed matter physics. A fascinating feature of BEC is that it provides the key to ultimate control over matter waves; this in turn has opened the door to a first generation of atom lasers demonstrating pulsed [10–12] and continuous output couplers [13]. All four types of atom lasers generate their output from a Bose–Einstein condensate. The condensate is prepared by evaporative cooling, which can be regarded as the pump mechanism of the atom laser.

In our experiment we prepare the Bose–Einstein condensates in a novel magnetic trap of unprecedented simplicity and efficiency [14]. The trap combines the quadrupole and the Ioffe configuration (QUIC-trap). A major advantage of our concept is that the magnetic trap is built inside a μ-metal shielding. This avoids magnetic field fluctuations of the environment and allows us to control the magnetic trapping potential with extreme precision. Stability of the magnetic fields in the sub-milligauss regime was a prerequisite for the demonstration of the atom laser with a continuous output-coupler.

The QUIC-trap consists of just three coils: two identical quadrupole coils and an additional Ioffe coil. A quadrupole trap is formed when the current flows through the quadrupole coils. This configuration is used to load atoms from the magneto-optical trap into the magnetic trap. By turning on the current through the Ioffe coil, the trap centre moves towards the Ioffe coil and the trapping potential is converted into an Ioffe-type geometry.

Our experimental setup consists of two magneto-optical traps mounted on top of each other with the trap centres 28 cm apart. The upper system is a standard vapour cell trap, which is pumped by a 2 l/s ion pump. A 5 cm long tube connects the vapour cell to the second magneto-optical trap (UHV-MOT) which is located in an ultra-high vacuum glass cell.

Both magneto-optical traps have the same laser beam geometry, which consists of three pairs of mutually counter-propagating beams. One of the beam axes is oriented along the horizontal x axis and the other two beam axes are oriented along the diagonals in the y-z plane. All laser beams are derived from grating stabilized diode lasers. We use the moving molasses technique [15, 16] to transfer the atoms from the vapour cell trap into the UHV-MOT. By multiply repeating the transfer we typically collect 10^9 atoms within 60 s. Before loading the atoms into the magnetic quadrupole trap the atomic cloud is compressed [17] and cooled by polarization-gradient cooling.

Then the atoms are optically pumped into a low-field-seeking spin state and loaded into the magnetic quadrupole potential. Subsequently the trap is converted into the Ioffe configuration.

Evaporative cooling of the atoms is performed in the Ioffe configuration by rf-induced spin flips. The velocity distribution of the cooled cloud is probed by absorption imaging. The cloud is released from the magnetic trap and expands ballistically before it is illuminated for typically 100 μs by a near resonant probe laser beam which is tuned to the $F = 2 \rightarrow F = 3$ transition of the D_2-line.

In the night of the 21st January 1998 we observed Bose–Einstein condensation in our laboratory: the black shadow of an elliptically shaped Bose–Einstein condensate, surrounded by thermal atoms appeared on the monitor. On the very next day we proudly presented our measurements to Theodor Hänsch and gave him the data file. A few hours later he gently knocked on our laboratory door and entered showing us the data rendered in three dimensions (see Fig. 2).

In August 1998 we started experimenting with radio-frequency output coupling from the Bose–Einstein condensate. We saw the first glimpse of an atom laser beam for an output coupling period of only 1 ms. On the image we already saw its shape, directed and collimated. Less than half an hour later we ran the experiment for a 10 ms output coupling period and obtained an image similar to that shown in Fig. 3.

Fig. 2. The images display the velocity distribution of the atomic cloud for decreasing temperatures. The left frame corresponds to a temperature just above condensation. In the middle frame the thermal cloud and the condensate can be seen, and the right frame shows an almost pure condensate. The field of view is 1 mm × 1 mm and the image of the distribution is taken after 17 ms time of flight. The image has been rendered by Theodor Hänsch

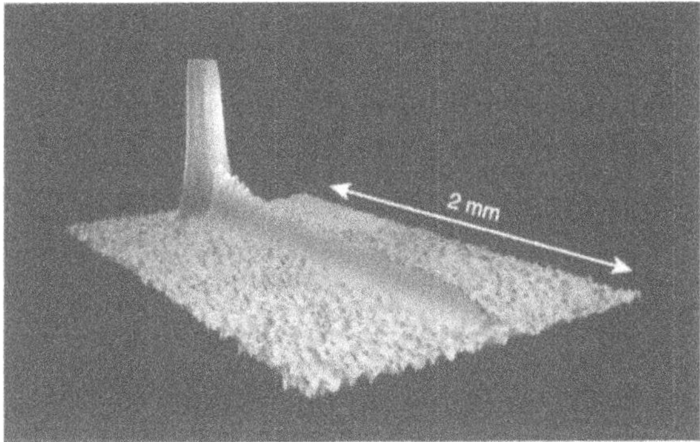

Fig. 3. Three dimensional graph of an atom laser beam extracted over a period of 13 ms from a Bose–Einstein condensate. The vertical axis corresponds to the axis with the length scale. At the top of the image the Bose–Einstein condensate appears saturated due to its high optical density

Continuous output coupling is accomplished with a monochromatic rf-field that resonantly transfers condensed atoms from the magnetically trapped into the untrapped state. The mechanism requires that the level of fluctuations in the magnetic trapping field is much less than the change of the magnetic trapping field over the spatial size of the condensate. Let us consider the geometry of the output coupling process in more detail. The geometry of the magnetic field of our trap gives rise to a harmonic potential which confines the condensate in the shape of a cigar, with its long axis oriented perpendicular to the gravitational force. The radial diameter of the condensate is 10 μm. Due to gravity, both the minimum of the trapping potential and the condensate are displaced by 13 μm relative to the minimum of the magnetic field. Atoms are predominantly transferred into the untrapped state at the intersection of the condensate with the shell that is determined by the electron spin resonance condition [13].

A unique, though non-technical, characteristic of the atom laser is that it drew enormous media attention. This culminated in the appearance of the atom laser and Theodor Hänsch in the main television news. The request for the interview came on the same day as it was broadcast.

It is difficult to predict in which direction the field of BEC and the atom laser will move in the next few years. A host of applications, such as clocks and atom microscopes, appear promising once a continuously pumped atom laser is available. Another direction might be the study of interactions between a BEC and a surface. Big challenges are also the observation of a BCS-transition in a fermionic gas or the observation of a quantum phase transition between a superfluid and and MOTT-insulator state of a BEC trapped in an optical lattice. Whatever way the field will move, one thing is certain: grating-stabilized diode lasers will be used in these experiments.

Acknowledgement

We would like to thank Theodor Hänsch for the loan of the original grating-stabilized diode laser setup for photographic purposes.

References

1. J. Camparo, Contemp. Phys. Methods **26**, 443 (1985)
2. R. Lang, K. Kobayashi, IEEE J. Quantum Electron. **QE-16**, 347 (1980)
3. R. Dahmani, L. Hollberg, R. Drullinger, Opt. Lett. **12**, 876 (1987)
4. K. MacAdam, A. Steinbach, C. Wieman, Am. J. Phys. **60**, 1098 (1992)
5. C. Wieman, L. Hollberg, Rev. Sci. Instrum. **62**, 1 (1991)
6. L. Ricci et al., Opt. Commun. **117**, 541 (1995)
7. M.H. Anderson, J.R. Ensher, M.R. Matthews, C.E. Wieman, E.A. Cornell, Science **269**, 198 (1995)
8. C.C. Bradley, C.A. Sackett, J.J. Tollett, R.G. Hulet, Phys. Rev. Lett. **75**, 1687 (1995); C.C. Bradley, C.A. Sackett, R.G. Hulet, Phys. Rev. Lett. **78**, 985 (1997)
9. K.B. Davis et al., Phys. Rev. Lett. **75**, 3969 (1995)
10. M.-O. Mewes et al., Phys. Rev. Lett. **78**, 582 (1997)
11. B. Anderson, M. Kasevich, Science **282**, 1686 (1998)
12. E. Hagley et al., Science **283**, 1706 (1999)
13. I. Bloch, T. Hänsch, T. Esslinger, Phys. Rev. Lett. **82**, 3008 (1999)
14. T. Esslinger, I. Bloch, T.W. Hänsch, Phys. Rev. A **58**, 2664 (1998)
15. M. Kasevich, D.S. Weiss, E. Riis, K. Moler, S. Kasapi, S. Chu, Phys. Rev. Lett. **66**, 2297 (1991)
16. K. Gibble, S. Chang, R. Legere, Phys. Rev. Lett. **75**, 2666 (1995)
17. W. Petrich, M.H. Anderson, J.R. Ensher, E.A. Cornell, J. Opt. Soc. Am. B **11**, 1332 (1994)

Optical Components for a Robust Bose–Einstein Condensation Experiment

Ennio Arimondo and Maria Allegrini

1 Introduction

Within the last thirty years spectacular progress has been made in the experimental ability to probe and manipulate atomic/molecular species through laser spectroscopy techniques. Several elements have driven such progress. The first one, strictly technical, was the availability of advanced optical elements to a large research community, such as highly reflective mirrors, solid-state lasers, acusto-optical elements. While those elements were developed for other communities, the basic research community was able to make prompt use of them. These advanced optical elements placed in the hands of ingenuous researchers stimulated the development of new configurations for the optical probing of atoms, molecules and solids. Simultaneously the easy access to the spectroscopic determination of the internal and external variables stimulated new ideas for the optical manipulation of those variables. This flexible use of laser sources allowed the production of samples with very special properties, i.e., laser cooled atoms/molecules and quantum degenerate gases.

Over these years Theodor Hänsch has contributed with a steady output of new ideas to develop the tools required to laser-probe and manipulate atoms in different surroundings. As a result of these original contributions, the development and spread of laser spectroscopy methods has been largely inspired by Hänsch's published works.

Nowadays a physics laboratory dealing with laser spectroscopy includes in its routine operation a large number of devices and tools originated by the progress briefly sketched above, and thus by the research work of Hänsch. Often for a newcomer, for instance an undergraduate or graduate student, several parts of the laboratory apparatus appear to be made out of standard components, hiding the large effort in foresight, ingenuity and trial required for the realization of those components. Thus our contribution for the 60th birthday of Theodor Hänsch aims to present the close connection existing between the work published by his group and several components installed in our laboratories in the areas of laser spectroscopy, laser cooling, and partially installed also in the near-field scanning optical microscopy/spectroscopy laboratory.

Several optical components and techniques currently used in our laboratories and linked to the publications of Hänsch and coworkers will be briefly described. We will comment where modifications in the apparatus or experimental approach have been applied with respect to the publications that inspired our work. The second section deals with laser sources, the design of extended cavity diode lasers, and the realization of narrow-band diode lasers on the D_1 line of cesium. Section 3 discusses the frequency locking of laser diodes to saturated absorption lines of atomic samples. Section 4 discusses laser cooling and presents some results for the temperature reached in laser cooling. All the elements described here are part of the present experiment in Pisa on the Bose–Einstein condensation of rubidium atoms.

2 Laser Sources

2.1 Diode Lasers

Present research in atomic/molecular spectroscopy requires spectrally narrow laser sources. Solid-state diode lasers represent an economic, compact and efficient solution. Diode lasers do not cover all spectral regions, because their commercial use is concentrated around specific wavelengths. Nevertheless the diode lasers available in the near-infrared region have stimulated attention on rubidium and cesium atomic investigations, and also in our group [1–7]. Moreover other transitions, for instance those of atomic and molecular oxygen, are covered by diode lasers [8], and near-infrared lasers have been used in the operation of optical tweezers [9].

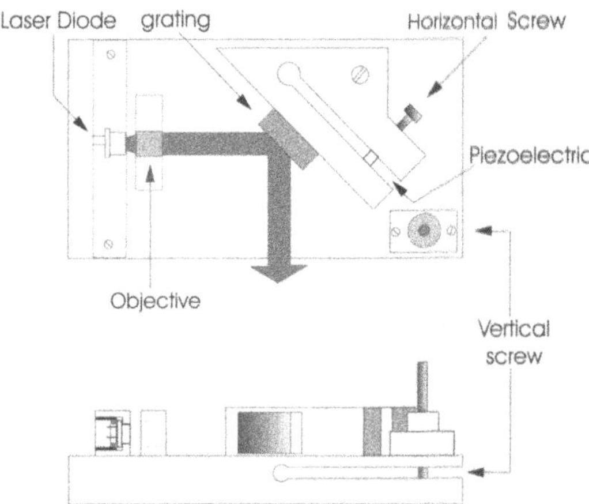

Fig. 1. Schematics of the mechanical setup of the grating stabilized diode laser system

The standard design producing diode lasers with the required frequency narrowing and long-term thermal stability is based on the work by Hänsch's group in 1995 [10]. It is composed of an external cavity containing an objective and a grating in the Littrow-mounting configuration. In our laboratory that design was reproduced, to a large extent, with the drawing of Fig. 1. Also the material the mounting is made of, the collimating lens, the piezoelectric transducer, the temperature sensor, and the Peltier elements for the temperature stabilization of the whole mounting have been chosen following the suggestions of the original reference [10]. Only the 1800 lines/mm grating from the Edmunds Company provides a cheap replacement for the better-quality model of the original design. We currently use diode lasers with an antireflection coating on the front face, offering a larger tuning range for the laser output frequency.

2.2 Cesium D_1 Laser

The development of laser spectroscopy has been strongly enhanced by the use of new laser sources. As an important example of such progress in 1995 the Hänsch group pointed out that a commercially available product, the C86135E/C86136E series laser diodes from EG&G Optoelectronics, could be tuned to the D_1 line of cesium [11]. That paper stimulated great attention within the laser cooling community, because the laser manipulation of cesium on the D_1 line opened new schemes for controlling or modifying the atomic momentum. As a consequence we also placed an order for those laser diodes and started experiments using that radiation. It was planned to apply D_1 laser radiation to laser cooled cesium atoms in order to produce a precisely controlled light shift within a scheme of one-dimensional velocity-selective coherent population trapping of cesium atoms [12]. At that point we realized that the laser diodes quoted above did not provide the laser power required for the planned experiment. Thus we started to search around between alternative commercially available products. When we realized that broad area lasers, with large output power, were available on the market, we tried to produce a single mode operation of the broad area laser by grating feedback and operation on an external cavity, i.e., operating the laser diode as in the Hänsch group papers [10, 11]. We reached our aim in [4] introducing some modifications into the standard external cavity design of Fig. 1 based on the Ricci et al. scheme [10]. In order to compensate for the astigmatism of the broad-area laser, we inserted into the external cavity a cylinder lens, beside the objective and the diffraction grating mounted in the Littrow configuration. The cylinder lens produces a long thin strip across the grating grooves and allowed the realization of a stable resonator in both the directions parallel and orthogonal to the laser junction. This configuration produced a reasonable single mode output power, but the quality of the output beam was not exceptional so that the coupling into the cesium cooled cloud was not very efficient. Since then we have injected the broad area laser diode using the radiation emitted from one EG&G diode, producing an output beam with moderate quality, but achieving a larger output power so that part of the power could even be wasted in the coupling with the cooled atoms.

3 Frequency Locking

In order to obtain the required-long term frequency stability, the diode lasers are locked to the narrow signal provided by the saturated absorption in a reference cell. Saturated absorption, produced by a weak laser probe counterpropagating through an absorbing medium together with a strong saturating laser beam, was originally introduced by Hänsch and coworkers in order to measure accurately the α-Balmer transition in hydrogen [13, 14].

At present, saturation spectroscopy is a standard technique accessible even to physics students in their laboratory courses. A simple set-up with counterpropagating pump and probe lasers, as described in standard textbooks, provides a signal-to-noise ratio large enough to generate the error signal required for frequency locking of the laser to an atomic line. Figure 2 shows a typical saturation spectroscopy signal obtained with cesium atoms on the D_2 resonance line for the hyperfine transitions $F_g = 4 \rightarrow F_e$. Both Doppler-free saturation signals and cross-cover signals are present in the figure. Optical pumping between hyperfine and Zeeman levels modifies the strength of the cross-over signals, as discussed in the original paper by Hänsch et al. [14], and again discussed in the framework of laser polarization spectroscopy [15–17]. Indeed we verified that the relative polarization of the pump and probe lasers modifies the strength of the cross-over signals.

Two saturated absorption set-ups are used in our laboratory to observe the Doppler-free signals, both shown in Fig. 3. In the upper scheme of Fig. 3, called Zeeman locking, the atomic cell is contained within a solenoid, around

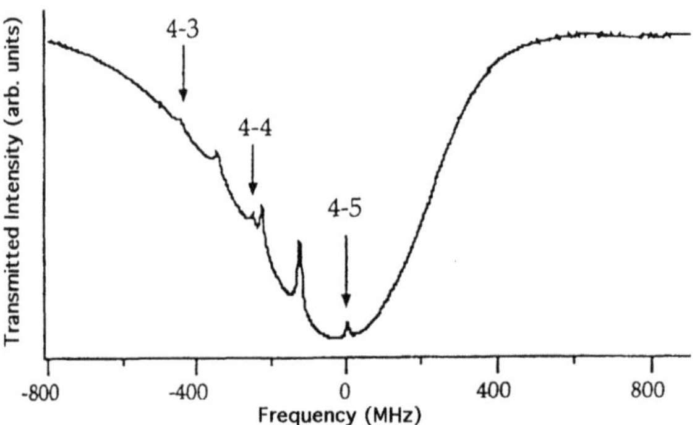

Fig. 2. Saturated absorption spectrum for the cesium hyperfine transitions $F_g = 4 \rightarrow F_e$ on the D_2 line. The marked peaks are produced by cross-over signals intermediate between the Doppler-free saturation signals

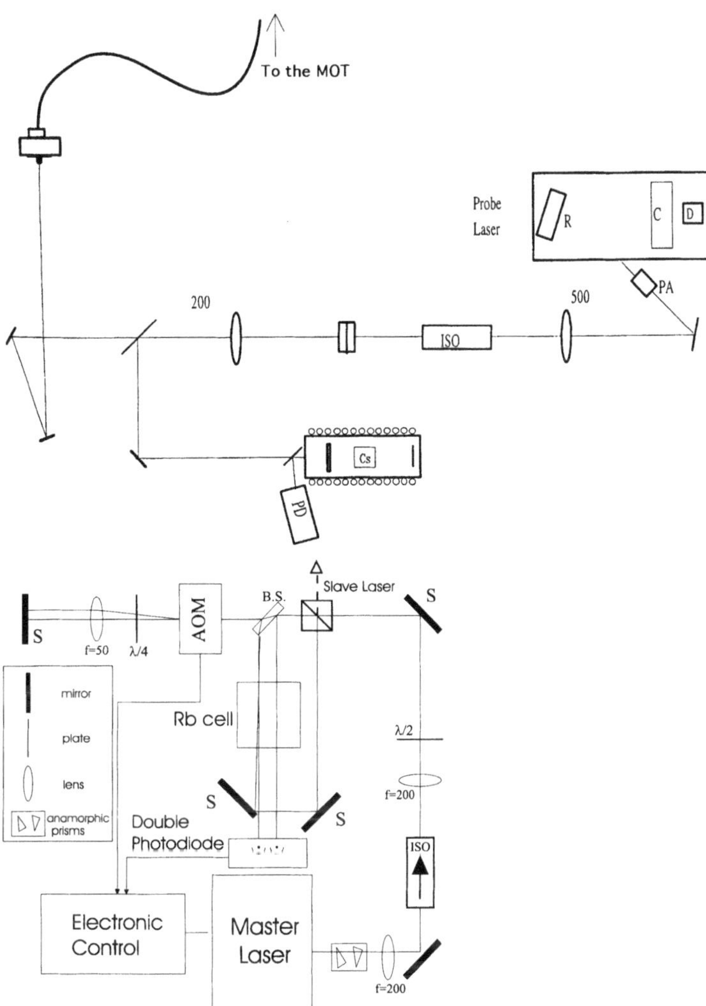

Fig. 3. *Upper scheme:* Zeeman-lock scheme for frequency locking the master laser to the saturated absorption peaks of cesium atoms contained inside a solenoid. *Lower scheme:* called frequency shift lock, the saturating laser beam is frequency shifted by an AOM operating in double pass

20 cm long, providing a weak magnetic field along the laser propagation direction. The solenoid is driven by a constant current furnishing a constant bias field, and by a modulating current, creating a modulating magnetic field. This field modulation produces a modulation on the probe signal monitored by the photodiode PD in the figure. The modulation signal, phase detected through a lock-in amplifier operating at the modulation frequency, generates an error signal applied to the piezoelectric device controlling the laser cavity length in

Fig. 1. No modulation signal, whence no error signal, is produced when the laser is in resonance with the atomic absorption within the magnetic field, thus keeping the laser frequency at the center of the absorption transition. The bias magnetic field shifts the frequency of the Doppler-free signals and allows one to tune the locked laser frequency around the resonance up to the value required for the cooling process.

In the lower scheme of Fig. 3 an acoustic-optical modulator (AOM) in double passage is inserted into the path of the saturating pump laser beam. As a consequence the saturation signals are obtained with a frequency offset given by the AOM frequency. A frequency modulation of the AOM generates, through lock-in detection, the error signal controlling the laser frequency. An additional AOM inserted between the frequency-locked master laser and the slave laser brings the slave laser to the frequency requested for the cooling process.

4 Laser Cooling

One of the most important contributions by Theodor Hänsch to atomic physics is represented by his work, in collaboration with Arthur L. Schawlow, proposing the cooling of gases by laser radiation [18]. That short and dense paper triggered in a few years, when lasers became more reliable, a flourish of activity on laser cooling, applying that technique to a large range of atoms and also simple molecules. [18] described in detail the characteristics of the laser cooling mechanism currently defined as Doppler cooling and representing the first stage of any current laser cooling experiment. Moreover that reference proposed to increase the efficiency of the cooling process by applying to the absorbing medium broadband laser radiation exciting half of the Doppler width. This laser cooling scheme has been applied in a few cases by implementing a frequency modulation to a single mode laser or by using a multimode laser.

In experiments dealing with alkali laser cooling and aimed at Bose–Einstein condensation, the Doppler cooling mechanism is masked by the polarization gradient mechanism, a process not forecast by the original work of Hänsch and Schawlow and understood only after its experimental observation. Also in our laser cooling investigations on cesium and rubidium, we easily reach the polarization gradient regime where temperatures lower than those associated to the Doppler limit are produced. However, in that regime the temperature depends on the number N of atoms contained in the laser cooled cloud, because the rescattering of spontaneously emitted photons limits the final temperature. This role of the rescattering produces a dependence of the laser cooling temperature T on the $N^{1/3}$ quantity as shown in the upper plot of Fig. 4. Moreover it has been experimentally verified that the temperature dependence on the atomic number N, the laser Rabi frequency Ω and the laser detuning $\delta < 0$ can be described through the unified scaling

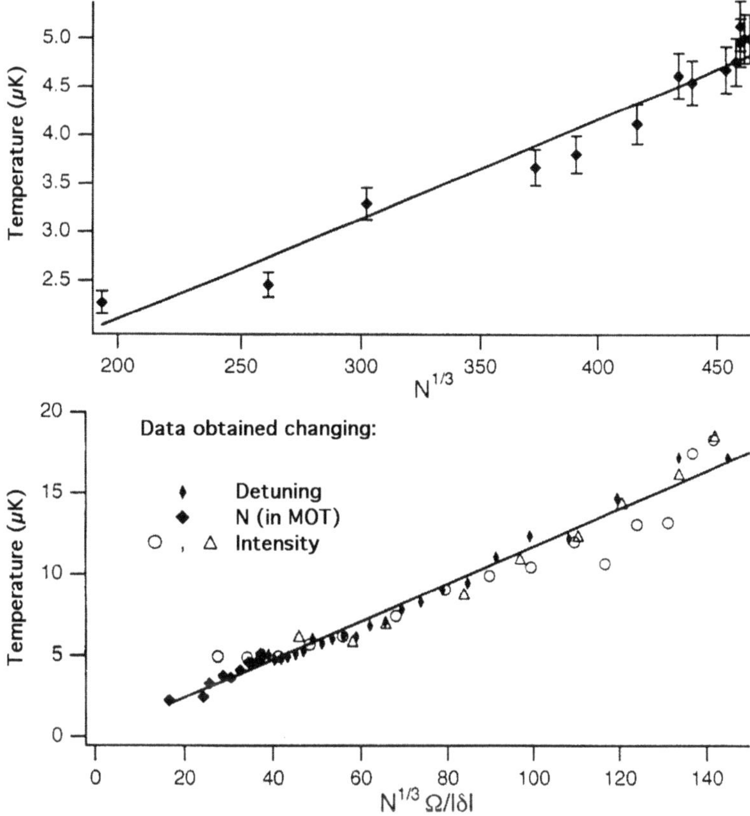

Fig. 4. *Upper plot:* the measured linear dependence of the cold cesium temperature T on the parameter $N^{1/3}$ at fixed Rabi frequency Ω and laser detuning δ. *Lower plot:* the linear dependence of the cold cesium temperature T on the scaling parameter $N^{1/3}\Omega/|\delta|$

parameter $N^{1/3}\Omega/|\delta|$, as shown by the data and the linear fit of the lower plot of Fig. 4. The temperatures of cesium atoms we achieved in different laser cooling schemes are discussed in [5].

5 Conclusions

We have presented technical schemes, optical configurations, experimental results of laser spectroscopy and laser cooling, all of them installed in the present set-up for the realization and investigation of the rubidium Bose–Einstein condensate. The optical configurations are simple to build, reliable to operate and offer the long-term stabilities required for measurements on the condensates. The experimental results of laser spectroscopy and laser cooling show that this apparatus allows the realization of very good control and manipulation of the atomic species we are dealing with. The results we have presented are strictly connected to previous research by Hänsch.

Acknowledgements

Our friendship with Theodor Hänsch started in 1969 at the Carberry Tower near Edinburgh (Scotland) for the Tenth Scottish Summer School, where the late Arthur Schawlow lectured on the early development of lasers. As a follow-up of that School, while Theodor started his post-doc and later splendid career at Stanford, we stopped in Reading (England) inspired by the lectures of the late George Series on optical pumping. In 1979, given the beautiful dye laser applications to nonlinear spectroscopy performed by Theodor Hänsch, we had the opportunity to spend a few months in Stanford. There while watching him play with a toy train along the corridors of the Varian building in order to develop a simple scheme for the wavemeter, we were asked by our little daughter to buy a similar toy! His return to Germany and the creation of the well equipped laboratory in Garching allowed us to establish a closer scientific collaboration between Pisa and Munich with exchanges in both directions. Recently his appointment as professor in Florence has provided new opportunities for his visits to Pisa. We have greatly appreciated Theodor Hänsch's scientific knowledge, but also his candid modesty and his willingness to contribute with original ideas and stimulus to other people's research work. We like to think of him as a modern "inventor".

The scientific results presented in this work are the collective effort of our students and coworkers quoted in the listed publications. We are grateful to all of them for their contributions.

References

1. S. Grego, M. Colla, A. Fioretti, J.H. Müller, P. Verkerk, E. Arimondo: Opt. Commun. **132**, 519 (1996)
2. C. Gabbanini, A. Evangelista, S. Gozzini, A. Lucchesini, A. Fioretti, J.H. Müller, M. Colla, E. Arimondo: Europhys. Lett. **37**, 251 (1997)
3. A. Fioretti, J.H. Müller, P. Verkerk, M. Allegrini, E. Arimondo, P.S. Julienne: Phys. Rev. A **55**, R3999 (1997)
4. D. Cassettari, E. Arimondo, P. Verkerk: Opt. Lett. **23**, 1135 (1998)
5. A. di Stefano, D. Wilkowski, J.H. Müller, E. Arimondo: Appl. Phys. B **69**, 263 (1999)
6. J.H. Müller, D. Ciampini, O. Morsch, G. Smirne, M. Fazzi, P. Verkerk, F. Fuso, E. Arimondo: J. Phys. B **33**, 4095 (2000)
7. G.D. Telles, W. Garcia, L.G. Marcassa, V. Bagnato, D. Ciampini, M. Fazzi, J.H. Müller, D. Wilkowski, E. Arimondo: Phys. Rev. **63**, 033406 (2001)
8. F. Cervelli, F. Fuso, M. Allegrini, E. Arimondo: Appl. Surf. Sci. **127/129**, 679 (1998)
9. S. Grego, E. Arimondo, C. Frediani: J. Biomed Phys. **2**, 332 (1997)

10. L. Ricci, M. Weidemüller, T. Esslinger, A. Hemmerich, C. Zimmermann, V. Vuletic, W. König, T.W. Hänsch: Opt. Commun. **117**, 541 (1995)
11. S.B. Ross, S.I. Kanorsky, A. Weis, T.W. Hänsch: Opt. Commun. **120**, 155 (1995)
12. C. Menotti, P. Horak, H. Ritsch, J.H. Müller, E. Arimondo: Phys. Rev. A **56**, 2123 (1997)
13. P.W. Smith, T.W. Hänsch: Phys. Rev. Lett. **26**, 740 (1971)
14. T.W. Hänsch, I.S. Shahin, A.L. Schawlow: Phys. Rev. Lett. **27**, 707 (1971)
15. C. Wieman, T.W. Hänsch: Phys. Rev. Lett. **36**, 1170 (1976)
16. R.E. Teets, F.V. Kowalski, W.T. Hill, N. Carlson, T.W. Hänsch: SPIE **113**, 80 (1977)
17. T.W. Hänsch: Nonlinear high-resolution spectroscopy of atoms and molecules, Proc. Int. School of Physics E. Fermi (North-Holland, Amsterdam 1977) p. 17
18. T.W. Hänsch, A.L. Schawlow: Opt. Commun. **13**, 68 (1975)

From Atoms to Single Biomolecules Through Bose–Einstein Condensates: Un Saluto da Firenze per Theodor

Francesco Cataliotti, Sven Burger, Paolo De Natale, Chiara Fort, Giovanni Giusfredi, Massimo Inguscio, Francesco Minardi, Pablo Cancio Pastor, and Francesco Pavone

The contribution of Theodore Hänsch to scientific progress in the field of the interaction between light and matter has been impressive, continuous and ubiquitous during the last 35 years. The oldest of us still remember the world impact of the first saturation and two-photon spectroscopy results which were only the beginning of a long series of inventions. We have now the privilege of benefiting from a stimulating interaction with a Theodor "fiorentino". Discussions have produced the birth and death of many ideas, but they have also reinforced the image of a scientist, curious and ready to enthusiastically follow new adventures.

In this work we present new results from recent experiments at LENS on a Bose–Einstein condensate interacting with laser beams, on the manipulation of single biomolecules with magnetic fields and on the laser spectroscopy of helium. We hope that Theodor's imagination will be attracted by the images of these new phenomenologies that we send him like a special "cartolina di auguri".

1 Condensates in an Optical Lattice

Bose–Einstein condensates (BEC) confined in optical periodic potentials have a remarkably wide spectrum of applications. Since the pioneering work of Anderson and Kasevich [1] exploring macroscopic quantum interference, such systems have been used to experimentally investigate phenomena ranging from the emergence of squeezed-like states [2], to the study of phase coherence [3, 4], to the investigation of superfluidity on a local scale [5] to the emergence of topological effects in the condensation process [6], to the demonstration of 1D Josephson junction arrays [7]. Many theoretical proposals have also been suggesting the use of these systems for the study of classical and quantum decoherence [8, 9], nonlinear propagation [10] and quantum logic [11]. The main reason for this broad interest lies in the possibility to vary in a continuous way the periodic potential strength from zero to much higher than any other energy scale in the problem, thus realizing an almost ideal tool to investigate phenomena peculiar to condensed matter physics as well as solid state physics.

Unlike in all other experiments of this kind we do not first produce a BEC in a magnetic trap and then load it into an optical potential. On the contrary, we directly condense inside the combined potential of a blue detuned optical standing wave superposed on a harmonic magnetic trap. This enables us to investigate the effect of the gas temperature on the new phenomenology emerging from the presence of the optical potential. In particular, here we will focus our attention on the behaviour of mixed atomic clouds where a condensate is initially in equilibrium with a substantial thermal cloud. We will show how the presence of the periodic potential allows the coherent nature of the BEC to dramatically appear both in the expansion of the ground state of the system as well as in the dynamical behaviour.

Our experimental apparatus has been described in detail elsewhere [5, 12]. ^{87}Rb atoms are loaded from a double magneto-optical trap apparatus into a cigar-shaped magnetic trap (incidentally, of the type developed in Theodor's group for BEC [13]), where they are evaporatively cooled to slightly above the BEC transition temperature. At this stage we switch on the optical lattice (the light comes from a Hänsch-type extended-cavity diode laser [14]) oriented along the magnetic-trap axis and continue the evaporation process to the desired temperature. In the experiments described here the optical potential height is much higher than the temperature of the atoms. This ensures that each well in the lattice is isolated from the adjacent ones except for tunneling. Condensation in our system therefore takes place almost independently in each well, while the non-negligible tunneling rate ensures that phase coherence is maintained over all sites [4, 6]. Releasing the atoms from the combined trap will lead to interference of the atomic ensembles emerging from the various wells. Unlike the case of the expansion of an atomic cloud confined in a double trap, for which interference fringes will show up even in absence of coherence among the trapped samples, in the multiple-well case we expect interference peaks to appear only if a well-defined phase relation exists among all the different clouds. The situation is analogous to the use of a diffraction grating as opposed to a double-slit experiment in optics. Indeed, in the expansion the condensed cloud will form interference peaks pulling out from each other with a velocity of $2h/m\lambda$ (λ is the laser wavelength and m the atomic mass). These correspond to the momentum components in which the modulated condensate wavefunction can be decomposed. In Fig. 1 we show a series of pictures of the expansion of the condensate from the mixed trap taken for different release times. After the first few milliseconds where the limited resolution of the imaging system does not allow us to observe a spatial pattern, we can observe the appearance of two side-peaks of condensed atoms carrying twice the momentum of the trapping laser, corresponding to the first diffraction orders. The central feature corresponding to the zero-order diffraction peak instead is composed of a mixed cloud. By fitting the Gaussian wings of this cloud we can deduce the temperature of the system. Note that due to the incoherent nature of the thermal cloud no thermal component is present around the diffracted orders. The interferogram shown in Fig. 1 provides us with all the relevant information about the ground state of the system. From the width of the diffracted peaks we can deduce the mutual coherence of the sub-condensates in each lattice site, while from the ratio of the intensities one can extract the sub-condensate sizes [4].

The expansion interferogram therefore shows a distinct difference between the behaviour of a coherent and an incoherent cloud of atoms. But perhaps the most spectacular manifestation of this difference appears in the dynamics of the system. Indeed, when the atoms are driven to oscillate in the harmonic potential through the optical lattice barriers we observe a complete separation of the coherent and incoherent parts of the cloud. The condensate atoms oscillate in the potential, coherently tunneling through the optical barriers,

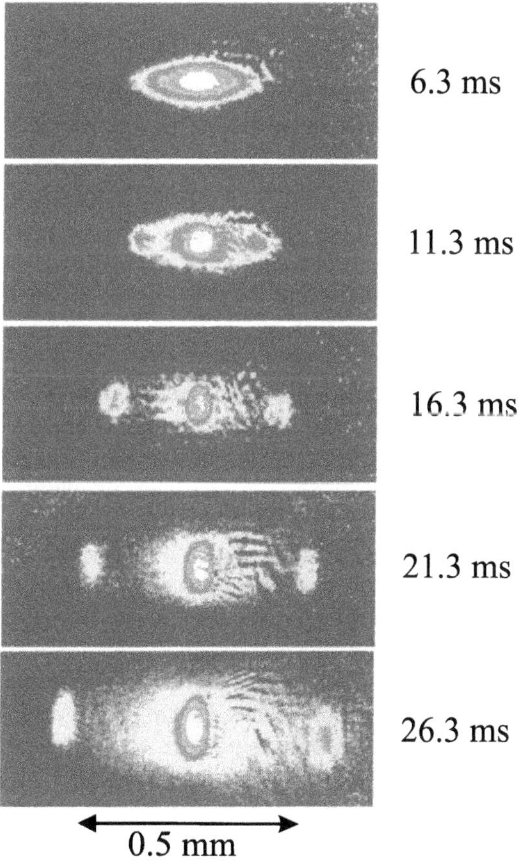

Fig. 1. Expansion of the Bose–Einstein condensate from the combined trap. Two interference peaks fly away from the original cloud at 11.4 mm/s (or twice the recoil velocity of one lattice photon $2h/m\lambda$). The Gaussian cloud around the central peak comes from non-condensate atoms and allows the measurement of temperatures

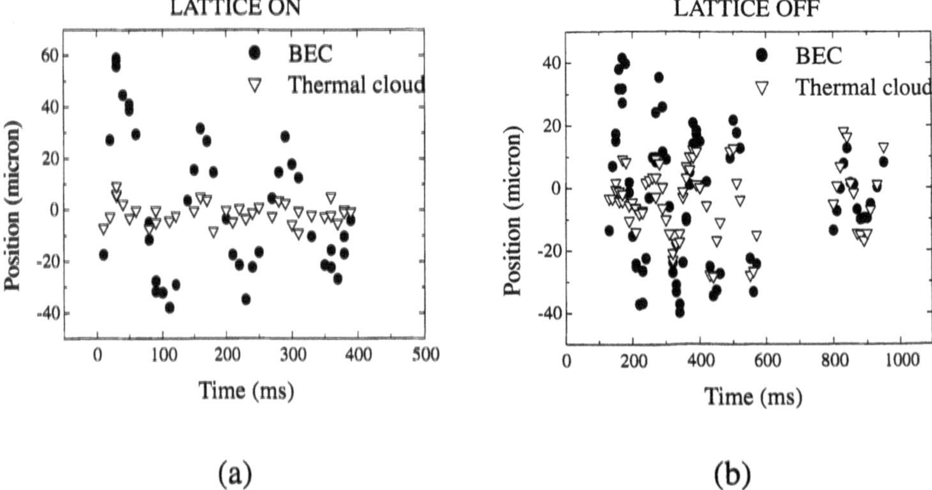

Fig. 2. (**a**) Center-of-mass oscillation in the combined trap of the BEC and of the thermal cloud after displacement of the BEC position. (**b**) Center-of-mass oscillation in the purely harmonic trap of the BEC and of the thermal cloud after displacement of the BEC position

while the thermal component remains trapped in the single wells and provides dissipation. We took advantage of this observation [7] to perform the following experiment. After we have created the mixed cloud in the combined trap we suddenly displace the magnetic trap along the optical-lattice axis, and then we wait half an oscillation cycle and bring the magnetic trap centre back to its original position. At the end of this procedure the condensate and the thermal cloud both have zero velocity but the condensate fraction now sits on the side of the magnetic potential while the thermal cloud is still at the bottom. In Fig. 2a we report the centre-of-mass position of the condensate and thermal clouds as a function of time spent in the trap after the displacing procedure is performed. The condensate starts to oscillate coherently, tunneling through the barriers of the optical lattice. However, there is dissipation due to collisions with the thermal atoms.

In Fig. 2b we again show the centre-of-mass positions of the condensate and thermal components during the oscillation, but this time the optical lattice is switched off after the displacing procedure is performed. Again, the condensate starts to oscillate and collides with the thermal component, but this time there is nothing to hold the thermal cloud fixed and the cloud is dragged along with the condensate. In essence the behaviour of the system reflects the fact that elastic s-wave collisions damp out the relative centre-of-mass motion of the two systems.

These experiments show that the optical lattice gives the possibility of independently controlling the dynamics of the coherent and incoherent part of an atomic ensemble. This would allow us, for example, to completely separate a thermal cloud from a condensate, obtaining a "zero temperature" sample. In practice one would have a new atom-optical element capable of filtering out the incoherent part from an atom laser output, thus realizing a totally new kind of output coupler.

2 Optical and Magnetic Manipulation of Single Biomolecules

Ashkin [15] demonstrated in 1986 the possibility of trapping a single dielectric particle by means of an extremely focused Gaussian laser beam (optical tweezers). The dipole interaction of the laser beam with a dielectric particle allows its trapping in the higher intensity region (centre of the focus) by means of a gradient force. The optical tweezers were used for the manipulation of micrometre and submicrometre dielectric particles attached to single biomolecules [16] (optical tweezers allow application of forces to dielectric spheres of micrometre sizes ranging from a few piconewtons to several hundred piconewtons). Also, Theodor Hänsch used the dipole-force interactions to trap neutral atoms in a periodic structure [17]. Here, two standing waves, oscillating with a $\pi/2$ time phase delay, confined the atoms by dipole forces to the intensity maxima, forming a two-dimensional collimated array of linear de Broglie waveguides, spaced by half an optical wavelength.

Magnet-based separation systems [18] (used often in biochemical protocols) have inspired researchers to use magnets to apply forces on magnetic beads in single biomolecule manipulation studies [19, 20]. Magnetic tweezers exert on magnetic beads weaker forces than do optical tweezers (typically, unless very powerful magnets are used, from a few tens of femtonewtons to a few tens of piconewtons); but, unlike optical tweezers, they are able to rotate the beads. This feature allows application of a torque on the molecule attached to the magnetic bead [20].

A very attractive and compact design of magnetic manipulator, developed by Theodor Hänsch et al. [21] for atom trapping, inspired us to develop a new hybrid system based on an integrated design of optical and magnetic tweezers. With this system we have rotated magnetic beads trapped by optical tweezers with a magnetic manipulator. The latter has been obtained following a new design inspired by the tip-trap geometries [21]. This magnetic manipulator (Fig. 4) allows the rotation of magnetic beads on each one of the three axes (3D rotation) while they are trapped by optical tweezers. Rotation of beads is obtained by means of a time orbital potential (TOP) operation similar to those used for the first time in atomic physics [22]. The position of the bead is detected by means of an interferometric scheme based on DIC (differential interference contrast) detection and allows sensitivities of nanometres in milliseconds [23].

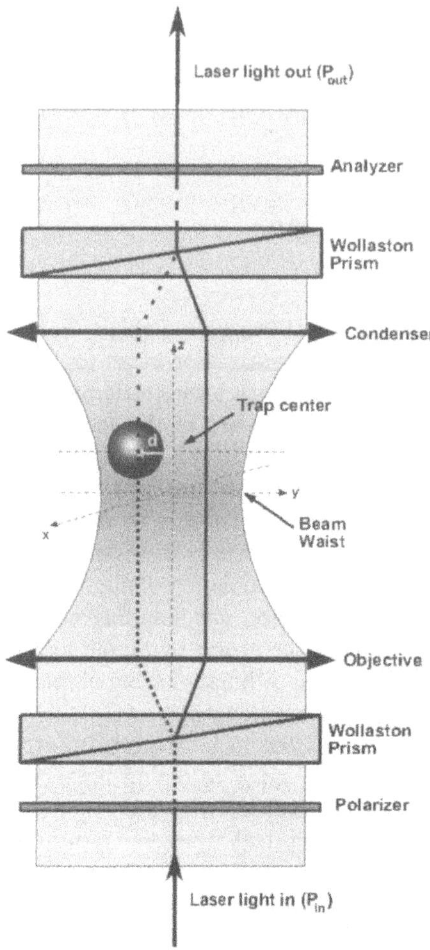

Fig. 3. DIC detection scheme

The working principle of the DIC position detection system is illustrated in Fig. 3. Polarization of laser light is initially selected by a polarizer to reach the first Wollaston prism with the right polarization angle. The first Wollaston prism splits the laser light into two beams with orthogonal polarization, then focused by the objective to two overlapping diffraction-limited spots (a few hundreds of nanometers apart) working as a single optical trap. The second Wollaston prism, after the passage through the specimen, recombines them into one single elliptically polarized beam (linear if no objects are present in the laser beams paths). An analyser with the polarization axis orthogonal with respect to the polarizer axis measures the two beams dephasing; the outgoing laser light power (P_{out}) will be zero when a bead is trapped in the middle of the two spots (equal optical paths for the two beams), but will be not zero when the bead is displaced from the trap center.

Detection of scattered light (P_{out}) through the DIC detector allows us to obtain the bead displacement (d) and the force exerted by the optical tweezer (F). For small displacements the force exerted by the tweezer is harmonic and the position detector response is linear: $F = -kd$, $d = \beta V$. Knowledge of the two parameters k (the tweezer spring constant) and β (the position detector conversion factor (PCDF)) is sufficient for a complete calibration of the optical tweezers. Different calibration methods can be used to obtain k and β (drag-force method, power spectrum method, equipartition theorem method, time-of-flight method, etc., see [24]), but they all make use of Stokes' law. Using Stokes' law, the force (F) due to a viscous fluid flow is proportional to the viscous coefficient (η), the bead radius (R) and its velocity relative to the fluid (v): $F = -6\pi\eta Rv$. The specimen is always immersed in a water solution inside a microscope slide, so it is possible to calibrate the tweezers force against the fluid viscous force.

The optical tweezer was integrated with a magnetic manipulator constituted by eight independent electromagnets,each with a tip-pole expansion (Fig. 4). Eight coils were mounted in an optimized parallel-piped structure. The mounting structure, the core of the coils and the tip-poles were made of nickel-plated soft iron, while the screws were made of zinc-plated iron. All the components were designed in order to keep constant the magnetic field flow along the structure (constant section). The trap ($44 \times 44 \times 49\,\mathrm{mm}^3$) was mounted, by means of an aluminum support (not shown in Fig. 2), on the microscope collimator in order to overlap the objective focal plane with the centre of the magnetic trap. Each coil (mean radius of about 3 mm) was made of 180 turns of copper wire (cross-section of about $0.2\,\mathrm{mm}^2$), giving a resistance of 350 mW and an inductance of 570 mH. Coils were driven by

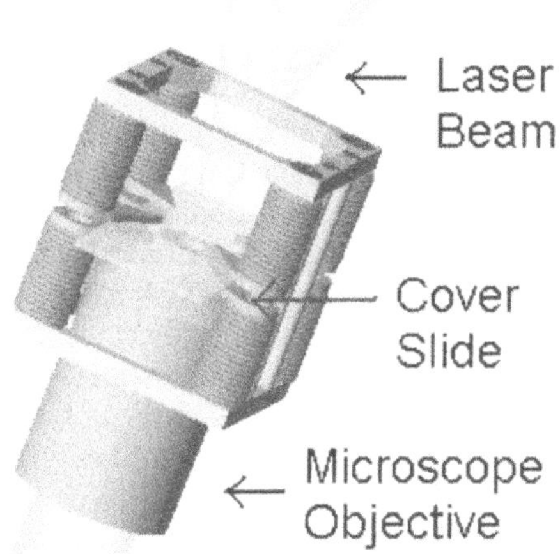

Fig. 4. The micro TOP manipulator

a personal computer through a current boost. The magnetic field generated in a typical configuration (current 2 A for each coil) was about 350 G near the tip-pole and about 70 G in the middle region of the trap.

We have trapped super-paramagnetic beads with diameters ranging from 0.4 μm to 2.6 μm by using the optical tweezers and obtaining higher forces with respect to the polystyrene beads. Using the viscous drag method calibration 9-10, we have measured a force of 230 pN with 150 mW of input power at 1064 nm (using super-paramagnetic beads R99-15 Merck Eurolab, France; mean diameter 2.6 μm). Similar measurements on polystyrene beads have shown slightly weaker forces (210 pN). We believe this difference may be related to the different index of refraction between the two kinds of beads. The use of a water-immersed objective allows a trapping depth inside the fluid chamber of more than 200 μm. We did not observe any apparent dramatic heating on the beads when trapped with laser radiation. By using the magnetic manipulator we have rotated the single beads and clusters of beads with a TOP (time orbital potential) operation. The TOP configuration was generated by creating a uniform field B_o parallel to the plane of rotation, and rotating it as $B = B_o e^{i\omega t}$. We have estimated a maximum magnetic force momentum G'' of 10 – 16 Nm (with 2.6 μm super-paramagnetic beads, magnetic mass susceptibility $= (100 \pm 25) \times 10^{-6}\,\mathrm{m^3/kg}$) and a maximum rotation frequency (100 Hz) limited by the inductance of the circuit. With optimum current configuration, it was possible in principle to choose an arbitrary axe of rotation. By using magnetic beads with some imperfections (irregularities in the spherical shape) in order to visualize the rotation, we have also observed single-bead rotation while optically trapped. The detection system allowed detection of the trapped bead position within the dimension of the laser focus; in particular we were limited to the Brownian motion of the bead, obtaining a sensitivity of a few nm in 1 ms (all particulars may be found in [25]).

Currently we are applying this technique to the study of DNA–protein interactions at the single-molecule level. In particular, we are working on the theme of lactose digestion in bacteria. For this purpose we have considered the lactose repressor (Lac repressor) molecule, which is a tetrameric protein (Molecular weight = 152 kDa). It participates in the lactose metabolism in bacteria: it regulates the expression of those genes, which allow for the production of all the substances, which makes possible the digestion of the lactose molecule. The regulation of the expression of bacterial genes is a fundamental process; it means that the information coded in the bacterial DNA has to be read and transformed into the production of proteins, but in a way controlled by some regulatory molecules like the Lac repressor. Regulation is achieved by formation of a loop in the DNA structure. The disposal of the lactose molecule inside the bacterium cell starts a mechanism in which expression of bacterial DNA is enhanced, while on the contrary the absence of lactose represses this

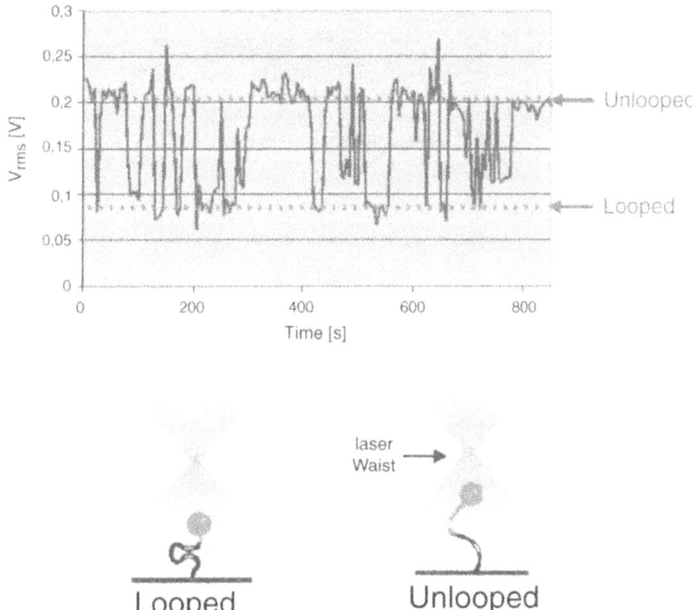

Fig. 5. RMS Signal vs time for a 1 µm length of DNA with a 400 nm diameter bead. Two levels are visible, corresponding to the looped and unlooped states respectively

expression. Bacterial repressors were the first regulatory proteins of any kind to be identified, and the Lac repressor was the first to be isolated.

The aim of our research is to study the influence of DNA twisting on the characteristic times of the interaction between the Lac repressor and DNA, at the single-molecule level. By tracking the bead's Brownian motion by means of laser light, we can distinguish the DNA looped and unlooped states. With the TOP trap we can twist the DNA, both in the positive and in the negative way, by twisting a magnetic bead attached to the DNA molecule. Figure 5 shows measurements of the looping-unlooping transitions as a function of time. The vertical axis corresponds to the RMS noise induced in the transmitted laser beam by the Brownian motion of the bead attached to the DNA molecule. The variations of the tether elongation, due to the attachment of the Lac Repressor, are monitored by a focused laser beam. By detecting the noise induced in the beam transmission, it is possible to detect the looping transitions.

3 Helium Spectroscopy

Hydrogen is the simplest atom and, as such, has always had a central role in the scientific research of Theodor Hänsch. Tests of quantum electrodynamics (QED) and determinations of fundamental constants, like the Rydberg

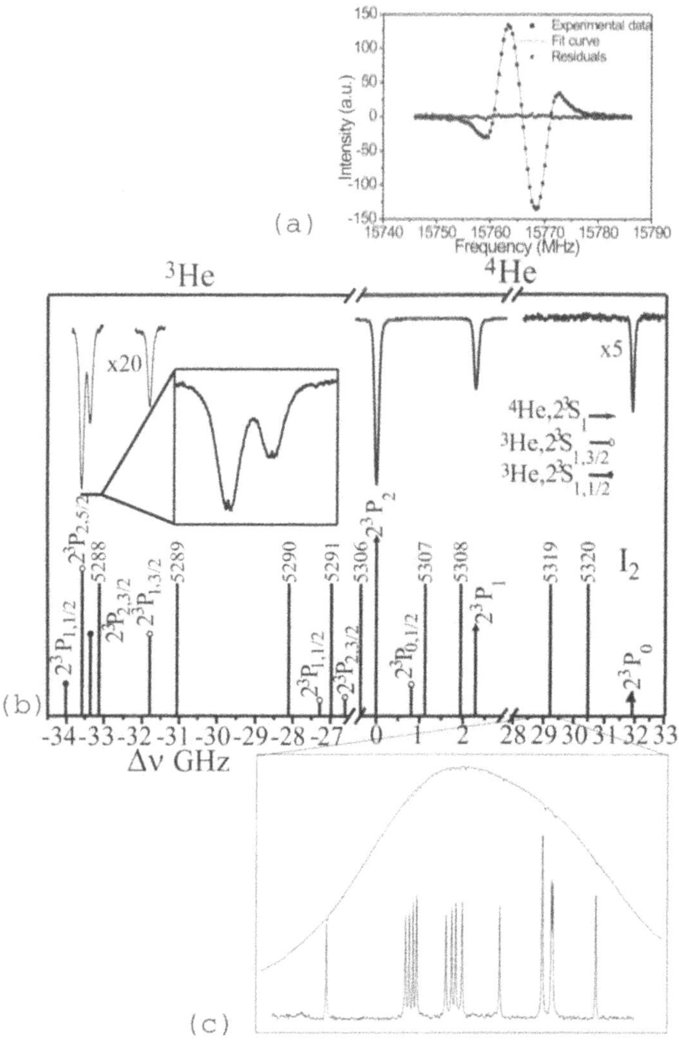

Fig. 6. (**a**) A typical third-derivative helium recording, obtained with the LENS apparatus; (**b**) the possible coincidences among the rich I_2 line spectrum and the lines of He_4 and He_3 that also have a hyperfine structure; (**c**) the hyperfine components of a I_2 line recorded both in direct absorption, along with the Doppler profile, and using a frequency-modulation (FM) technique

constant, have become more and more accurate thanks to the spectroscopic measurement of hydrogen transition wavelengths or frequencies. Work on the hydrogen atom inspired the more recent work performed at LENS in Firenze on the "second" simple system, helium. Experimental research on helium has widely benefitted from advances in atomic theories, especially from an extension of Dirac theory to deal with its spectrum. Helium has therefore become the prototypical many-body system and a unique playground to test, for instance, calculations of the electron–electron Lamb shift which, of course, cannot be found in hydrogen. In 1993, soon after the first pure frequency measurement had been performed on hydrogen, by using a Schottky diode as mixer [26], the first frequency measurement was performed on helium. The absolute frequency of the $2\,^3S_1$–$3\,^3P_0$ transition at 389 nm was measured to 2.4 parts in 10^{10} and the Lamb shift of the metastable $2\,^3S_1$ level was determined with an accuracy of 1.9 parts in 10^4, by using as a frequency reference a previously measured ^{87}Rb two-photon transition that was about 82 GHz from the helium frequency [27].

For measuring the atomic fine structure constant α, helium was early recognized as a better candidate than hydrogen, due to its larger fine-structure splitting and longer lifetimes. In particular, for the helium $2P$ state, the ratio fine-structure interval/natural linewidth is about 2 orders of magnitude higher than in hydrogen. An experiment was therefore planned at LENS, with the aim of measuring the fine-structure energy splittings, in the $2\,^3P_\mathrm{J}$ manifold, as the difference of two optical transition frequencies, by exciting an atomic beam of helium atoms to the $2\,^3S_1$ metastable state, using a DC discharge. At that time, diode lasers were finding increasing application in hydrogen spectroscopy, especially in Hänsch's group in Munich [28]. Hänsch's group in particular developed frequency-interval dividers using grating-tuned diode lasers. Studies at LENS were started to test the suitability of diode lasers with a DBR structure, which had recently become available, to excite the helium transitions at a wavelength of 1083 nm. An extended cavity configuration, that reduced the laser linewidth to less than 200 kHz and a frequency-offset phase-locked technique [29] were found useful for scanning the helium lines. A 2 kHz accurate determination of the $2\,^3P_0$–$2\,^3P_1$ energy splitting in helium was thus obtained [30]. When this experiment was running, an important line-centre frequency shift was also observed and studied, due to mechanical effects of light in the saturation-spectroscopy configuration [31]. One of the limiting factors to improving the accuracy of these frequency measurements was the lack of accurate and stable frequency standards in the near-infrared region. Recently, advances in nonlinear optical techniques have made it possible to build a new frequency reference, based on a diode laser, a Yb^+ fiber amplifier and a periodically-poled $KTiOPO_4$ crystal [32]. The underlying idea was to double the frequency of diode lasers at 1083 nm wavelength to access the rich and metrological spectrum of I_2 in the green region for getting an improved stability and to make possible absolute frequency measurements of He transitions. To build this new device a significant effort was made in Firenze to test the suitability of Yb^+-doped fiber amplifiers and doubling techniques for spectroscopic applications [33]. Measurements performed on helium using this frequency reference have been performed, showing a stability of 1.9 in 10^{12} at 1 s, improving to 4.1 in 10^{13} at 300 s [34].

This new set-up enables us to make an a posteriori absolute frequency measurement of the I_2 component used as reference and, thus, of He transitions, with respect to a frequency standard. To this purpose, the new technique of femtosecond laser comb generation [35], introduced by Theodor Hänsch, would be perfectly suited and its latest astonishing result, i.e. one entire octave spectral coverage, could even directly refer the $2\,^3S_1$–$2\,^3P_J$ transitions of atomic helium to the cesium primary standard.

References

1. B.P. Anderson, M.A. Kasevich, Science **282**, 1686 (1998)
2. C. Orzel, A.K. Tuchman, M.L. Fenselau, M. Yasuda, M.A. Kasevich, Science **291**, 2386 (2001)
3. M. Greiner, I. Bloch, O. Mandel, T.W. Hänsch, T. Esslinger, cond-mat/0105105
4. P. Pedri, L.P. Pitaevskii, S. Stringari, C. Fort, S. Burger, F.S. Cataliotti, P. Maddaloni, F. Minardi, M. Inguscio, cond-mat/0108004
5. S. Burger, F.S. Cataliotti, C. Fort, F. Minardi, M. Inguscio, M.L. Chiofalo, M. Tosi, Phys. Rev. Lett. **86**, 4447 (2001)
6. S. Burger, F.S. Cataliotti, C. Fort, P. Maddaloni, F. Minardi, M. Inguscio, cond-mat/0108037
7. F.S. Cataliotti, S. Burger, C. Fort, P. Maddaloni, F. Minardi, A. Trombettoni, A. Smerzi, M. Inguscio, Science **293**, 843, (2001)
8. L. Pitaevskii, S. Stringari, cond-mat/0104458
9. A. Cuccoli, A. Fubini, V. Tognetti, R. Vaia, cond-mat/0107387
10. F.Kh. Abdullaev, B.B. Baizakov, S.A. Darmanyan, V.V. Konotop, M. Salerno, cond-mat/0106042
11. R. Burioni et al., cond-mat/0004100
12. C. Fort et al., Europhys. Lett., **49**, 8, (2000)
13. T. Esslinger, I. Bloch, T.W. Hänsch, Phys. Rev. A **58**, R2664, (1998)
14. L. Ricci, M. Weidemüller, T. Esslinger, A. Hemmerich, C. Zimmermann, V. Vuletic, W. König, T.W. Hänsch, Opt. Commun. **117**, 541 (1995)
15. A. Ashkin, J.M. Dziedzik, J.E. Bjorkholm, S. Chu, Opt. Lett **11**, 288 (1986)
16. A. Ashkin, J.M. Dziedzik, Science **235**, 1517 (1987)
17. A. Hemmerich, T.W. Hansch, Phys. Rev. Lett., **70**, 410 (1993)
18. R.R. Birss, R. Gerber, M.R. Parker, IEEE Trans. Magn. **12**, 829 (1976); J.H.P. Watson, A.S. Bashaj, IEEE Trans. Magn. **25**, 3803 (1989)
19. S.B. Smith, L. Finzi, C. Bustamante, Science, **258**, 1122 (1992)
20. T.R. Strick, J.-F. Allemand, D. Bensimon, A. Bensimon, V. Croquette, Science, **271**, 1835, (1996)
21. V. Vuletic, T.W. Hänsch, C. Zimmermann, Europhys. Lett., **36**, 349 (1996)
22. W. Petrich, M.H. Anderson, J.R. Ensher, E.A. Cornell, Phys. Rev. Lett. **74**, 3352 (1995)

23. K. Svoboda, C.F. Schmidt, B.J. Schnapp, S. Block, Nature **365**, 721 (1993)
24. M. Capitanio, G. Romano, R. Ballerini, D. Dunlap, L. Finzi, M. Giuntini, F.S. Pavone, submitted to Rev. Sci. Instrum.
25. L. Sacconi, G. Romano, R. Ballerini, M. Capitanio, M. de Pas, D. Dunlap, M. Giuntini, L. Finzi, F.S. Pavone, in press for Opt. Lett, Sept. 2001
26. F. Nez, M.D. Plimmer, S. Bourzeix, L. Julien, F. Biraben, R. Felder, Y. Millerioux, P. De Natale, Europhys. Lett. **24**, 635 (1993)
27. F.S. Pavone, F. Marin, P. De Natale, M. Inguscio, F. Biraben, Phys. Rev. Lett. **73**, 42 (1994)
28. Th. Udem, A. Huber, B. Gross, J. Reichert, M. Prevedelli, M. Weitz, T.W. Hänsch, Phys. Rev. Lett. **79**, 2646 (1997)
29. M. Prevedelli, P. Cancio, G. Giusfredi, F.S. Pavone, M. Inguscio, Opt. Commun. **125**, 231 (1996)
30. F. Minardi, G. Bianchini, P. Cancio, G. Giusfredi, F.S. Pavone, M. Inguscio, Phys. Rev. Lett. **82**, 1112 (1999)
31. F. Minardi, M. Artoni, P. Cancio, M. Inguscio, G. Giusfredi, I. Carusotto, Phys. Rev. A **60**, 4164 (1999)
32. P. Cancio, P. Zeppini, A. Arie, P. De Natale, G. Giusfredi, G. Rosenman, M. Inguscio, Opt. Commun. **176**, 453 (2000)
33. R. Paschotta, D.C. Hanna, P. De Natale, G. Modugno, M. Inguscio, P. Laporta, Opt. Commun. **136**, 243 (1997); D.J.E. Knight, F. Minardi, P. De Natale, P. Laporta, Eur. Phys. J. D **3**, 211 (1998); P. Cancio, P. Zeppini, P. De Natale, S. Taccheo, P. Laporta, Appl. Phys. B **70**, 763 (2000)
34. N. Picqué, P. Cancio, G. Giusfredi, P. De Natale, J. Opt. Soc. Am. B, **18**, 692 (2001)
35. S.A. Diddams, D.J. Jones, J. Ye, S.T. Cundiff, J.L. Hall, J.K. Ranka, R.S. Windeler, R. Holzwarth, T. Udem, T.W. Hänsch, Phys. Rev. Lett. **84**, 5102 (2000)

Cavity Cooling with a Hot Cavity

Vladan Vuletić

Cavity cooling of arbitrary particles by coherent scattering is significantly improved inside resonators with optical gain. The friction force arises from frequency-selective amplification of the radiation emitted by the particle. As possible implementations of active cavity cooling we discuss a laser below threshold and an injection-locked laser system.

1 Introduction

Whereas from a classical point of view the emission of radiation by an oscillating atomic dipole is a local property determined by the magnitude of the dipole moment and the angular frequency of oscillation ω, quantum electrodynamics introduces the more comprehensive concept of delocalized electromagnetic modes to describe spontaneous emission or scattering. In classical electrodynamics, the ω^4-dependence of the spontaneously emitted or Rayleigh scattered power arises from the Larmor formula for the emission by an accelerated point charge. From a quantum mechanical perspective, the spontaneous emission and the scattering rate are both proportional to the density of electromagnetic modes at the emission frequency. In free space, the mode density scales as ω^2, and multiplication by the square of the matrix element for the atom light coupling, scaling as $\omega/\hbar$, and the photon energy $\hbar\omega$ then yields the classical expression. The strong ω^4-frequency dependence is responsible not only for the blue sky, but also explains why the edible laser [1], rather than the edible maser, was experimentally realized.

If we change the density of electromagnetic modes available to the atom, e.g. by placing a resonator around it, then the emission rate is modified accordingly [2–4]. In this sense spontaneous emission is as much an atomic property as that of the electromagnetic vacuum surrounding the atom. Since the density of electromagnetic modes is a nonlocal feature determined by boundary conditions away from the atom, it is possible to influence the atomic emission rate by manipulating spatially extended electromagnetic modes.

Conventional Doppler cooling [5], which makes use of the conservation of momentum in scattering, was such a fantastic idea that for some time even its inventors had doubts as to its feasibility. Can you imagine simply shining light into a vapor cell and observing atoms near $-273\,^\circ\mathrm{C}$ moving at the speed of an ant? Then how long before we can count the frequency of light, perform precision spectroscopy of antihydrogen, implement magnetic motors to move Bose–Einstein condensates, realize an atomic laser, or maybe build time machines?

Doppler cooling relies on the anisotropic absorption of light by a moving two-level atom, where cooling events are favored over heating events because the incident light is tuned below the atomic resonance by an amount on the order of the Doppler effect. Conventional Doppler cooling therefore requires a closed two-level system. Alternatively, by utilizing the frequency variation of the electromagnetic mode density in resonators, it is possible to devise optical cooling schemes that arise from an asymmetry in emission, rather than in absorption [6–9]. The frequency-dependent modification of emission rates inside a resonator implies a corresponding change of the light-induced mechanical forces on the atom. Since under certain circumstances the sign of the dissipative force depends exclusively on the detuning between incident light and resonator, without reference to the atomic level structure [9], this cavity cooling technique holds promise for generalizing laser cooling to arbitrary light scatterers, including atoms with a complicated level structure, molecules, or even mesoscopic particles.

Several authors have studied how to use a modification of the boundary conditions for the electromagnetic field, i.e. a passive optical resonator, to laser cool atoms in various situations, either by means of spontaneous emission from a two-level system [6–8, 10], or by means of classical (coherent) Rayleigh scattering [8–12]. In this chapter we analyze how a resonator with an intracavity gain medium, i.e. an active cavity, can be used to significantly improve the performance of cavity cooling, resulting in a larger cooling force and a lower temperature than those attainable with passive resonators.

2 Cavity Cooling with an Intracavity Gain Medium

The basic idea of cavity Doppler cooling or cavity sideband cooling in a passive resonator is to enhance the coherent (classical) scattering of blue photons over that of red photons, thereby extracting energy and entropy from the scatterer [9, 12]. By analogy to conventional Doppler cooling, the maximum cooling force is of order

$$f_{\max} = \hbar k \Gamma_0', \tag{1}$$

where $\hbar k$ is the photon momentum, and Γ_0' the on-resonance scattering rate into the cavity. Therefore the figure-of-merit is the ratio η_0 of Γ_0' and the scattering rate Γ_{sc} into free space, and is given by [12]

$$\eta_0 = \frac{\Gamma_0'}{\Gamma_{sc}} = 2E_c \frac{\Delta\Omega}{4\pi}. \tag{2}$$

η_0 is proportional to the solid angle $\Delta\Omega$ that is subtended by the simultaneously resonant cavity modes, and to the intensity enhancement factor $E_c = F/\pi$, where F is the cavity finesse. The cavity-to-free-space scattering ratio η_0 determines the magnitude of the cavity cooling force relative to that for conventional Doppler cooling at the same scattering rate.

The scattering rate into the resonator strongly depends on the detuning Δ of the emitted light relative to the cavity resonance. For not too large detuning Δ, the frequency-dependent scattering ratio $\eta_c(\Delta) = \Gamma_c'(\Delta)/\Gamma_{sc}$ is a Lorentzian,

$$\eta_c(\Delta) = \eta_0 \frac{\gamma_c^2}{\gamma_c^2 + \Delta^2} = \frac{2E_c}{1 + (\Delta/\gamma_c)^2} \frac{\Delta\Omega}{4\pi}, \tag{3}$$

where γ_c is the amplitude decay rate constant for the cold (passive) cavity. The enhancement factor is inversely proportional to the cavity linewidth, and in terms of the cavity round trip time τ is given by $E_c = (\gamma_c \tau)^{-1}$.

Cooling is achieved by tuning the cavity resonance to the blue side of the atomic emission spectrum, such that the emission of blue-detuned motional sidebands is enhanced, and the emission of red-detuned motional sidebands is suppressed. This leads to a cooling force that is proportional to the Lorentzian cavity-to-free-space scattering ratio $\eta_c(\Delta)$, and results in a final temperature that is inversely proportional to its slope. A narrower cavity linewidth therefore has the advantages of a larger cooling force, and a lower temperature of order $k_B T \approx \hbar\gamma_c/\eta_0$ for $\eta_0 < 1$ and $k_B T \approx \hbar\gamma_c$ for $\eta_0 > 1$ [9, 12].

Although current 'supermirrors' with very low scattering and absorption losses can sustain large enhancement factors $E_c > 10^4$ [13], even larger values will be necessary to attain scattering ratios $\eta \gg 1$ in confocal resonators, which have the largest cooling volume [12]. In addition, the cooling performance would be improved if one were able to vary E and consequently the cavity linewidth in real time in order to maximize the velocity capture range in the beginning, and minimize the temperature at the end of the cooling. Since in active resonators the linewidth is a function of the gain and much smaller than in passive resonators, the use of active resonators for cavity cooling represents a promising alternative.

For an atom placed inside an active resonator a large and strongly frequency-dependent scattering rate can be implemented in two ways. One is to use the resonator below laser threshold as a regenerative amplifier [14] for the emitted light. In this case the bandwidth and the enhancement factor are determined by the value of the regenerative gain. An alternative is to cool inside the slave laser of an injection-locked master–slave system [14], where the injecting field is tuned close to the low-frequency boundary of the locking range. Then for the light emitted by the atom the amplification and the bandwidth are determined by the proximity of the system to the boundary of the locking range. The cooling parameters are then conveniently controlled inside a feedback loop with the locking phase angle serving as the error signal.

2.1 Cavity Cooling Inside a Regenerative Amplifier

Since the intensity enhancement factor and the cooling force in a passive resonator are inversely proportional to the cavity loss, it should be possible to improve the cooling performance by compensating part of the round-trip loss with intracavity gain. In this scenario the system must be kept below laser threshold in order not to saturate the gain. Although spontaneous-emission noise in the gain medium will prevent operation of the system arbitrarily close to threshold, a significant bandwidth reduction can still be obtained [14].

To calculate the power that an atom radiates inside a linear regenerative amplifier, we write the field emitted by the atomic dipole into the cavity mode(s) under consideration as $\mathsf{E}_{\text{atom}}(t)\mathrm{e}^{\mathrm{i}kx-\mathrm{i}\omega t}$, where $\mathsf{E}_{\text{atom}}(t)$ is a slowly varying field envelope. As long as the scattering rate into the cavity does not exceed the cavity decay rate, the envelope of the field circulating inside the cavity can be approximated by its steady-state value $\mathsf{E}_{\text{st}}(t)$, which is governed by

$$\mathsf{E}_{\text{st}}(t) = g^2 r^2 \mathrm{e}^{2\mathrm{i}kL} \mathsf{E}_{\text{st}}(t) + \mathsf{E}_{\text{atom}}(t). \tag{4}$$

Here $L = \frac{1}{2}c\tau$ is the resonator length, r is the field reflection coefficient for each mirror, and g the single-pass amplitude gain of the intracavity gain medium. Introducing the detuning $\Delta = \omega - \omega_{\text{c}}$ relative to the nearest cavity resonance frequency ω_{c}, the intracavity field can be simply written as

$$\mathsf{E}_{\text{st}}(t) = \frac{\mathsf{E}_{\text{atom}}(t)}{1 - g^2 r^2 \mathrm{e}^{\mathrm{i}\Delta\tau}}. \tag{5}$$

Below laser threshold the denominator is non-zero and we can define the width of the regenerative amplifier ("hot-cavity bandwidth") $2\gamma_{\text{h}}$ by setting

$$\gamma_{\text{h}}\tau = 1 - g^2 r^2. \tag{6}$$

Then (5) close to resonance ($\Delta\tau \ll 1$) reduces to the Lorentzian form

$$\mathsf{E}_{\text{st}} = \frac{\mathsf{E}_{\text{atom}}}{\gamma_{\text{h}}\tau + \mathrm{i}\Delta\tau}. \tag{7}$$

The modified emission rate, i.e. the change in emitted power, inside the cavity can be understood as being due to the interference between the steady-state field (7) and the primary field E_{atom} of the oscillating atomic dipole. The power emitted by the atom into the resonator is proportional to $|\mathsf{E}_{\text{st}} + \mathsf{E}_{\text{atom}}|^2 - |\mathsf{E}_{\text{st}}|^2$, rather than to $|\mathsf{E}_{\text{atom}}|^2$ as in free space. Therefore the cavity-to-free-space scattering ratio is given by

$$\eta_{\text{h}}(\Delta) = \frac{\Gamma'_{\text{h}}(\Delta)}{\Gamma_{\text{sc}}} = \frac{2E_{\text{h}}}{1 + (\Delta/\gamma_{\text{h}})^2} \frac{\Delta\Omega}{4\pi}, \tag{8}$$

where $E_h = (\gamma_h \tau)^{-1}$ is the enhancement factor for the regenerative amplifier. This result is formally identical to that for the passive cavity (3) if one replaces the width of the cold cavity $2\gamma_c = 2(1 - r^2)/\tau$ by the smaller gain bandwidth of the regenerative amplifier $2\gamma_h = 2(1 - g^2 r^2)/\tau$. The larger enhancement factor $E_h = (\gamma_h \tau)^{-1}$ can be attributed to the effective reduction of cavity round trip loss from $1 - r^2$ to $1 - g^2 r^2$ due to the intracavity gain g. A reduction of the hot-cavity bandwidth γ_h by several orders of magnitude compared to γ_c has been observed [14]. Inside a regenerative amplifier it should therefore be possible to increase the scattering ratio η and the cavity cooling force substantially beyond values attainable even with supermirrors.

2.2 Cavity Cooling Inside an Injection-Locked Laser

The above analysis is concerned with an active resonator below threshold that regeneratively amplifies the radiation emitted by the atom. Cooling is achieved by frequency-selective amplification of the higher-energy motional sidebands. The amplified field is fed back phase coherently onto the atom via the resonator, and extracts a larger scattered power from the oscillating dipole by means of constructive interference. It is interesting to ask whether the same type of frequency-dependent emission enhancement is also available inside a laser. In this case the laser field itself could serve as the incident field that is being scattered in the cooling process.

Here we propose using an injection-locked system [14] to realize frequency dependent amplification of the scattered light. The particle to be cooled is placed inside the slave resonator, where it is irradiated by the slave field, whose frequency is equal to the master frequency ω_m. If ω_m is tuned near the low-energy edge of the injection-locking region, the slave will display gain peaked at its (higher) free-running frequency ω_c whose value and bandwidth depend on the proximity of the system to the injection-locking boundary. Then light scattered by the atom on blue motional sidebands near ω_c is amplified and fed back onto the atom, which results in cooling.

The only difference between cooling inside a regenerative amplifier and inside an injection-locked laser is that in the former the round-trip gain for the radiation emitted by the atom is determined by the small-signal gain g, while in the latter the single-pass gain is saturated to a value g_s that depends on the power and detuning of the master field. To analyze cavity cooling inside an injection-locked laser we merely need to find the saturated gain g_s as a function of the injection parameters. Just as for the regenerative amplifier, the cooling is then simply characterized by a Lorentzian cavity-to-free-space scattering ratio of the form (8), with a saturated hot-cavity bandwidth γ_s defined in analogy to (6) by

$$\gamma_s \tau = 1 - g_s^2 r^2. \tag{9}$$

To calculate g_s, we assume that the master field $\mathsf{E}_m e^{ik_m x - i\omega_m t}$ with time-independent real envelope E_m is incident onto the slave laser with free-running frequency ω_c. Both slave laser mirrors have amplitude transmission and reflection coefficients q and r, respectively. The steady-state condition (4) with the atomic source field E_{atom} replaced by the master field inside the cavity $q\mathsf{E}_m$ in the frame rotating at ω_m then reads

$$\mathsf{E}_{st} = g_s^2 r^2 e^{i\Delta_m \tau} \mathsf{E}_{st} + q\mathsf{E}_m. \tag{10}$$

Here $\Delta_m = \omega_m - \omega_c$ is the detuning of the master field relative to the free-running slave laser. Writing the steady-state field as $\mathsf{E}_{st} = \mathsf{E}_0 e^{i\phi}$ with real E_0, we find the following relation between g_s and the locking angle ϕ:

$$\tan\phi = \frac{g_s^2 r^2 \sin \Delta_m \tau}{1 - g_s^2 r^2 \cos \Delta_m \tau}. \tag{11}$$

For not too large detuning $\Delta_m \tau \ll 1$ we can express the saturated gain half-width γ_s as

$$\gamma_s \tau = 1 - g_s^2 r^2 = \frac{\Delta_m \tau}{\Delta_m \tau - \tan\phi}. \tag{12}$$

As the master is tuned towards the edges of the locking range, the locking phase ϕ approaches the values $\pm\pi/2$ [14], and the saturated bandwidth γ_s is reduced towards zero, which according to (3) or (8) leads to a large enhancement factor $E_s = (\gamma_s \tau)^{-1}$. The injection-locking width $\pm\Delta_l$ is given by [14]

$$\Delta_l = \frac{\mathsf{E}_m}{q\mathsf{E}_0} \gamma_c. \tag{13}$$

Here E_m and $q\mathsf{E}_0$ are the master and slave laser amplitude, respectively, measured outside the slave cavity, and $2\gamma_c$ is the cold-cavity width of the slave laser. In terms of the locking bandwidth Δ_l the enhancement factor for cooling inside an injection-locked laser can be approximately written as

$$E_s = (\gamma_s \tau)^{-1} \approx \frac{|\Delta_m|}{\Delta_l - |\Delta_m|}. \tag{14}$$

The cavity cooling force becomes very large as the master detuning $|\Delta_m|$ approaches Δ_l. This behavior is analogous to that of the regenerative amplifier near laser threshold, since for $|\Delta_m| \geq \Delta_l$ the slave reverts to its free-running frequency.

Compared to cooling with a regenerative amplifier, cooling inside an injection-locked laser offers the advantage that the gain is easily controlled and stabilized in a feedback loop using the locking phase as an error signal. For instance, a standard Pound–Drever [15] or Hänsch–Couillaud lock [16] can be used to stabilize the locking angle to a value very close to $\pi/2$. This should result in reliable operation of hot-cavity cooling, while ensuring a very large cooling force and low final temperatures. Furthermore, the high intensities inside a laser cavity should allow one to cool at very large detuning from atomic or molecular resonances, while maintaining a reasonably large scattering rate.

2.3 Limitations to Cavity Cooling with Intracavity Gain Due to Spontaneous-Emission Noise

Both for passive and for active resonators the product of the enhancement factor E and the gain half-width γ is constant and equal to the resonator free spectral range τ^{-1}. From the relation between amplification bandwidth and round-trip gain (6) it may then appear that arbitrarily large enhancement factors can be obtained by adjusting the round-trip gain g^2r^2 to a value very close to unity. However, at very small round-trip loss the amplification of photons that are spontaneously emitted in the gain medium will prevent stable operation, either by triggering laser action and saturating the gain in the regenerative amplifier, or by reverting the slave to its free-running frequency in the injection-locked system.

The limit due to amplified spontaneous noise is easily estimated by noting that the same mechanism is responsible for the Schawlow–Townes linewidth $\Delta\omega_{\mathrm{ST}}$ of a laser [17]. The smallest possible value for the gain bandwidth $2\gamma_{\mathrm{h}}$ of the regenerative amplifier or $2\gamma_{\mathrm{s}}$ of the injection-locked system is given by $\Delta\omega_{\mathrm{ST}}$ at threshold,

$$\gamma_{\mathrm{h,s}} \geq \frac{1}{2}\Delta\omega_{\mathrm{ST}}. \tag{15}$$

Consequently the maximum enhancement factor for cavity cooling with gain is given by the ratio of the free spectral range τ^{-1} and $\Delta\omega_{\mathrm{ST}}$,

$$E_{\mathrm{s,h}} = \frac{1}{\gamma_{\mathrm{s,h}}\tau} \leq \frac{2}{\Delta\omega_{\mathrm{ST}}\tau}. \tag{16}$$

In general, $\Delta\omega_{\mathrm{ST}}$ is much smaller than the cold-cavity linewidth, which should result in significantly improved cooling performance compared to passive resonators.

3 Conclusion

Resonators with gain constitute a promising possibility to improve cavity cooling. The gain bandwidth and therefore the final temperature are limited by the Schawlow–Townes linewidth. Enhancement factors larger than 10^6 appear feasible, which even for small mode solid angles would result in a cavity-to-free-space scattering ratio $\eta \gg 1$. Then the cooling of new atomic species and perhaps even of selected molecules directly from a background vapor [12] may be within reach.

Considering the title of this volume, it would be quite appropriate if the temperature of arbitrary laser-cooled light scatterers were finally limited by the fundamental linewidth of lasers. An implementation of cavity cooling with gain would also represent just another step on this amazingly successful and surprising journey that began some 25 years ago with an improbable idea [5] about how to freeze particles with that strange hot and cold, amazing and beautiful state of light. If there is one thing that Prof. Theodor Hänsch is teaching us again and again, and that I am deeply grateful for, then it is not so much how science can become great fun (I suspected that already when as an undergraduate I first heard him talk about light forces, after which no other field of physics could compete), but how playfulness can become great science.

References

1. T.W. Hänsch, A.L. Schawlow, IEEE J. Quant. Electron. **7**, 45 (1971)
2. E.M. Purcell, Phys. Rev. **69**, 681 (1946)
3. D. Kleppner, Phys. Rev. Lett. **47**, 233 (1981)
4. P. Goy, J.M. Raimond, M. Gross, S. Haroche, Phys. Rev. Lett. **50**, 1903 (1983); R.G. Hulet, E.S. Hilfer, D. Kleppner, *ibid.* **55**, 2137 (1985); W. Jhe, A. Anderson, E.A. Hinds, D. Meschede, L. Moi, S. Haroche, *ibid.* **58**, 666 (1987); D.J. Heinzen, J.J. Childs, J.E. Thomas, M.S. Feld, *ibid.* **58**, 1320 (1987); D.J. Heinzen, M.S. Feld, *ibid.* **59**, 2623 (1987)
5. T.W. Hänsch, A.L. Schawlow, Opt. Comm. **13**, 68 (1975)
6. T.W. Mossberg, M. Lewenstein, D.J. Gauthier, Phys. Rev. Lett. **67**, 1723 (1991); M. Lewenstein, L. Roso, Phys. Rev. A **47**, 3385, (1993)
7. J.I. Cirac, A.S. Parkins, R. Blatt, P. Zoller, Opt. Commun. **97**, 353 (1993); J.I. Cirac, M. Lewenstein, P. Zoller, Phys. Rev. A **51**, 1650 (1995)
8. P. Horak, G. Hechenblaikner, K.M. Gheri, H. Stecher, H. Ritsch, Phys. Rev. Lett. **79**, 4974 (1997)
9. V. Vuletić, S. Chu, Phys. Rev. Lett. **84**, 3787 (2000)
10. G. Hechenblaikner, M. Gangl, P. Horak, H. Ritsch, Phys. Rev. A **58**, 3030 (1998); M. Gangl, H. Ritsch, Phys. Rev. A **61**, 011402 (1999); M. Gangl, H. Ritsch, Eur. J. Phys. D **8**, 29 (2000); P. Domokos, P. Horak, and H. Ritsch, J. Phys. B (2001); M. Gangl, H. Ritsch, Phys. Rev. A **61**, 043405 (2000); M. Gangl, H. Ritsch, J. Mod. Opt **47**, 2741 (2000)
11. V. Vuletić, A.J. Kerman, C. Chin, S. Chu, in *Proceedings of the 17th International Conference on Atomic Physics*, ed. by E. Arimondo, P. de Natale, M. Inguscio (American Institute of Physics, Melville, New York 2001) pp. 356–366
12. V. Vuletić, H.W. Chan, A.T. Black, Phys. Rev. A **64**, 033405 (2001)
13. P. Münstermann, T. Fischer, P. Maunz, P.W.H. Pinkse, G. Rempe, Phys. Rev. Lett. **82**, 3791 (1999); J. Ye, D.W. Vernooy, H.J. Kimble, *ibid.* **83**, 4987 (1999)
14. A.E. Siegman, *Lasers* (University Science Books, Mill Valley 1986) and references therein
15. R.W.P. Drever, J.L. Hall, F.V. Kowalski, J. Hough, G.M. Ford, A.J. Munley, H. Ward, Appl. Phys. B **31**, 97 (1983)
16. T.W. Hänsch, B. Couillaud, Opt. Commun. **35**, 441 (1980)
17. A.L. Schawlow, C.H. Townes, Phys. Rev. **112**, 1940 (1958)

Zeeman-Tuned Slowing: Surfing the Resonance Wave

David H. McIntyre, Shannon K. Mayer, Nancy S. Minarik, and Mark H. Shroyer

Laser cooling is an elegant technique that has had a profound influence on atomic physics in the last 25 years. It is also simple enough to explain to an introductory physics class using Newtonian concepts. These attributes of elegance and simplicity are hallmarks of Theodor Hänsch's approach to physics. Moreover, Theodor thoroughly enjoys doing physics and his enthusiasm is contagious. He is able to explain complex phenomena in such simple terms that they become glaringly obvious and you are left asking why you had not thought of them earlier. When Art Schawlow and Theodor Hänsch first came up with the idea of laser cooling, Theodor was concerned that it was so obvious that they would not be able to publish a paper on the idea without demonstrating it experimentally [1]. They did publish the idea [2], and were gracious enough to leave the experimental demonstration to others.

Like any graduate student in the Schawlow–Hänsch group, I (DHM) quickly learned that physics was fun and that what we were doing was simple at heart. At group meetings, discussion was steered around to the simplest approach. Anyone who tried to explain an idea using something as sophisticated as contour integration in the complex plane, for example, could expect a volley of questions from Theodor Hänsch and Art Schawlow probing the basic concepts of the idea. I vividly recall Theodor telling me that if I couldn't explain something simply, then I did not truly understand it. That idea has stayed with me ever since, and has guided both my research and my teaching. Theodor Hänsch's enthusiasm for physics was especially evident to me when I joined his group in Munich. We were busy building the hydrogen experiment and setting up several new experiments. These opportunities brought forth a flurry of ideas, ranging from schemes to increase the precision of the hydrogen $1S$–$2S$ measurements to designs for vacuum equipment. As always, Theodor Hänsch's ideas were elegant and simple. For example, his plan to divide optical frequencies down to measurable microwave frequencies [3] seems so obvious in hindsight, and it has now become an important component of the march to ever-increasing precision of the Rydberg constant [4].

1 Introduction

In our contribution to this volume honoring Theodor Hänsch and his remarkable scientific career, we wish to briefly describe some work we have done in laser cooling, stressing the fun and simple aspects. We have developed a well-collimated, intense source of low-velocity atoms using an atomic funnel [5]. The funnel relies on magneto-optic trapping of atoms in a two-dimensional magnetic field [6]. The atoms are pushed along the axis of the field to form a slow beam. We originally used chirped laser slowing to slow atoms from a thermal beam and load them into the trap. In order to increase the flux of atoms, we have since changed to Zeeman-tuned slowing of the thermal beam. In designing, building, and testing the Zeeman slower, we learned many details about the fundamentals of Zeeman slowing, in particular regarding the differences between σ^+ and σ^- slowing.

2 Zeeman-Tuned Slowing Basics

One-dimensional slowing of an atomic beam relies on the radiation pressure from a counterpropagating laser beam. The deceleration of the atoms is proportional to the scattering rate, which has the usual Lorentzian resonance shape. The resonance condition is that the Doppler-shifted laser frequency $\omega_\mathrm{L} + k\nu$ equal the atomic frequency $\omega_0 + \Delta\omega_\mathrm{field}$, which may include shifts due to external fields. Efficient slowing requires that the resonance condition be nearly satisfied in order to maximize the deceleration. Unfortunately, as atoms are slowed, the changing Doppler shift takes the atoms away from resonance. The solution is to change the parameters at one's disposal, which means modifying either the laser frequency or the atomic perturbation. In either case, the effect is to move the resonance curve in velocity space to keep up with the slowing atoms. A surfer riding a wave provides a simple mechanical analogy of this effect, and is illustrated in Fig. 1.

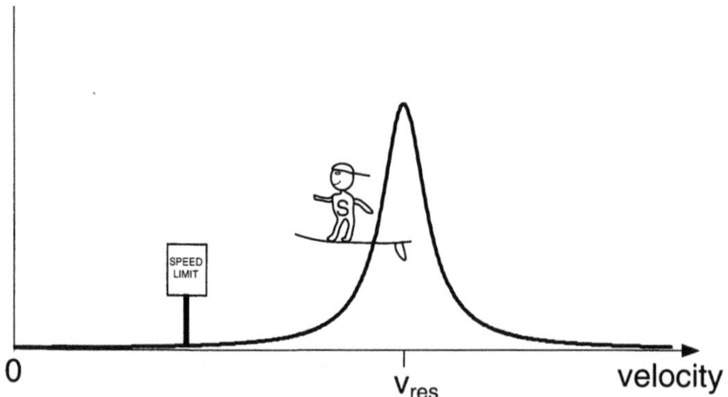

Fig. 1. A surfer on a wave provides a simple mechanical analogy for an atom experiencing resonant deceleration. The wave moves in velocity space toward zero velocity and the surfer rides the wave at approximately a constant position on the wave

The surfer analogy is useful in illustrating the self-stabilizing nature of the Doppler slowing mechanism. If the wave moves too fast, the surfer ends up near the steep top and so picks up speed to return to the optimum point on the wave. If the wave moves too slowly, the surfer ends up out front where the wave is shallow, slows down, and waits for the wave to catch up. This self-stabilization works only on the front of the wave, which corresponds to the atomic velocity being less than the resonance velocity. This stability has been derived theoretically for Zeeman slowing [7].

In Doppler slowing there is a maximum acceleration imposed by the finite lifetime of the upper excited state. That is, the average time for the atom to return to the ground state and be ready for another absorption limits the scattering rate. In the surfer analogy, the wave moves in velocity space, so the acceleration limit on the atoms in position space corresponds to a speed limit on the wave-riding surfer in velocity space. Efficient slowing requires that the resonance curve speed in velocity space approach but not exceed this speed limit.

In chirped slowing, the laser beam is frequency tuned to stay in resonance with the slowing atoms [8]. Modern techniques allow precise control of the laser frequency as a function of time, so it is generally a simple matter to impose the linear frequency chirp required for constant deceleration ($\nu = \nu_0 + at$ requires $\mathrm{d}\omega_\mathrm{L}/\mathrm{d}t = ka$). In addition, the laser frequency can be abruptly changed to turn off the slowing process and produce atoms at any desired final velocity. Unfortunately, chirped slowing addresses only atoms that are in synch with the chirp, and a pulsed beam results, with a small duty cycle.

Zeeman-tuned slowing uses the Zeeman shift provided by a spatially varying magnetic field to keep the slowing atoms in resonance with a fixed-frequency laser beam [9]. Since Zeeman-tuned slowing (as well as Stark tuned slowing [10]) operates in the spatial domain, it produces a continuous beam of slowed atoms. Circularly polarized light is used to drive a transition between extreme hyperfine energy levels, giving a Zeeman shift for the transition of

$$\Delta\omega_\mathrm{Zeeman} = \pm\frac{\mu_\mathrm{B}}{\hbar}B\,. \tag{1}$$

The resonance condition can then be written as

$$\Delta + k\nu_\mathrm{res} \mp \frac{\mu_\mathrm{B}}{\hbar}B = 0\,, \tag{2}$$

where Δ is the laser detuning ($\omega_\mathrm{L} - \omega_0$). Throughout this article we follow the convention that upper (lower) signs apply to the $\sigma^+(\sigma^-)$ configuration. Since the sign of the shift is explicit, the magnetic field B is positive for both cases.

For constant deceleration a_0, the velocity of a slowing atom is

$$\nu^2 = \nu_0^2 - 2a_0 z\,, \tag{3}$$

and an atom will be brought to rest after travelling a distance $z_0 = \frac{\nu_0^2}{2a_0}$. Satisfying the resonance condition for these slowing atoms requires a magnetic field with a spatial profile given by

$$B(z) = B_\mathrm{b} \pm B_\mathrm{t}\sqrt{1 - \frac{z}{z_0}}\,, \tag{4}$$

where the bias field $B_\mathrm{b} = \pm\frac{\hbar\Delta_0}{\mu_\mathrm{B}}$ and the taper field $B_\mathrm{t} = \frac{\hbar k \nu_0}{\mu_\mathrm{B}}$. The designed detuning Δ_0 is that required to have atoms at rest at z_0 be resonant. Later we discuss the situation where the detuning Δ differs from the designed detuning Δ_0. The magnetic field profiles of (4) are shown in Fig. 2. For the σ^+ configuration, the magnetic field decreases along the length of the solenoid, while for the σ^- case, the magnetic field increases. Due to the sign difference in the Zeeman shift, both cases result in the resonance curve in Fig. 1 moving toward lower velocities as it must for slowing. Note that the designed detuning Δ_0 is positive (or zero) for the σ^+ case and negative for the σ^- case.

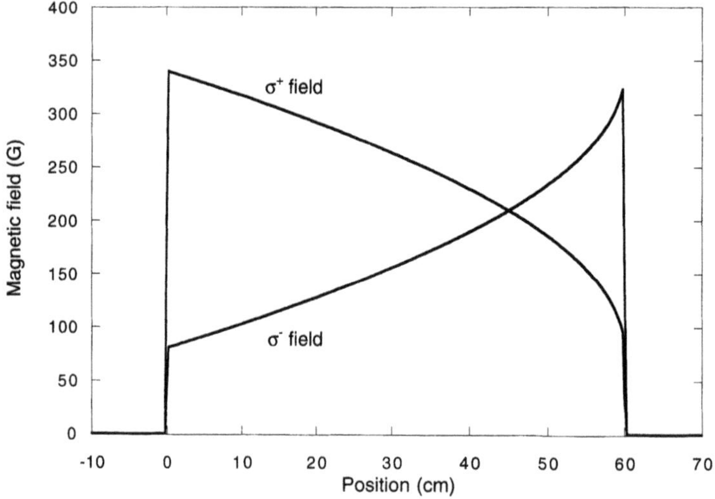

Fig. 2. Ideal magnetic field profiles for Zeeman-tuned slowing. The field for σ^+ slowing decreases with position and the σ^- field increases

3 Zeeman-Tuned Slowing Experiments

Zeeman-tuned slowing was first demonstrated by the Phillips group at NIST, who used a fixed-frequency, σ^+ polarized laser beam and a decreasing strength magnetic field to slow [9] and later stop [11] a thermal sodium beam. While large changes in velocity were achieved using σ^+ Zeeman-tuned slowing, it was observed that atoms slowed to low velocity (less than ~100 m/s) never went out of resonance with the slowing laser, and were thus brought to rest or turned around within the solenoid. For applications requiring a slow atomic beam, extracting the slow atoms from the solenoid posed a technical challenge. To overcome this problem Metcalf and Phillips turned off the slowing laser and allowed the atoms to drift out of the solenoid [9]. This yielded a pulse of slow atoms rather than a continuous beam. Bagnato et al. utilized a special magnetic field design that allowed the atoms to be optically pumped out of the cycling transition near the exit end of the solenoid [12]. The atoms, no longer resonant with the slowing laser, drifted out of the solenoid.

Barrett et al. overcame the difficulty associated with σ^+ Zeeman-tuned slowing more simply by demonstrating a modified version of the Zeeman slower [13]. In contrast with earlier work, they employed σ^- polarized light and a magnetic field of increasing strength. This simple modification resulted in a significantly more accessible continuous beam of slow atoms.

4 Zeeman-Tuned Slowing Simulations

To understand the differences between σ^- and σ^+ slowing, we have performed simulations of the atomic motion. Following Theodor Hänsch's lead again, we have used a spreadsheet to perform these calculations. We perform a simple numerical integration of the classical equation of motion, given the deceleration

$$a = \frac{\hbar k \Gamma}{2M} \frac{s}{1 + s + \frac{4}{\Gamma^2}\left(\Delta + k\nu \mp \frac{\mu_B}{\hbar} B\right)^2}, \tag{5}$$

where the saturation parameter $s = I/I_{\text{sat}}$, I is the laser intensity, and I_{sat} is the saturation intensity. The ideal magnetic field of (4) is designed to bring atoms to rest at the end of the field region at z_0. The simulation shown in Fig. 3 confirms this for the case of σ^+ slowing. To make the resonant nature of the slowing obvious, all three terms in the resonance condition are plotted: the laser detuning, the Doppler shift, and the Zeeman shift. Since the two vertical axes are offset by the laser detuning, the resonance condition is satisfied when the Doppler shift and Zeeman shift curves overlap. During the slowing, the Doppler curve lies beneath the Zeeman curve, which corresponds to the stable side of resonance (i.e., the surfer is on the front of the wave). In this ideal case, the acceleration is constant throughout the slowing for the chosen detuning. The results of this simulation for the ideal field are identical for σ^+ and σ^- slowing, other than the different detunings required.

The simulation of Fig. 3 corresponds to a specific detuning, nominally equal to the designed detuning Δ_0 for bringing atoms to rest. In practice, we want atoms at finite velocities so they can leave the field and be useful. Rather than redesigning the field, consider how a field designed to stop atoms

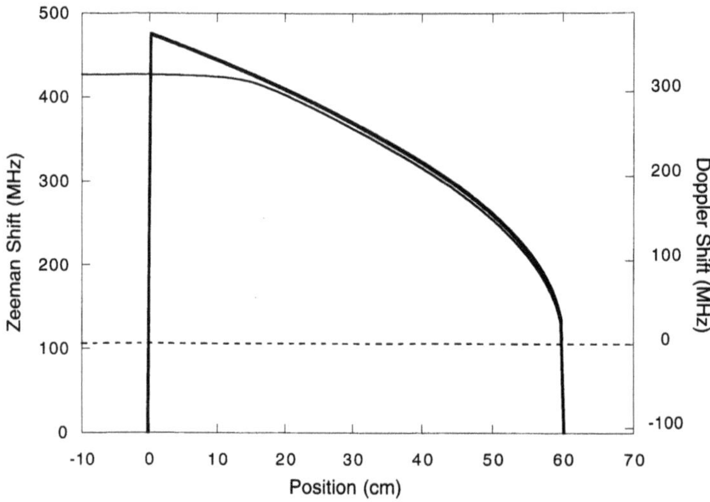

Fig. 3. Simulation of an atom moving in a σ^+ Zeeman slower with an ideal magnetic field. The plot shows the Zeeman shift (*thickest line*), Doppler shift (*solid line*), and laser detuning (*dashed line*). The two *vertical axes* have been offset by the laser frequency detuning, so that the resonance condition is satisfied when the Zeeman shift and Doppler shift curves overlap

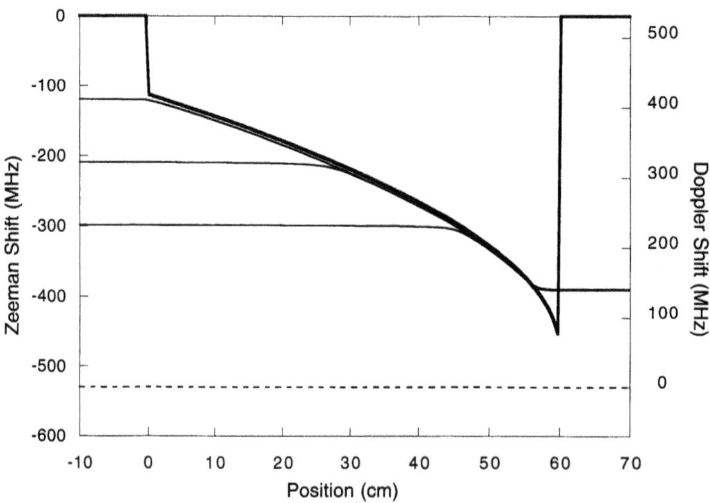

Fig. 4. Simulation of an atom moving in a σ^- Zeeman slower with an ideal magnetic field. Atoms with three different initial velocities are slowed to the same final velocity

with a detuning Δ_0 works when the detuning is changed. From the resonance condition we know that increased velocities require decreased detunings (i.e., less positive for σ^+ and more negative for σ^-) at the same field strength. So to produce a larger final velocity would require a detuning below the value used for Fig. 3. Such a case is shown in Fig. 4 for σ^- slowing, again using an ideal field. For this detuning, atoms with differing initial velocities are slowed to the same, non-zero velocity, and stop decelerating before the end of the field. The final velocity as a function of detuning for this field is shown in Fig. 5. These results are still the same for σ^+ and σ^- slowing.

To understand how the final velocity changes with detuning, consider how the atom arrives at its final velocity, again by referring to the surfer analogy. To deposit the surfer at a particular location in velocity space requires either that (1) the wave speeds ahead toward lower resonance velocities faster than the allowed speed limit for the surfer, or (2) the wave stops and retreats back toward higher resonance velocities. In both cases, the surfer is left in flat water at the final location in velocity space. For σ^+ slowing, the magnetic field decreases from B_b+B_t to B_b during the slowing, and then decreases further to zero as the atom leaves the field, which corresponds to the resonance velocity always decreasing. For σ^- slowing, the magnetic field increases from $B_b - B_t$ to B_b during the slowing, and then decreases to zero as the atom leaves the field, which corresponds to the resonance velocity decreasing during the slowing and then increasing as the atom leaves the field. Thus we see that

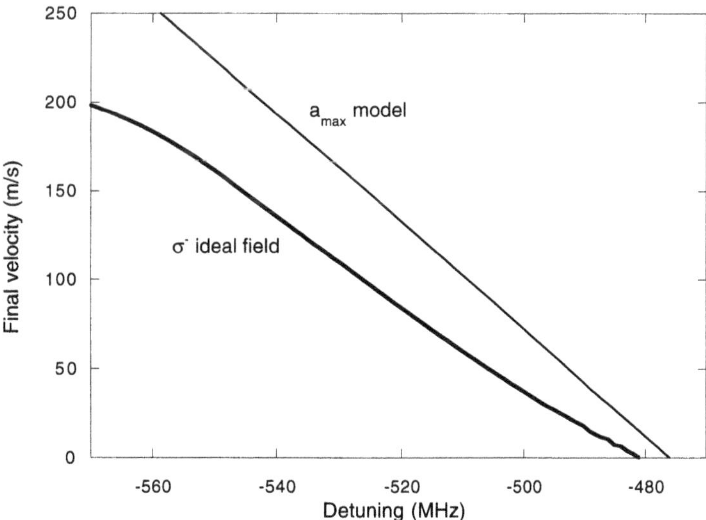

Fig. 5. Final velocity of slowed atoms as a function of laser detuning. The *lower, thick line* shows the result of the simulation using the field of Fig. 4. The *upper, thinner line* is the result of the model wherein the slowing stops when the maximum acceleration is exceeded

of the two possibilities for turning off the slowing mechanism to reach a final velocity, only the first is possible for σ^+ slowing, while both are possible for σ^- slowing. This distinction is the essence of the difference between the two types of Zeeman slowing.

In the first case where the wave speeds ahead of the surfer, the atom would have to accelerate faster than the allowed acceleration in order to stay in resonance. To see how the final velocity is reached, consider the condition that the acceleration of a resonant atom must be less than or equal to the maximum allowed acceleration: $a_{\mathrm{res}} \le a_{\mathrm{max}}$. The acceleration of a fully saturated ($s \gg 1$) transition in Doppler slowing is $a_{\mathrm{D}} = \frac{\hbar k \Gamma}{2M}$ and the maximum acceleration for a finite laser intensity is

$$a_{\mathrm{max}} = a_{\mathrm{D}} \frac{s}{1+s} . \tag{6}$$

For Zeeman slowing, the acceleration of a resonant atom can be expressed in terms of the magnetic field gradient:

$$\begin{aligned} a_{\mathrm{res}} &= \frac{\mathrm{d}\nu_{\mathrm{res}}}{\mathrm{d}t} = \frac{\mathrm{d}}{\mathrm{d}t}\left[-\frac{\Delta}{k} \pm \frac{\mu_{\mathrm{B}}}{\hbar k} B(z)\right] \\ &= \pm \frac{\mu_{\mathrm{B}}}{\hbar k} \frac{\mathrm{d}B}{\mathrm{d}z} \frac{\mathrm{d}z}{\mathrm{d}t} , \end{aligned} \tag{7}$$

yielding the condition [7]

$$\frac{\mu_{\mathrm{B}}}{\hbar k} \left|\frac{\mathrm{d}B}{\mathrm{d}z}\right| \nu_{\mathrm{res}} \le a_{\mathrm{max}} . \tag{8}$$

From this condition, it is clear that slower atoms can tolerate a larger magnetic field gradient than faster atoms. For the ideal fields above, only atoms at zero velocity can tolerate the infinite gradient at the end of the solenoid. Faster atoms can only tolerate a smaller gradient, which occurs sooner along the atom's path. Beyond that point, the atoms cannot keep up with the resonance curve and decouple from the slowing process. Thus as the laser detuning is decreased in order to increase the desired final velocity, the position where the atoms find the largest tolerable gradient occurs farther from the exit end of the solenoid, as seen in Fig. 4.

In the case of the ideal fields, the acceleration of a resonant atom can be written as [14]

$$a_{\mathrm{res}} = a_0 \left(1 + \frac{\Delta_0 - \Delta}{k\nu_0\sqrt{1 - z/z_0}}\right) . \tag{9}$$

This relation shows that the acceleration is equal to the designed acceleration a_0 only for the detuning Δ equal to the designed detuning Δ_0. For detunings less than that, the acceleration increases as the atom moves down the solenoid and exceeds a_{max} before it reaches the end. The position where this occurs is found by solving the equality of (8), noting that the velocity of a resonant atom can be written as

$$k\nu_{\mathrm{res}} = \Delta_0 - \Delta + k\nu_0\sqrt{1 - z/z_0}\,. \tag{10}$$

Assuming that the atom decouples quickly from the field after this point, the final velocity of the atom is the resonant velocity at that point, giving

$$k\nu_{\mathrm{final}} = \frac{\Delta_0 - \Delta}{1 - a_0/a_{\mathrm{max}}}\,. \tag{11}$$

We see that if the magnetic field is designed to have an acceleration a_0 nearly equal to the maximum allowed acceleration a_{max}, then small changes in the detuning result in large changes in the final velocity. In the limit where $a_0 = a_{\mathrm{max}}$, it is not possible to slow atoms to a final velocity other than the designed zero velocity because any change in the detuning increases the resonance acceleration beyond the allowed value. This model for the final velocity is plotted in Fig. 5 with the simulation results. The final velocity in the simulation is always less than our simple model since the slowing does not turn off instantly, but otherwise the agreement is good.

Up to this point in the analysis, σ^+ and σ^- slowing are still identical, with the final velocity determined by the acceleration limit. To see how the two cases differ we must consider more realistic fields, which allows for the possibility of the other mechanism for turning off the slowing. More realistic fields can be modeled by multiplying the ideal fields by the function

$$f(z) = \frac{1}{2}\left(\frac{z}{\sqrt{z^2 + r^2}} + \frac{z_0 - z}{\sqrt{(z - z_0)^2 + r^2}}\right), \tag{12}$$

which effectively softens the sharp discontinuities at the ends of the solenoid of radius r. Outside the solenoid, $f(z)$ is multiplied by the field at the closest end. The effect of this on the ideal fields used above is shown in Figs. 6 and 7 along with simulations of the motion. For σ^+ slowing, the acceleration limit still determines the final velocity. For the σ^- case, the slowing is now turned off by the field reversing direction, i.e., the atomic resonance has retreated to higher velocities. As indicated in Fig. 7, the final velocity is approximately the resonance velocity at the maximum field:

$$k\nu_{\mathrm{final}} = \frac{\mu_{\mathrm{B}}}{\hbar}B_{\mathrm{max}} - \Delta\,. \tag{13}$$

If we generalize our definition of the designed detuning Δ_0 to be $\Delta_0 = \mu_{\mathrm{B}}\frac{B_{\mathrm{max}}}{\hbar}$, then we can write

$$k\nu_{\mathrm{final}} = \Delta_0 - \Delta\,. \tag{14}$$

for easy comparison with the earlier result.

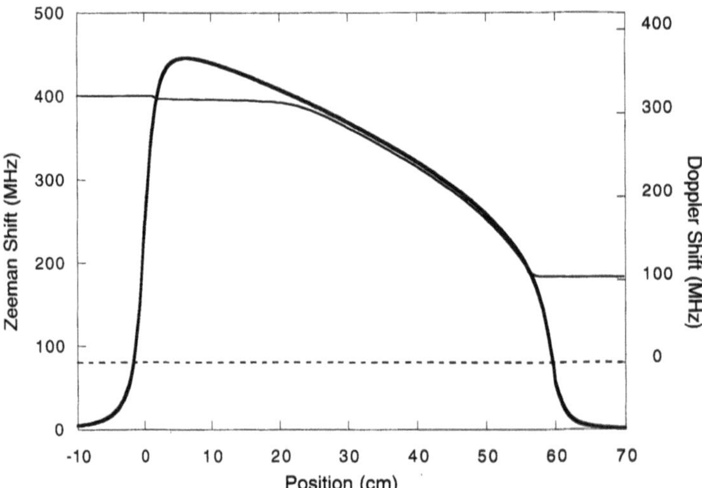

Fig. 6. Simulation of an atom moving in a σ^+ Zeeman slower with a realistic magnetic field. This magnetic field is the same ideal field used in Fig. 3, modified by (12)

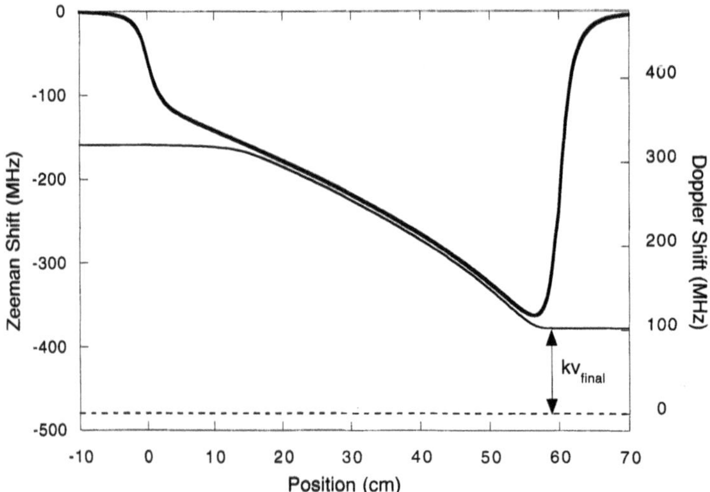

Fig. 7. Simulation of an atom moving in a σ^- Zeeman slower with a realistic magnetic field. This magnetic field is the same ideal field used in Fig. 4, modified by (12)

The detuning dependence of the final velocity for the two simulations of Figs. 6 and 7 is shown in Fig. 8. The σ^+ simulation agrees well with the model that the slowing stops when a_{max} is exceeded and the σ^- simulation agrees well with the model that the slowing stops when the maximum in the field

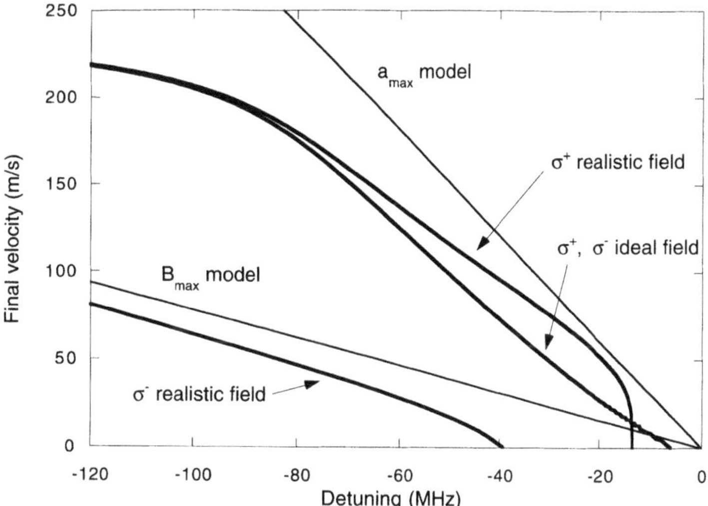

Fig. 8. Final velocity of slowed atoms as a function of laser detuning. The *thicker lines* show the results of simulations in the denoted cases. The *thinner lines* show the results of the two models for how the slowing is turned off. All results have been shifted horizontally such that zero detuning corresponds to stationary atoms at the end of the field being resonant

$B_{\max}$ is reached. The slope of the σ^- simulation is close to the expected value of λ, which is 0.78 m/s/MHz for rubidium. This slope is small enough that extraction of low-velocity atoms is possible. For the σ^+ case, the slope of the final velocity simulation curve gets steeper at low velocities when the field is modeled more realistically, and is so steep near zero velocity that it becomes impossible to extract atoms from the field in a well-defined velocity class. Since the realistic σ^+ magnetic field doesn't have the infinite gradient at the end, slow atoms do not find a steep enough gradient to decouple from the slowing process. This model is consistent with problems common to Zeeman slowing in the σ^+ configuration.

5 Experiment

We have studied these ideas in our Zeeman slowing experiment. A schematic diagram of the experimental apparatus is shown in Fig. 9. The Zeeman slower is a 60 cm long solenoid with layers of wire wound around a constant-diameter water-cooled cylinder. For maximum flexibility, the 60 cm long magnet was built with independently controllable bias and taper magnetic fields. The constant bias magnetic field consists of wire wound at a constant pitch. The taper magnetic field, constructed by winding wire on a lathe at increasing

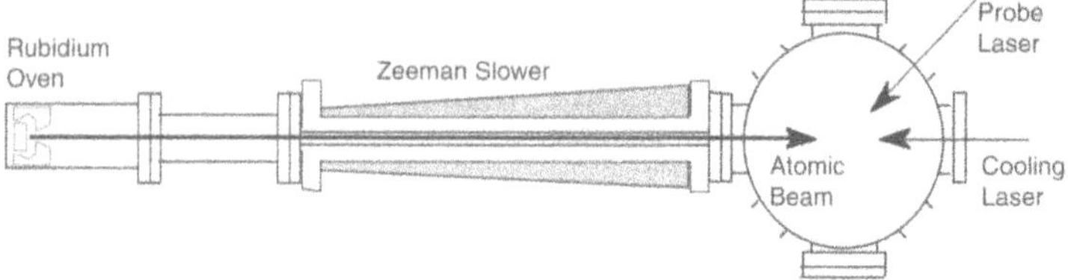

Fig. 9. Schematic diagram of Zeeman slowing experiment

pitch, approximates a field of

$$B = 360\,\mathrm{G} - 330\,\mathrm{G}\sqrt{1 - \frac{z}{60\,\mathrm{cm}}} \,. \tag{15}$$

To generate an increasing strength magnetic field for σ^- slowing, the currents in the bias and taper coils are in the same direction such that the fields add together. A decreasing strength field for σ^+ slowing is achieved by running current in opposite directions in the two coils.

Within the Zeeman slower, the atomic beam is slowed by the scattering force from a counterpropagating laser beam ($I \approx 4I_{\mathrm{sat}}$) appropriately detuned from the $F = 3$ to $F' = 4$ hyperfine transition of the D_2 line of ^{85}Rb. After exiting the solenoid, the atoms propagate to the detection region where the probe laser beam is aligned to intersect the atomic beam at an angle of 45° in order to measure the velocity distribution of the slowed beam. Fluorescence from the atomic beam is measured using a photomultiplier tube. The lasers used in this experiment are grating stabilized diode lasers operating near 780 nm [15], and are based upon work done in Theodor Hänsch's group in Munich [16].

The slow atomic beam is loaded into a two-dimensional magneto-optic trap (MOT) [5]. Efficient loading of the MOT requires a final velocity of approximately 20 m/s; atoms with larger velocities cannot be trapped and atoms with smaller velocities tend to miss the trap due to their transverse velocity acquired during one-dimensional Zeeman slowing. For our application, a length of 60 cm maximizes the change in atomic velocity while assuring that the transverse size of the atomic beam remains less than the size of the MOT.

The slowing of atoms within the solenoid is depicted in the simulation of Fig. 10, showing that atoms with initial velocities less than 250 m/s should be slowed to 20 m/s. Experimental measurements of the final velocity as a function of laser detuning for both σ^- and σ^+ slowing are shown in Fig. 11. For σ^- slowing, the final velocity can be well controlled with the detuning over a large range. The data agree well with the simulation using the measured magnetic field. For detunings below −950 MHz, two slow atom peaks are observed, and the data for the faster peak agrees well with the simulation using atoms with larger initial velocities. We conclude that the two slowed

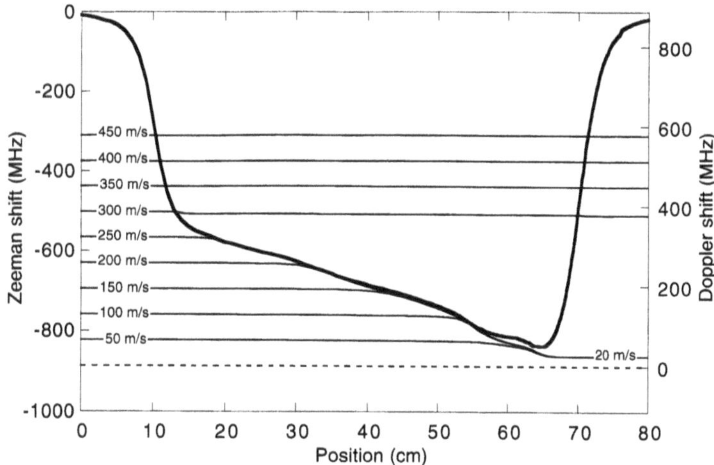

Fig. 10. Zeeman shift (*thick line*) and simulated Doppler shift as a function of position in the experimental magnetic field. The Doppler shift curves are labeled with the initial atomic velocity and the final velocity is indicated. The laser detuning is shown by the *dashed line*

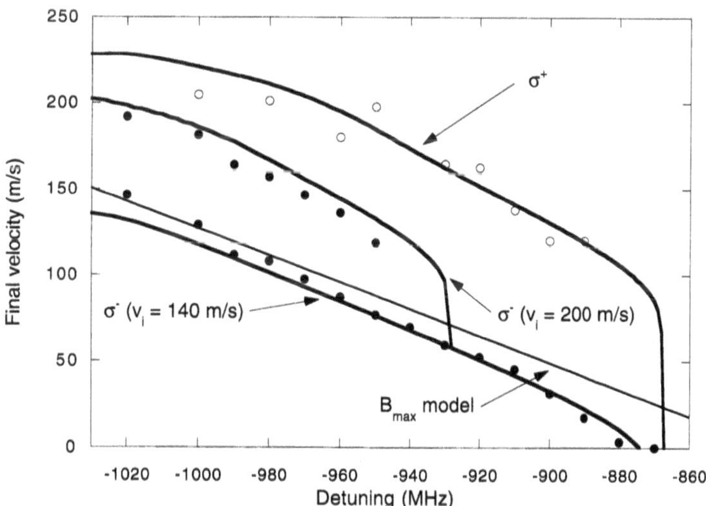

Fig. 11. Experimental data for final velocity of slowed atoms as a function of laser detuning. The filled (*open*) circles are measurements for σ^- (σ^+) slowing. For σ^- slowing, two slow atom peaks are found for detunings below −950 MHz. Simulations are shown as *solid thick lines* with labels. The *thin line* shows the final velocity expected for the simple model where the slowing is turned off at the maximum field. To aid comparison, the σ^+ data and simulation were shifted −960 MHz to make the curves coincide at zero velocity

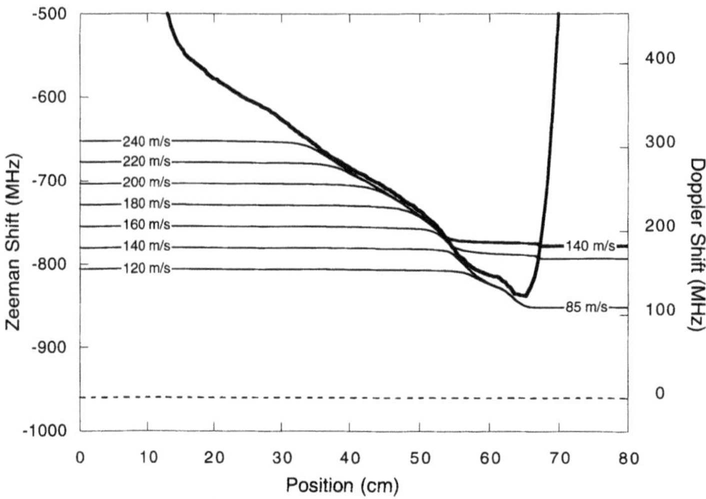

Fig. 12. Zeeman shift (*thick line*) and simulated Doppler shift as a function of position in the experimental magnetic field for the case where multiple final velocities are possible. The Doppler shift curves are labeled with the initial atomic velocity and the final velocities are indicated. The laser detuning is shown by the *dashed line*

atom peaks are caused by faster atoms decoupling from the field at a fixed point in the field where the gradient is too large due to an irregularity in the field. Simulations for differing initial velocities shown in Fig. 12 confirm this. Since these atoms decouple at a fixed point, the slope of the final velocity versus detuning curve is the same as for the atoms that decouple at the maximum field.

For σ^+ slowing, the slow atom peak was negligible for final velocities less than $\approx$120 m/s. The simulation shows a near-infinite slope at low velocities and the same slope as σ^- slowing for higher velocities, indicating that the fast atoms in this case again decouple from the field at a fixed point where the gradient is too large due to an irregularity in the field. As others have found, atoms at low velocities tend to continue accelerating and be turned around inside the σ^+ slower [9].

6 Conclusion

The experimental data agree well with the simple model for how Zeeman-tuned slowing works, in particular with the models for how the slowing is turned off to reach the final velocity. In σ^+ slowing, the atoms decouple from the field when the magnetic field gradient becomes larger than the allowable value determined by the maximum acceleration. In σ^- slowing, the atoms decouple from the field at the maximum field location since the resonance retreats to higher velocities after that. These important differences are particularly manifested in real magnetic fields (as opposed to ideal fields) and make it difficult if not impossible to extract slow atoms from a σ^+ slower. It should be noted that this problem has been solved by others by adding extra field coils to cause the field to increase to turn off the slowing, with the necessary decrease occurring at much larger gradients made with smaller coils [14, 17].

This work is an outgrowth of ideas and techniques that began in Theodor Hänsch's fertile imagination. It is a pleasure and a privilege to contribute to this volume honoring his achievements, and we are sure he will continue to demonstrate his ability to make physics elegant, simple, and fun.

Acknowledgement

Support for this work was provided in part by the Office of Naval Research.

References

1. T.W. Hänsch, *Lasers, Spectroscopy and New Ideas: A Tribute to Arthur L. Schawlow*, W.M. Yen, M.D. Levenson (eds.) (Springer-Verlag, Berlin, Heidelberg 1987), p. 3
2. T.W. Hänsch, A.L. Schawlow, Opt. Commun. **13**, 68 (1975)
3. H.R. Telle, D. Meschede, T.W. Hänsch, Opt. Lett. **15**, 532 (1990)
4. M. Niering, R. Holzwarth, J. Reichert, P. Pokasov, Th. Udem, M. Weitz, T.W. Hänsch, P. Lemonde, G. Santarelli, M. Abgrall, P. Laurent, C. Salomon, A. Clairon, Phys. Rev. Lett. **84**, 5496 (2000)
5. T.B. Swanson, N.J. Silva, S.K. Mayer, J.J. Maki, D.H. McIntyre, J. Opt. Soc. Am. B **13**, 1833 (1996)
6. E. Riis, D.S. Weiss, K.A. Moler, S. Chu, Phys. Rev. Lett. **64**, 1658 (1990)
7. V.S. Bagnato, A. Aspect, S.C. Zilio, Opt. Commun. **72**, 76 (1989)
8. W. Ertmer, R. Blatt, J.L. Hall, M. Zhu, Phys. Rev. Lett. **54**, 996 (1985)
9. W.D. Phillips and H. Metcalf, Phys. Rev. Lett. **48**, 596 (1982)
10. R. Gaggl, L. Windholz, C. Umfer, C. Neureiter, Phys. Rev. A **49**, 1119 (1994)
11. J. Prodan, A. Migdall, W.D. Phillips, I. So, H. Metcalf, J. Dalibard, Phys. Rev. Lett. **54**, 992 (1985)
12. V.S. Bagnato, C. Salomon, E. Marega Jr., S.C. Zilio, J. Opt. Soc. Am. B **8**, 497 (1991)
13. T.E. Barrett, S.W. Dapore-Schwartz, M.D. Ray, G.P. Lafyatis, Phys. Rev. Lett. **67**, 3483 (1991)
14. P.A. Molenaar, P. van der Straten, H.G.M. Heideman, H. Metcalf, Phys. Rev. A **55**, 605 (1997)
15. J.J. Maki, N.S. Campbell, C.M. Grande, R.P. Knorpp, D.H. McIntyre, Opt. Commun. **102**, 251 (1993)
16. L. Ricci, M. Weidemüller, T. Esslinger, A. Hemmerich, C. Zimmermann, V. Vuletic, W. König, T. Hänsch, Opt. Commun. **117**, 541 (1995)
17. M.E. Firminio, C.A. Faria Leite, S.C. Zilio, V.S. Bagnato, Phys. Rev. A **41**, 4070 (1990)

A New Approach for Laser Cooling of Calcium

Andreas Hemmerich

We explore an efficient method for preparing large samples of ultracold calcium atoms. An optimized conventional (Doppler-limited) magneto-optical trap collects atoms from a Zeeman cooled atomic beam using a strong dipole transition within the singlet system. This transition is not completely closed thus yielding an intense flux of atoms into the metastable triplet state $^3P_2$. We obtain a flux of above 10^{10} atoms/s into the $^3P_2$ state. We find that our MOT lifetime of 23 ms is mainly limited by this loss channel and thus the $^3P_2$ production is not hampered by inelasic collisions. A second magneto-optical trap sharing the same magnetic field gradient is superimposed which captures and further cools the metastables using the narrow-band infrared transition $^3P_2 \rightarrow ^3D_3$. In our present experiment we were able to prepare 3×10^8 atoms at temperatures below 20 microkelvin within 250 ms. Minor technical improvements of our setup promise to yield above 10^{10} atoms at submicrokelvin temperatures within 1 s.

1 Introduction

Laser cooling of neutral atoms was first proposed by Theodor Hänsch and Arthur Schawlow in the 1970s [1]. The detailed development of their far-seeing proposal and its application to the alkali group has culminated in the first observation of Bose–Einstein condensation (BEC) twenty years later [2,3]. Selecting alkali atoms for this venture proved successful due to the simplicity of their hydrogen-like energy level schemes which provide strong closed cycle transitions. The implementation of efficient laser cooling schemes for more complex atomic species remains a challenge which promises a wealth of new physics. Earth-alkali atoms offer narrow optical transition lines because of their two valence electrons yielding spin zero singlet states and spin one triplet states. These intercombination lines open up new possibilities for improved laser cooling schemes with the promise of reaching particularly low temperatures close to or even below the critical temperature for BEC [4–6] on a millisecond timescale. Novel scenarios for reaching quantum degeneracy, as for example matter wave amplification by optical pumping [7], may become feasible with earth alkaline atoms [8]. The simple structure of their zero spin ground state allows for precise comparisons between theory and experiments in cold collision studies [9, 10]. Ultracold earth alkaline samples should also boost the development of novel optical time and length standards [11, 12]. Unfortunately, there are obstacles that impede the success of too simple-minded laser cooling strategies. Because of the absence of ground state Zeeman structure, conventional laser cooling is limited by the Doppler limit which amounts to temperatures in the millikelvin range in contrast to microkelvin temperatures usually obtained for alkalies. Moreover, the principal cooling transition (for all earth alkalis with the exception of magnesium) has a built-in loss channel that can be only incompletely closed by an additional repumping laser thus limiting the size of samples that can be captured.

Calcium with its 408 Hz intercombination line at 657 nm offers a particularly attractive combination of readily accessible transition wavelengths and high potential spectroscopic precision. Our approach to produce large, dense samples of ultracold calcium atoms involves the realization of a magneto-optical trap (MOT) for long-lived (118 min [13]) metastable 4^3P_2 atoms [14]. This MOT (referred to as 4^3P_2 MOT in the following) operates on the closed cycle $4^3P_2 \rightarrow 3^3D_3$ transition at 1978 nm (cf. Fig. 1). Because of its narrow linewidth of 60 kHz this infrared transition provides a Doppler temperature of only 1.3 microkelvin. In addition, its Zeeman substructure implies the presence of polarization gradient cooling with the promise of temperatures approaching the recoil limit. The softness of the 2 micron photons provides a particularly low value of this fundamental cooling limit of only 122 nK.

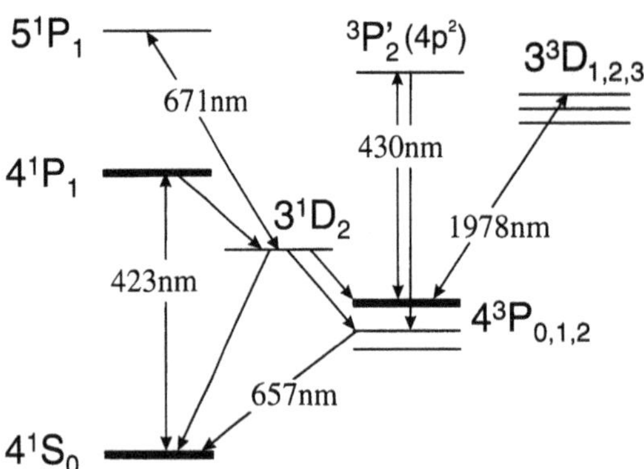

Fig. 1. Relevant energy levels of ^{40}Ca including the strong cooling and trapping transition at 423 nm, the repumping transition at 671 nm, and the cooling and probing transitions for metastable atoms at 1978 nm and 430 nm respectively

2 Production of Precooled Metastable Atoms

In order to load a 4^3P_2 MOT, precooled 4^3P_2 atoms have to be produced at a high rate. We obtain a continuous flux of more than 10^{10} cold 4^3P_2 atoms per second from a MOT operating on the $4^1S_0 \rightarrow 4^1P_1$ transition at 423 nm (natural linewidth $\Gamma_{nat}/2\pi = 34.6$ MHz), referred to as 4^1S_0 MOT in the following. We utilize the fact that there is a weak decay channel for excited 4^1P_1 atoms into the 3^1D_2 state occuring at a rate of 2180 s^{-1} [15–17]. About 78 % of the 3^1D_2 atoms return to the ground state in approximately 3 ms either directly in a quadrupole transition or via the 4^3P_1 state which decays via the 657 nm intercombination line. These atoms are not subject to trapping forces during this time and will move out of the trap at an average speed of 1 m/s. Thus the illuminated capture volume of the 4^1S_0 MOT has to be sufficiently large (Ø = 1 cm) to completely recapture them once they returned to the ground state. The remaining 22 % are transfered to the desired metastable 4^3P_2 state. These atoms exhibit the Doppler-limited 4^1S_0 MOT temperature of typically 1–2 mK, which appears ideal as a starting point for further cooling.

3 Experimental Apparatus

A detailed description of the 4^1S_0 MOT setup is given in [8]. This MOT is loaded from a thermal calcium oven at 650 °C through a conventional decreasing field Zeeman slower operated with a slowing laser tuned 270 MHz below resonance. The appropriate laser light at 423 nm is produced by a frequency-doubled Ti:sapphire laser pumped with 10 W pumping power at 532 nm from a commercial solid state laser system. This results in about 300 mW of blue radiation of which 2/3 can be used for the experiment, having to be shared upon the Zeeman cooler, a two-dimensional transverse cooling stage at its exit and the MOT itself. The MOT beams are about 10 mm in diameter yielding a saturation parameter of 1/2. For optical pumping experiments (repumping 3^1D_2 atoms) described below, we employ an external cavity diode laser system at 671 nm. The 10 mW diode provides 3.6 mW output at the $3^1D_2 \rightarrow 5^1P_1$ transition wavelength of 671 nm. A fraction of 1.3 mW is available in the trap after beam shaping. For direct detection of metastable 4^3P_2 atoms we use a frequency-doubled external cavity diode laser at 860 nm with 30 mW usable output power. After resonant frequency doubling in $KNbO_3$ and beam shaping we have 5 mW available resonant with the 4^3P_2 (4s4p)→ $^3P_2$ ($4p^2$) transition at 430 nm. As a frequency reference for this laser we have combined a DC discharge with a calcium heatpipe. Inside a 15 cm glass tube calcium is heated to 600 °C in a small steel cup which represents the cathode of a DC discharge operated with 1 kV in 2 torr of neon. Using Doppler-free polarization spectroscopy we obtain dispersive signals with signal to noise above 50, well suited for frequency stabilization of the laser. For analyzing our system we can observe independently the fluorescence at 657 nm with a photomultiplier tube and at 423 nm with a calibrated photodiode. In order to operate the 4^3P_2 MOT we have constructed a compact Tm:YAG laser emitting two longitudinal modes separated by ca. 1 GHz.

Pumped with 500 mW power from a Ti:sapphire laser at 786 nm the output is above 50 mW at the desired wavelength of 1978 nm. We stabilize this laser near the $^3P_2 \rightarrow ^3D_3$ transition with the help of Doppler-free polarization spectroscopy in the discharge heatpipe described above.

4 Flux of Metastable Atoms

We can determine the flux of metastable atoms indirectly by measuring the number of 4^1P_1 atoms in our 4^1S_0 MOT, which can be achieved by observing the 423 nm fluorescence. In the steady state we find a 4^1P_1-population $N(4^1P_1) = 4 \cdot 10^7$ which translates into a flux of

$$N(4^1P_1) \cdot 0.22 \cdot 2180 \text{ s}^{-1} = 2 \cdot 10^{10} \text{ s}^{-1}$$

into the 4^3P_2 state. We can check whether $N(4^1P_1)$ is limited by imperfect recycling of atoms decaying from 3^1D_2 to the ground state by means of a lifetime measurement of the 4^1S_0 MOT. Such a measurement is done by shutting off the Zeeman cooler, thus terminating further loading and observing the decaying 423 nm fluorescence. The result in Fig. 2 shows a clean exponential decay with a time constant of 22.6 ms. Using the rate equation model discussed in [8], cf. (3), with the saturation parameter of 0.5 in our 4^1S_0 MOT we can verify 100 % recapture efficiency.

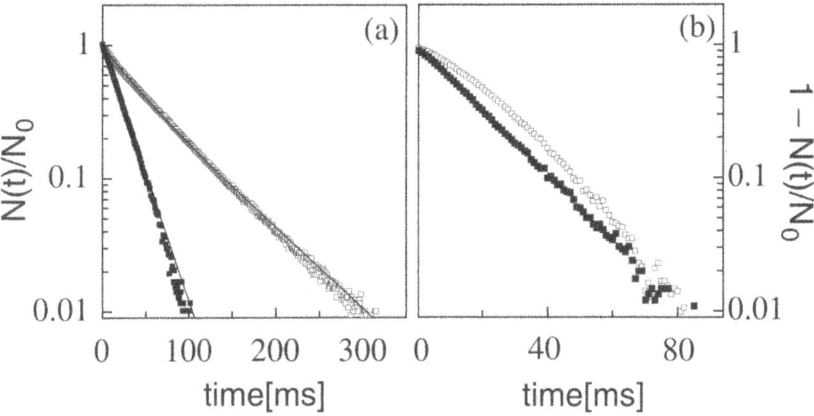

Fig. 2. Decay (**a**) and loading (**b**) of the trap with (□) and without (■) application of a repumping laser acting on the $^1D_2$ population

In order to observe the flux of atoms into the 4^3P_2-state in a direct fashion we have employed an additional laser beam at 430 nm which optically pumps all 4^3P_2 atoms into the 4^3P_1 state with a rate above 10^8 s^{-1}. These 4^3P_1 atoms add to those produced by decay from 3^1D_2 thus increasing the 657 nm fluorescence emerging from the 4^1S_0 MOT, when they decay to the ground state. As illustrated in Fig. 3 we observe an increase of the red fluorescence level by 1/8 when the optical pumping laser is active. Under the realistic assumption that all 4^3P_2 atoms are optically pumped to 4^3P_1 instantaneously with respect to the timescale of their production our observation suggests that the $3^1D_2 \rightarrow 4^3P_2$ decay rate is 1/8 times that of the $3^1D_2 \rightarrow 4^3P_1$ decay rate which is to be compared with the theoretical branching ratio of 1/3. This discrepancy is resolved by noting that the $3^1D_2 \rightarrow 4^3P_2$ decay takes 10 ms on average and thus these atoms have travelled about 1 cm, three times more than the 4^3P_1 atoms. As a consequence the 4^3P_1 atoms remain entirely in the region mapped onto the photomultiplier, while this is only the case for about 40 % of the 4^3P_2 atoms.

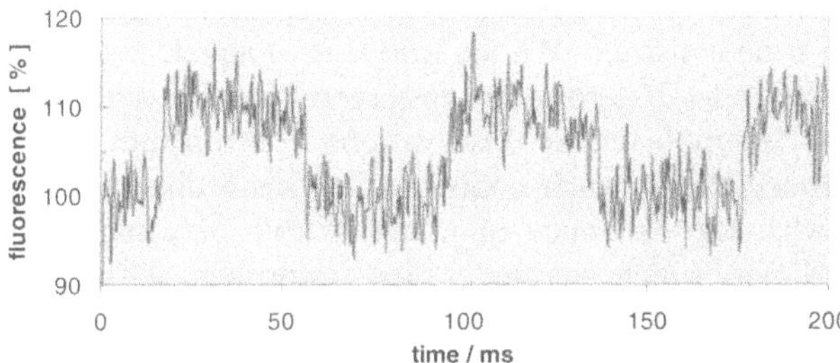

Fig. 3. When the 430 nm probing laser is pulsed on while the 4^1S_0 MOT is operating the red fluorescence at 657 nm is increased by 12 % because metastable $^3P_2$ atoms are transferred to $^3P_1$ and decay to the ground state via the intercombination line

5 Possible Limitations by Cold Collisions

The purely exponential trap decay seen in Fig. 2a (filled squares) suggests that at the present capture rate of the 4^1S_0 MOT collisional loss is not yet a limiting loss mechanism comparable to the desired loss due to the production of metastables. As a consequence, in our case of perfect recycling the production rate of 4^3P_2 atoms equals the capture rate of the 4^1S_0 MOT, i.e. all atoms captured from the Zeeman cooler are eventually transferred into the desired metastable state. It appears natural to ask whether a further increase of the 4^1S_0 MOT capture rate by improvement of the Zeeman cooler will yield a further increase of 4^3P_2-production, or rather take us into the regime where inelastic collisions contribute a comparable loss. In order to investigate this question we have decreased the loss channel which results from the production of metastables by repumping the 3^1D_2 atoms with a 671 nm laser back into the 4^1S_0 MOT cycling transition [18]. In this case, we observe a more than three-fold increase of the trap decay time to 72 ms (cf. Fig. 2a, unfilled squares) in combination with a twofold increased 423 nm fluorescence level. Moreover, the decay acquires a clearly nonexponential character yielding a slight curvature in the logarithmic representation in Fig. 2a (unfilled squares). We have also observed the loading process of the 4^1S_0 MOT by suddenly turning on the Zeeman cooler and recording the 423 nm fluorescence (Fig. 2b). It appears that in contrast to the observations regarding the trap decay there is only a very small increase of the loading time constant if we activate the 671 nm repumper. These observations strongly indicate the presence of an additional density-dependent loss channel occurring for the repumped 4^1S_0 MOT. In the absence of such an additional loss mechanism our trap should be limited by the linear loss resulting from collisions with hot background atoms, which for our imperfect vacuum conditions would still suggest a more than tenfold increase of the lifetime and the fluorescence level. Moreover, for linear loss the loading and decay should be purely exponential with the same time constant. We assume that inelastic two-body collisions between cold 4^1S_0 and 4^1P_1 atoms are responsible for the additional trap loss. This hypothesis is supported by observations in strontium, where a comparably large cross-section for such collisions has been observed and theoretically explained with the existence of a metastable $^1\Pi_g$ molecular state for the strontium dimer, which enables a close approach of the 4^1S_0 and 4^1P_1 collision partners without spontaneous emission [19].

The rate equation for the trapping dynamics of the 4^1S_0 MOT including two-body collisions reads

$$\dot{\mathrm{N}} = R - \Gamma \mathrm{N} - \beta \int n^2(r)\,\mathrm{d}^3 r \tag{1}$$

where R is the 4^1S_0 MOT capture rate, Γ accounts for the linear loss due to collisions with hot background atoms, β is the coefficient for inelastic collisions and $n(r) = n\,\mathrm{e}^{-(r/a)^2}$ is the Gaussian distribution of the atomic density. The solution of this differential equation for the condition of terminated loading (trap decay) is

$$\frac{N}{N_0} = \frac{\mathrm{e}^{-\Gamma t}}{1 + \frac{\xi}{1-\xi}\left(1 - \mathrm{e}^{-\Gamma t}\right)} \tag{2}$$

where $N_0 = n_0\left(\sqrt{\pi}a\right)^3$ denotes the steady state population,

$$n_0 = \sqrt{\frac{2\Gamma^2}{\beta^2} + \frac{8R}{\beta}\left(\sqrt{2\pi}a\right)^3} - \frac{\sqrt{8}\Gamma}{2\beta} \tag{3}$$

is the corresponding steady state peak density in the trap and

$$\xi = \frac{\beta n_0}{\sqrt{8}\Gamma + \beta n_0} = \frac{\sqrt{1 + \frac{4R\beta}{\Gamma^2\left(\sqrt{2\pi}a\right)^3}} - 1}{\sqrt{1 + \frac{4R\beta}{\Gamma^2\left(\sqrt{2\pi}a\right)^3}} + 1} \tag{4}$$

is a parameter in the interval [0, 1] which denotes the fraction of quadratic loss relative to the total trap loss. In the case of trap loading the trapped population increases according to

$$1-\frac{N}{N_0}=\frac{\mathrm{e}^{-\gamma t}}{\frac{1}{1+\xi}+\frac{\xi}{1+\xi}\mathrm{e}^{-\gamma t}} \quad , \quad \gamma=\frac{1+\xi}{1-\xi}\Gamma . \tag{5}$$

While the trap decay involves the linear loss rate Γ loading occurs with an increased rate γ. The reason for this behavior is that the loading is abruptly terminated once the quadratic regime is reached leading to shorter loading times as compared to what is expected due to linear losses. In contrast to that, the linear loss time constant dominates the trap decay because the initial steady state density is never too far inside the quadratic loss regime. This effect is clearly observed in our experiments. The above solutions are used to fit the decay and loading data shown in Fig. 2. The relevant time constants were 22.6 ms without repumping and 71.8 ms with repumping. The parameter ξ turned out to be negligible without repumping ($\xi = 0.001$) but takes the value $\xi = 0.32$ when repumping is applied. The explicit expression for the parameter ξ in (4) shows that a diminishment of the linear loss rate Γ by a factor of 3.1 introduced by the 671 nm repumper yields the same increase in ξ as an enlargement of the capture rate R by a factor of $(3.1)^2 \approx 10$. We can thus conclude that an order of magnitude increase of the capture rate would yield a loss of about 32 % of the atoms due to inelastic collisions rather than a transfer to the desired metastable state.

6 Magneto-Optical Trap for Metastable Atoms

The $^3P_2$ atoms are produced at a temperature of about 2 mK slightly above the Doppler cooling limit for the 423 nm cooling line of 0.8 mK. This is sufficiently low to capture them in a $^3P_2$ MOT employing the closed $^3P_2 \rightarrow {}^3D_3$ transition at 1978 nm. The narrow linewidth of this transition of about 57 kHz [20] together with its large wavelength provide a very low Doppler limit of only 1.3 μK and a recoil limit of 122 nK. Moreover, the $J = 2 \rightarrow J = 3$ level structure allows for polarization gradient cooling (PGC) which should allow for temperatures approaching the recoil limit, i.e. well below a microkelvin. In our experiment we use the same magnetic field gradient of 15 G/cm for both MOTs in order to let us operate our experiment in an entirely continuous mode. This represents a compromise between the best value for the $^1S_0$ MOT of 60 G/cm – which optimizes the capturing of atoms from the atomic beam – and a value below 1 G/cm which would account for the narrow linewidth of the 1978 nm transition. In order to enable operation of the $^3P_2$ MOT at 15 G/cm and to extend its velocity capture range, we strongly saturate the $^3P_2 \rightarrow {}^3D_3$ transition. We use about 5 mW in each of the infrared beams of about 8 mm diameter, yielding an intensity I which is 1.6×10^4 times the saturation intensity of $I_S = 1.28\,\mu W\,cm^{-2}$. For this mode of operation the Doppler limit amounts to $T_D = \hbar\Gamma_I/2k_B = 170\,\mu K$, where $\Gamma_I = \Gamma\sqrt{1+I/I_S}$ denotes the power-broadened transition linewidth. This appears to be an undesirably high temperature, however the presence of PGC should allow us to go significantly beyond the Doppler limit. Further temperature decrease into the nanokelvin range should be possible by ramping down the magnetic field as well as the laser intensity and detuning.

A typical loading and detection cycle is shown in Fig. 4. Before $t = 43$ ms both lasers at 423 nm and 1978 nm are activated for 250 ms loading 3×10^8 atoms into the $^3P_2$ state. The operation of the $^1S_0$ MOT is monitored by the red fluorescence at 657 nm resulting from atoms pumped into the $^3P_1$ state which decays within 0.4 ms to the ground state. At $t = 43$ ms the $^1S_0$ MOT

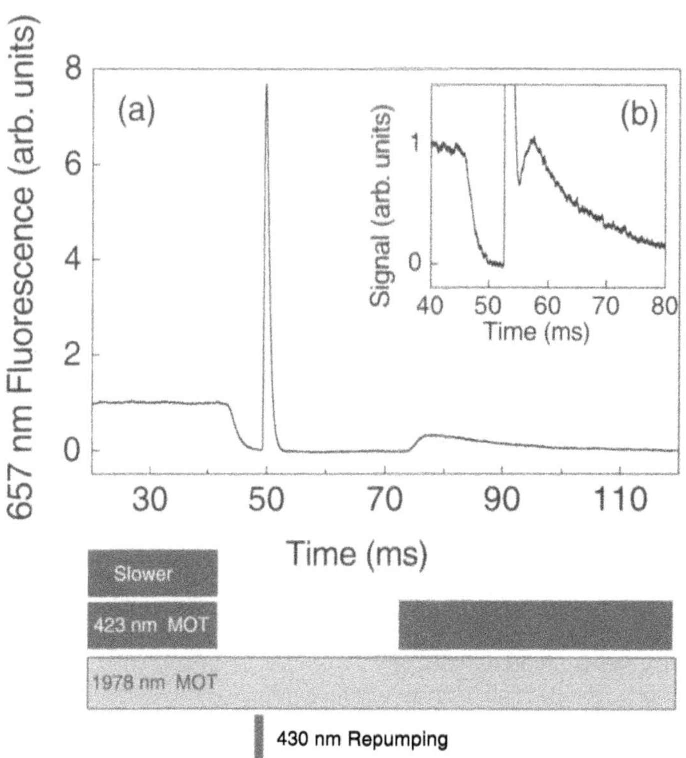

Fig. 4a,b. A typical loading and detection cycle. The gray bars below the time axis indicate which of the laser beams are active. The $^1S_0$ MOT and slower beams are turned off after both traps have been loaded for 250 ms. Following a variable delay time τ_1 a pulse of 430 nm light optically pumps all $^3P_2$ atoms into the $^3P_1$ state, yielding a burst of 657 nm fluorescence. The atomic sample is now subject to gravity and ballistic expansion during a time interval τ_2. Finally, the $^1S_0$ MOT beams are turned on again recapturing a fraction of these atoms, which leads to a revival of 657 nm fluorescence and Zeeman slower beams are blocked, i.e. only the $^3P_2$ MOT continues to operate and the red fluorescence terminates within a few ms (limited by the lifetime of the singlet D state $^1D_2$). After a variable delay time τ_1 which amounts to 5 ms in the specific case of Fig. 4, i.e. at $t = 48$ ms, a 2 ms long pulse of 430 nm light is applied which optically pumps all $^3P_2$ atoms into the $^3P_1$ state, yielding a burst of 657 nm fluorescence. The atomic sample is now in the ground state and subject to gravity and ballistic expansion. After a variable delay time τ_2 (25 ms in Fig. 4) the $^1S_0$ MOT beams are turned on again (at $t = 73$ ms in Fig. 4) recapturing a fraction of these atoms, which leads to a revival of 657 nm fluorescence. This fluorescence finally dies out in accordance with the 22 ms lifetime of the $^1S_0$ MOT resulting from the transfer of the atoms into the $^3P_2$ state again. The detail (b) shows the case when the free flight duration is only $\tau_2 = 10$ ms. Note that the number of recaptured atoms exceeds the number of atoms initially trapped in the $^1S_0$ MOT. Measurements as shown in Fig. 4 allow us to determine the trap lifetime of the $^3P_2$ MOT if we vary τ_1 and observe the size of the 657 nm fluorescence peak. In Fig. 5a we show the resulting lifetime of $\tau_{\mathrm{IR}} = 261$ ms which we attribute to our imperfect vacuum of only 10^{-8} mbar.

At the present precision that we can achieve in such measurements (which is limited by laser frequency drifts) we do not recognize any nonexponential decay component, that would indicate collisional loss. We can also estimate the temperature via a simple recapture measurement, as shown in Fig. 5b, where we have varied τ_2 and recorded the size of the recapture signal. The temperature of the atomic sample is expected to be due to PGC in the center of the trap. The region where PGC can operate is roughly determined by the condition that the Zeeman shift should not exceed the light shift. For the parameters used in our experiment this region is expected to be on the order of 2 mm in diameter. In fact, by recording the optical pumping efficiency of the 430 nm laser beam versus its diameter, we find that about 50 % of the atomic population is located outside a sphere of 2 mm diameter. This indicates the possible existence of a non-Gaussian velocity distribution consisting of a colder core with a temperature determined by PGC and a hotter halo exhibiting the Doppler temperature increased by power broadening. This complicates the interpretation of recapture data. In order to obtain a rough estimate of the temperature we can resort to the somewhat oversimplified picture saying that the decay rate observed in Fig. 5b is given by the mean velocity $\bar{v}$ of the atoms divided by the radius $r = 4$ mm of the capture volume of the $^1S_0$ MOT. With $\bar{v} = (2/\sqrt{\pi})\sqrt{(2k_{\mathrm{B}}T)/m}$ this yields a temperature of about 50 µK. We have to employ a slightly more subtle model in order to better understand our data. We assume a bimodal thermal distribution of the $^3P_2$ atoms with different widths and temperatures and let each fraction expand ballistically. We use the initial width σ_{h}, temperature T_{h}, and relative particle number N_{h} of the hot fraction, and the temperature of the cold fraction T_{c} as fit parameters. In Fig. 5b a best

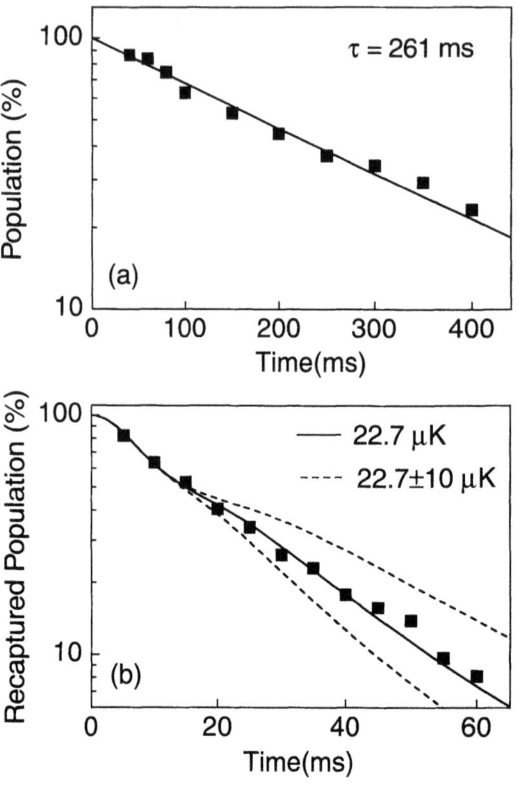

Fig. 5. (**a**) Lifetime measurement of the $^3P_2$ MOT. The *solid line* shows an exponential fit. (**b**) The number of atoms recaptured in the $^1S_0$ MOT is plotted versus the ballistic expansion time τ_2. The *solid line* shows a theoretical fit using a bimodal thermal distribution as explained in the text. We find a temperature of 22.7 μK for the cold fraction of atoms cooled by PGC and a temperature of 409 μK for atoms only subjected to Doppler cooling. For comparison the *dashed lines* show the theoretical fits for $22.7 \pm 10\,\mu$K

fit is obtained for $\sigma_h = 6.1\,\text{mm}, T_h = 409\,\mu\text{K}$, which amounts to 2.5 times the Doppler temperature corresponding to our strongly saturated mode of operation, $N_h = 2.2$, and $T_c = 22.7\,\mu\text{K}$. These results nicely confirm our expectations of a bimodal thermal distribution with a cold core prepared by PGC surrounded by a hotter cloud only subject to Doppler cooling. The dashed curves show the predictions when T_c is increased or decreased by 10 μK. The optical pumping transition at 430 nm connects the $^3P_2$ and $^3P_1$ states to a higher lying $^3P_2$ ($4p^2$) state with a line strengths ratio of 3:1. Thus, in the optical pumping of $^3P_2$ atoms into $^3P_1$ on average four photons are absorbed from the 430 nm laser beam and, together with a 657 nm photon, spontaneously emitted. This leads to a temperature increase of 3.8 μK and a center of mass velocity change of 9 cm/s. In a time interval of 20 ms this amounts to a 1.8 mm horizontal displacement of the atomic cloud, together with a vertical displacement of 2 mm due to gravity. These effects have not been taken into account in our model such that our temperature estimations are to be referred to as upper bounds. The value $T_c = 22.7\,\mu\mathrm{K}$ found for the recaptured ground state atoms indicates an upper bound of 19 μK for the temperature of the $^3P_2$ MOT.

We can determine our transfer rate from the $^1S_0$ MOT into the $^3P_2$ MOT as follows. By setting τ_2 to zero in Fig. 4 we recapture about 1.5 times the original $^1S_0$ MOT population, which amounts to $N_{\mathrm{IR}} = 3 \times 10^8$ atoms. Thus, the loading rate of the $^3P_2$ MOT is $N_{\mathrm{IR}} \times \tau_{\mathrm{IR}}^{-1} = 1.1 \times 10^9$ atoms/s. This is about one order of magnitude lower than our production rate of $^3P_2$ metastables $R = 2 \times 10^{10}$ atoms/s, i.e. only about 6 % of the metastables are captured. The reason is that the infrared laser beams due to limited access to the vacuum chamber are only 8 mm in diameter whereas the cloud of metastables produced via decay from the $^1D_2$ state is expected to have a width of around 20 mm, because this decay takes on average 10 ms and the 2 mK cold atoms can fly 10 mm during this time. This mismatch of the capture volume fully accounts for the 6 % capture efficiency. Employing slightly larger laser beams should allow us to make use of the full rate R.

The continuous mode of operation of our trap described above is optimized for collecting cold metastable atoms at a high rate at the price of not fully exploiting the cooling potential of our scheme. In order to obtain this high production rate at a temperature in the nanokelvin range, in forthcoming experiments we will combine a capture phase as described above with a duration of the lifetime of the $^3P_2$ MOT with a several 10 ms long cooling phase where the magnetic field gradient and the infrared laser intensity and detuning are synchronously ramped down. In a further step we plan to transfer the atoms into an optical dipole trap operating near 1064 nm, which yields comparable light shifts for the $^1S_0$ and $^3P_1$, $m = \pm 1$ states. This should allow us to apply resolved sideband cooling on the intercombination line with the promise of all-optical (and therefore fast) access to the quantum degenerate regime.

7 Conclusion

We have demonstrated a scheme for producing large samples of very cold calcium atoms. While at present we can produce about 10^9 atoms/s at a temperature below 20 μK, minor technical improvements promise rates above 10^{10} atoms/s. A principal upper limit due to light-assisted two-body collisions is expected only at above 10^{11} atoms/s. Our experiment can be optimized for far lower temperatures (below 1 μK) when operated in a pulsed mode including a phase of high capture efficiency as described in this chapter and a short cooling phase, where the magnetic field gradient as well as the laser intensity and detuning are reduced.

8 Acknowledgements

I would like to acknowledge my team at the Universität Hamburg and in particular Jan Grünert whose PhD work forms the basis of this chapter. This work has been partly supported by the Deutsche Forschungsgemeinschaft under contract number He2334/2.3, DAAD 415 probral/bu, and the European research network "Cold Atoms and Ultra-Precise Atomic Clocks" (CAUAC). We also acknowledge future support of the Schwerpunktprogramm 1116.

9 Epilog

This book is dedicated to Theodor Hänsch on the occasion of his 60th birthday. It is a pleasure to take this opportunity to recall a few scenes from the early days of his engagement in Munich. At the end of 1990 I had finished my PhD thesis at Theodor's research group at the Max-Planck Institut für Quantenoptik (MPQ) in Garching, when he offered me the opportunity to set up new labs in his second group at the university in the city of Munich. This group was still small, consisting of Wolfgang Heckel and his two PhD students Martin Specht and Johannes Pedarnig who were pursuing activities in the field of tunneling and near field microscopy in close neighborhood to the IBM research group of nobel laureate Gerd Binnig. Setting up labs was a known issue for PhD students during these pioneer times at MPQ and I had tasted the outstanding scientific freedom and the informal and international spirit that was connected with working with Theodor. About three years before he had just arrived from Stanford starting a large research group at MPQ with all its practical implications. Despite my specialization in mathematical physics he courageously invited me to become number three of his PhD students in Garching. Shortly before, Reinald Kallenbach and Claus Zimmermann had been employed to set up new records in hydrogen spectroscopy that would outdate those obtained by Theodor's former group at Stanford. The royal challenges of precision spectroscopy were thus distributed and the way seemed wide open for defining new exciting research goals. Many topics were under consideration and I got to know Theodor's broad and unprejudiced enthusiasm for physics. I believe that if I had insisted on elaborating on something strange like particle physics or cosmology, he would not have argued against it as long as he could recognize a payable, novel and exciting physical perspective. I thus accepted Theodor's new offer with no hesitation with the plan in mind to try a more serious second excursion into the world of light forces that I had stumbled into during my PhD, and I began to collect a small group of enthusiastic young physicists. Tilman Esslinger, who had worked with me for his diploma thesis, joined in as a PhD student and slightly later Matthias Weidemüller who returned from a one year stay at the Ecole Normale and finally in 1994 Axel Görlitz. At the same time Theodor had invited Claus Zimmermann, who was interested in magnetic traps, to start a small research team one floor above us. Leo Ricci and slightly later Vladan Vuletic joined him as PhD students and thus the "city gang" was complete. The university institute was located in the center of Schwabing, Munich's former artist quarter, with plenty of bookstores, cafes and Italian restaurants around. The peculiar charm of the location appeared to make up well for the better technical basis we had enjoyed at MPQ at the price of its less favored location out in the meadows of the northern Munich suburbs. A number of exciting years were ahead, much inspired by the peculiar academic spirit Theodor taught us. Theodor's mistrust in the readily practiced use of quantum mechanical explanations, if immediate intuition is not available, became a guiding principle for us. Many lively discussions, sometimes acquiring a slightly sophistic character, were dedicated to taking classical interpretations to the limits. It often turned out surprising how little quantum mechanics was needed in most cases. A second holy principle was KISS, standing for "keep it simple stupid". Physical explanations that could not be boiled down into intuitive pictures but rather relied on large bodies of mathematical formalism were certainly not in accordance with KISS. However, more importantly there was an experimental version of that principle. Inspired by Theodor's appreciation for small but smart technical toys, simple and elegant technical solutions in the lab appeared to us as prerequisites for exploring the world of the really small. During these privileged years as an assistant in Theodor's university group I have learned that curiosity and courage to question the obvious in combination with a sense of playfulness is a fruitful approach to physics and that striving for elegance and simplicity can be a quite useful guideline in the search for experimental solutions. Until today a never running dry stream of novel and influential ideas and their realizations has arised from Theodor's research activities and I believe that still many will follow.

References

1. T. Hänsch, A. Schawlow, Opt. Commun. **13**, 68 (1975)
2. M.H. Anderson *et al.*, Science **269**, 198 (1995)
3. K. Davis *et al.*, Phys. Rev. Lett. **75**, 3969 (1995)
4. T. Ido, Y. Isoya, H. Katori, Phys. Rev. A **61**, R061403 (2000)
5. E.A. Curtis, C.W. Oates, L. Hollberg, arXiv:physics 0104061 (2001)
6. T. Binnewies, G. Wilpers, U. Sterr, F. Riehle, J. Helmcke, T.E. Mehlstäubler, E.M. Rasel, W. Ertmer: arXiv:physics/0105069 (2001)
7. R. Spreeuw, T. Pfau, U. Janicke, M. Wilkens, Europhys. Lett. **32**, 469 (1995)
8. J. Grünert, G. Quehl, V. Elman, A. Hemmerich, J. Mod. Opt. **47**, 2733 (2000)
9. T.P. Dinneen, K.R. Vogel, E. Arimondo, J.L. Hall, A. Gallagher, Phys. Rev. A **59**, 1216 (1999)
10. M. Machholm, P.S. Julienne, K.-A. Suominen, arXiv:physics 0103059 (2001)
11. T. Kisters, K. Zeiske, F. Riehle, J. Helmcke, J. Appl. Phys. B **59**, 89 (1994)
12. F. Ruschewitz *et al.*, Phys. Rev. Lett. **80**, 3173 (1998)
13. A. Derevianko, Phys. Rev. Lett. **87**, 023002 (2001)

14. J. Grünert, S. Ritter, A. Hemmerich, submitted to Phys. Rev. Lett. (2001)
15. F. Strumia, in *Laser Science and Technology*, ed. by A.N. Chester, S. Marellucci (Plenum, New York 1988) pp. 367–401
16. N. Beverini, F. Giammanco, E. Maccioni, F. Strumia, G. Vissani, J. Opt. Soc. Am. B **6**, 2188 (1989)
17. T. Kurosu, F. Shimizu, Jpn. J. Appl. Phys. **29**, L2127 (1990)
18. C. Oates, F. Bondu, R. Fox, L. Hollberg, Eur. Phys. J. D. **7**, 449 (1999)
19. K.R. Vogel, T.P. Dinneen, A. Gallagher, J.L. Hall, IEEE Trans. Instrum. Meas., **48**, 618 (1999)
20. R.L. Kurucz, B. Bell, *1995 Atomic Line Data* (Smithsonian Astrophysical Observatory, Cambridge, Mass., 1995) CD–ROM No. **23**

Part V

Nonlinear Optics and Spectroscopy

The Design of Enhancement Cavities for Second Harmonic Generation

Tim Freegarde and Claus Zimmermann

1 Introduction

The use of resonant cavities to enhance the intensity of fundamental radiation has allowed second harmonic generation, originally confined to large-frame pulsed lasers, to be applied to low power continuous-wave devices. Theodor Hänsch and his co-workers, driven in particular by the need for stable ultraviolet sources for hydrogen spectroscopy, have from the beginning been in the vanguard in developing these techniques: resonant cavities enhanced the frequency doubling [1] and mixing [2] of c.w. dye lasers; crucial active [3] and passive [4] stabilization techniques were contributed; and a doubly-resonant scheme was investigated [5]. Frequency doubling is now regularly achieved with remarkable efficiency using tiny laser diodes and, taking the harmonic itself as the source for a second doubling stage, even quadrupling of their frequency has become practical [6].

Cavity design in theory requires only textbook physics, but it is complicated in practice by off-axis astigmatism and Brewster-angle refraction. The use of off-axis spherical mirrors in resonators containing other astigmatism-inducing components was considered in the early days of the laser by Kogelnik et al. [7], whose description of the physics would be hard to surpass. Dunn and Dunn [8] subsequently extended the treatment to the case of a ring laser with an intra-cavity frequency-doubling crystal at the Brewster angle, and others have since considered related arrangements in some detail [9, 10]. Recently, Sun et al. addressed the calculation of beam parameters within a given cavity specifically for the generation of diode laser harmonics [11].

Many researchers simplify the exercise by seeking solutions with a certain symmetry, such as a circular beam waist at the primary or secondary focus, and either accept imperfect mode-matching or introduce external beam-shaping components. We describe instead the design of cavities that match an arbitrarily elliptical beam such as that emitted by a laser diode, via an off-axis spherical mirror, to the optimal beam for second harmonic generation, which in general is also elliptical. Starting from the ellipticity of the fundamental laser beam, an arbitrary choice of mirror curvature and the desired beam profile in the crystal, whose determination we have addressed in a previous paper [12], we provide a recipe by which the dimensions of a suitable resonator may be calculated.

2 Resonator Geometry

A typical four-mirror “bow-tie” enhancement cavity is shown in Fig. 1. The nonlinear crystal is enclosed at the ‘primary’ beam waist midway between a pair of off-axis spherical concave mirrors, whose inclination defines the tangential (azimuthal) coordinates (x, z) that are locally aligned with z parallel to the ray axis. Two plane mirrors, of which one is usually the input coupler, complete the ring cavity. In Fig. 1, the crystal is inclined in the tangential plane.

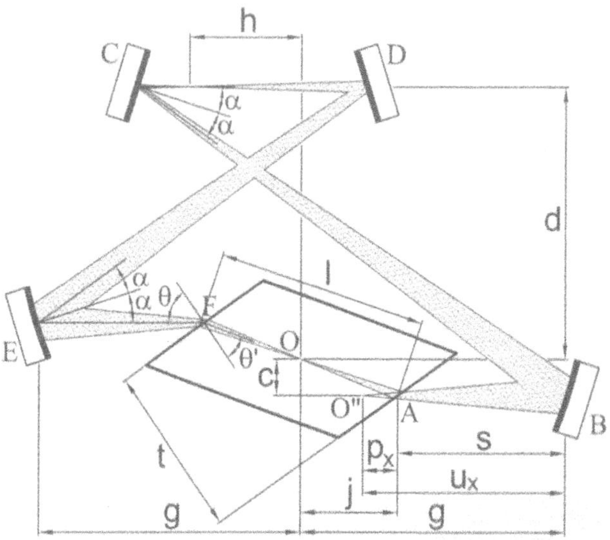

Fig. 1. Schematic layout of the enhancement cavity shown in plan view. The crystal is here shown inclined at Brewster's angle in the tangential orientation. O$''$ is the apparent position of the focus when viewed from outside the crystal

The crystal, of thickness t measured perpendicular to its faces, is presumed to have a refractive index η and may be inclined at the Brewster angle to the ray axis. The concave mirrors, of normal focal length f, make an angle α with the ray axis. The positions of the two plane mirrors are unimportant so long as they complete the cavity and allow the appropriate round-trip distance, but in the geometry shown they are also inclined at an angle α.

The crystal may alternatively be inclined in the sagittal (y, z) direction. Strictly, this renders the resonator non-planar, not least because the modes are elliptically polarized [13], but in many practical cases the distortions introduced will be small. The relevant dimensions are shown in Fig. 2.

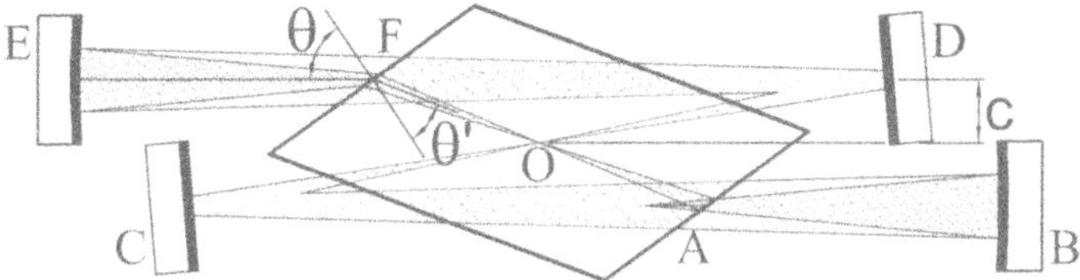

Fig. 2. Side view of the enhancement cavity showing dimensions appropriate when the crystal is inclined sagittally

3 Design Procedure

The design of the cavity starts with the choice of crystal material, dimensions and cut, which in turn define the best Gaussian beam for harmonic generation. Knowing this beam within the crystal, we calculate the corresponding external form and the apparent optical thickness of the crystal. These parameters, together with the ellipticity of the original laser beam, are the target solutions for a pair of equivalent free-space cavities corresponding to the tangential and sagittal directions and, with the addition of an assumed mirror curvature, suffice to establish the resonator dimensions by numerical solution. Finally, we convert the optical dimensions to the physical positions and orientations of the mirrors.

3.1 Selecting the Nonlinear Crystal

Crystals are generally chosen on the basis of their transparency and damage thresholds at the fundamental and harmonic wavelengths, the requirement of phase-matching, the likely conversion efficiency and, at high powers, the susceptibility to damage and thermally-induced dephasing. The efficiency, whose theoretical calculation has been described elsewhere [12, 14], depends upon the nonlinearity and Poynting vector walk-off together with the crystal dimensions, which are usually constrained by availability and cost. Whether or not a Brewster-angle crystal cut is chosen will probably depend upon the availability of anti-reflection coatings for the normal incidence case.

To calculate the optimal beam for harmonic generation, we need to know the optical path length within the crystal. For a plate of thickness t and refractive index η inclined at the Brewster angle, the ray length between the two faces is

$$l = \frac{t\sqrt{\eta^2 + 1}}{\eta}, \tag{1}$$

while at normal incidence the corresponding length is simply $l = t$.

3.2 Determining the Required Resonator Beam

Having chosen the crystal, we calculate the optimal fundamental beam [12, 14]. Generally elliptical, this is characterized in the tangential and sagittal planes by the complex Gaussian beam parameters $q_{x,y}(z)$, defined at the foci by

$$\frac{2}{\mathrm{i}} q_{x,y}(0) = b_{x,y} = \frac{2\pi w_{x,y}^2 \eta}{\lambda_0}, \tag{2}$$

where the confocal parameters $b_{x,y}$ are functions of the waist radii $w_{x,y}$, the vacuum wavelength λ_0 and the refractive index η. The chosen crystal parameters define the optimal beam through the ratio $\xi_{x,y} = l/b_{x,y}$.

The beam undergoes refraction at the crystal faces, and hence from outside appears to have differently sized and positioned foci. We now calculate the apparent confocal parameters $b''_{x,y}$ and focal positions for both planes. If the beam waist at the crystal centre is described by $q = \mathrm{i}b/2$, and a matrix with elements A, B, C and D describes propagation from the crystal centre to its face and subsequent refraction, then the beam leaving the crystal face, characterized by $q' = (Aq/\eta + B)/(Cq/\eta + D)$, could equivalently result from the free propagation of a beam whose waist is described by $q'' = \mathrm{i}b''/2$ through a distance p to the same point. Since $q' = q'' + p$, we find

$$p = \mathrm{Re}(q'), \quad b'' = \frac{2}{\mathrm{i}} \mathrm{Im}(q'). \tag{3}$$

For given true beam parameters b_x and b_y at the crystal focus, we hence derive equivalent free-space parameters b''_x and b''_y and the corresponding apparent focus to crystal-face distances p_x and p_y, which for later convenience we write in terms of the mean p and astigmatic focal shift z as $p_x = p + z/2$ and $p_y = p - z/2$. The mean distance u between the concave mirror and the apparent crystal foci is given by $u = p + s$, where s is the distance from the mirror to the crystal face; the tangential and sagittal crystal focus distances are thus $u_x = u + z/2$ and $u_y = u - z/2$, respectively. Expressions for $b''_{x,y}$, p and z for our three specific configurations are given in Table 1.

The cavity, which is now defined by the mirror focal length f, incidence angle α, astigmatism z and the mean distance between the mirror and the crystal focus u, is now examined as if the crystal were absent. Our aim is to find the conditions under which the images of the chosen beam radii coincide

Table 1. Effective beam parameters and positions for normal incidence and for tangential and sagittal inclination at the Brewster angle

	b''_x	p	z	b''_y
normal incidence	b_x/η	$t/2\eta$	0	b_y/η
Brewster (tangential)	b_x/η^3	$\frac{t}{4\eta^4}\left(1+\eta^2\right)^{3/2}$	$-\frac{t}{2\eta^4}\sqrt{1+\eta^2}\left(\eta^2-1\right)$	b_y/η
Brewster (sagittal)	b_x/η	$\frac{t}{4\eta^4}\left(1+\eta^2\right)^{3/2}$	$+\frac{t}{2\eta^4}\sqrt{1+\eta^2}\left(\eta^2-1\right)$	b_y/η^3

at a distance v from the mirror and are in the same proportion as the axes of the elliptical laser beam. To simplify the calculation, we convert the object and image distances and confocal parameters to dimensionless quantities by scaling according to the focal length. Thus $b''_{x,y}$ become $X, Y = b''_{x,y}/2f$, z becomes $Z = z/2f$ and u and v are rewritten as $U = u/f$ and $V = v/f$. We also introduce the parameter $\phi = \cos\alpha$. Beam waists are already scaled in terms of the wavelength λ_0 by our use of the confocal parameter b.

3.3 Solving the Resonator Equations

To find the dimensions (u, v, α) of a cavity giving the required parameters b''_x, b''_y, z, mirror focal length f and secondary waist ellipticity parameter $e_2 = w_{y2}/w_{x2}$ (where $w_{x2,y2}$ are the secondary waist radii) we refer to the function

$$\mathcal{F}(u, v) = \left(\frac{u-1}{v-1}\right)(1 - (u-1)(v-1)), \tag{4}$$

which is shown in Fig. 3.

This is the equation which, for normal incidence and in the absence of astigmatism, defines the radius w of the waist that lies a distance d_1 from a mirror of focal length f when the secondary waist distance is required to

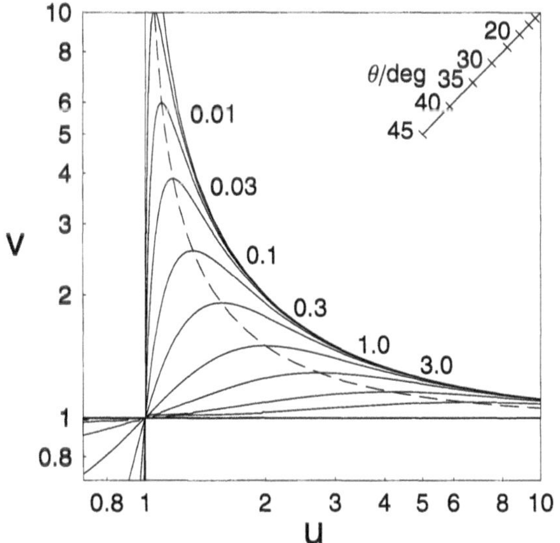

Fig. 3. Logarithmic plot of $\mathcal{F}(u, v)$ for graphical solution of astigmatic cavity design, showing the normalized principal waist size as a function of cavity geometry. The *dashed line* is the locus of the principal mode maxima

be d_2:

$$\left(\frac{\pi w^2}{\lambda_0 f}\right)^2 = \mathcal{F}\left(\frac{d_1}{f}, \frac{d_2}{f}\right). \tag{5}$$

In terms of the function $\mathcal{F}(u,v)$, our aim is the simultaneous solution of

$$\mathcal{F}\left(\frac{U+Z}{\phi}, \frac{V}{\phi}\right) = \frac{X^2}{\phi^2} \tag{6}$$

$$\mathcal{F}\left((U-Z)\,\phi, V\phi\right) = Y^2\phi^2 \tag{7}$$

$$\frac{\mathcal{F}\left(V/\phi, (U+Z)\,/\phi\right)}{\mathcal{F}\left(V\phi, (U-Z)\,\phi\right)} = \frac{1}{e_2^4\phi^4}, \tag{8}$$

which respectively require specific values for the crystal waist in the tangential and sagittal directions and the secondary waist ellipticity. Given values for X, Y, Z and e_2, these equations allow us to find values of U, V and ϕ.

A first approximation can sometimes be found graphically. For the example of normal incidence ($Z=0$), the solution to the first two equations lies at the intersection of the appropriate contours, displaced according to α along the 45° line shown. Solution of the third equation involves evaluation of the function at the reflection in this 45° line of the intersection point, the process being repeated for a range of ϕ until the required value is found.

In the more general case, numerical solution is necessary. For given values of X, Y, Z and ϕ – when we have defined the target crystal beam waists and astigmatism and have made an arbitrary choice of mirror focal length – (6) and (7) may be combined to yield a quartic equation for U, as a function of the mirror incidence angle via $\phi = \cos\alpha$:

$$\begin{aligned}
0 = {} & U^4\left(\phi^3-\phi\right) + U^3\left(1-\phi^4\right) \\
& + U^2\left(\phi^4(X+Z)+\phi^3\left(1-2Z^2\right)-\phi^2(X-Y+4Z)-\phi\left(1-2Z^2\right)-Y+Z\right) \\
& + U\left(\phi^4 Z\left(-2X+Z\right) - \phi^3\left(2X+Y-2Z\right) + 2\phi^2\left(X+Y\right)Z \right.\\
& \qquad \left. +\phi\left(X+2\left(Y+Z\right)\right) - Z\left(2Y+Z\right)\right) \\
& + \left(\phi^4\left(X-Z\right)Z^2 + \phi^3\left(X\left(Y+2Z\right) + Z\left(-Y-3Z+Z^3\right)\right)\right. \\
& \qquad +\phi^2\left(X-Y-2Z-XZ^2+YZ^2+4Z^3\right) \\
& \qquad \left. -\phi\left(X\left(Y+Z\right)+Z\left(-2Y-3Z+Z^3\right)\right) - Z^2\left(Y+Z\right)\right).
\end{aligned} \tag{9}$$

The solution may be inserted into (8) and ϕ adjusted until the required secondary waist ellipticity is obtained. In most practical cases, there will be a large range of f (and a corresponding range of α) for which solutions are possible.

3.4 Extracting the Cavity Dimensions

Having found values for the scaled parameters U, V and ϕ, we may immediately write down the dimensions u, v and α upon which they are based.

For practical application, however, these must be converted into the real dimensions g, c, d and h shown in Figs. 1 and 4.

The distance from the crystal centre to the concave mirror is given by $g = s + j = u + j - p$, where p is given in Table 1. For the case of the Brewster angle, we have $j = t/\sqrt{\eta^2+1}$ and hence

$$g = u + t\frac{\left(3\eta^4 - 2\eta^2 - 1\right)}{4\eta^4\sqrt{\eta^2+1}}, \tag{10}$$

while at normal incidence we have simply $2j = l = t$ and thus

$$g = u + t\frac{\eta - 1}{2\eta}. \tag{11}$$

The lateral displacement c of the external rays from the crystal centre is given in both the Brewster-angle cases by

$$c = \frac{t}{2}\frac{\eta^2 - 1}{\eta\sqrt{\eta^2+1}}. \tag{12}$$

The transverse displacement of the plane mirrors (and thus the secondary beam waist) from the crystal centre is

$$d = \frac{1}{\sin 2\alpha}\left(v - g\cos 2\alpha - |v\cos 2\alpha - g|\right). \tag{13}$$

This expression, which is also valid when the resonator forms a trapezium ($v\cos 2\alpha < g$), becomes simply $d = (v + g)\tan\alpha$ for the bow-tie cavity of Fig. 1. If $v < g$, then the plane mirrors will be closest to the crystal, and the concave mirrors will thus enclose the secondary focus. In the configuration shown, the plane containing this secondary beam waist will be displaced from that containing the crystal centre by a distance

$$h = \frac{c}{\tan\alpha}. \tag{14}$$

3.5 Mirror Curvature and Angle of Incidence

The freedom so far to make an arbitrary choice for either the mirror focal length or angle of incidence allows us to take into account additional effects such as coma. This has been treated using geometrical optics by Dunn and Ferguson [15], who found that, for $b \ll l$, coma due to the off-axis mirror and a crystal face inclined tangentially at the Brewster angle cancel when

$$f\sin\alpha = t\frac{\left(\eta^2+1\right)^{\frac{1}{2}}\left(\eta^4 - 1\right)}{\eta^7} \tag{15}$$

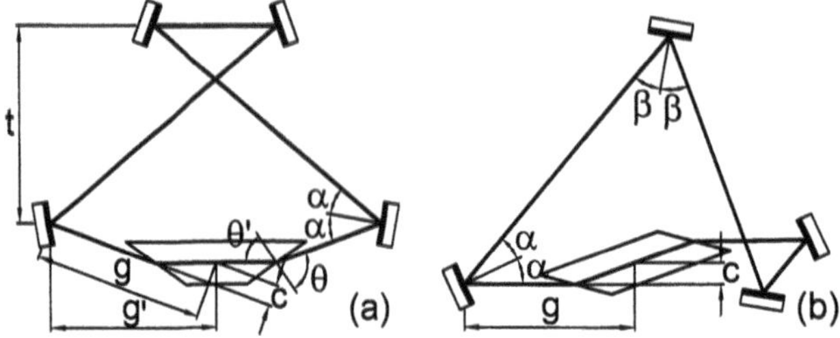

Fig. 4. Alternative ring cavities which permit compensation of coma: (**a**) ring resonator with Brewster-angle crystal faces oppositely inclined; (**b**) ring resonator with concave mirrors oppositely inclined

provided that the crystal face is inclined in the direction that takes it closer to being parallel to the mirror. This latter requirement is not satisfied by the common arrangement in which a crystal with parallel faces is placed within the bow-tie resonator of Fig. 1, but may indeed be met by the alternative ring cavities shown in Fig. 4.

In the cavity of Fig. 4a, the crystal faces are inclined in opposite directions [16, 17]; g and c are as given in (10) and (12), while $h = 0$ and

$$d = (v + g') \tan\left(\alpha - (\theta - \theta')/2\right) , \tag{16}$$

where $g' = g\cos(\theta - \theta') + c\sin(\theta - \theta')$. Alternatively, as in the cavity of Fig. 4b, it may be the mirrors that are oppositely inclined [18–20]; g and c are again unchanged, and the plane mirror angle of incidence is given by

$$\beta = \tan^{-1} \frac{c\cos 2\alpha - g\sin 2\alpha}{c\sin 2\alpha + g\cos 2\alpha - v} . \tag{17}$$

If coma may be neglected and the secondary waist ellipticity is unimportant, then both f and $\phi = \cos\alpha$ will be free parameters. It is apparent from Fig. 3 that the beam waist may be adjusted, by varying u, up to a maximum value determined by v [4]. At these maxima, the mode size is least sensitive to mirror position and, for some researchers, simultaneous maxima of tangential and sagittal mode sizes therefore provide the preferred configuration [21].

Contours of constant $\mathcal{F}(u, v)$, corresponding to a given waist size, have a common form in a linear plot, being scaled about the point $(1, 1)$ by a factor of $\sqrt{\mathcal{F}}$ in u and its reciprocal in the v-direction. The turning points of these curves, $\partial\mathcal{F}/\partial u)_v = 0$, describe a locus, shown dashed in Fig. 3,

$$(u - 1)(v - 1) = \frac{1}{2} \tag{18}$$

with $\mathcal{F} = (u - 1)^2$. For the u and v coordinates to coincide for both the tangential and sagittal directions, we find

$$2Z + Y - X = \phi - \frac{1}{\phi} \tag{19}$$

$$\frac{1}{2X}\phi^4 + \phi^3 - \phi - \frac{1}{2Y} = 0. \tag{20}$$

Substituting for X, Y and Z, we seek a focal length f that will simultaneously satisfy these equations. If both cavities lie in the upper right quadrant of Fig. 3 ($u, v > 1$), this will be given by

$$f^2 = \frac{4(2z+b''_y-b''_x)b''_x b''_y z \pm \sqrt{b''_x b''_y (b''_x+b''_y)^2 (2z+b''_y-b''_x)^3 (2z+b''_x-b''_y)}}{4(b''_x-b''_y)^2}. \tag{21}$$

The sign of $b''_{x,y}$ should be changed if the corresponding cavity instead lies in the opposite quadrant ($u, v < 1$). The mirror incidence angle may then be found by substituting this value back into (19) to obtain ϕ.

References

1. L.A. Bloomfield, B. Couillaud, Ph. Dabkiewicz, H. Gerhardt, T.W. Hänsch Phys. Rev. A **26**, 713 (1982)
2. B. Couillaud, Ph. Dabkiewicz, L.A. Bloomfield, T.W. Hänsch, Opt. Lett. **7**, 265 (1982); B. Couillaud, L.A. Bloomfield, T.W. Hänsch, Opt. Lett. **8**, 259 (1983)
3. T.W. Hänsch, B. Couillaud, Opt. Commun. **35**, 441 (1980)
4. A. Hemmerich, D.H. McIntyre, C. Zimmermann, T.W. Hänsch, Opt. Lett. **15**, 372 (1990)
5. C. Zimmermann, R. Kallenbach, T.W. Hänsch, J. Sandberg, Opt. Commun. **71**, 229 (1989)
6. C. Zimmermann, V. Vuletic, A. Hemmerich, T.W. Hänsch, Appl. Phys. Lett. **66**, 2318 (1995)
7. H.W. Kogelnik, E.P. Ippen, A. Dienes, C.V. Shank, IEEE J. Quantum Electron. **8**, 373 (1972)
8. C.E. Dunn, M.H. Dunn, Optica Acta **28**, 1413 (1981)
9. D.M. Kane, Opt. Commun. **71**, 113 (1989)
10. J.-P. Weber, IEEE J. Quantum Electron. **30**, 2407 (1994)
11. X.G. Sun, G.W. Switzer, J.L. Carlsten, Appl. Phys. Lett. **76**, 955 (2000)
12. T. Freegarde, J. Coutts, J. Walz, D. Leibfried, T.W. Hänsch, J. Opt. Soc. Am. B **14**, 2010 (1997)
13. A.E. Siegman, *Lasers* (University Science Books, California 1986)
14. G.D. Boyd, D.A. Kleinman, J. Appl. Phys. **39**, 3597 (1968)
15. M.H. Dunn, A.I. Ferguson, Opt. Commun. **20**, 214 (1977)
16. J.S. Nielsen, Opt. Lett. **20**, 840 (1995)
17. J.D. Bhawalkar, Y. Mao, H. Po, A.K. Goyal, P. Gavrilovic, Y. Conturie, S. Singh, Opt. Lett. **24**, 823 (1999)
18. J.C. Bergquist, H. Hemmati, W.M. Itano, Opt. Commun. **43**, 437 (1982)
19. A. Steinbach, M. Rauner, F.C. Cruz, J.C. Bergquist, Opt. Commun. **123**, 207 (1996)
20. T. Kaing, M. Houssin, Opt. Commun. **157**, 155 (1998)
21. F. Schmidt-Kaler, D. Leibfried, S. Seel, C. Zimmermann, W. König, M. Weitz, T.W. Hänsch, Phys. Rev. A **51**, 2789 (1995)

Raman Technique for Femtosecond Pulse Generation

Stephen E. Harris, Alexei V. Sokolov, David R. Walker, Deniz D. Yavuz, and Guang-Yu Yin

It is an honor to have been asked to write an article for this volume celebrating Theodor Hänsch's pre-eminent and towering contributions to laser physics and, at the same time, honoring his 60th birthday. I (Steve Harris) have known Theodor Hänsch since the late 1960s when he came to Stanford to work with Professor Arthur Schawlow. Since then we have collaborated several times and he has helped me many times. It is a pleasure to look forward to our continuing interactions in the years ahead.

Since his early years, Theodor has been interested in precision optical spectroscopy. Within the last few years he has shown that the frequency comb which is generated by a femtosecond mode-locked laser may be broadened to more than an octave of optical bandwidth. This provides an unprecedented frequency chain linking MHz radio frequencies, in one step, to the optical region [1]. The present paper is about a different type of nonlinear laser light source which produces a comb of sidebands with many octaves of optical bandwidth. The work of this paper is also related to my interactions with Theodor Hänsch in another way. In 1993, we wrote a paper suggesting the possibility of using high-order harmonic generation as an approach to attosecond timescale laser sources [2]. As part of this work Theodor suggested that one must include diffraction in the consideration of the temporal shape of the optical waveform. I will return to this subject later in the paper.

Our approach to the generation of femtosecond timescale radiation is an outgrowth of our work on electromagnetically induced transparency (EIT). The essential idea is to establish a vibrational or rotational molecular eigenstate with a coherence which has the maximum value $|\rho_{\mathrm{ab}}| = 0.5$. When this is the case, the nonlinear portion of the dipole moment is of the same magnitude as the linear portion. Phasematching plays a negligible role and the generation of a very broad spectrum of sidebands becomes possible.

1 Generation by Phased and Antiphased Molecular States

We consider the experimental situation denoted in Fig. 1. We apply two frequencies, a pump frequency E_0 (Ti:sapphire) and a Stokes frequency E_{-1} (Nd:YAG). The molecular medium is deuterium with a fundamental vibration frequency of 2994 cm^{-1}.

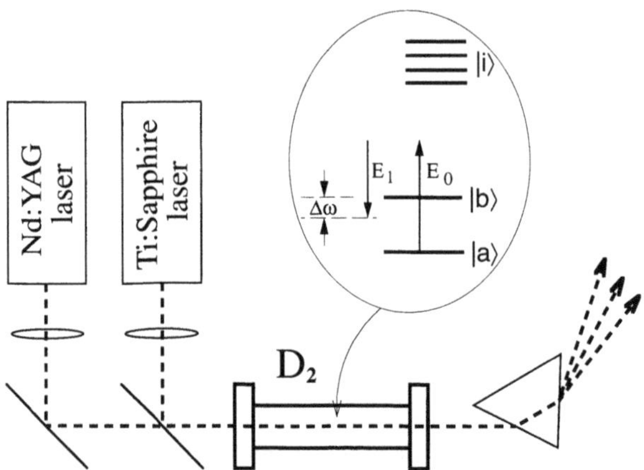

Fig. 1. Experimental setup and energy level diagram for coherent molecular excitation and collinear Raman generation

The detuning, $\Delta\omega$, from the molecular resonance establishes the sign of the molecular coherence. When tuned below resonance, as shown, the molecular coherence is in phase with the two-photon driving field. When above resonance, the molecular coherence is in antiphase with this field. At resonance it is in quadrature. To establish a particular eigenstate one sets the detuning, $\Delta\omega$, to be greater than the linewidth of the molecular transition. One then increases the two-photon Rabi frequency until it is large compared to $\Delta\omega$. When this is the case, the coherence is, in essence, saturated at its maximum value, and there is no advantage to further increase of the driving laser power [3]. For the deuterium vibrational system described here, this saturation occurs at a laser power density of about 10^{10} W/cm^2.

The equation governing the propagation [3] of the individual Raman sidebands is

$$\frac{\partial E_q}{\partial z} = -j\eta\hbar\omega_q N \left(a_q \rho_{\mathrm{aa}} E_q + d_q \rho_{\mathrm{bb}} E_q + b_q \rho_{\mathrm{ab}} E_{q-1} + c_q \rho_{\mathrm{ab}}^* E_{q+1} \right) . \quad (1)$$

The quantities ρ_{ij} are the elements of the 2×2 density matrix of the effective Hamiltonian. N is the number of molecules per volume and $\eta = (\mu/\epsilon_0)^{1/2}$. The quantities a_q and d_q determine the Stark shift of the molecular states $|a\rangle$ and $|b\rangle$ (Fig. 1). The drive strength is determined by the coupling constants b_q and its conjugate c_q. The importance of the condition of maximum coherence becomes clear: since a_q, b_q, c_q, and d_q are of the same order, then, as $|\rho_{\mathrm{ab}}|$ approaches its maximum value of 0.5, the driving terms in (1) are almost as big as the dispersive terms. Phasematching then plays a negligible role in the generation of the Raman sidebands. By making explicit use of the expressions for the dispersive constants [3], we obtain conservation conditions for photons and power:

$$\frac{\partial}{\partial z}\left(\sum_q \frac{1}{\hbar\omega_q}\frac{|E_q|^2}{2\eta}\right) = 0$$

$$\frac{\partial}{\partial z}\left(\sum_q \frac{|E_q|^2}{2\eta}\right) = -\frac{N\hbar}{2}\left(\omega_{\mathrm{b}} - \omega_{\mathrm{a}}\right)\left[\frac{\partial}{\partial t}\left(\rho_{\mathrm{bb}} - \rho_{\mathrm{aa}}\right)\right] . \quad (2)$$

In the experiments described here we use a Q-switch injection-seeded Nd:YAG laser with a per pulse energy of 100 mJ, a pulse length of 12 ns, and a repetition rate of 10 Hz. The Ti:sapphire laser has an energy of about 75 mJ and a pulse length of 16 ns. The laser beams are focussed to a nearly diffraction-limited spot in the center of the D_2 cell. The spot sizes at the focus for the 106 nm and 807 nm wavelength lasers are 460 μm and 395 μm, respectively. The deuterium cell is cooled to liquid nitrogen temperature with a cooled region 50 cm long. At this temperature, the Doppler width is 260 MHz and the population of the ground rotational state of D_2 is 60% of the total population.

We find that detuning to the low-frequency side of resonance produces the largest number of anti-Stokes sidebands with the most energy. These results, taken from [4], are shown in Fig. 2. Note that generated pulse energies exceed 0.1 mJ out to the tenth anti-Stokes sideband. The totality of 17 sidebands,

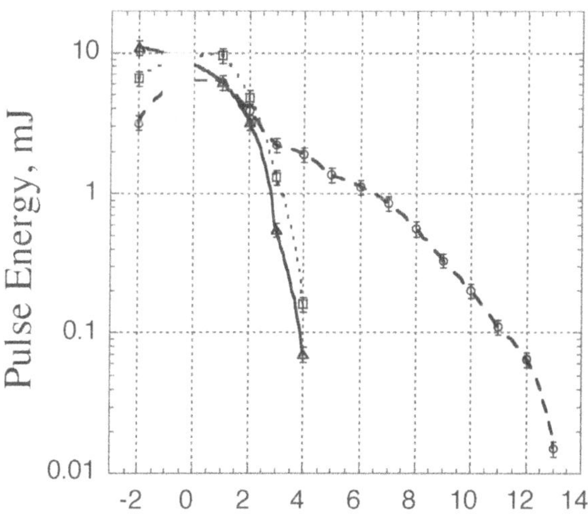

Fig. 2. Pulse energies generated at $P = 72$ torr. The *triangles* correspond to on-resonance ($\Delta\omega = 0$), and the *circles* and *squares* correspond to generation by phased ($\Delta\omega = 500$ MHz) and antiphased ($\Delta\omega = -200$ MHz) states of D_2, respectively. (From Sokolov et al. [4])

spaced by about 2994 cm^{-1}, extends from 2.94 μm in the infrared to 195 nm in the vacuum ultraviolet and has an overall width of about four octaves.

2 Phase Control

2.1 Light Modulation at Molecular Frequencies

To demonstrate that the sidebands which we generate are coherent, both over their spatial profile and over their temporal extent of about 20 ns, we have used these sidebands to generate both amplitude-modulated (AM) and frequency-modulated (FM) light with a modulation frequency of 2994 cm^{-1}. The experimental setup follows an idea of Sokolov [5] and is shown in Fig. 3. In this work, three sidebands (1.06 μm, 807 nm, and 650 nm) are separated out from the Raman comb. The phase of the 650 nm beam is varied by tilting a gas plate (thickness 1.0 mm). The beams are then retroreflected and recombined into a new region of the D_2 cell. We measure the intensity of the generated anti-Stokes sideband at 544 nm as a function of the phase of the returning 650 nm beam. Since the Raman medium only responds to intensity variations in the driving beam, there will be no anti-Stokes radiation produced when the incident sidebands have FM phases. Conversely, maximum generation occurs with AM phases. The experimental results are shown in Fig. 4.

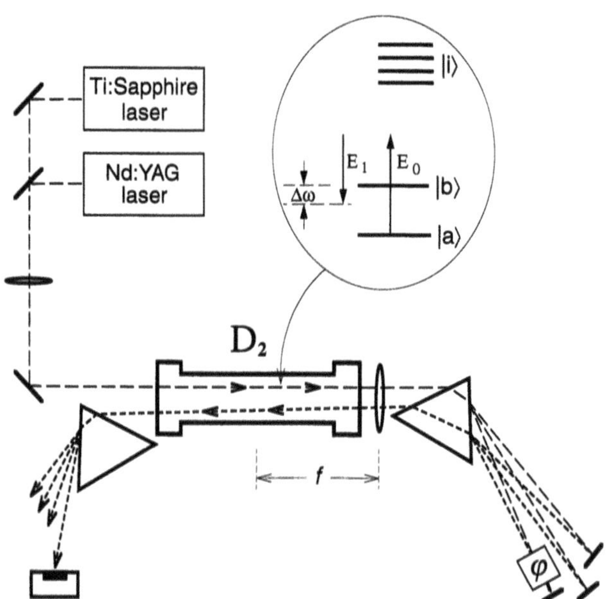

Fig. 3. Experimental setup and energy level diagram for collinear Raman generation, synthesis of FM and AM light, and detection of the modulation

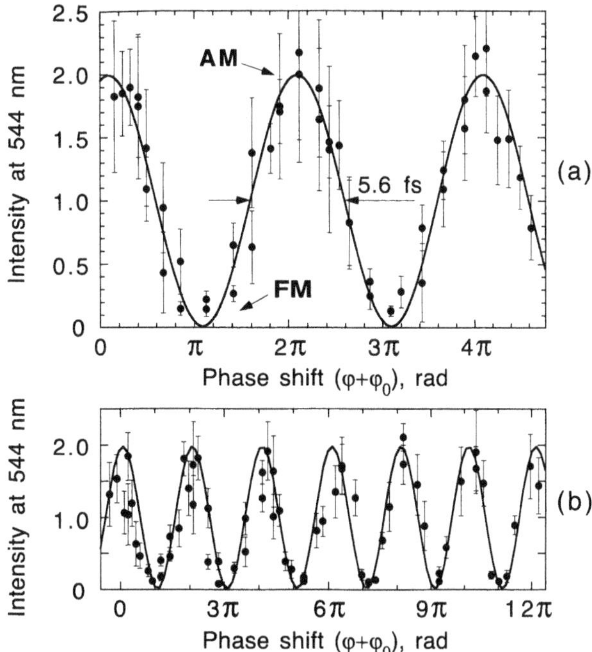

Fig. 4a,b. Anti-Stokes generation at 544 nm (arbitrary units) as a function of the relative phase φ of the three sidebands (at 1.06 µm, 807 nm, and 650 nm). (**a**) Data set 1; (**b**) Data set 2, showing that additional phase retardation does not reduce the contrast. (From Sokolov et al. [5])

We remark that, in earlier years, there has been significant work on a technique called anti-Stokes coherent Raman spectroscopy (CARS). A CARS experiment is much like this one, except that only two sidebands, rather than three, are incident on the Raman cell. Because two sidebands are present, the modulation is both AM and FM and the excitation of the medium does not depend on their relative phase. Conversely, as shown here, three sidebands with particular relative phases will either maximize or minimize the excitation of the medium.

2.2 Phase-Controlled Multiphoton Ionization

Our next experiment involves a more ambitious demonstration of phase control. We examine multiphoton ionization of Xe using five Raman sidebands at wavelengths of 1.06 µm, 806 nm, 650 nm, 544 nm, and 468 nm. A schematic of the experiment [6] is shown in Fig. 5. It is of interest that multiphoton ionization of Xe requires eleven photons of the longest wavelength of 1.06 µm, but only five photons of the shortest wavelength of 468 nm.

We control the phases of these sidebands both by tilting mirrors or by varying the pressure of an N_2 cell inserted for this purpose. The sideband

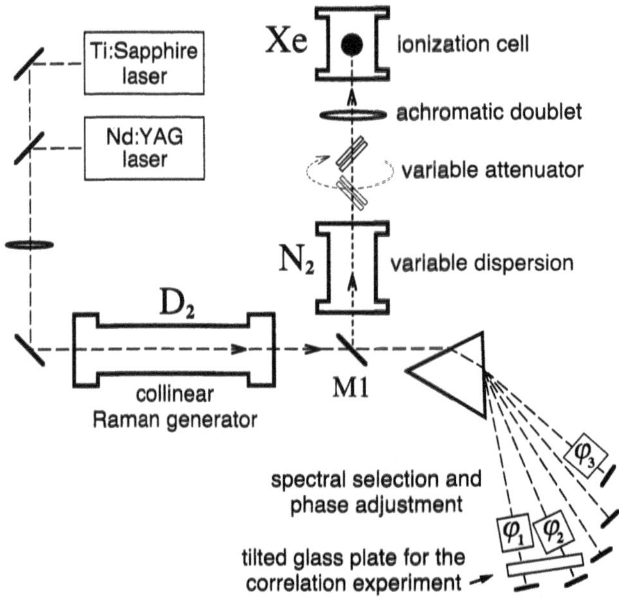

Fig. 5. Experimental setup for multiphoton ionization of Xe with five Raman sidebands (at 1.06 μm, 807 nm, 650 nm, 544 nm, and 468 nm). The mirror M1, which is displaced in the vertical plane, picks off the slightly offset retroreflected beam. We vary sideband phases independently with tilted glass plates, or by varying dispersion in the N_2 pressure cell

energies in the beam which is incident on the target cell are 33 mJ at 1.06 μm, 28 mJ at 807 nm, 4.2 mJ at 650 nm, 0.9 mJ at 544 nm, and 0.2 mJ at 468 nm. The pressure of the target Xe gas is 3×10^{-5} torr and the number of ions is measured with a Channeltron detector. In a first experiment, we measured the dependence of the Xe-ion signal on the total pulse energy for three different sets of relative phases, i.e. for three different pulse shapes. Although the absolute magnitude of the signal depended on the particular pulse shape, somewhat to our surprise, its slope did not. For each choice of phase the Xe-ion signal vs. pulse energy was approximately fitted by a sixth-order power law.

In the next set of experiments, we used the Xe detector to characterize the Raman light source. This was done by a cross-correlation measurement where we varied the delay of the 544 nm and 468 nm beams with respect to the 650 nm, 807 nm, and 1.06 μm beams. The result of this procedure is shown by the solid line in Fig. 6. Here, the intensities of the monochromatic components are adjusted to equalize their contribution to the total ionization. The dashed line shows the response of a detector obeying a sixth-order power law for five spectral components of equal amplitude. The dotted line in this figure shows the result of a perturbation calculation using the known energy levels

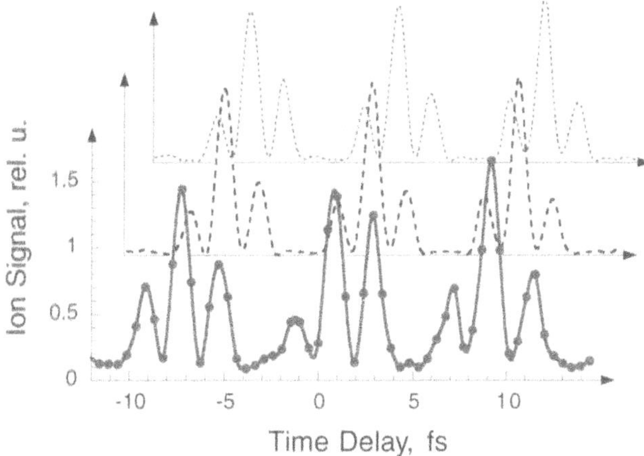

Fig. 6. Correlation of waveforms, synthesized by subsets of Raman sidebands: experiment (*solid line*); time-domain calculation, $\int \{\sum \cos [\omega_m(t + \tau_m)]\}^{12}\, \mathrm{d}t$ (*dashed line*); and perturbative frequency domain calculation (*dotted line*). (From Sokolov et al. [6])

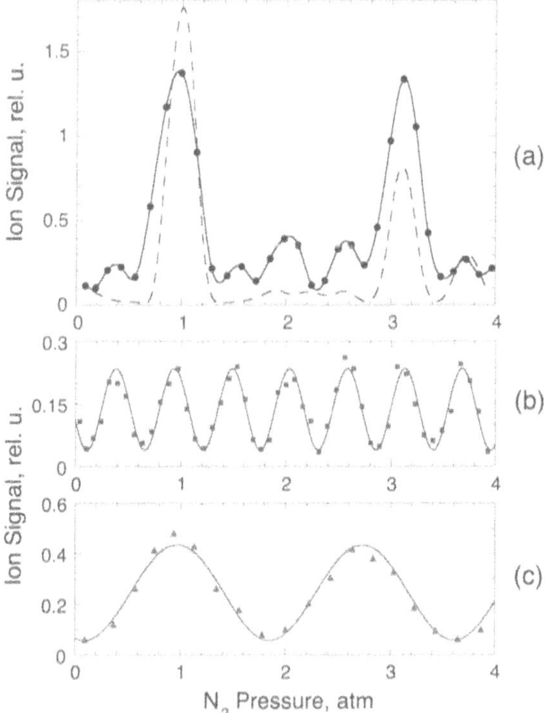

Fig. 7. Number of Xe ions produced in the setup of Fig. 1 versus pressure in the N_2 cell. For part (**a**) all five sidebands are present. For parts (**b**) and (**c**) we leave only three equidistant sidebands [1.06 µm, 650 nm, and 468 nm for part (**b**), and 650 nm, 544 nm, and 468 nm for part (**c**)]. The solid line is a cubic spline interpolation of the data in part (**a**) and a sinusoidal fit to the data in parts (**b**) and (**c**). The dashed line shows $\int [\sum \cos (\omega_m t + \varphi_m)]^{12}\, dt$, where φ_m are detemined by the N_2 dispersion. (From Sokolov et al. [6])

of Xe and assuming that the matrix elements between all levels are equal and are normalized to an oscillator strength of unity. We observe reasonable agreement between the calculations and experiment.

To further test the mutual coherence among the sidebands, we put a 34-cm-long cell of N_2 into the beam. We first adjust the sideband phases such that the ion signal maximizes for one atmosphere pressure and then count the number of Xe ions per laser shot as a function of the N_2 pressure (Fig. 7). We first observe a sharp decrease in the ion signal as a function of N_2 pressure. As the N_2 pressure is further increased to about 3.1 atm, the dispersion is such that the sidebands rephase and the ion signal again increases. We repeat this experiment for subsets of three equally-spaced sidebands and, in this case, observe nearly sinusoidal dependence of the ion signal on the N_2 pressure.

The results of Figs. 6 and 7 establish evidence for good mutual coherence among the frequency components of the Raman source. Since these frequency components extend over an octave of optical bandwidth, we infer that we are able to synthesize a train of single-cycle pulses with a repetition period of 11 fs and a pulse duration of about 2 fs.

3 Multiplicative Technique

As in traditional mode locking, the synthesis of well-formed femtosecond and single-cycle optical pulses will require phase correction of as many modes as can be obtained. Present liquid crystal and acoustic techniques allow in excess of 1000 resolvable pixels and it is desirable to have a technique for generating a total number of sidebands which are on this order. This section suggests a method for accomplishing this objective. A schematic of the proposed technique [7] is shown in Fig. 8. Two Raman generators with transition frequencies ω_a and ω_b are placed in series. The cells are driven by three lasers whose frequencies are ω_0, $(\omega_0 - \omega_a)$, and $(\omega_0 - \omega_b)$. Two of these lasers drive the coherence of the ω_a transition which, in turn, modulates each of the incident laser beams. At the output of the first cell the generated spectrum consists of two sets of sidebands which are offset from each other by ω_b. These sidebands enter the next cell and, acting in unison, drive the ω_b coherence. This coherence imposes additional FM modulation on each of the entering sidebands. The net result is a Fourier series with angular frequencies

$$w_{q,r} = \omega_0 + q\omega_a + r\omega_b \ . \tag{3}$$

For each pulse the driving lasers are randomly phased and the spectrum of generated sidebands carries these three random phases. The set of generated sidebands is phase corrected and the recombined beam is incident on a target cell where the temporal structure and peak power are to be studied. The phase correction is set by assuming that the driving lasers have fixed (non-random) phases.

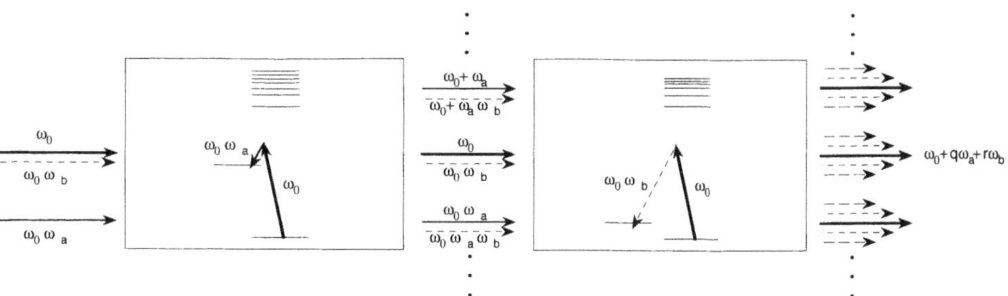

Fig. 8. Schematic of the suggested technique. Three randomly phased laser beams with frequencies denoted by ω_0, $(\omega_0 - \omega_a)$, and $(\omega_0 - \omega_b)$ drive series Raman generators with incommensurate transition frequencies ω_a and ω_b. The generated spectrum is a Fourier series with terms of the form $\omega_0 + q\omega_a + r\omega_b$. With phase correction the series synthesizes to a random train of pulses with a peak power enhancement which is equal to the product of the number of sidebands of the series generators, if alone, and an average pulse periodicity given by (4). (From Harris et al. [7])

We have found that in the limit of a large number of modes, irrespective of the initial random phases, the Fourier transform of the phase-corrected comb corresponds to a train of optical pulses with a peak power which is enhanced relative to the average driving power by a factor equal to the product of the modes of the independent cells. Individual pulses occur at random times with an average periodicity T. If the Raman transition frequency and number of modes of each (independent) cell is denoted by ω_a and N_a, and ω_b and N_b; and if the ratio of ω_a to ω_b is irrational, then the peak power enhancement, average periodicity T, and pulse width of the random pulse train are:

$$\begin{aligned} \text{peak/av} &= N_a\, N_b \\ T &= (2\pi)\,\frac{N_a\, N_b}{N_a\omega_a + N_b\omega_b} \\ \delta T &= (2\pi)\,\frac{1}{N_a\omega_a + N_b\omega_b}\,. \end{aligned} \tag{4}$$

It does not matter which of the two cells of Fig. 8 is placed ahead of the other and, in fact, the two molecular species may be mixed in a single cell. Peak-to-average power enhancements of a factor of about 1000 should be possible. Also, according to (4), the average periodicity of the randomly timed pulses should be extendable into the picosecond range.

4 Propagation

We return to the question of how a spectrum which is many octaves wide will propagate in free space. Since the radiation source has a finite size, each side-band will have a different diffraction pattern. The temporal waveform of the generated electric field will therefore depend on the distance of observation from the source. If one assumes a uniform Gaussian aperture with a beam waist w_0 at $z = 0$ and an electric field over this aperture of $E(t, 0)$, then, as noted in [2], the electric field for all t and z is given by

$$E(t,z) = E(t - z/c,\ z = 0) - \frac{2cz}{w_0^2} \int_0^\infty E(t - \tau - z/c,\ z = 0) \exp\left[-(2cz/w_0^2)\tau\right]\ d\tau\ . \quad (5)$$

The temporal waveform $E(t, z)$ has lowest and highest harmonics ω_i with significant Fourier content and, with these harmonics, we may associate minimum and maximum confocal parameters $b_i = (\omega_i/c)w_0^2$. For z much less than $b_{\min}$ or much greater than $b_{\max}$, (5) has the limits

$$\begin{aligned} E(t,z) &= E(t - z/c,\ 0), \quad z \ll b_{\min}\ , \\ &= \frac{w_0^2}{2cz} \frac{\partial E}{\partial t}(t - z/c,\ z = 0), \quad z \gg b_{\max}\ . \end{aligned} \quad (6)$$

In the Gaussian near field, the temporal waveform replicates the atomic currents (i.e. the velocity of the charges). As the beam propagates to the far field, the relative amplitudes and phases of its Fourier components change and the on-axis temporal waveform is differentiated and replicates the acceleration.

Most experiments will be performed at the focus of a lens or mirror, and the optical pulses which should be designed are the integral of the pulse which is desired at the focus. Propagation, i.e. differentiation, enhances the higher-frequency sidebands relative to those at lower frequency. For example, in order to obtain a single-cycle pulse whose spectrum is independent of frequency, one would use a spectrum which falls off as the square of frequency. Since Raman sources of the type described here fall off at the high end of the spectrum, this effect will make it easier to obtain multi-octave, single-cycle pulses.

5 Connections to Other Work

Of particular pertinence to the subject of this paper is the work of Korn and colleagues, who have impulsively excited a vibrational mode of SF_6 and used the resulting time-varying refractive index to produce a sequence of compressed femtosecond pulses [8]. In 1993, Losev and Lutsenko [9] suggested and demonstrated on-resonance two-color pumping, and Yoshikawa and Imasaka [10] suggested the application of multi-component Raman spectra to femtosecond timescale pulses. Rotational Raman spectra extending from the infrared to the ultraviolet have been demonstrated [11]. Hakuta and colleagues have used solid hydrogen for collinear Raman sideband generation [12]. Possible application of Raman sidebands to the synthesis of ultra-broadband solitons have been suggested by Kaplan [13] and by Yavuz et al. [14].

Acknowledgements

I thank Theodor Hänsch for many years of insightful physics and inspiration, and for his continuing friendship. This work was supported by the U.S. Air Force Office of Scientific Research, the U.S. Army Research Office, and the U.S. Office of Naval Research.

References

1. R. Holzwarth, Th. Udem, T.W. Hänsch, J.C. Knight, W.J. Wadsworth, P.St.J. Russell, Phys. Rev. Lett. **85**, 2264 (2000)
2. S.E. Harris, J.J. Macklin, T.W. Hänsch, Opt. Commun. **100**, 487 (1993)
3. S.E. Harris, A.V. Sokolov, Phys. Rev. A **55**, R4019 (1997)
4. A.V. Sokolov, D.R. Walker, D.D. Yavuz, G.Y. Yin, S.E. Harris, Phys. Rev. Lett. **85**, 562 (2000)
5. A.V. Sokolov, D.D. Yavuz, D.R. Walker, G.Y. Yin, S.E. Harris, Phys. Rev. A **63**, R051801 (2001)
6. A.V. Sokolov, D.R. Walker, D.D. Yavuz, G.Y. Yin, S.E. Harris, Phys. Rev. Lett. **87**, 033402-1/033402-4 (2001)
7. S.E. Harris, D.R. Walker, D.D. Yavuz, Raman Technique for Single-Cycle Pulses. (submitted for publication)
8. A. Nazarkin, G. Korn, M. Wittmann, T. Elsaesser, Phys. Rev. Lett. **83**, 2560 (1999); M. Wittmann, A. Nazarkin, G. Korn, Phys. Rev. Lett. **84**, 5508 (2000); M. Wittmann, A. Nazarkin, G. Korn, Opt. Lett. **26**, 298 (2001)
9. L.L. Losev, A.P. Lutsenko, Quantum Electron. **23**, 919 (1993); L.L. Losev, A.P. Lutsenko, Opt. Commun. **132**, 489 (1996)
10. S. Yoshikawa, T. Imasaka, Opt. Commun. **96**, 94 (1993)
11. H. Kawano, Y. Hirakawa, T. Imasaka, IEEE J. Quantum. Electron. **34**, 260, (1998)
12. K. Hakuta, M. Suzuki, M. Katsuragawa, J.Z. Li, Phys. Rev. Lett. **79**, 209 (1997)
13. A.E. Kaplan, Phys. Rev. Lett. **73**, 1243 (1994)
14. D.D. Yavuz, A.V. Sokolov, S.E. Harris, Phys. Rev. Lett. **84**, 75 (2000)

High-Order Harmonics and White Light: Looking for Fringes and Finding Much More

Marco Bellini

1 Introduction

Complex and expensive experimental systems are not always necessary to perform interesting research and to shed light on novel physical phenomena. Very simple experiments often provide great food for thought and can sometimes give new, deep, physical insight when applied in different contexts from those they were initially designed for.

It certainly takes an unusual dose of ingenuity and creativity to think of such simple but decisive experiments but, perhaps more importantly, it takes an exceptionally open mind to look at the results of these experiments from different, and apparently distant, points of view, in order to learn something completely new.

Since 1996, I have been lucky enough to work with Theodor Hänsch in the framework of his collaboration with the University of Firenze. During this time, I have had the opportunity to appreciate such rare qualities in him and to be inspired by his enthusiasm and curiosity towards the beauty and the challenges of physics.

In the following I will give a short description of a couple of key experiments we have been performing together at LENS, the European Laboratory for Nonlinear Spectroscopy of Firenze. In particular, I will focus on the studies of the radiation produced in the extreme ultraviolet by the process of high-order harmonic generation, and on the white-light supercontinuum obtained when very intense laser pulses impinge on transparent materials.

I will show how, using very simple experimental ideas to investigate the coherence properties of these two different kinds of "extreme" light sources, we were able to obtain very interesting and unexpected results. I will also try to show how, by looking at the outcome of such experiments from a wider perspective, it was possible to find new and surprisingly fruitful results in fields that could initially appear as almost completely uncorrelated, like strong-field atomic physics and high-precision optical metrology.

2 Spectroscopy with Sequences of Pulses

The driving idea for our experiments was the possibility of using ultrashort light pulses of various kinds to perform high-resolution spectroscopy. This is one of the experimentalist's dreams and would allow us to exploit the extremely high peak intensities characteristic of the short pulses to excite multiphoton processes, or to drive highly nonlinear phenomena (and generate new wavelengths, for example), while maintaining the high resolution characteristic of CW sources to investigate very narrow spectral structures of atoms or molecules.

Of course, the two conditions of ultrashort pulse duration and high spectral resolution normally appear in striking contrast, since short pulses invariantly correspond to broad spectral bandwidths that limit the frequency resolution to the inverse of the pulse duration. However, if pairs of time-delayed and phase-locked pulses (like those generated by splitting a single laser pulse by means of a Michelson interferometer) are used, a simple Fourier transformation shows that the corresponding spectrum maintains the broad bell-shaped envelope, but also acquires a sinusoidal modulation with a spectral period given by the inverse of the temporal separation between the two pulses. It is this fringe period that now sets the instrumental resolution and allows one, in principle, to investigate very fine spectral features if long time delays are available.

We demonstrated that this was indeed the case in 1996, during my first collaboration with Theodor Hänsch, when we showed that it was possible to measure line splittings (the hyperfine separation of the $8S_{1/2}$ state in cesium in that case) in a two-photon transition with a spectral resolution much better than that given by the single-pulse spectral width [1].

The idea of using a pair of phase-locked pulses in order to achieve a better spectral resolution can also be extended by the use of longer sequences of equally time-delayed and phase-locked pulses. The spectrum that one obtains in this case still presents the broad bandwidth connected to the short pulse duration, but is now modulated in a sharper and sharper fashion as longer pulse sequences are used (see Fig. 1). In the ideal limit of an infinite train of phase-locked and equally-spaced pulses, the resulting spectrum essentially consists of a "comb" of infinitely sharp lines, equally separated by a frequency spacing corresponding to the inverse of the interpulse period.

The potential advantages of such a peculiar spectral distribution are evident: if such a comb is realized in practice, it can be used as a precise ruler to measure unknown frequency intervals in a relatively simple way. By locking two laser lines to two different "teeth" of the comb, and by counting the integer number of interposed teeth, one can immediately obtain the unknown frequency gap if the separation between the teeth is well known.

A mode-locked laser is a natural way for generating such an ideally infinite sequence of time-delayed pulses with a well-defined phase relationship [2]. Its spectrum (given by the set of equally spaced longitudinal modes of the cavity) is a broad comb of frequencies with a mode separation equal to the measurable and controllable pulse repetition rate. The largest frequency gap that can be bridged with such a comb is determined by the inverse of the

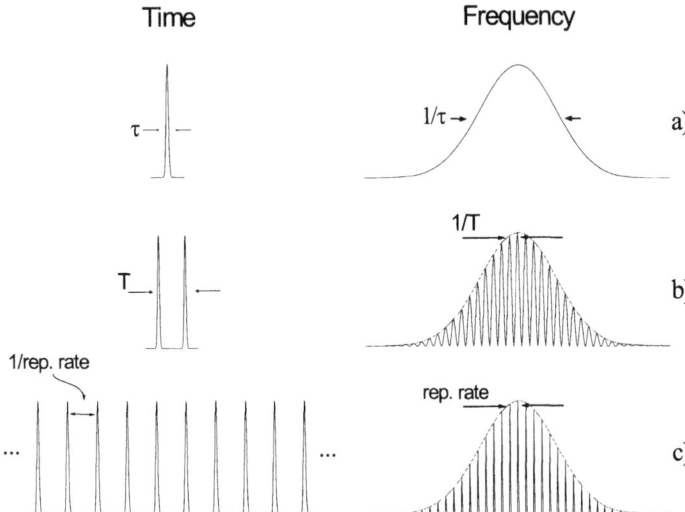

Fig. 1. A single laser pulse has a frequency bandwidth which scales with the inverse of its duration τ. If one uses a pair of phase-locked pulses delayed by a time T, the resulting spectrum maintains the broad envelope of width $1/\tau$ but with a sinusoidal modulation of period $1/T$. This width sets the new instrumental resolution. If one uses an infinite sequence of pulses locked in phase and equally delayed in time of a time interval T, the spectrum breaks up in a "comb" of very narrow lines (the "teeth") equally spaced by a frequency interval $1/T$

pulse duration, but it can be widely extended if nonlinear interactions are used to broaden the spectrum.

These ideas are some of the best examples of the importance of having such a clear and wide perspective on physics as to imagine useful connections between normally separated worlds. As a matter of fact, people dealing with ultrafast lasers usually think of their applications only in the temporal domain, while spectroscopists generally believe that ultrastable lasers are necessary to do their job and only speak in terms of frequency. It is instead easy (once someone has shown you the way) to get the best of both worlds and to do so without a growth in complexity, but rather with an enormous increase in the range of possible applications.

In any case, one of the essential requisites for these novel spectroscopic techniques to work is that the phase coherence between the pulses in the sequence is accurately preserved. The experiments described below were mainly aimed at checking this coherence preservation in particular highly nonlinear processes, but their results went far beyond the initial scope.

3 High-Order Harmonics

At the time of the two-pulse experiments on cesium, we had already developed at LENS a reliable source for the generation of high-order harmonic pulses and we were still mainly characterizing this radiation while trying to think of clever ways to use it for spectroscopy in the vacuum and extreme ultraviolet (VUV and XUV) [3].

Pulses with frequencies which are odd-order harmonics of the fundamental laser frequency are generated by the interaction of short and intense laser pulses with the atoms of a supersonic jet. Depending on the wavelength, duration and peak intensity of the pulses, and on the ionization potential of the atoms, very high orders can be efficiently generated at wavelengths down to the XUV or to the soft X-ray regions.

Though harmonic sources are extremely appealing due to the lack of other easily accessible alternatives in these spectral regions, it looked like no true spectroscopy would have ever been done with them, due to the extremely broad bandwidth associated with the short duration of the harmonic pulses. In fact, even if some low-order harmonics can be generated with pump pulses in the picosecond range, allowing one to keep an acceptable spectral resolution for selected applications, higher-order harmonics can only be created at intensities above $10^{13}\,\mathrm{W/cm^2}$ by ultrashort laser pulses; and a 100 fs pulse is already characterized by a spectral width in the THz range.

We immediately thought of overcoming this limit with the application of the two-pulse technique to the harmonic radiation, by splitting and delaying the XUV pulses by means of a Michelson interferometer before sending them to the samples under study. Unfortunately, the use of this technique with harmonic pulses is far from straightforward, mainly because good interferometers cannot be built to work in the XUV due to the lack of suitable optics.

A possible solution to the problem was to invert the two steps of harmonic generation and pulse splitting, by moving the interferometer in the path of the laser beam, in order to create two phase-locked and time-delayed pump pulses that would have generated equally phase-related XUV pulses. The question at this point was about the preservation of the phase lock in the generated pulses: if the process of harmonic generation was an incoherent one, no phase relationship would have been preserved between the XUV pulses and the whole scheme for spectroscopy with harmonics would have been useless.

A simple way to test the mutual phase coherence between the harmonic pulses was to generate them in two separate spatial regions and look for interference fringes in the far field: the existence of interference would have shown that the two secondary sources have preserved a memory of the phases of their parent pulses.

We first performed a preliminary experiment with the third harmonic generated in air and, though it was not possible to directly extrapolate the results to higher orders, it gave very interesting and encouraging indications [4]. At the same time, Theodor Hänsch discussed these ideas with Anne L'Huillier, one of the major world experts of high-order harmonics, and, though she initially suspected that the generation process would have completely messed up the phases of the pulses destroying any interference, we agreed to do a joint experiment to test these ideas.

The first experiment was carried out in Lund, where harmonics were generated by focusing 30 ps laser pulses into an argon gas jet. After the focusing lens, the pulses coming from the laser were split and given a slight fixed displacement (both in space and time) thanks to the walk-off in a birefringent plate. A common polarization component was selected before entering the interaction region. Harmonics were then generated in two different positions in the gas jet and we looked for interference fringes in the far field after the spectral dispersion operated by a grating. We unexpectedly observed nice, stable and highly contrasted fringes, indicating that the generation process was not as phase-destructive as we initially thought [5], and demonstrating that harmonic generation was a suitable source for the high-resolution spectroscopic technique we had in mind.

The second, more accurate, experiment was done at LENS (see Fig. 2). Here we used much shorter (100 fs) laser pulses and we split and delayed the pulses by means of a Michelson interferometer, so that their temporal and spatial separation in the focus could be carefully adjusted and controlled.

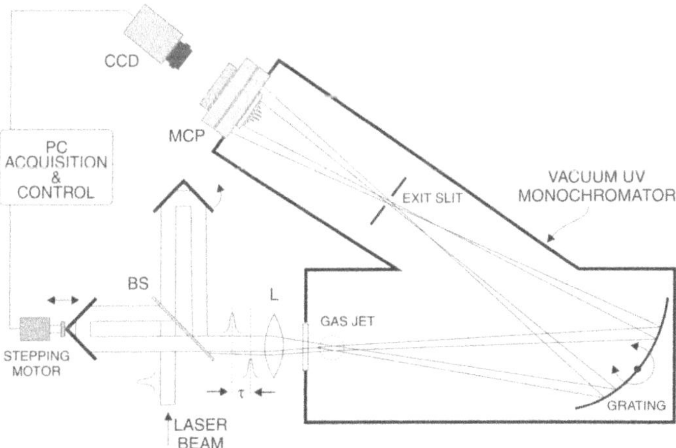

Fig. 2. Scheme of the experimental setup used at LENS for the observation of the interference fringes and for the measurement of the temporal coherence of high-order harmonic pulses. The pump laser pulses are split and delayed, and their output directions are slightly tilted by means of a Michelson interferometer, so that the two beams are focused in different positions of the gas jet and generate harmonics independently. We looked for interference fringes on a Micro Channel Plate detector placed after the exit slit of a vacuum monochromator

Again, we saw very clear interference fringes on the Micro Channel Plate detector placed beyond the exit slit of our vacuum monochromator, and we were also able to measure the temporal coherence of the harmonics by observing the decay of the fringe visibility as a function of the delay [6].

We demonstrated that one can generate almost transform-limited XUV pulses with coherence times of the order of the expected duration of the harmonics themselves (about 40 fs), showing that not only is the phase not scrambled in the process, but also that a negligible frequency chirp is imparted to the secondary pulses.

While doing this we also discovered something that was quite a mystery at the beginning: the presence of two clearly distinct spatial regions in the pattern of harmonic emission, with drastically different coherence properties (see Fig. 3). An inner region with the long coherence times described above, and a diffuse outer halo, containing more than half of the total emitted flux, with an extremely short temporal coherence, of the order of few femtoseconds.

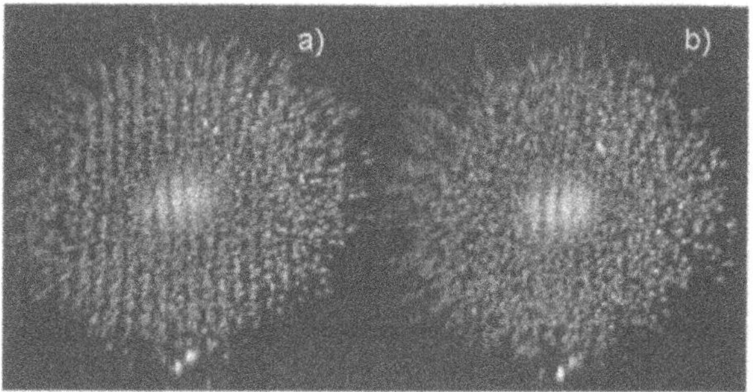

Fig. 3a,b. Snapshots of the interference fringes produced by the 15th harmonic generated in argon. In (**a**), taken at a delay of 0 fs, fringes appear all over the image with good visibility, while they disappear in the outer region of (**b**), taken at a delay of 15 fs. The diffuse halo surrounding the central bright spot has a much shorter coherence time than the inner region

We unexpectedly found that this was one of the most direct proofs of the validity of current theoretical models used to describe the microscopic physical mechanisms involved in the process of harmonic generation.

According to the standard picture of the high-order harmonic generation process [7], every half optical cycle of the laser pulse, electrons undergo tunnel ionization through the potential barrier formed by the atomic potential and by the electric field potential of the laser; after being accelerated in the ionization continuum by the field, they may come back to the ion core and finally recombine to emit harmonic photons that release the accumulated kinetic and ionization energy. Single-atom models also predict that harmonics are emitted with an intensity-dependent phase, proportional to the amount of time spent in the continuum by the generating electrons.

Simple calculations show that the highest harmonic orders (in the so-called *cutoff*) can only derive from electrons which have been released at a well-defined time in each half optical cycle of the laser. On the other hand, it is equally easy to see that lower-order harmonics (in the so-called *plateau*) essentially come from two different classes of electrons which are emitted at different moments, spend different amounts of time in the continuum following different trajectories, but nevertheless come back to the ion with the same correct kinetic energy to generate that given harmonic (see Fig. 4).

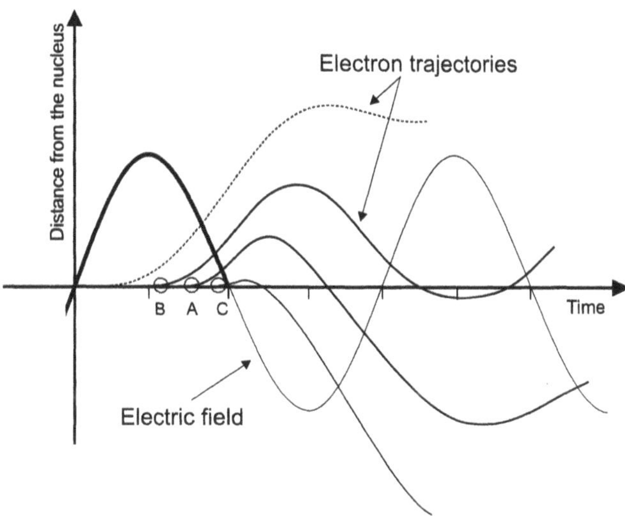

Fig. 4. Schematic representation of the possible classic electronic trajectories leading to the emission of harmonic photons. Only a half optical cycle of the laser is considered here since everything repeats with this periodicity. Electrons which are ionized when the field amplitude grows are driven away from the ion and never recombine to generate harmonics (*dashed curve*). Electrons escaping with a decreasing field will leave the ion, oscillate in the continuum and come back with some supplementary kinetic energy that can be converted into photon energy (*solid curves*). The maximum return energy (corresponding to the highest harmonic orders generated) is carried by electrons ionized in A. A given lower harmonic can be generated by electrons released either in B or in C, since both trajectories correspond to the same final return velocity (same angle of intersection with the horizontal axis). Electrons "born" in B suffer a stronger intensity-dependent phase modulation since they spend a much longer time in the continuum

For these harmonics, the "short" trajectory imposes a phase that does not vary much with the laser intensity, whereas the phase corresponding to the "long" trajectory varies rapidly with the laser intensity. Such a "long" electronic trajectory gives rise to strongly divergent angular emission because the rapid spatial variation of the phase with the focused laser intensity leads to a strong curvature of the phase front. This radiation also has a very short coherence time, since the harmonic pulse is strongly chirped due to the rapid temporal variation of the phase during the pulse. In contrast, the phase variation corresponding to the "short" trajectory is much less important. The emitted radiation has a long coherence time and is much more collimated (for a more detailed discussion see [8]).

Our simple experiment allowed us to directly observe these effects for the first time. It was somehow amazing that, just from the observation of the appearance and behavior of interference fringes, we were able to gain such a deep insight on the inner dynamics of atoms in the presence of strong laser fields.

Moreover, some of the most interesting latest developments in the field of high-order harmonics have been a more or less direct consequence of these results: harmonic sources have been used for interferometric studies of metals and plasmas in the XUV [9], and the possibility of using pairs of harmonic pulses for high-resolution spectroscopy has also been recently demonstrated [10]. Finally, our proof of the role of the different electron trajectories in the process of generation is giving a substantial push in the race towards the attosecond barrier.

4 White Light

In 1997, while we were still performing the experiments on the coherence of high-order harmonics, we decided to use our simple experimental apparatus to test the mutual phase coherence of the light pulses generated in a different kind of extremely nonlinear process. We just replaced the system for harmonic generation with a plain plate of calcium fluoride (CaF_2) to generate pulses of white light.

The process of white-light continuum generation provides a simple and efficient way to achieve extreme spectral broadening, by focusing intense enough laser pulses into transparent materials [11, 12]. The dominant process leading to spectral super-broadening is the self-phase-modulation of the pulse due to an intensity dependent refractive index of the medium, but a number of other linear and nonlinear effects play a role as well, including self-focusing, parametric four-photon mixing, stimulated Raman and Brillouin scattering and shock-wave formation. The generation of the continuum is then the result of a very complex interplay between competing processes and the characteristics of the output beam appear strongly dependent on the exact initial conditions of the interaction and hardly predictable. In particular, one is led to expect that the white-light pulses produced by phase-locked pump pulses might loose any precise phase relationship in the generation process.

Again, like in the case of harmonics, we were rather pessimistic about the possibility of a residual phase coherence between the resulting pulses but, since we had been lucky the first time, and since it was a very simple

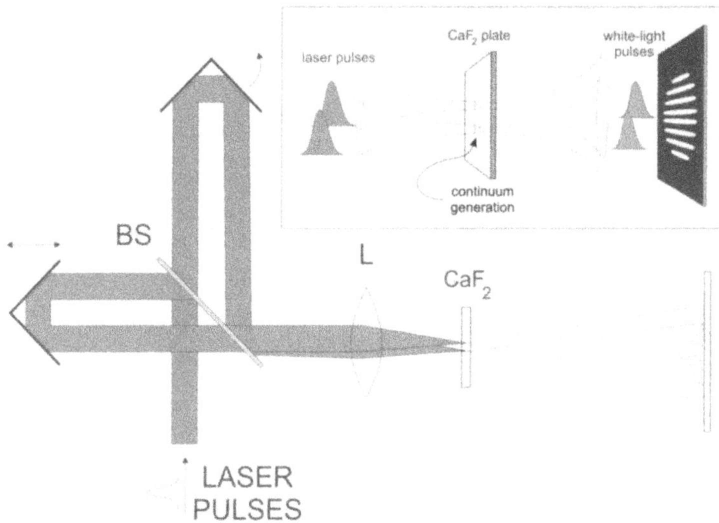

Fig. 5. Scheme of the experiment for the observation of white-light interference fringes. The two pump pulses are generated by the same laser system and by the same Michelson interferometer described above. Two independent supercontinuum pulses are generated in the material and then propagate and overlap on a screen

experiment to perform with our existing apparatus, we just decided to give it a try.

A 2 mm thick CaF_2 plate was placed in the focal plane of a lens after the Michelson interferometer as shown in Fig. 5, so that the two phase-locked laser pulses could independently produce two white light pulses. After the interaction zone and after the two diverging continua had propagated in air for some distance, they were finally overlapped on a screen where we hoped to observe interference fringes.

Our results were once again unexpected and intriguing: when the two pump pulses were properly balanced in intensity and adjusted for zero relative delay, the two white-light continua that they generated separately showed the surprisingly clear and stable white interference fringes shown in Fig. 6, indicating that we were dealing with highly phase-correlated secondary sources [13].

It is instructive to emphasize that there is a substantial difference between this and a simple Young's or Michelson's type experiment: in such cases two spatial portions of the same beam, or two time-delayed replicas of the same pulse are recombined to give interference. In our experiments on harmonics and white light, on the contrary, the interference fringes appeared because of the spatio-temporal superposition of two secondary light pulses independently generated in two separate positions of the medium. For the complex and apparently unpredictable characteristics of the generation processes at play, one could expect such pulses to be highly uncorrelated.

Fig. 6. Snapshot of the observed interference pattern. This picture was taken with an exposure time of about 2 seconds and with a laser repetition rate of 1 kHz. The very good visibility of the fringes indicates that the phase lock is not only preserved on a shot-to-shot basis, but that a constant phase relationship is maintained over at least thousands of shots

In general, in order not to damage the crystals by destructive optical breakdown and to avoid multiple filamentation in the medium, we limited the energy of each 100 fs pulse coming from the interferometer to less than 3 μJ. Under such conditions, a single filament was generally created by each focused beam in the transparent material, with an estimated peak intensity of the order of 10^{12-13} W/cm^2. Nonetheless, we observed that even in the case of multiple filaments and depending on their number, more or less complicated interference structures appeared, indicating that the mutual coherence among the resulting continua was conserved.

After the first appearance of the white-light fringes, we also tried to generate similar fringe patterns from different materials: quartz plates and plain glass microscope slides proved effective to produce stable interference, as well as water cells and Plexiglas plates. In general, any transparent material that we could find in the laboratory was able to produce nicely interfering white light at various degrees of efficiency.

We obtained an even more visual demonstration of the effective phase-lock over the whole visible spectrum by recording the spectrally dispersed interference pattern. We used a planar diffraction grating and a cylindrical lens, with the two focal spots aligned in a direction parallel to the grating grooves. In such a configuration, the interference fringes are parallel to the direction of the spectral dispersion and can be followed as they change color. Also in this case it was possible to produce stable fringe patterns of high contrast, as shown in Fig. 7. This was another clear indication that the two continua were mutually phase-coherent over the entire visible spectrum.

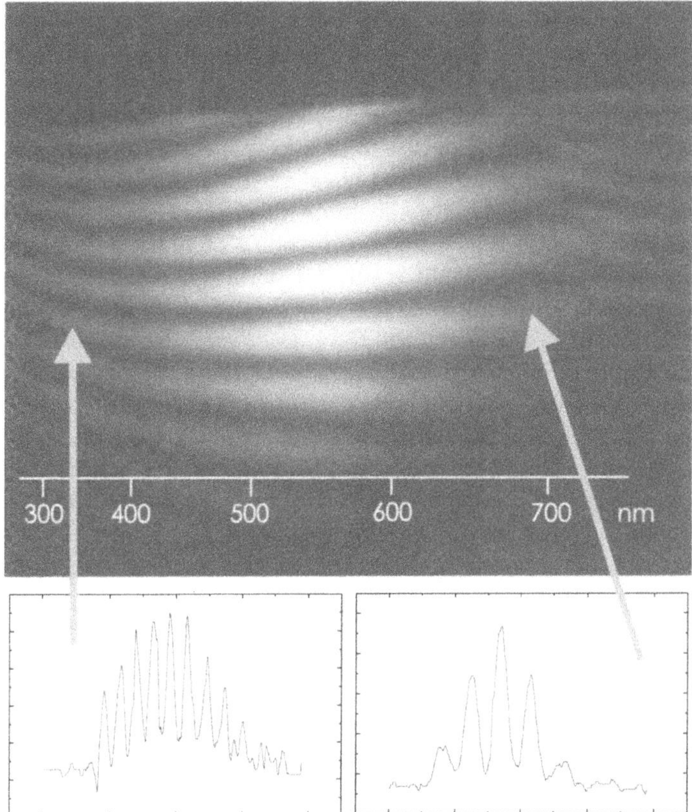

Fig. 7. Spectrally dispersed white-light fringes (note that in the original picture the color of the fringes changes from violet to red when moving from the left to the right side of the screen). Clear and well-defined fringes indicate that a stable phase relationship is conserved across all the generated visible spectrum. Also reported are two vertical lineouts showing that more than one optical octave is spanned by the continuum: the fringes in the right graph have a period which is more than twice as large as that of the fringes in the left graph

Though the experiment was of an extreme simplicity (it took just one day to see the first white fringes), it nonetheless bore very important consequences. After the first surprise at the look of the nice interference patterns, it soon became evident that, by demonstrating the phase preservation in the process of supercontinuum generation, we had also shown that it was possible to generate a sequence of phase-locked white-light pulses from a sequence of phase-locked pump pulses. In other words, we had shown that it was in principle possible to build a white-light frequency comb.

The fantastic possibilities of such a device were already clear to Theodor (not quite so clear to me at the beginning!): one could build an extremely "long" and precise ruler to measure optical frequencies throughout the whole visible range in a single step, doing away with clumsy laser frequency chains, divider stages, and so on.

The ideal path to the generation of a white-light comb involved the use of a train of phase-locked infrared pulses from a mode-locked laser to produce a train of phase-locked, white-light pulses after the interaction with a medium. Only two main obstacles could hinder the practical realization of such a wide comb: the first one was the possibility of phase scrambling between successive pulses in the output train, but, with our experiment, we had demonstrated that this was not the case. The second one was of a more technical nature, and depended on the limited pulse energy of mode-locked lasers. At the nanojoule-level energy characteristic of these high-repetition-rate systems, it was absolutely impossible to generate a supercontinuum and, unfortunately, no solution to this problem was available at that time. So, although these experiments had been performed in 1997, we kept all the nice and colorful pictures we had recorded aside for quite a long time because, without the possibility of such an application in sight, we just did not know what to do with them, apart from showing them around among the general amazement.

Finally, in 1999, a new kind of optical fiber came out [14] with new, exceptional properties: these "photonic crystal fibers" or "holey fibers" are essentially constituted by a very small silica core surrounded by a regular structure of "holes". They have the very interesting characteristic of supporting the propagation of highly spatio-temporally confined visible pulses over long distances. This allows for high nonlinearities to build up along the fiber and can give rise to the generation of a supercontinuum already from nanojoule-level pump pulses.

The results of our experiment and the introduction of these new fibers boosted the research toward the realization of white-light combs. Such devices are now a reality and are revolutionizing the whole field of optical metrology and spectroscopy. Femtosecond frequency combs are now replacing old-style frequency chains wherever a precise measurement of an optical frequency is required. With the realization of combs so wide as to extend over more than one optical octave, this technique now constitutes a self-referencing method allowing the measurement of absolute optical frequencies with extremely high accuracy in a single step from the frequency standard [15].

5 Conclusions

To conclude, I hope I have been able to give an impression of the kind of nice work that can be produced by the interaction with such a source of an amazing number of clever ideas as Theodor Hänsch. Ranging from the apparently simple ones (like the suggestion he once gave us on how to time-share the pulses from our laser among different simultaneous experiments: why had we never thought about it before?), to those that have really revolutionized the history of laser physics, there is always a good and unexpected idea waiting to pop out, thanks to his immense enthusiasm and curiosity towards science. It has been, and I hope it will be for a long time to come, a great pleasure to contribute to the realization of some of these ideas.

References

1. M. Bellini, A. Bartoli, T.W. Hänsch, Opt. Lett. **22**, 540 (1997)
2. J.N. Eckstein, A.I. Ferguson, T.W. Hänsch, Phys. Rev. Lett. **40**, 847 (1978)
3. C. de Lisio, C. Altucci, R. Bruzzese, S. Solimeno, F. Vigilante, M. Bellini, P. Foggi, Opt. Commun. **121**, 73 (1995)
4. M. Bellini, T.W. Hänsch, Appl. Phys. B **65**, 677 (1997)
5. R. Zerne, C. Altucci, M. Bellini, M.B. Gaarde, T.W. Hänsch, A. L'Huillier, C. Lyngå, C.-G. Wahlström, Phys. Rev. Lett. **79**, 1006 (1997)
6. M. Bellini, C. Lyngå, A. Tozzi, M.B. Gaarde, T.W. Hänsch, A. L'Huillier, C.-G. Wahlström, Phys. Rev. Lett. **81**, 297 (1998)
7. P.B. Corkum, Phys. Rev. Lett. **71**, 1994 (1993)
8. C. Lyngå, M.B. Gaarde, C. Delfin, M. Bellini, T.W. Hänsch, A. L'Huillier, C.-G. Wahlström, Phys. Rev. A **60**, 4823 (1999)
9. D. Descamps, C. Lyngå, J. Norin, A. L'Huillier, C.-G. Wahlström, J.-F. Hergott, H. Merdji, P. Salières, M. Bellini, T.W. Hänsch, Opt. Lett. **25**, 135 (2000)
10. M. Bellini, C. Cavalieri, C. Corsi, M. Materazzi, Opt. Lett. **26**, 1010 (2001)
11. R.L. Fork, C.V. Shank, C. Hirlimann, R. Yen, Opt. Lett. **8**, 1 (1983)
12. P.B. Corkum, C. Rolland, T. Srinivasan-Rao, Phys. Rev. Lett. **57**, 2268 (1986)
13. M. Bellini, T.W. Hänsch, Opt. Lett. **25**, 1049 (2000)
14. J.K. Ranka, R.S. Windeler, A.J. Stentz, Opt. Lett. **25**, 25 (2000)
15. S.A. Diddams, D.J. Jones, J. Ye, S.T. Cundiff, J.L. Hall, J.K. Ranka, R.S. Windeler, R. Holzwarth, T. Udem, T.W. Hänsch, Phys. Rev. Lett. **84**, 5102 (2000)

MeV Electrons and Positrons from a Femtosecond Table-Top Laser System

Klaus J. Witte, George D. Tsakiris, Christoph Gahn, and Georg Pretzler

We experimentally demonstrate the feasibility of generating a well collimated multi-MeV electron jet utilizing 790 nm/120 fs/1.2 TW pulses from a titanium:sapphire laser running at 10 Hz. The method uses the process of relativistic self-channeling in an underdense plasma of almost critical density. In a second step, we have succeeded in producing 2×10^8 positrons/sec by letting the electron jet impinge on a 2 mm thick lead converter. The scheme exhibits a favorable scaling to higher laser intensities.

1 Introduction

The last decade has witnessed an almost explosive-like progress in short-pulse laser technology which has resulted in a remarkable increase in laser powers and intensities. Two achievements have been responsible for this development. Firstly, through the mechanism of Kerr-lens modelocking, it has become relatively easy to reliably generate femtosecond pulses. Secondly, the new technique of chirped pulse amplification (CPA) has made it possible to amplify these pulses. A particularly attractive amplifying material is titanium:sapphire emitting at 800 nm. It enables table-top laser systems to be built with TW peak powers running at high repetition rates. The ATLAS (Advanced Titanium:Sapphire Laser) facility at the Max-Planck-Institut für Quantenphysik (MPQ) belongs to this class of lasers. It is a two-table-top system running at 10 Hz repetition rate and releasing 120 fs pulses with energies of up to 250 mJ and focusable to intensities of up to 4×10^{18} W/cm^2. The experiments reported below were carried out with this laser.

Intensities beyond 10^{18} W/cm^2 give access to a new regime of laser–matter interaction identified as high-intensity physics [1]. Present activities cover the generation of MeV γ-rays using solid targets, XUV radiation in the form of harmonics of the fundamental laser frequency, generation of fast ion jets from the rear side of thin foils, and fusion neutrons employing either deuterated planar targets or deuterium clusters. Another focus of interest deals with the propagation of femtosecond pulses through an underdense fully ionized plasma and the accompanying process of electron acceleration. Besides being of fundamental importance due to the occurrence of new phenomena such as giant magnetic fields and currents at the Alven limit, there are three other reasons for investigating this subject: (a) it is intimately related to the physics of plasma-based accelerators that can sustain extremely large acceleration gradients of the order of 1 GV/cm. Because this is three orders of magnitude higher than the value achieved in conventional accelerators, the plasma-based accelerators hold out the promise to replace the large-scale RF linacs by more compact devices; (b) in the advanced inertial confinement scheme known as "fast ignitor" [2], the laser-accelerated electrons play the central role in igniting the hot spot by depositing their energy in this fuel zone; (c) laser-based production of γ-rays, neutrons, positrons, π-mesons, μ-leptons, and laser-induced nuclear reactions require large fluxes of multi-MeV electrons.

In this paper, we describe the work we have recently performed with ATLAS pulses focused to $>10^{18}$ W/cm^2 on the generation of a relativistic electron beam [3] and its application for the production of 10^6 positrons (e^+) per shot with a mean energy of $\sim$2 MeV [4]. The e-beam is created under previously unexplored conditions, characterized by pulses of relatively *low energy* ($\leq$200 mJ) interacting with a *high-density* gas jet providing electron densities, $n_e \sim 0.3\, n_c$, where $n_c = \epsilon_0 m\omega_L^2/e^2$ is the plasma critical density corresponding to the laser pulse center frequency ω_L; ϵ_0 is the permittivity of free space, m is the electron mass, and e is the electron charge. Our work is the first demonstration of a *switcheable* positron source providing 2×10^8 e^+/s or an equivalent activity of 2×10^8 Bq.

After a short review on the interaction physics of TW laser pulses with gas-jet targets and the electron acceleration mechanisms in Sect. 2, we present the experimental setup used by us along with the main experimental results in Sect. 3. The positron generation and detection is described in Sect. 4. The conclusions and some potential prospects are given in Sect. 5.

2 High-Intensity Laser Pulse Interaction with an Underdense Plasma

2.1 Self-Focusing and Channel Formation

Hydrogen or helium gas jets are best suited for the generation of a homogeneous underdense plasma. This is for two reasons. Firstly, the density in the neutral gas jets can be precisely measured interferometrically. Secondly, at a peak intensity of $>10^{18}$ W/cm^2, the intensity far into the wings of the pulse, both in the propagation direction and transversally, is strong enough to completely ionize the gas, which requires $\sim10^{14}$ W/cm^2 for hydrogen and $\sim10^{15}$ W/cm^2 for helium. Defocusing ionization gradients are hence absent and the actual pulse then interacts with a *fully ionized* plasma of well-known electron density whose refractive index is given by $\eta_e = 1 - n_e/(2\gamma n_c)$ for pulses many light periods long. The factor $\gamma = (1 + a_L^2)^{1/2}$ accounts for the relativistic mass increase of the electrons executing a quiver motion in the linearly polarized laser field. The quantity $a_{\mathrm{L}} = eA_{\mathrm{L}}/(mc^2)$ is the normalized amplitude of the vector potential, A_{L}, which in terms of the peak electric field, E_{L}, can be expressed as $A_{\mathrm{L}} = cE_{\mathrm{L}}/\omega_{\mathrm{L}}$. Since the intensity is highest on axis, and hence so is γ, the relativistic mass increase produces an on-axis hump in the refractive-index profile. In addition, electrons are expelled from this region due to the *transverse* gradient of the ponderomotive potential $\propto |E_{\mathrm{L}}|^2$. Both effects self-focus the beam which is counteracted by diffraction. When the pulse power, P_{L}, exceeds the critical value, $P_{\mathrm{c}}[\mathrm{GW}] \approx 16.4\,\mathrm{n_c}/\mathrm{n_e}$, self-focusing dominates over diffraction [5].

Theoretical analysis has shown that depending on the initial laser and plasma conditions an intense short pulse with $P_{\mathrm{L}} > P_{\mathrm{c}}$ can propagate either in a stable single-channel mode or can break up into many filaments, thus exhibiting unstable behavior [6]. Recently, the transition from whole beam self-focusing to the strong filamentation regime has been observed [7]. More insight into the filamentation process has been gained by 3D PIC (Particle-In-Cell) simulations [8]. The filamentation is thought to be seeded by the Weibel instability occurring in the flow of counterstreaming electrons. In the present case, fast electrons are initially produced by stimulated Raman scattering and propagate with the pulse around its propagation-axis. The slower counterstreaming electrons move outside this region. The break-up of this scenario leads not only to current filamentation but also to light filamentation. Each filament is accompanied by a strong quasi-static magnetic field. However, the 3D PIC simulations have revealed a new phenomenon: the beam first passes through an unstable filamentary phase and then all filaments merge into a single one, the super-channel. The coalescence of the current and light filaments into a single filament or one plasma channel for $a_{\mathrm{L}} > 1$ is attributed to reconnection of the azimuthal magnetic field lines and is accompanied by a factor of > 10 increase of the on-axis light intensity. The refractive index profile in the plasma channel is similar to that in an ordinary fiber so that light can be transported over considerable distances [9].

2.2 Electron Acceleration

Concomitant with the channel formation is the generation of relativistic electrons with energies of up to several tens of MeVs. These energies greatly exceed the maximum quiver energy, $a_{\mathrm{L}}^2 mc^2$, acquired by a single electron in the focused laser beam, which for $a_{\mathrm{L}} \approx 1$ is less than 1 MeV . Two mechanisms due to collective effects associated with the plasma inside the channel are thought to be responsible for the observed relativistic electron acceleration: Laser Wakefield Acceleration (LWFA) and Direct Laser Acceleration (DLA).

In LWFA, a short laser pulse excites an electron plasma wave by expelling the electrons from the position they hold in the absence of the laser field. This is due to the axial gradient of the ponderomotive potential and works best when the pulse-length, $c\tau_{\mathrm{L}}$, matches the plasma wavelength, $\lambda_{\mathrm{p}} = 2\pi/\omega_{\mathrm{p}}$. For longer pulses, a break-up instability provides pulse portions with the required length [10]. The electron displacement and the restoring force due to ion inertia leads to the creation of a longitudinal plasma wake field propagating with a phase velocity equal to the group velocity of the laser pulse. In the potential well of this wave, background electrons can be trapped and accelerated to high speeds in the laser direction. When the oscillation amplitude of the electrons responsible for the plasma wake field comes close to the plasma wavelength, λ_{p}, wave-breaking occurs. The trapped electrons then damp the wake field irreversibly and appear as a collimated relativistic electron beam [11]. The maximal energy, $\gamma_{\max} = 2\omega_{\mathrm{L}}^2/\omega_{\mathrm{p}}^2$, a trapped electron can reach corresponds to the situation where it has become so fast that it has moved ahead of the accelerating flank of the potential well and starts decelerating [12]. Obviously, high plasma densities do not favor a high energy gain.

Direct Laser Acceleration, the second mechanism, presumes the existence of strong quasi-static electric and magnetic fields succeeding the channel formation. A part of the electrons initially occupying the channel volume is kicked out of this region because of the radial gradient of the ponderomotive potential. As a result of this charge separation, a strong radial electric field is generated. Another part of these initial electrons is axially accelerated by the excitation of a plasma wave in the leading edge of the laser pulse via forward Raman scattering. These electrons represent an electric current surrounded by an azimuthal magnetic field. A relativistic channel electron moving in the combined radial electric and azimuthal magnetic fields executes a motion which is similar to the one in the wiggler field of a free-electron laser [13]. It oscillates in the transverse direction at the so-called betatron frequency, $\omega_{\mathrm{b}} \approx \omega_{\mathrm{p}}/(2\gamma)^{1/2}$, while drifting along the channel with the velocity v. When ω_{b} coincides with the Doppler shifted laser frequency, $\omega_{\mathrm{L}} - k_{\mathrm{L}} v$, as seen by the electron, a resonance occurs leading to an effective energy exchange between the laser electric field and the electron. The $\boldsymbol{v} \times \boldsymbol{B}$ action of the laser magnetic field converts the energy gained in the transverse direction into longitudinal acceleration. This mechanism [14] is similar to that in an inverse free-electron laser [15], where the wiggler field is replaced by the self-generated quasi-static electric and magnetic fields in the channel. In contrast to LWFA, in DLA it follows from the resonance condition, $\omega_{\mathrm{b}} \approx \omega_{\mathrm{L}} - k_{\mathrm{L}} v$ that γ increases with n_{e}, i. e. the higher n_{e} the more energy the electrons can gain.

3 Generation and Characterization of the Relativistic Electron Beam

3.1 The Experimental Setup

In order to avoid-beam quality degradation resulting from self-phase modulation and small-scale self-focusing in air due to the high intensity of $\sim$100 GW/cm^2, the ATLAS pulses have to be transported through evacuated tubes to the target chamber. An $F_{\#} = 3$ high-quality off-axis parabolic mirror then focuses the pulses onto the edge of the free-expansion gas jet at intensities beyond 10^{18} W/cm^2 (see Fig. 1). The jet is generated by a high-pressure gas nozzle with a circular orifice of 500 μm diameter. The gas density profile was measured interferometrically. The density profiles at distances of 100 to 400 μm from the nozzle orifice have a bell-shaped radial profile and an exponential fall-off with distance from the nozzle orifice. In the selected interaction region located ~100 μm above the nozzle exit, the radial density profile is Gaussian with half the peak density at the edge of the orifice. The peak molecular density at this position is linearly proportional to the backing pressure, p, up to a maximal value of 4×10^{20} /cm^3 requiring p ~50 bar. Self-focusing and channel formation is diagnosed by means of a side-scattering imaging system at 90° to the laser beam direction. A narrow-band interference filter ensures that only laser light scattered by channel electrons is used for time-integrated image recording. The spatial resolution is 5 μm.

Two different diagnostics for the electron jet were employed. The first one is shown schematically in Fig. 1 and consists of a phosphorescent screen placed behind a 100 μm thick aluminum foil that blocked the laser and plasma light as well as electrons with energies below 200 keV. The electron jet could thereby be visualized on the screen and qualitatively optimized. At an elec-

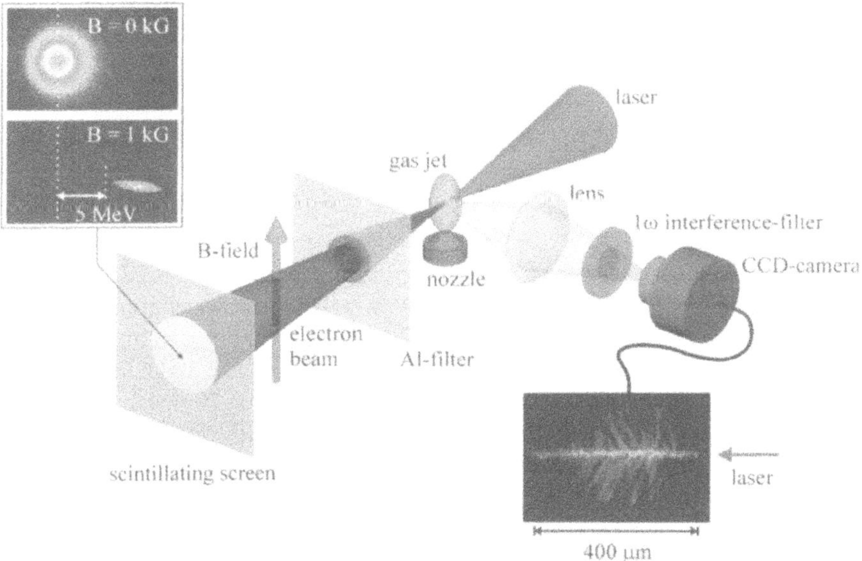

Fig. 1. Schematics of the experimental setup showing two of the diagnostics used. The scintillating screen visualizes the e-beam and enables its qualitative optimization. The appearance of MeV electrons is demonstrated by utilizing a magnetic field deflecting the electrons according to their energy (*upper left*). The CCD camera records the channel formation (*lower right*). The depicted channel was observed at optimal plasma conditions

tron density of 3×10^{19} /cm^3, a faint round spot appeared on the screen centered on the laser axis; with increasing density, the spot became brighter. When a static magnetic field of 1 kG was applied between filter and screen, the spot moved in the direction expected for an electron beam. From the spot, the FWHM divergence of the e-beam was estimated as $\sim 15°$. This is an upper limit because the low-energy electrons suffer multi-Coulomb scattering in the aluminum foil and hence increase the spot size.

The main electron diagnostic was a compact 45-channel magnetic spectrometer [16], which replaced the filter and screen after beam optimization and with which quantitative measurements of the electron energy spectrum in the range from 0.5 to 12.5 MeV were performed. The spectrometer was absolutely calibrated using various β-emitters. It was located 14 cm away from the laser focus, resulting in a 1 msr collection angle. The spectrometer could be rotated around the e-beam axis up to an angle of 10°, thus enabling an angularly resolved measurement of the electron energy spectrum.

3.2 Experimental Results

A typical energy spectrum measured in a helium plasma along the laser beam direction at maximal laser intensity and an electron density of 2×10^{20} /cm^3 is shown in Fig. 2. The fast electrons with energies >1 MeV exhibit a Boltzmann-like distribution with an effective temperature of $T_{\rm eff} = 5$ MeV. The largest energy recorded was 12.5 MeV, which was the spectrometer limit. About 10^8 electrons per msr with a mean kinetic energy of 5 MeV can be readily generated at the rate of 10 Hz. Noteworthy is the good agreement between the experimental results and 3D PIC simulations [2] which could precisely model the electron density profile and the laser pulse in space and time.

For a given laser power and frequency, the threshold for relativistic self-focusing, $P_{\rm RSF} = P_{\rm c}[{\rm GW}] \approx 16.4 n_{\rm c}/n_{\rm e}$, was transversed by changing the gas density at constant laser power. The onset of self-focusing was monitored by the channel images mediated through side-scattered laser light (see Fig. 1). No channel was observed below an electron density of 2×10^{19} /cm^3 for a laser power of $\sim$1.2 TW, in accordance with the above equation. As the plasma density was increased, the self-focusing power threshold dropped and a larger portion of the laser beam was trapped in the channel, whereby its length grew. At the electron density of 2×10^{20} /cm^3, we found an optimum at which most of the laser energy was trapped in the channel. Both channel length and $T_{\rm eff}$ are maximized here, having values of 0.4 mm (see Fig. 1) and 5 MeV (see Fig. 2), respectively. At higher densities, the number of electrons still increases, but the temperature and channel length decrease. This is attributed to higher energy losses due to electron heating as $n_{\rm e}$ approaches $n_{\rm c}$.

Angularly resolved measurements of the energy spectrum revealed that the beam is azimuthally symmetric, confirming the result obtained with the

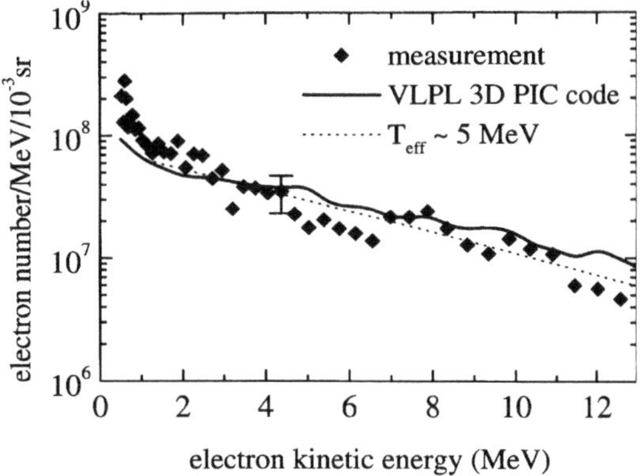

Fig. 2. Typical electron spectrum measured in a helium plasma along the laser beam direction within a solid angle of 1 msr at the optimal electron density of $n_e = 2 \times 10^{20}\,/\text{cm}^3$. A similar spectrum was observed for hydrogen at the same electron density

phosphorescent screen (see Fig. 1). At $n_e = 2 \times 10^{20}\,/\text{cm}^3$, the hottest electrons with $T_{\text{eff}} = 5\,\text{MeV}$ are in the center of the beam at 0° and colder ones appear in the outer parts with $T_{\text{eff}} = 4\,\text{MeV}$ at 5° and finally with $T_{\text{eff}} = 2\,\text{MeV}$ at 10°. The spectra integrated over a solid angle of 100 msr corresponding to a cone angle of 20° yield a T_{eff} of only 3.3 MeV, since the outer parts of the beam are colder than its center. We also measured the total number of all MeV-electrons within the 20°-cone, N_{tot}, as a function of the plasma density. When changing n_e from 4×10^{19} to $2 \times 10^{20}\,/\text{cm}^3$, N_{tot} grew from 4×10^7 to 2×10^{10}. This sharp increase in N_{tot} as well as the concomittant rise in temperature expected from the resonance condition, $\omega_b = \omega_L - k_L v$, supports the assertion that the mechanism accelerating the electrons is dominated by DLA and not by LWFA. For $N_{\text{tot}} = 2 \times 10^{10}$, the conversion efficiency of laser pulse energy into energy of a collimated jet of MeV-electrons amounts to 5%. Assuming that the electron emission lasts for about 200 fs (the laser pulse duration plus the difference between the transit times of a 10 MeV and an 1 MeV electron over the channel length), the corresponding current is ~15 kA, which is compatible with the Alven limit amounting to 350 kA for 10 MeV electrons.

The scaling of T_{eff} with the focal laser intensity is an important piece of information for two reasons: (a) it helps to further identify the role of the two mechanisms envisaged for electron acceleration, LWFA and DLA, and (b) provides insight into the application potential of relativistic electron beams in situations where a large number of MeV electrons is required.

As to (a), keeping the electron density at its optimal value of $2 \times 10^{20}\,/\text{cm}^3$, we observe a clear scaling of T_{eff} with intensity of the form $\propto I^{1/2} \sim a_L$, i.e. T_{eff} scales linearly with the electric field of the laser pulse. This finding is in excellent agreement with the results obtained by Pukhov with his 3D PIC code [17] for our experimental conditions. A detailed code-based analysis of the electron dynamics reveals further that the laser-induced wake field exists only for a short time at the head of the laser pulse and that most electrons gain their energy via DLA and not via LWFA. This remains true even for plasma densities much lower than those investigated by us experimentally. Only for very short laser pulses ($\leq$10 fs) and very thin plasmas with $n_e \ll n_c$, is LWFA unequivocally the dominant mechanism [18]. As to (b), since T_{eff} scales with $I^{1/2}$ and N_{tot} also increases when I increases (less than linearly), the collimated electron jet turns out to be an attractive source for γ-rays, positrons and neutrons via the giant resonance mechanism, and for π- and μ-particles. However, the neutron generation and in particular the π- and μ-generation require stronger lasers than ATLAS. The upgraded version of ATLAS providing 10-TW pulses will be just sufficient for the neutron production. The π- and μ-production needs pulses of several hundred TW. Such lasers with high repetition rates will be available in a few years. It remains an open question whether the electron jet will be useful for fast ignition as envisaged in [2]. A major uncertainty refers to the propagation of the jet through the highly overdense plasma surrounding the hot spot. The issue of disintegration of the jet by filamentation is not yet settled.

4 Pair Creation and Positron Detection

As long as the electron beam produced in the gaseous target comprises electrons with kinetic energies >1 MeV, there is a finite probability of generating electron–positron pairs in a high-Z converter. The question that arises is: Can the number of created pairs unequivocally be detected? There is a distinct peculiarity associated with this new source of nuclear radiation. Unlike common radioactive sources which undergo disintegration at a given rate and emit continuously over a long period of time, the laser-based source emits a burst of nuclear radiation within a very short time interval. Although this is in principle advantageous, since the source can be turned "on" and "off" at will, the known coincidence techniques normally employed for the detection of nuclear radiation are not applicable in this case.

The scheme employed by us for the generation of positrons from high-Z converters is analogous to that in linear electron accelerators. The MeV-electrons emerging from the gas jet hit a 2 mm thick lead slab, thereby generating bremsstrahlung or γ-photons, which in turn produce electron–positron pairs via interaction with the lead nuclei. The trident process in which the electrons directly generate electron–positron pairs via electron–nucleus collisions is negligible under our conditions [4]. Positron emission has also been reported from the direct interaction of PW (PetaWatt) laser pulses with solid gold targets [19]. However, such pulses can presently be provided only by huge single-shot facilities.

The experimental setup for the positron detection is straight forward and is schematically depicted in Fig. 3. To suppress the background signal due to stray γ-photons in favor of the weak positron signal, the detector had to be carefully shielded by an appropriate arrangement of lead bricks (not shown in Fig. 3). In addition, the primary electron jet was confined to a diameter of 1 cm by a corresponding bore in the plastic block surrounding the 2 mm thick lead converter. Plastic is a low-Z material and can hence stop electrons without producing undue bremsstrahlung. The distance between the converter and the gas jet amounted to 16 cm. The beam confinement reduces the number of MeV-electrons from a total of 2×10^{10} to $(8 \pm 1.7) \times 10^{8}$. This reduction was made in order to be able to perform a clean demonstration experiment. The positrons emanating from the converter have a quasi-isotropic distribution [20]. Those travelling in the e-beam direction have to pass through another 2 cm in the plastic bore before they enter the region where a magnetic field of 150 mT provided by two permanent magnets is present. Due to the magnetic field, the positrons describe an orbit of 180° and are then detected by a light-tight, 1.5 cm thick plastic scintillator coupled to a photomultiplier tube. The absolutely calibrated detector covers the positron energy range of (2 ± 0.08) MeV and subtends a solid angle of 7 msr to the converter.

At the beginning of the experiment, the electron energy spectrum was carefully measured and its reproducibility was established. The spectrum (Fig. 4) can be closely approximated by a Boltzmann distribution with $T_{\mathrm{eff}} = (2.7 \pm 0.1)$ MeV. The feasibility of the positron detection was then checked as follows. For the measured electron spectrum and the employed lead converter, the energy distribution of the γ-photons was calculated (see Fig. 4). Assuming each e^{+}–e^{-} pair shares the energy of the γ-photon or the electron, respectively, which generated it, we arrive at an estimate of the expected

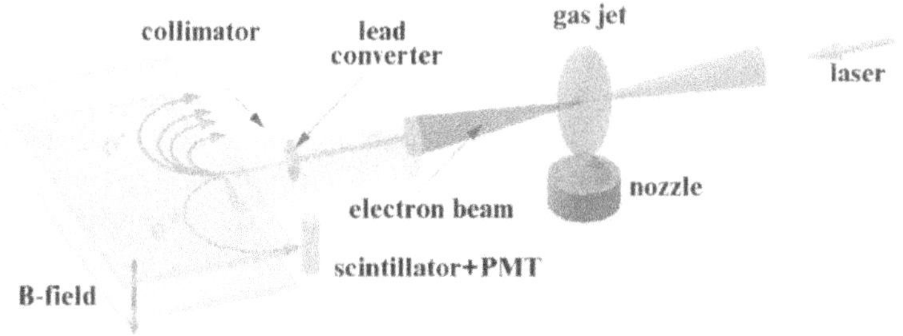

Fig. 3. Schematics of the miniaturized arrangement used for the production of positrons. A fraction of the MeV electrons created in the gas jet enter the 1 cm bore in the plastic collimator and produce γ-photons in the 2 mm thick Pb converter. These in turn generate e^{+}-e^{-}-pairs. The positrons are separated from the primary and secondary electrons via a static magnetic field and detected by the scintillator–photomultiplier tube combination

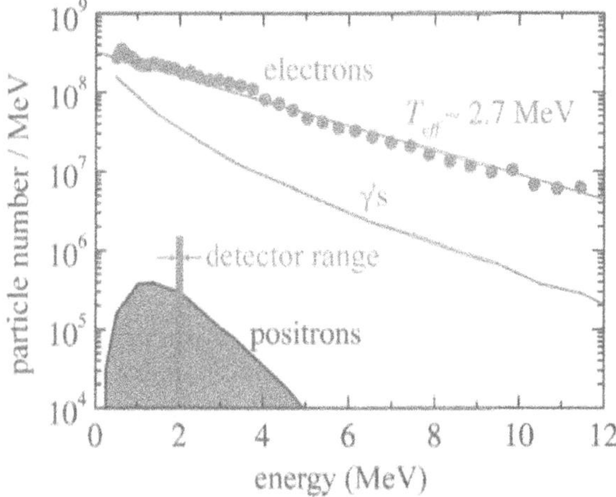

Fig. 4. Measured energy distribution of the primary electrons (full circles, exponential fit as dashed line) used to produce positrons. The calculated spectrum of the generated γ-photons and the expected positron spectrum is also shown. The calculation assumes a 2 mm thick lead converter. The vertical stripe denotes the energy range covered by the detector. It encompasses 5% of the total positron number

positron spectrum resulting from both contributing processes. As shown in Fig. 4, the e^+-spectrum peaks at $\sim$1.5 MeV and then drops rapidly at higher energies. This is due to the chosen thickness of the converter, which gets more transmissive with increasing energy of the γ-photons or electrons, respectively. The number of positrons, N_{pos}, per laser shot can now be estimated from the value $\Delta N_{\mathrm{pos}}/\Delta E = 3 \times 10^5/\mathrm{MeV}$ Fig. 4 for 2 MeV positrons as $N_{\mathrm{pos}} \approx (\Delta N_{\mathrm{pos}}/\Delta E)(\Delta \Omega_{\mathrm{pos}}/2\pi)\Delta E \sim 27e^+/\mathrm{shot}$. This result is confirmed by the measurements, which yield a value of 30 ± 15 positrons per shot.

In order to further substantiate our experimental result, we have performed detailed Monte-Carlo-type simulations using the code GEANT [21]. This code allows us to exactly simulate the experimental setup, i.e. collimator, converter, lead shielding, magnet, vacuum chamber wall, and detector. Two cases were systematically investigated. Firstly, for a fixed value of $T_{\mathrm{eff}} = 3\,\mathrm{MeV}$, the converter thickness was varied, leading to an optimal value of $l_{\mathrm{opt}} = 2\,\mathrm{mm}$. Secondly for the converter thickness fixed at $l_{\mathrm{opt}}, T_{\mathrm{eff}}$ was varied between 2 and 5 MeV. The simulation results for selected values of T_{eff} along with the experimental value for $T_{\mathrm{eff}} = (2.7 \pm 0.1)\,\mathrm{MeV}$ are presented in Fig. 5, where the number of positrons expected within the $(2 \pm 0.08)\,\mathrm{MeV}$ channel is given. As can be seen, the simulations predict the experimentally measured number of positrons for the geometry used.

Scaling the number of positrons detected within the energy interval of 0.16 MeV and the solid angle of 7.0 msr to the full energy range and the solid angle of 4π, a total number of 10^6 positrons per laser shot is obtained.

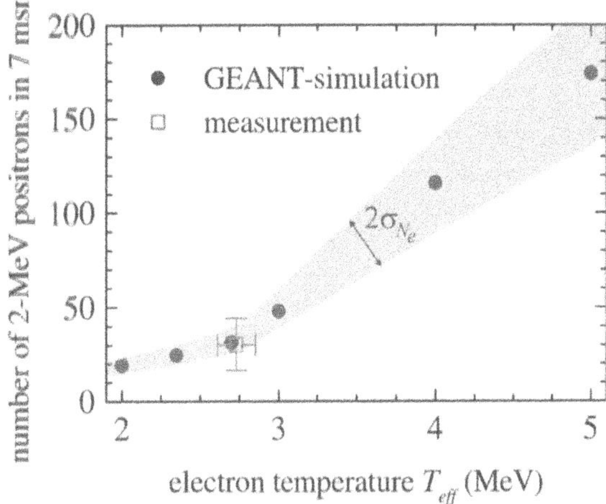

Fig. 5. Number of positrons seen by the detector as a function of the effective electron temperature, T_{eff}. The shaded area indicates the uncertainty $2\sigma_{\mathrm{Ne}}$ associated with the fluctuations in the total number of measured electrons

Using the full electron beam gives a positron number of $\sim 2 \times 10^7$ per laser shot, which for 10 Hz operation corresponds to an activity of 2×10^8 Bq. These values pertain to the ATLAS facility running in these experiments at a power level of ~1.2 TW/pulse. However, laser systems delivering pulses with $\sim$ 10 times more power and hence enabling correspondingly higher in-

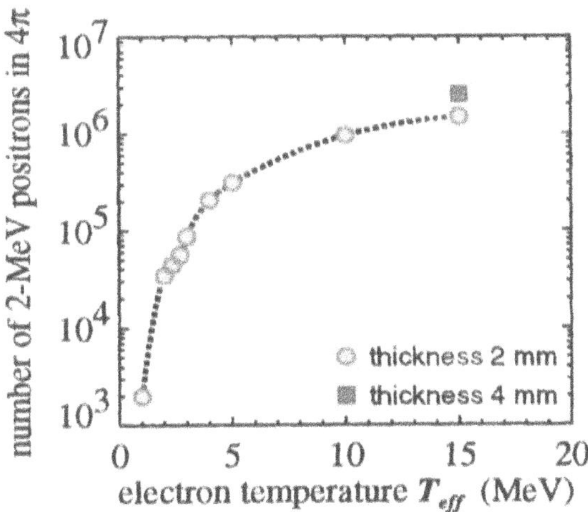

Fig. 6. GEANT simulation for the number of 2 MeV positrons produced within the solid angle of 4π as a function of the effective electron temperature, T_{eff}, for a 2 mm thick lead converter (circles). For a 4 mm thick converter and $T_{\mathrm{eff}} = 15$ MeV, the number of positrons nearly doubles

tensities and effective electron temperatures are available. How this increases the positron yield is shown in Fig. 6, where predictions of the GEANT code for higher values of T_{eff} and the same type of converter are given. For $T_{\mathrm{eff}} = 10\,\mathrm{MeV}$, a 20-fold increase in positron output is expected. At $T_{\mathrm{eff}} \sim 15\,\mathrm{MeV}$, saturation occurs. This is due to the higher energy of the γ-photons or electrons, respectively, which are less well absorbed by the converter. Increasing the converter thickness then improves the positron yield. Activities of 10^{10} Bq appear feasible when for a given electron temperature the converter thickness is optimized.

5 Conclusions

In recent years, positrons have played an important role in applications in diverse fields of physics, e.g. surface physics [22], positronium spectroscopy [23], and electron–positron plasmas [24]. Although for most of these applications conventional radioactive sources like ^{22}Na or ^{58}Co are presently used, the distinct possibility exists to replace them by laser-driven positron sources. Our results demonstrate that already today an appreciable number of electron–positron pairs can be created with a small sized laser system. Given the prodigious advances in laser technology, it is almost certain that in the near future there will be laser systems delivering 100 TW pulses at repetition rates approaching 1 kHz. The positron yield that can then be expected would be comparable to those from electron linear accelerators but from a facility orders of magnitude smaller in size. Other obvious advantages of laser-driven positron sources are that they can be turned on when they are needed, thus considerably reducing the radiation hazard associated with large-sized radioactive sources. It is for this reason that replacing the ^{22}Na source by a laser-driven positron source in the planned production of cold antihydrogen [25] is worth considering. The comparison of 1s–2s transition frequencies measured for antihydrogen and hydrogen would be a test of CPT invariance that involves the charges and masses of leptons and baryons at an unprecedented precision and might reveal conceivable small differences between the gravitational forces acting on matter and antimatter. Prof Hänsch and his group have achieved a precision in hydrogen spectroscopy [26] sufficient to contemplate this comparison to 1 part in 10^{15}. The pulsed character of the laser-driven positron source can also be exploited in positron-annihilation spectroscopy to eliminate the cumbersome timing electronics since the start signal is well defined by the laser pulse. It is beyond doubt that the laser-based pulsed positron source would be of advantage for other applications as well.

References

1. G. Mourou, C. Barty, M. Perry, Phys. Today **51**, 22 (1998)
2. M. Tabak, J. Hammer, M. Glinsky, W. Kruer, S. Wilks, J. Woodworth, E. Campbell, M. Perry, R. Mason, Phys. Plasmas **1**, 1626 (1994)
3. C. Gahn, G. Tsakiris, A. Pukhov, J. Meyer-ter-Vehn, G. Petzler, P. Thirolf, D. Habs, K. Witte, Phys. Rev. Lett. **83**, 4772 (1999)
4. C. Gahn, G. Tsakiris, G. Pretzler, K. Witte, C. Delfin, C. Wahlström, and D. Habs, Appl. Phys. Lett. **77**, 2662 (2000)
5. G. Sun, E. Ott, Y. Lee, P. Guzdar, Phys. Fluids **30**, 526 (1984)
6. A. Borisov, O. Shiryaev, A. McPherson, K. Boyer, C. Rhodes, Plasma Phys. Control. Fusion **37**, 569 (1995)
7. X. Wang, M. Krishnan, N. Saleh, H. Wang, D. Umstadter, Phys. Rev. Lett. **84**, 5324 (2000)
8. A. Pukhov, J. Meyer-ter-Vehn, Phys. Rev. Lett. **76**, 3975 (1996); Phys. Plasmas **5**, 1880 (1998)
9. M. Borghesi, A. McKinnon, L. Barringer, R. Gaillard, L. Gizzi, C. Meyer, O. Willi, Phys. Rev. Lett. **78**, 879 (1997)
10. P. Sprangle, E. Esarey, J. Krall, Phys. Plasmas **3**, 2183 (1996)
11. R. Wagner, S. Chen, A. Maksimchuk, D. Umstad, Phys. Rev. Lett. **78**, 3125 (1997)
12. T. Tajima, J. Dawson, Phys. Rev. Lett. **43**, 267 (1979)
13. G. Tsakiris, C. Gahn, V. Tripathi, Phys. Plasmas **7**, 3017 (2000)
14. A. Pukhov, Z. Sheng, J. Meyer-ter-Vehn, Phys. Plasmas **6**, 2847 (1999)
15. E. Courant, C. Pellegrini, W. Zakowicz, Phys. Rev. A **32**, 2813 (1985)
16. C. Gahn, G. Tsakiris, K. Witte, P. Thirolf, D. Habs, Rev. Sci. Instrum. **71**, 1642 (2000)
17. A. Pukhov, J. Plasma Phys. **61**, 425 (1999)
18. A. Pukhov, private communication
19. T. Cowan, M. Perry, M. Key, T. Ditmire, S. Hatchett, E. Henry, J. Moody, M. Moran, D. Pennington, T. Phillips, T. Sangster, J. Sefchik, M. Singh, R. Snavely, M. Stoyer, S. Wilks, P. Young, Y. Takahashi, B. Dong, W. Fountain, T. Parnell, J. Johnson, A. Hunt, T. Kühl, Laser Part. Beams **17**, 773 (1999)
20. R. Evans, *The Atomic Nucleus* (McGraw-Hill, New York 1995) pp. 701–710
21. R. Brun et al., *EANT User's Guide* CERN report DD/EE/82 (1982)
22. C. Szeles, K. Lynn, in *Encyclopedia of Applied Physics*, Vol. 14, ed. by G. Trigg (VCH, New York 1966) pp. 607–632
23. A. Rich, Rev. Mod. Phys. **53**, 127 (1981)
24. R. Greaves, C. Surko, Phys. Plasmas **4**, 1528 (1997)
25. G. Gabrielse, T. Roach, J. Estrada, D. Hall, P. Yesley, H. Kalinowsky, T. Hänsch, K. Eikema, J. Walz, W. Oelert, D. Grzonka, Th. Sefzick, T. Hijmans, W. Phillips, S. Rolston, J. Walraven, W. Jhe, D. Wineland, J. Bollinger, ATRAP Proposal SPSC 97-8/P306 *The Production and Study of Cold Antihydrogen*, presented to the CERN SPSLC on 25 March 1997
26. M. Niering, R. Holzwarth, J. Reichert, Th. Udem, M. Weitz, T. Hänsch, P. Lemonde, G. Santarelli, M. Abgrall, P. Laurent, C. Salomon, A. Clairon, Phys. Rev. Lett. **84**, 5496 (2000)

Small Molecules in Intense Laser Fields – Dissociation and Stabilization

Hartmut Figger, Domagoj Pavicic, and Karsten Sändig

Small molecules such as H_2^+ and D_2^+ in an ion beam were investigated in high-intensity laser fields with intensities between 10^{13} and 5×10^{15} W/cm^2. This new research field is characterized by the terms multiphoton absorption, nonlinear processes, light-induced molecular potentials and photon-dressed states. In our ion beam experiment the kinetic energy spectra of the fragments H and H^+ were detected with high resolution. The two fragmentation channels observed here exhibit an interesting fine structure not detected before. It can be easily associated in the first one, the dissociation channel, to the vibrational levels of the parent molecules H_2^+ and D_2^+, respectively. In the second fragmentation channel, namely in the multiphoton ionization process leading to Coulomb explosion, it was interpreted as charge resonance enhanced ionization (CREI) resulting in the preferred (critical) distances clearly observed here for the first time.

1 Introduction

Much of the research in Theodor Hänsch's group at the Max-Planck-Institute for Quantum Optics since 1987 was focused on high-precision measurements on elementary atomic systems like H and D. It was our (H.F.) wish to supplement it by accurate measurements on the most elementary molecular systems such as H_2^+, D_2^+ and HD^+ at the high light intensity limit.

New exciting effects like above-threshold ionization (ATI), above-threshold dissociation (ATD), or high harmonic generation (HHG) were discovered when atoms and small molecules were exposed to pulsed laser fields of femtosecond lengths and intensities of 10^{12} to 10^{16} W/cm^2 [1]. These effects depend nonlinearly on the intensity and are mostly initiated by multiphoton absorption. Much of the experimental work starts with the H_2 molecule while the theoretical work was concentrated on H_2^+, which is the smallest molecule in nature [1,2]. Assuming the Born–Oppenheimer approximation, its Schrödinger equation is solvable in a closed form, i.e analytically. It has a relatively simple two-state potential curve system with a bonding ground state $1s\,^2\Sigma_g$ and a repulsive excited state $2p\,^2\Sigma_u$, both being connected by a charge resonance transition, Fig. 1. The next excited states are at least 11.5 eV higher in energy and are also unbound.

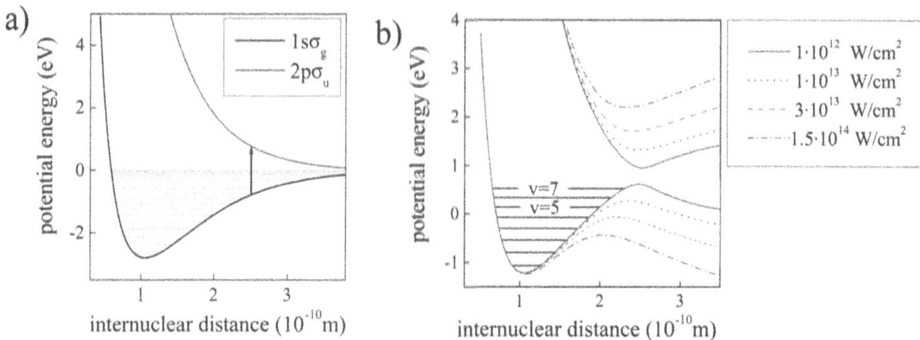

Fig. 1. (**a**) Lowest potential curves of H_2^+ with some transitions which can be induced by the Ti:sapphire laser. (**b**) Dressed potentials at 785 nm

In interpreting the experiments on H_2 it has usually been assumed that this molecule is ionized at the leading edge of the laser pulse and that a vibrational population distribution in the H_2^+ ion corresponding to the Frank–Condon factors is thereby achieved. Subsequently, this molecular ion is fragmentized [2–4]. More recent work shows that this assumption is inadequate [4]. More probably, the ionization takes place rather late at the top of the laser pulse and this process then interferes with the fragmentation of the molecule.

In order to exclude such problems, in this work a fast mass-selected, highly collimated ion beam of H_2^+ or of isotopes D_2^+ and HD^+ for the first time was exposed to high-intensity laser pulses, rather than H_2 gas leaking into a high-vacuum vessel [5–8]. The H and H^+ fragments are then detected on a 2-dimensional detector with very high energy resolution, Figs. 2, 3. This setup had already been introduced to molecule-intense laser field physics in an experiment on Ar_2^+ by Wunderlich et al. [9–11] from 1996 onwards. Recently an ion beam experiment on H_2^+ using femtosecond laser pulses that is similar and complementary in many points to ours was published by Williams et al. [12].

Such a setup provides a great variety of ways of preparing the beam: it can be prepared in a few vibrational levels as already demonstrated in the Ar_2^+ experiment [10, 11], it can be aligned by external fields, and it can even be neutralized. The state population distribution can be determined before it is exposed to the femtosecond laser pulse. Finally, the setup as described can be used in future to investigate a multitude of small molecules in this way.

This experiment allows a high energy resolution of the H and H^+ fragments of slightly better than 1%, which is an improvement by a factor of 5 to 10 or more than hitherto achieved in all other experiments previously done in this field on H_2 and H_2^+. Consequently, we resolve for the first time the structure of the two (three in the case of two-photon absorption) broad maxima having eV widths in the translational energy spectra, Figs. 4, 5, 7a. These broad maxima were already found to be due to the two possible fragmentation channels: (1) dissociation by one (and two respectively) photons and (2) photoionization (Coulomb explosion) [2–4], according to

$$H_2^+ + nh\nu \longrightarrow H + H^+ \qquad (a) \tag{1a}$$

$$H_2^+ + nh\nu \longrightarrow H^+ + H^+ + e^- \qquad (b) \tag{1b}$$

where n is the number of absorbed photons and hν their energy, and e^- is the ionized electron. In the first channel the two fragments escape with a kinetic energy of up to 1 eV, while in the second one, the Coulomb explosion channel, they escape with a kinetic energy of 1 to 4.5 eV [3, 4]. An overview momentum distribution of both channels in (1a,b) is shown in Fig. 2.

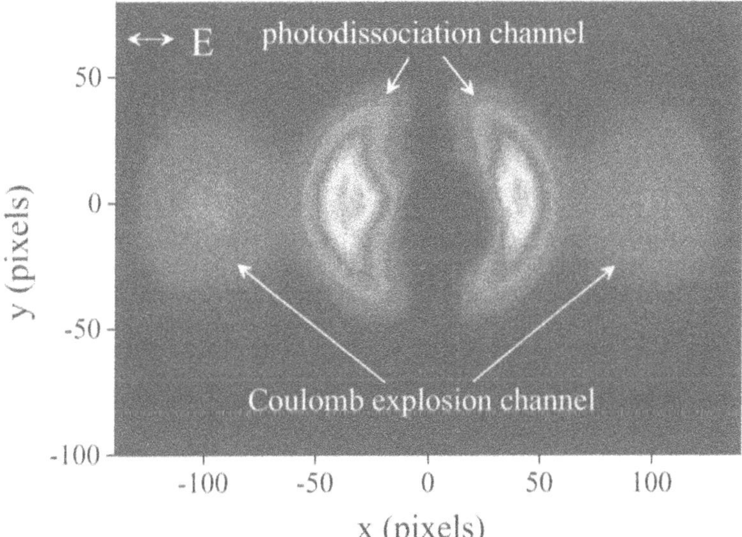

Fig. 2. Overview momentum spectrum of the fragmental protons. The parent beam of H_2^+ is blocked and would hit the MCP plate at the zero eV point. The laser intensity I was $I = 3 \times 10^{14}$ W/cm^2, the pulse duration $\tau = 135$ fs. The two fragmentation channels, dissociation and Coulomb explosion, are well discernible, the first one being dominant at this lower intensity. Furthermore, the very different angular distributions of the fragments of the two channels are noteworthy

Both broad maxima are resolved in this work into a multi-line system showing in more detail the physics behind them: so the resolution of the first maximum now shows the fate of most of the vibrational levels populated in the H_2^+ or D_2^+ beam with increasing laser pulse intensity (Figs. 4, 5). This is reported in the first part of Sect. 3.1.

Very recently we were also able to resolve the second maximum which is due to Coulomb explosion, for the intensities close to the threshold of this process, (1b); Figs. 2, 7a. This translational energy spectrum can be converted to the protons distances by means of the Coulomb law (Fig. 7b). The peaks here are the critical distances from which the Coulomb explosion preferentially starts and are to our knowledge experimentally clearly observed here for the first time. This is described in Sect. 3.2.

2 Measurement Principle and Experimental Setup

The molecular ion beam of H_2^+ was generated in a dc-discharge by a so-called duoplasmatron (Fig. 3; [5–11]). The ions were accelerated to 11.1 kV, mass-selected in a sector magnetic field, compressed by two ion lenses and collimated to a beam with a cross-section as small as $25 \times 300\,\mu\text{m}^2$. In that way it was matched to the diameter of the pulsed focused laser beam which crossed the ion beam at 90°. A rather small variation of the laser intensity over the ion beam was thus realized, leading to a well-defined spectrum.

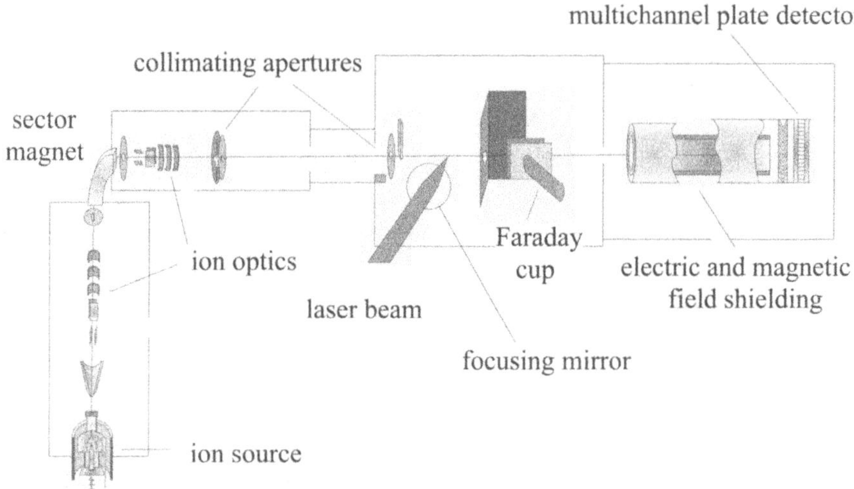

Fig. 3. Setup of the molecular ion beam experiment using lasers with pulse widths down to 80 femtoseconds. The fragments are detected by a MCP detector combined with a CCD camera

The H and H^+ fragments coming from one rovibrational level of H_2^+ with a dissociation energy $E_{v\mathrm{J}}$ have a translational energy of $1/2(\hbar\omega_\mathrm{L} - E_{v\mathrm{J}})$. This means that because of the dissociation probability $P_\mathrm{diss} \sim (\boldsymbol{d} \times \boldsymbol{E})^2 \cos^2\theta$, and the 4π spatial distribution of the molecules' axes the fragments lie on the caps of a virtual sphere. Here θ is the angle between the electric component E of the laser field, i.e. the polarization direction is perpendicular to the ion beam and the dipole moment d, which is along the molecular axis. Because of the many rovibrational levels populated in the H_2^+ beam, up to $v = 13$ we have a concentric system of virtual spheres which expand when moving towards the detector. Since the polarization of the laser was perpendicular to the ion beam, all the fragments have a velocity component also in this direction when they are projected towards the 2D detector. This consists of a multichannel plate (MCP) standing perpendicular to the beam. The multiplied electrons emerging from the channels hit a phosphorescent screen and the light flash generated there by them is read out by a CCD camera recording its local position, (La Vision system). The projected pole caps of the spheres generate a rather sharp concentric ring system on the screen as shown in Figs. 2, 4, 5, 7. The inner circles on the screen belong to the fragments from the lower vibrational levels because of their greater dissociation energy and hence lower kinetic energy and velocity perpendicular to the beam. The original 3-dimensional spherical velocity distribution can be reconstructed from the projection using the Abel transformation, as is shown in Fig. 4, resulting in a sharper ring system.

This setup allows a kinetic energy resolution of the fragments of slightly better than 1 per cent in the inner part of the screen. It is determined by the cross-section of the ion beam, the distance between the laser crossing point with the H_2^+ beam from the detector, the distance of the channels of the MCP plate and the spatial resolution of the CCD camera. The commercial femtosecond CPA("chirped pulse amplifier") – laser system ("Spectra Physics" system) has a repetition rate of up to 1000 pulses/s and has a maximum power of 2.5 mJ per pulse and a shortest pulse length of 80 fs. The laser media of the oscillator and of the amplifiers are Ti:sapphire crystals. Therefore, the laser wavelength was variable between 700 and 900 nm. The pump laser was an intracavity-doubled flashlamp-pumped Nd-YLF laser with about 20 W of output power. Great constancy of the intensity over the laser beam cross-section, and, in particular, small variation of the pulse heights of better than 3% for about 5 h of measuring time were necessary for the success of the experiments described below. Additionally, the ion beam current had to be constant to better than 2% during the same time.

2.1 Population Distribution

In order to have well-known molecular beam conditions, i.e. a well known vibrational distribution in the H_2^+ beam, the latter was determined before the high-intensity experiment. For this purpose the ion beam was excited by a low-intensity long-pulse ns laser, so that Fermi's golden rule (FGR) is valid, i.e. the dissociation rate is proportional to the light intensity and FC factors. A cut through the 2-dimensional image of the fragment distribution along the polarization direction of the laser was taken similar to that in Fig. 5, where the single vibrational levels are resolved as peaks. From their heights and FC factors, the relative populations of the vibrational levels were determined for the single vibrational levels with $v > 4$. In practice the populations of the H_2^+ vibrational levels were determined by a computer simulation of the experimental fragment distribution by varying the populations. Here the rotational temperatures for the different vibrational levels were tentatively taken from Koot et al. [14].

3 Measurements with High Laser Intensity

This section describes and interprets the finer structures of the two eV-broad maxima in the translational energy spectrum of the H^+ and H fragments, respectively, of H_2^+ in intense pulsed laser fields. The experimental overview momentum distribution as observed on the 2-dimensional detector is given in Fig. 2. This shows the distribution of the protonic fragments according to their projected momenta, which are proportional to the square root of the kinetic energy, as given in the Fig. 2. The polarization direction of the laser light is horizontal as indicated by the arrow $\boldsymbol{E}$ in Fig. 2. The parent H_2^+ beam is perpendicular to the detector plane and hits the detector at the centre, i.e. at the zero point, but is blocked in Fig. 2 by a Faraday cup. The distribution is symmetrical to the origin, which is also due to the homogeneous 4π distribution of the molecular axes in space. The two fragmentation channels, viz. the dissociation (1a) up to about 1.0 eV and the photoionization (1b) between 1 and 4 eV, are found here to be clearly separated.

At this comparatively low intensity of 3×10^{14} W/cm^2 in Fig. 2 the dissociation channel (1a) is the dominant one of the two, with its characteristic kidney shape, a very broad almost $\cos^2$-shaped angular distribution for the higher vibrational levels around 0.6 eV and a much narrower distribution for the low vibrational levels situated at lower translational energies around 0.25 eV.

A much narrower and very differently shaped angular distribution is detected for the second, the so-called Coulomb explosion channel, (1b), which is caused by multiphoton ionization of the bonding electron. It becomes more and more the dominant channel of the two when the laser intensity I increases to 10^{15} W/cm^2 and higher.

3.1 Dissociation

In the following, first the results for the first channel, the dissociation channel (a) of (1a), are discussed [5–8]. A high-resolution measurement of the neutral H fragments as projected on the MCP detector is shown in Fig. 4 for a pulse energy of 1 mJ and two different pulse lengths of 575 and 135 fs. These correspond to intensities in the foci on the ion beam of 3.5×10^{13} and 1.5×10^{14} W/cm^2, respectively. In these plots H fragments from the single vibrational levels are detected as partial circles for the first time in such high-intensity measurements [5–8]. Going from Fig. 4a to Fig. 4b, corresponding to amplification of the intensity of a factor of about 4.5 at constant pulse energy,

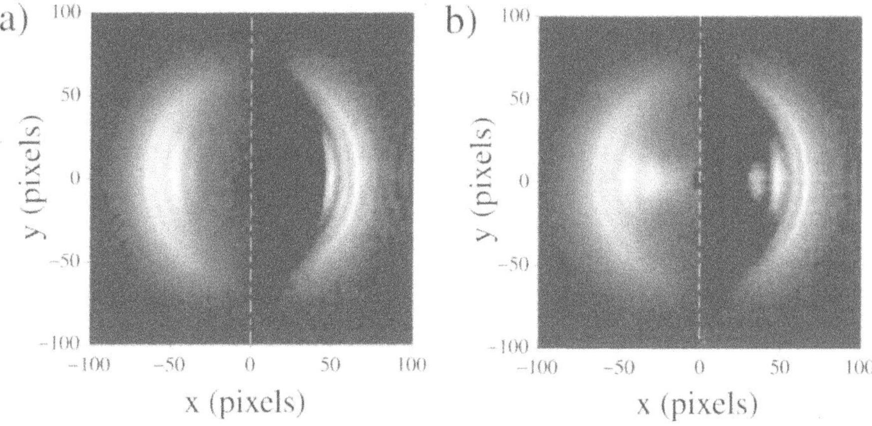

Fig. 4. On the *left-hand side* of figures (**a**) and (**b**): projection of the neutral H fragments on the 2-dimensional MCP detector at a pulse energy of 1 mJ, a laser wavelength of 785 nm and pulse length of (**a**) 575 fs, $I_0 = 3.5 \times 10^{13}\,\mathrm{W/cm^2}$, (**b**) 135 fs, $I_0 = 1.5 \times 10^{14}\,\mathrm{W/cm^2}$. The distance along the x-axis is proportional to the momentum of the fragments. The semicircle-type structure describes the angular distribution of the fragments from single vibrational levels of H_2^+.On the right-hand sides of (**a**) and (**b**): the Abel transformations illustrated by the distribution on a big circle

one notes a very strong change: lower vibrational levels now dissociate (in Fig. 4b) with a much narrower angular distribution while the higher levels remain almost unchanged. To get a more quantitative picture, cuts are made through the distributions along the polarization direction of the laser light, as shown in Fig. 5. All the *neutral* fragments here came from H_2^+ molecules, which experienced the maximum laser intensity since their axes were all directed parallel to the polarization direction here. In both experiments in Fig. 5 the pulse energy was 0.3 mJ, and the pulse lengths were 690 and 130 fs, respectively, corresponding to intensities of 4.7×10^{12} and $2.5 \times 10^{13}\,\mathrm{W/cm^2}$, i.e. the pulse energy was compressed to a shorter time interval in experiment b.

The greatest effect observed in Fig. 5 is the exploding of the peaks belonging to the $v = 7$ and $v = 8$ levels, when the intensity is increased by a factor of 5.5. This effect is called bond softening in the literature [1–4]. Furthermore, a striking downshift in the translational energy of the $v = 7$ peak corresponding to a downshift of the energy of this vibrational level concerned is observed. The peaks with higher v, in particular the $v = 9$ peak, remain rather constant in height and angular distribution when going to higher laser intensities. There is one striking exception: the $v = 11$ peak *decreases* with increasing intensity ("trapping"). The decrease of $v = 11$ is found in all other experiments to be very sensitive to power and intensity.

These effects, called above bond softening, level shifting and trapping, are given a satisfying explanation by the so-called **L**ight **I**nduced **P**otential

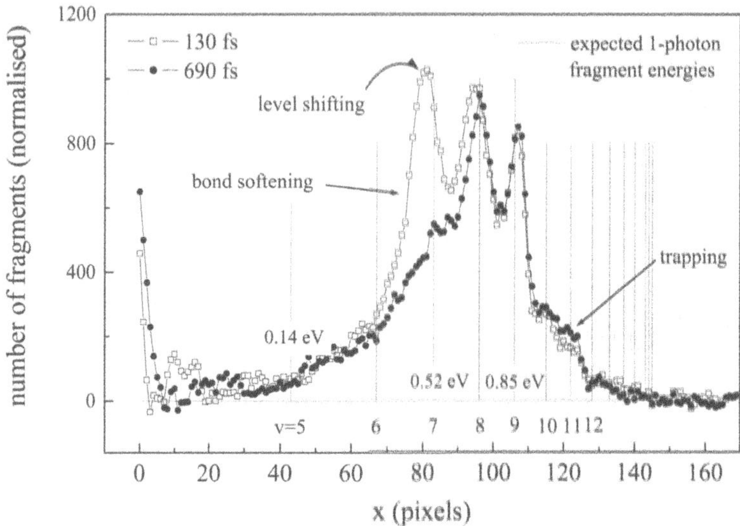

Fig. 5. Cut through the two-dimensional dissociation momentum spectrum along the polarization direction. The laser parameters were $\lambda = 785\,\mathrm{nm}$, $E_{\mathrm{pulse}} = 0.3\,\mathrm{mJ}$, the intensities at the focus were $I = 2.5 \times 10^{13}\,\mathrm{W/cm^2}$ and $I = 4.7 \times 10^{12}\,\mathrm{W/cm^2}$ respectively

(LIP) theory. This has to be applied when the molecules are exposed to such an intense light field that it cannot be treated any longer as a perturbation in the Hamiltonian. It will be briefly discussed in the following:

In Fig. 1a, part of the potential curve system of the H_2^+ molecular ion is shown. The ground state $1s\sigma_g$ is bonding, while the first excited state $2p\sigma_u$ is repulsive. The corresponding potential curves are called E_1 and E_2 in the following. Both are degenerate in energy at a large nuclear distance R and can be strongly coupled by a charge resonance transition whose transition probability is proportional to $R/2$. This system is rather similar to that of Ar_2^+, which was experimentally investigated in high laser fields by our group a few years ago, as mentioned above [9–11].

As long as the light intensity to which the molecule is exposed is small, its interaction energy can be treated as a perturbation to the Hamiltonian of the molecule. The molecule's dissociation can then be described by Fermi's golden rule (FGR), i.e. the dissociation probability is linear in the light intensity [1, 15].

The coupling of the two potential curves E_1 and E_2 by the light field in Fig. 1a is conveniently described by the magnitude of the Rabi frequency ω_R relative to the vibrational frequency ω of the molecule. The Rabi frequency ω_R is proportional to $\boldsymbol{E} \cdot \boldsymbol{D}$, where $\boldsymbol{E}$ is the electric component of the laser field and $\boldsymbol{D}$ is the transition dipole moment between the two states considered. In the case of $\omega_R > \omega$, that is when the field E is very high, the molecule can switch back and forth between the two curves during one molecular vibration, corresponding to a strong coupling of the two curves. In this case, the Hamiltonian H of the whole system "molecule in the laser field" has to be supplemented by the Hamiltonian of the light field, H_{light}. Then one has the Schrödinger equation [1]:

$$H\,|\Phi\pm\rangle = (H_{\text{electron}}(r) + V_{\text{proton}}(R) + H_{\text{light}} + V_{\text{int}})\,|\Phi\pm\rangle = E\pm\,|\Phi\pm\rangle\,. \quad (2)$$

Here H_{electron} is the Hamiltonian of the sum of the kinetic and electrostatic energies of the electron, V_{proton} the electrostatic potential of the two protons, R is he distance of the protons, and V_{int} the Hamiltonian of the light-molecule interaction. The kinetic energy of the nuclei is omitted to get the new potential curves $E \pm (R)$ as eigenvalues. The eigenfunctions $|\Phi\pm\rangle$ are linear combinations of a basis set consisting of product states of the electronic eigenstates of the molecule and the Fock states of the light.

Solving (2) by diagonalization yields the new light-induced potential curves $E\pm$ shown in Fig. 1b, [1]:

$$\begin{aligned} E\pm &= 1/2[E_1(R) + \hbar\omega_{\text{L}} + E_2(R)] \\ &\pm 1/2\sqrt{\{[E_1(R) + \hbar\omega_{\text{L}} - E_2(R)]^2 + \hbar^2\omega_{\text{Rabi}}^2\}}\,. \end{aligned} \quad (3)$$

The first term in (3) has the effect that after the lower curve $E_1(R)$ is shifted up by $\hbar\omega_{\text{L}}$ the two curves cross each other. The last sum term has the effect of avoiding the crossing. The gap there opens with increasing ω_{Rabi}, i.e. with the laser intensity I as is shown in Fig. 1b, where the new potential curves are drawn for increasing intensities. The effect of the widening gap is that the vibrational levels below the crossing, viz. $v = 8, 7$, shift down in energy before they become unbound ("level shifting", "bond softening"). We note that very similar potential curves to those in Fig. 1b were derived for the dc-Stark effect of H_2^+ [16].

The $v = 9$ level plays an outstanding role: it lies in the middle of the gap and is therefore unbound at all intensities, i.e. the dissociation rate should be almost unity for all light intensities in LIP theory. Considering classical FGR theory, one finds the Frank–Condon factor to the continuum maximal here. In our experiment, the peak height of $v = 9$ is almost constant with increasing laser intensities at constant pulse energy. Of course, also the strong saturation effects have to be taken into account here. The constancy of the peak height is effected by the big FC (Frank–Condon) factor in FGR in the classical picture as well as by high dissociation rates in LIP, both leading early to saturation [17]. These arguments are similarly but less valid for the higher levels with $v = 10, 12, 13$.

The angular distributions of the fragments of single vibrational levels can also be studied in this experiment by measuring the intensity along the semi-circular bright traces on the MCP detector. Surprisingly, one finds for all the v-levels in and above the crossing an almost $\cos^2$ distribution, while those below the crossing are very narrow with about a $\cos^{12}$ angular distribution, as already mentioned before. Since in general one only finds a $\cos^2$ distribution when the dissociation rate is linear in the intensity, this finding is consistent with the constancy of the peak heights found for the higher vibrational peaks with growing laser intensity. Here the pulse lengths were compressed and the integral $\int I(t)\,\mathrm{d}t$ thereby kept constant. According to FGR, in this case the dissociation rate per pulse should not change. However, saturation effects could be dominant in this intensity range as mentioned before and are therefore expected to contribute appreciably to the constancy of the peak heights and angular distributions.

Surely the most exciting effect observed in our measurements is the decreasing of the dissociation rate of levels 11, 12 with increasing intensity in Fig. 5. This effect is given a clear explanation inside LIP by the new bonding potential formed by the avoided crossing, where the levels $v = 11, 12$ now can be trapped. With increasing light intensity the trapping disappears again, which is explained by the potential minimum becoming shallower.

3.2 Coulomb Explosion after Photoionization

Very recently, we also investigated corresponding to channel (b) of (1b) called the Coulomb explosion after photoionization. It opens up at about $5 \times 10^{13}\,\mathrm{W/cm^2}$ and becomes the dominant channel when the intensity is increased to $10^{15}\,\mathrm{W/cm^2}$. The experiments here were performed with pulse intensities in that intensity range and with pulse lengths down to 80 fs.

The two channels are easily discernible in Fig. 2, the first one peaking at about 0.15 eV while the second one peaks at about 2.0 eV in a cut through the measured distribution along the polarization direction not shown here. The spatial H^+ distribution of channel (b), which was completely measured here for the first time (Fig. 2) on the MCP-plate differs characteristically from that of channel (a): from 10^{14} to $1.5 \times 10^{15}\,\mathrm{W/cm^2}$ its angular distribution is extremely narrow and can be approximated by a cos18 function at about 2 eV, as compared with about $\cos^{12}$ of the dissociation channel (a) for $v = 6 - 8$. This suggests an alignment of H_2^+ by the laser field along the polarization direction before its fragmentation.

Reducing the laser intensity to about $1.4 \times 10^{14}\,\mathrm{W/cm^2}$, which is just above the threshold for this channel, and at a pulse length of 80 fs produces clear structure in the fragment distribution cut along the polarization direction (Fig. 7a). The structure shows about 7 peaks with three prominent ones at 1.3, 1.7 and 2.2 eV of kinetic energy. It alters very sensitively with laser intensity and ion beam fluctuations. In a series of measurements where the intensity was increased in steps of about $0.2 \times 10^{14}\,\mathrm{W/cm^2}$ while all the other conditions were kept constant, it was found that the structure completely disappears at about $2.3 \times 10^{14}\,\mathrm{W/cm^2}$. It washes out to a broad continuum covering the energy range of the structure seen before having a sharp peak

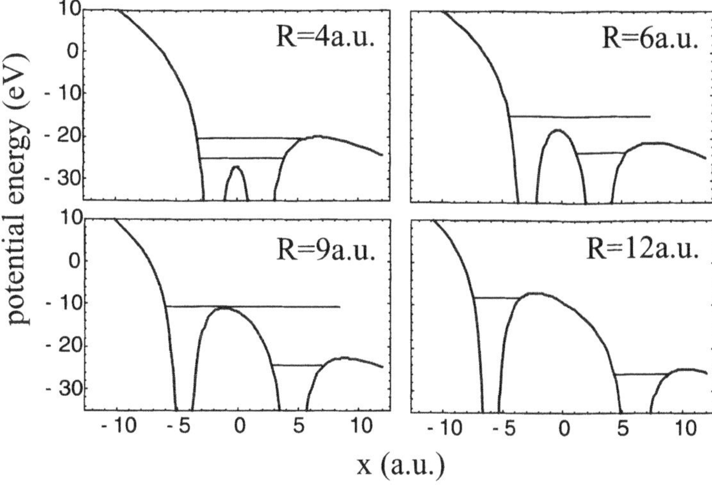

Fig. 6. Potentials seen by the electron of H_2^+ in a high electric field, shown for 4 different proton distances. The horizontal lines illustrate the two orbitals of the electron whose distance is $\boldsymbol{E} \cdot \boldsymbol{R}$. Indeed, the ionization rate is enhanced in this nuclear distance range

at about 1.7 eV. This shifts to about 2 eV when the laser power is increased to 10^{15} W/cm^2.

There are strong arguments for assuming that the fragmentation process here can be approximated by two steps: in the first step the electron is moved from the mid-point between the protons to one or other of the two protons by the charge resonance transition. Then the newly formed H atom and the proton depart from each other with a kinetic energy depending on the vibrational level, with an average energy of about 0.26 eV as we already learned in the dissociation experiment (process (a) of (1a)). The electron in the H atom not only feels the field of the protons but in addition the very strong electric component of the laser field, which tilts the whole electric potential as shown in Fig. 6 where this potential is shown when varying the proton distance R. Indeed, for R between 4 and 12 a.u. which means overstretching, the electron can easily fall out of this sum potential. The field of the protons is then no longer shielded by the electron's charge and they feel the full Coulomb repulsion.

Converting the measured kinetic energy of the protons in Fig. 7a on the x-axis to the protons' distance, using the Coulomb law after subtracting the average dissociation energy (0.26 eV) they gained in the first (dissociation) step, one obtains the plot in Fig. 7b. One now realizes that the repulsion indeed happens between $R = 4$ and 13 a.u., as predicted by the model in Fig. 6. In early model calculations by Zuo and Bandrauk in 1995 [18] using the Schrödinger equation, two critical distances namely 7 and 10 a.u. were

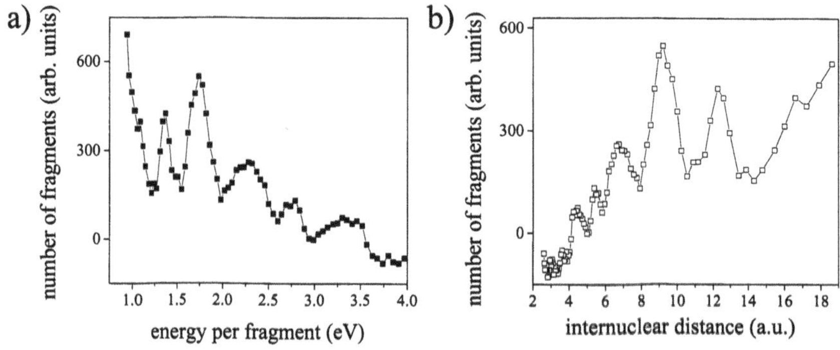

Fig. 7. (**a**) Cut through the momentum distribution of the Coulomb explosion channel along the polarization direction of the laser. Intensity $I = 1.4 \times 10^{14}\,\mathrm{W/cm^2}$, pulse length $\tau = 80\,\mathrm{fs}$. (**b**) Conversion of the abscissa of (**a**) to the internuclear distances R using Coulomb's law. From the kinetic energy of the fragments, the energy these had gained before by dissociation (0.26 eV) was subtracted

found as preferred distances from where the repulsion should take place. There was also a measurement started on H_2 which seemed to confirm this calculation [19]. Our plot in Fig. 7b displays peaks at approximately these distances; however, there are also additional ones.

The same authors as in [18] together with others improved their theoretical investigations by more and more sophisticated calculations until 1998. In fact, the last and most advanced results ("non-Born–Oppenheimer electronic-nuclear wave packet 1D model" at 800 nm) [20] reproduces our measurement in Fig. 7a much better. The bigger peaks are almost correctly reproduced when using here the intensity of approximately $1.3 \times 10^{14}\,\mathrm{W/cm^2}$. Their calculations also show the enormous sensitivity of the peak heights and their R-positions with the laser intensity. It is again surprising that as in the measurements on the first (dissociation) channel, the structure in the experiments is much deeper and clearer than the existing calculations predict.

4 Conclusion

Application of femtosecond laser pulses directly to H_2^+ molecules in an ion beam for the first time reveals a fine structure in the kinetic energy spectrum of the fragments of the dissociation channel as well as in the photoionization–Coulomb explosion channel. This affords a detailed understanding of the physical mechanism of the decay of H_2^+ in the two channels mentioned. For the first channel, all the processes such as bond softening, level shifting, trapping, etc., as predicted by LIP theory become apparent for single vibrational levels, while for the second channel the critical distances (CREI) are realized for the first time.

To explain why the fine structure we found was not observed in the many experiments done before, we point out that by starting the experiment on neutral H_2, as has mostly been done so far [e.g. 20, 21], two or three very intensity dependent processes described above take place here one after the other. Because of the light intensity variation from pulse to pulse this could lead to the structure being wiped out.

Acknowledgements

Dedicated to Theodor W. Hänsch on the occasion of his 60th birthday. We are grateful to Dr. J. Posthumus and Prof. H. Schüssler for their critical comments on the manuscript. H. Figger wants to express his gratitude to Theodor Hänsch for the free and relaxed atmosphere in our group, Theodor's constructive interest in our work, and his generous support over the many years.

References

1. A.D. Bandrauk (Ed.), *Molecules in Laser Fields* (Marcel Dekker, New York 1994)
2. A. Giusti Suzor, F.H. Mies, L.F. DiMauro, E. Charron, B. Yang, Topical Review, Dynamics of H_2^+ in Intense Laser Fields, J. Phys. B **28**, 309 (1995)
3. T.D.G. Walsh, F.A. Ilkow, S.L. Chin, J. Phys. B: At. Mol. Opt. Phys. **30**, 2167 (1997)
4. T.D.G. Walsh, F.A. Ilkow, S.L. Chin, F. Chateauneuf, T.T. Nguyen-Dang, S. Chelkowski, A.D. Bandrauk, O. Atabek, Phys. Rev. A **58**, 3922 (1998)
5. K. Sändig, H. Figger, T.W. Hänsch, Verh. Dtsch. Phys. Ges. 2/98 (1998)
6. K. Sändig, H. Figger, Ch. Wunderlich, T.W. Hänsch, in *Proc. XIV Int. Conf. Laser Spectroscopy, Innsbruck, Austria, 1999*, ed. by R. Blatt, J. Eschner, D. Leibfried, F. Schmidt-Kaler, (World Scientific, Singapore 1999)
7. K. Sändig, H. Figger, T.W. Hänsch, in *ICOMP VIII, Monterey, CA, 1999*, AIP Conf. Proc. **525** (Am. Inst. Phys., New York 1999)
8. K. Sändig, H. Figger, T.W. Hänsch, Phys. Rev. Lett. **85**, 4876 (2000)
9. Ch. Wunderlich, H. Figger, T.W. Hänsch, Chem. Phys. Lett. **256**, 43(1996
10. Ch. Wunderlich, E. Kobler, H. Figger, T.W. Hänsch, Phys. Rev. Lett. **78**, 2333 (1997)
11. Ch. Wunderlich, H. Figger, T.W. Hänsch, Phys. Rev. A **62**, 023401-1(2000)
12. I.D. Williams, P. McKenna, B. Srigengan, M.G. Johnston, W.A. Bryan, J.H. Sanderson, A. el-Zein, T. Goodworth, W.R. Newell, P.F. Taday, A.J. Langley, J.Phys. B., At. Mol. Opt. Phys. **33**, 1(2000)
13. T. Zuo, A.D. Bandrauk, Phys. Rev. A **52**, R2511 (1995)
14. W. Koot, W.J. van der Zande, D.P. de Bruijn, Chem. Phys. **115**, 297 (1987)
15. R. Schinke, *Photodissociation Dynamics* (Cambridge University Press, Cambridge 1993) p. 49

16. A. Saenz, Phys. Rev. A **61**, 051402(R) (1999)
17. F. Rebentrost, private communication
18. T. Zuo, A.D. Bandrauk, Phys. Rev. A **52**, R2511 (1995)
19. G.N. Gibson, M. Li, C. Guo, J. Neira, Phys. Rev. Lett. **79**, 2022 (1997)
20. T.D.G. Walsh, F.A. Ilkow, S.L. Chin, F. Chateauneuf, T.T. Nguyen-Dang, S. Chelkowski, A. Bandrauk, O. Atabek, Phys. Rev. A **58**, 3922 (1998)
21. C. Trump, H. Rottke, M. Wittmann, G. Korn, W. Sandner, M. Lein, V. Engel, Phys. Rev. A **62**, 063402 (2000)

Linear and Nonlinear Raman Spectroscopy of Gases

Heinz W. Schrötter

The results of experimental investigations by spontaneous and Coherent Anti-Stokes Raman Spectroscopy (CARS) in the laboratory of the author since the appointment of Theodor Hänsch to the Chair of Physics at the University of Munich are reviewed.

1 Introduction

When J. Brandmüller, who had held a Chair of Physics at the Sektion Physik der Ludwig–Maximilians–Universität München for about 20 years, announced to the Faculty two years in advance his wish to retire at the age of 65, namely the end of March 1986, activities to fill this position began immediately. In particular H. Walther proposed to try to win Theodor Hänsch from Stanford for this Chair of Physics together with the position of Director at the Max-Planck-Institute of Quantum Optics in Garching. After some discussions the Faculty nominated Theodor Hänsch for the first position on a list of three as Brandmüller's successor, and the Bavarian Ministry of Science and Art invited him for negotiations. In December 1985 he was in Munich and gave a talk at the Colloquium of Munich Physicists. Finally on March 29, 1986, two days before Brandmüller's retirement, he agreed to fill the chair as soon as possible and was appointed on April 30, 1986. To my knowledge this was the shortest vacancy of a full professorship of physics in our faculty.

The field of research in Brandmüller's laboratory had been Raman spectroscopy of molecules and crystals from various aspects [1]; a complete list of references up to 1985 is given there. In the group of the author Raman spectra of gases were investigated; in particular Raman scattering cross-sections were measured [2]. Both linear and non-linear techniques were used, in particular Coherent Anti-Stokes Raman Spectroscopy (CARS). This work was continued after the arrival of Theodor Hänsch and in this article the results obtained in this period are reviewed.

2 Linear Raman Spectroscopy

Spontaneous Raman spectra were excited by cw argon ion lasers and photoelectrically recorded with a Jarrell–Ash dual monochromator equipped with holographic gratings [3]. The high sensitivity of this spectrometer made it possible to detect weak second-order Raman spectra of overtones and combination bands of benzene, hexadeuterobenzene [4], and *sym*-trideuterobenzene [5] in the gaseous phase and to deduce anharmonicity constants. The rovibrational Raman spectrum of dimethylacetylene (but-2-yne) was recorded and depolarization ratios and relative normalized scattering cross-sections were measured [6]. In order to make the high laser power in the cavity of the argon ion ring laser used for CARS spectroscopy [7] (see next section) available also for linear Raman spectroscopy, an arrangement for the transfer of Raman scattered light by fiber optics was built and tested [8]. The scheme of the spectrometer including this arrangement is shown in Fig. 1, it was

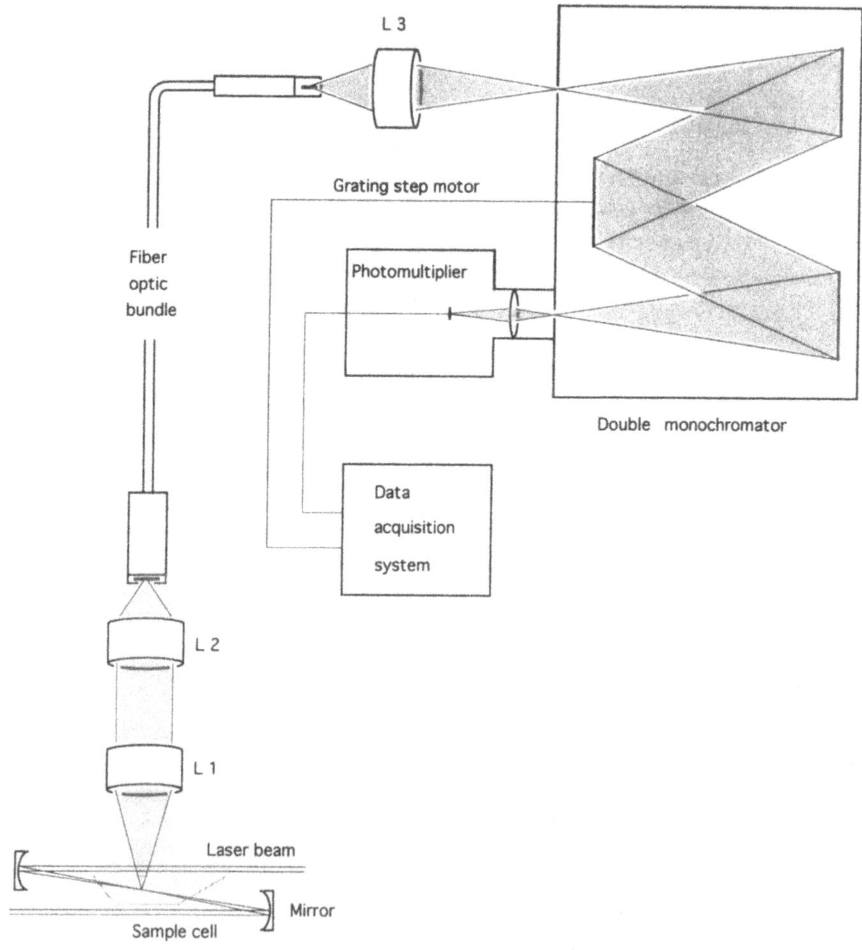

Fig. 1. Schematic arrangement of the experimental setup for linear Raman spectroscopy. L1, L2, L3: lens systems. From [9], with permission by Wiley, Chichester

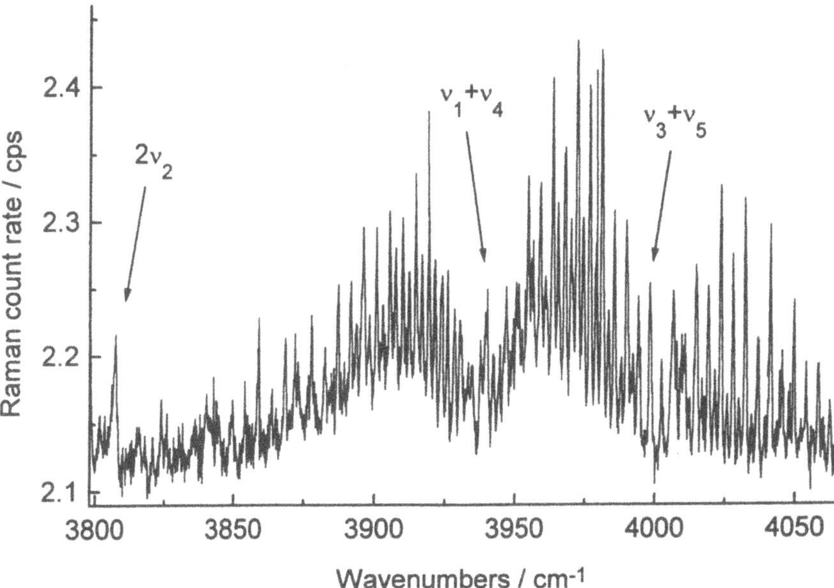

Fig. 2. Rovibrational Raman spectrum of the $2\nu_2, \nu_1 + \nu_4$ and $\nu_3 + \nu_5$ bands of $^{13}C_2H_2$. From [10], with permission by Wiley, Chichester

used for an investigation of the Raman spectrum of toluene vapour [9]. After the experiments in my laboratory in Munich were terminated, the sample arrangement with a multiple reflection system originally constructed by Finsterhölzl et al. [3] was transferred to the European Laboratory for Non-Linear Spectroscopy (LENS) in Florence and served there for recording the vibration-rotation Raman spectrum of $^{13}C_2H_2$ acetylene [10]. In addition to the fundamental bands, overtone and combination bands were detected. Figure 2 shows as an example the combination band $\nu_1 + \nu_4$; in this region also $\nu_3 + \nu_5$ and the overtone $2\nu_2$ can be clearly distinguished, in contrast to the same band system in $^{12}C_2H_2$ [3], where an elaborate analysis was necessary to find the positions of these second-order bands. Further tentative positions for the dyad $\nu_1 + \nu_2$ and $2\nu_2 + 2\nu_5$ of $^{13}C_2H_2$ were found [10], which were confirmed later by high-resolution stimulated Raman spectroscopy from vibrationally excited states by Bermejo et al. [11].

3 Nonlinear Raman Spectroscopy

When the high-resolution capability of nonlinear Raman techniques for vibration-rotation bands of gases became apparent [12, 13], we decided to construct a cw-CARS spectrometer using intracavity excitation [14] in the resonator of an argon ion ring laser [7]. With this set-up vibration-rotation bands of methane CH_4 [15], hydrogen sulfide H_2S [16], ammonia NH_3 [17], and nitrogen N_2 [7] were investigated. Figure 3 shows the last version of the cw-CARS spectrometer including the provisions for frequency stabilization of the lasers and frequency determination [18]. It was used to record CARS spectra of *sym*-trideuterobenzene [5], ethane [18], and various isotopomers of ammonia [19–21]. For $^{14}NH_3$ the experimental spectra of the Q-branch of the ν_1 band [19, 21] could be fitted using pressure broadening coefficients determined from infrared spectra [22], whereas for $^{15}NH_3$ it was necessary to reduce the pressure broadening coefficients in order to obtain satisfactory agreement with the experimental line profiles [19]. A part of the CARS spectrum of the Q-branch of the ν_1 band is shown in Fig. 4. In an attempt to clear up the linewidth effects also the CARS and infrared spectra of the ν_1 bands of $^{14}ND_3$ and $^{15}ND_3$ were recorded and assigned [20]. For the deuterated species the lines are much more closely spaced and therefore the assignment is more difficult than for the other isotopomers. In spite of the recent availability of high-resolution FTIR spectra [23] the agreement between the experimental CARS spectra and those simulated from infrared data is not yet completely satisfactory and calls for further investigations.

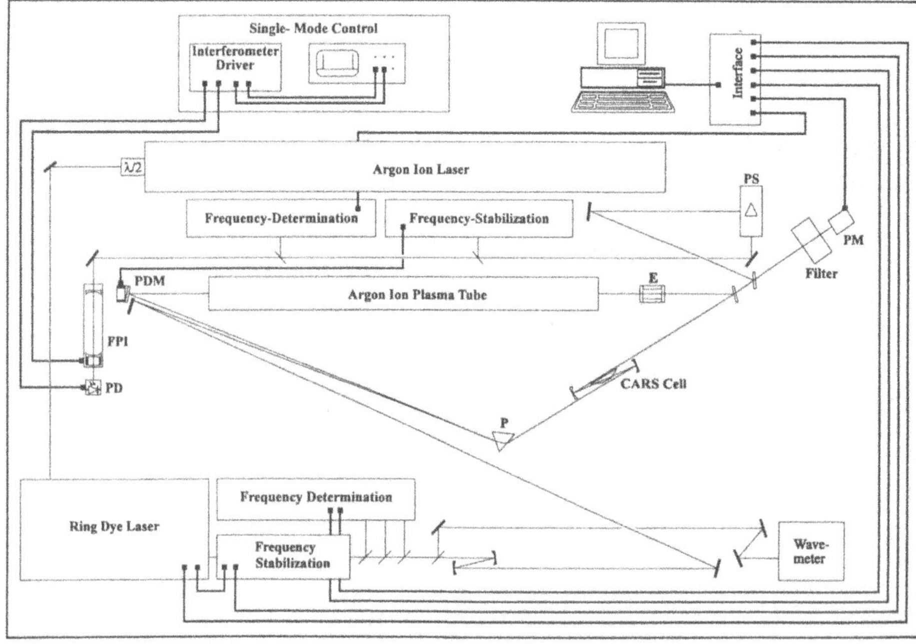

Fig. 3. Schematic arrangement of the high-resolution cw-CARS spectrometer. E = air-spaced etalon; FPI = Fabry–Pérot interferometer; PS = prism; PD = photodiode; PM = photomultiplier; PDM = piezo-mounted mirror. From [18], with permission by Wiley, Chichester

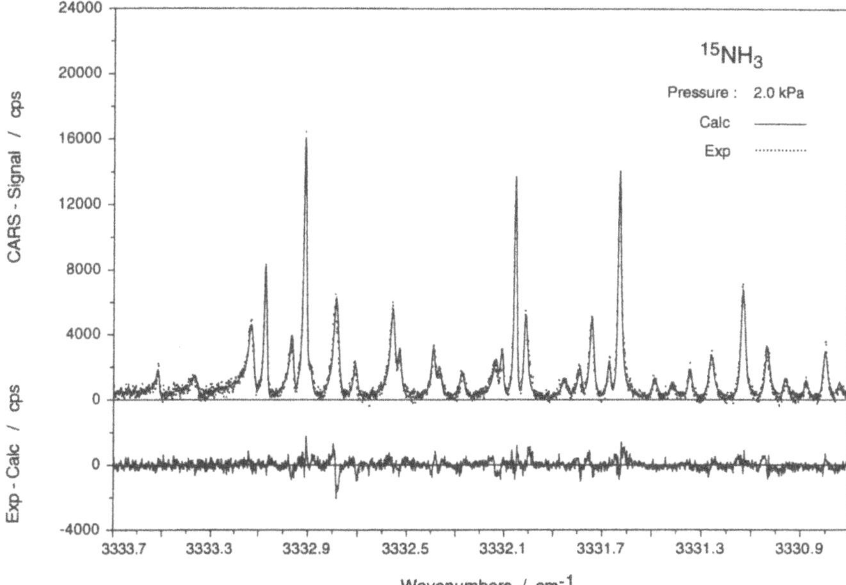

Fig. 4. Comparison of experimental (*dotted line*) and calculated (*solid line*) CARS spectrum of the Q-branch of the ν_1 band of $^{15}NH_3$ at a pressure of 2.0 kPa. *Lower curve*: residuals (exp – calc). From [21], with permission by Wiley, Chichister

4 Conclusion and Acknowledgement

Much of the work reviewed here and also other papers [24–28] and reviews [29–38] were the result of close cooperation with colleagues in Bologna [10], Bratislava [6, 19–21, 28], Calcutta [4], Dijon [15, 24, 26, 29–36], Florence [10], Giessen [20], Madrid [25, 26], Orsay [15, 16], Ottawa [16], Prague [17, 20], Ulm [5], and Zagreb [9, 27]. I thank them for the productive collaboration and Theodor Hänsch for the friendly cooperation in the ten years of my activities at his Chair of Physics.

References

1. I. Dietrich, W. Kiefer, H.W. Schrötter, J. Raman Spectrosc. **17**, 1 (1986)
2. H.W. Schrötter, H.W. Klöckner, Raman Scattering Cross Sections in Gases and Liquids. In: *Raman Spectroscopy of Gases and Liquids*, *Topics Curr. Phys.*, Vol. 11, ed. by A. Weber (Springer, Berlin, Heidelberg 1979) pp. 123-166
3. H. Finsterhölzl, H.W. Schrötter, G. Strey, J. Raman Spectrosc. **11**, 375 (1981)
4. S. Eppinger, A. Rabold, T.N. Misra, S.K. Nandy, H.W. Schrötter, J. Mol. Struct. **266**, 389 (1992)
5. S. Zeindl, K. Sailer, F. Bauer, H.W. Schrötter, K. Ruoff, J. Mol. Struct. **350**, 185 (1995)

6. P. Meßler, H.W. Schrötter, K. Sarka, J. Raman Spectrosc. **25**, 647 (1994)
7. H. Frunder, L. Matziol, H. Finsterhölzl, A. Beckmann, H.W. Schrötter, J. Raman Spectrosc. **17**, 143 (1986)
8. M. Schopp, H.W. Schrötter, N. Douklias, Appl. Spectrosc. **44**, 562 (1990)
9. Th.S. Bican, H.W. Schrötter, V. Mohaček Grošev, J. Raman Spectrosc. **26**, 787 (1995)
10. M. Becucci, E. Castellucci, L. Fusina, G. Di Lonardo, H.W. Schrötter, J. Raman Spectrosc. **29**, 237 (1998)
11. D. Bermejo, R.Z. Martinez, G. DiLonardo, L. Fusina, J. Chem. Phys. **111**, 519 (1999)
12. V.I. Fabelinsky, B.B. Krynetsky, L.A. Kulevsky, V.A. Mishin, A.M. Prokhorov, A.D. Savel'ev, V.V. Smirnov, Opt. Commun. **20**, 389 (1977)
13. A. Owyoung, C.W. Patterson, R.S. McDowell, Chem. Phys. Lett. **59**, 156 (1978); Erratum: Chem. Phys. Lett. **61**, 636 (1979)
14. A. Hirth, K. Vollrath, Opt. Commun. **18**, 213 (1976)
15. H. Frunder, D. Illig, H. Finsterhölzl, H.W. Schrötter, B. Lavorel, G. Roussel, J.C. Hilico, J.P. Champion, G. Pierre, G. Poussigue, E. Pascaud, Chem. Phys. Lett. **100**, 110 (1983)
16. H. Frunder, R. Angstl, D. Illig, H.W. Schrötter, L. Lechuga-Fossat, J.M. Flaud, C. Camy-Peyret, W.F. Murphy, Can. J. Phys. **63**, 1189 (1985)
17. R. Angstl, H. Finsterhölzl, H. Frunder, D. Illig, D. Papoušek, P. Pracna, K.N. Rao, H.W. Schrötter, Š. Urban, J. Mol. Spectrosc. **114**, 454 (1985)
18. T.S. Bican, J. Jonuscheit, U. Lehner, H.W. Schrötter, J. Raman Spectrosc. **26**, 699 (1995)
19. J. Jonuscheit, U. Lehner, D. Illig, S. Anders, K. Sarka, H.W. Schrötter, J. Mol. Struct. **349**, 389 (1995)
20. M. Stamova, S. Anders, J. Jonuscheit, H.W. Schrötter, P. Pracna, Š. Urban, S. Klee, M. Winnewisser, K. Sarka, J. Mol. Struct. **482–483**, 481 (1999)
21. S. Anders, J. Jonuscheit, U. Lehner, K. Sarka, H.W. Schrötter, J. Raman Spectrosc. **31**, 711 (2000)
22. V.N. Markov, A.S. Pine, G. Buffa, O. Tarrini, J. Quantum Spectrosc. Radiat. Transfer **50**, 167 (1993)
23. M. Snels, L. Fusina, H. Hollenstein, M. Quack, Mol. Phys. **98**, 837 (2000)
24. Y. Ouazzany, J.P. Boquillon, H.W. Schrötter, Mol. Phys. **63**, 769 (1988)
25. F. Liebrecht, H. Finsterhölzl, H.W. Schrötter, S. Montero, Indian J. Pure Appl. Phys. **26**, 51 (1988)
26. J. Plíva, M. Terki-Hasseine, B. Lavorel, R. Saint-Loup, J. Santos, H.W. Schrötter, H. Berger, J. Mol. Spectrosc. **133**, 157 (1989)
27. V. Mohaček Grošev, H.W. Schrötter, J. Jonuscheit, J. Raman Spectrosc. **26**, 137 (1995)
28. K. Sarka, H.W. Schrötter, J. Mol. Spectrosc. **179**, 195 (1996)
29. H.W. Schrötter, H. Berger, B. Lavorel, J. Mol. Struct. **141**, 195 (1986)
30. H.W. Schrötter, B. Lavorel, Pure Appl. Chem. **59**, 1301 (1987)
31. H.W. Schrötter, H. Frunder, H. Berger, J.P. Boquillon, B. Lavorel, G. Millot, High Resolution CARS and Inverse Raman Spectroscopy. In: *Advances in Nonlinear Spectroscopy*, Vol. 15, ed. by R.J.H. Clark, R.E. Hester (Wiley, Chichester 1988) pp. 97–147
32. H.W. Schrötter, H. Berger, J.P. Boquillon, B. Lavorel, G. Millot, Croatica Chem. Acta **61**, 487 (1988)

33. H.W. Schrötter, J.P. Boquillon, State-of the Art on High-Resolution CARS. In: *Raman Spectroscopy: Sixty Years On, Vibrational Spectra and Structure*, Vol. 17B, ed. by H.D. Bist, J.R. Durig, J.F. Sullivan (Elsevier, Amsterdam 1989) pp. 1–14
34. H.W. Schrötter, J.P. Boquillon, High-Resolution Coherent Anti-Stokes Raman Spectroscopy of Small Molecules. In: *Recent Trends in Raman Spectroscopy*, ed. by S.B. Banerjee, S.S. Jha (World Scientific, Singapore 1989) pp. 113–126
35. H.W. Schrötter, H. Berger, J.P. Boquillon, B. Lavorel, G. Millot, J. Raman Spectrosc. **21**, 781 (1990)
36. B. Lavorel, G. Millot, M. Rotger, G. Rouillé, H. Berger, H.W. Schrötter, J. Mol. Struct. **273**, 49 (1992)
37. H.M. Heise, H.W. Schrötter, Rotation-Vibration Spectra of Gases. In: *Infrared and Raman Spectroscopy, Methods and Applications*, ed. by B. Schrader (VCH, Weinheim 1995) pp. 253–297, Chap. 4.3
38. H.W. Schrötter, Raman Spectra of Gases. In: *Handbook of Raman Spectroscopy*, ed. by I. Lewis, H.G.M. Edwards (Marcel Dekker, New York 2001) pp. 307–350, Chap. 8

Nonlinear Properties of Laser-Generated Giant Surface Acoustic Wave Pulses in Solid Materials

Hans A. Schüssler and Alexandre A. Kolomenskii

The nonlinear evolution (compression and stretching) of the profile of very high-amplitude surface acoustic wave (SAW) pulses with acoustic Mach numbers M$\sim$0.01 and the formation of shock fronts were studied in solid materials. The excitation of such giant SAW pulses was performed with nanosecond laser pulses through a strongly absorbing layer and used for the determination of the nonlinear acoustic constants. The mechanical forces associated with the large surface acceleration in high-amplitude SAWs detach particles from the surface. For nanosecond SAW pulses the limit of the surface acceleration of about 10^{10} m/s^2 is set by the fracture of the material and corresponds to the removal of all particles larger than about 0.1 µm. This technique can also be used for the determination of the Hamaker constant of the adhesion force.

1 Introduction

Surface acoustic waves (SAWs) propagate along the surface of a solid and produce periodic strains of the material, decaying exponentially from the surface inward to the medium. They can exhibit large energy densities due to their confinement to the narrow subsurface region, which is of the order of one wavelength in depth. When strongly excited, SAWs will drive the medium into the nonlinear elastic regime. Therefore, as an intense SAW propagates the evolution of the wave shape in time allows us to glean information on the higher-order nonlinear acoustic parameters, and in this way the nonlinear elastic constants of the surface material. In the last few years such studies [1, 2] have emerged mainly by employing efficient ways to couple the energy of the laser pulse to the surface by optimizing the absorption of pulsed laser radiation and by simultaneously reducing other effects such as reflection, optical breakdown, and ablation. Other SAW excitation mechanisms [3] such as by transducers or particle beams exist, but do not easily lend themselves to nonlinear wave generation.

There is a wide range of interest in SAWs, spanning from, on the laboratory scale, novel methods for elastic materials characterization to, on a global scale, contributing in seismology to earthquake research. SAWs have also found applications in signal-processing devices. In addition, a novel surface-cleaning technique [4] applicable to microcircuits makes use of the large accelerations of SAWs to shake off small particles from silicon wafers. Finally, crystalline materials are anisotropic and the propagation of SAWs in them leads to such important effects as phonon focusing [5].

2 Characterization of Solid Materials

2.1 Experimental Arrangement

The SAWs of this experiment were produced dominantly by absorption in a thin covering layer and resulted in surface pressures of up to several GPa. Therefore, we labeled the resulting waves as giant SAWs. To minimize geometrical SAW diffraction, we focused the laser pulse onto the surface in a narrow line about 10 µm in width and 8 µm in length. In the experiment a Q-switched Nd:YAG laser with a variable pulse energy of up to 100 mJ, 26 ns pulse duration, and wavelength of 1.06 µm was used for excitation. For detection a dual-probe-beam technique was employed, to enable us to observe nonlinear transformations in a single laser-excited SAW pulse.

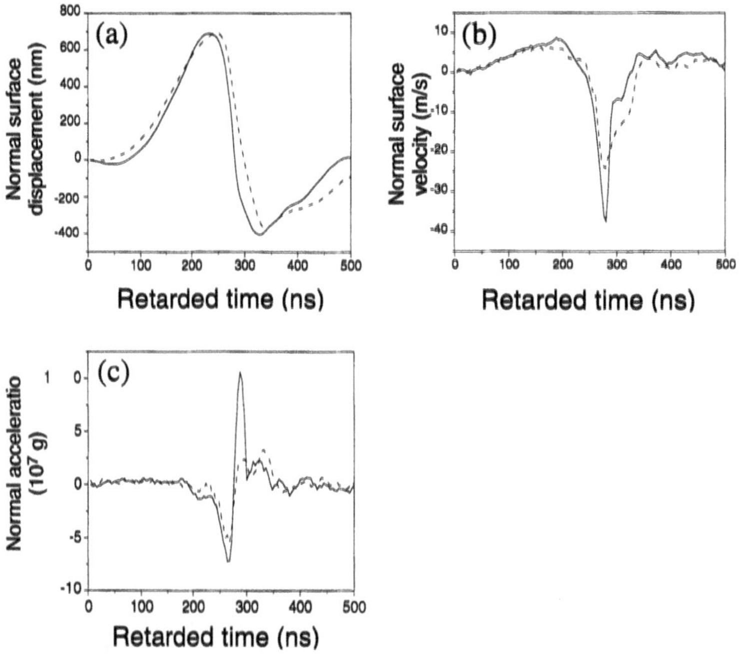

Fig. 1. SAW waveforms in aluminum for two propagation distances $x_1 = 13.8$ mm (*dashed line*) and $x_2 = 26.4$ mm (*solid line*): (**a**) normal surface displacement, (**b**) normal surface velocity, (**c**) normal surface acceleration

2.2 Measurements

The detailed registration of the nonlinear changes in the surface particle velocities of aluminum, copper, and fused silicon is the main result of our observations. As an example we show in Fig. 1 the time development of nonlinear compression in a SAW pulse for polycrystalline aluminum [6]. The corresponding frequency spectrum (Fig. 2) of the vertical particle velocity becomes also richer in higher harmonics, demonstrating frequency up-conversion. A different behavior, namely nonlinear wave stretching and additional down-conversion of the frequency resulting in two sharp peaks in the vertical surface velocity and two corresponding shock fronts for the in-plane surface velocity, was observed for fused silica.

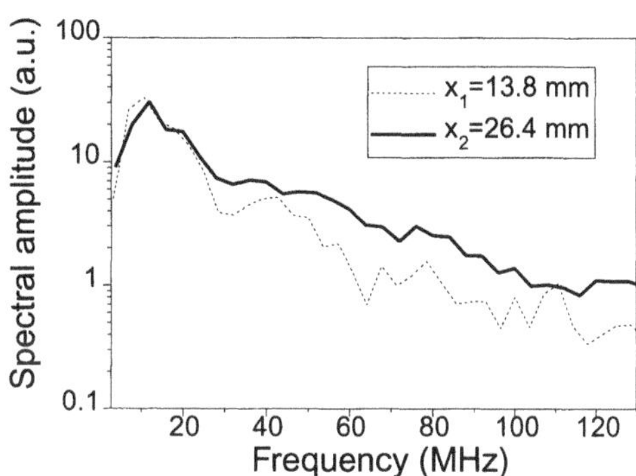

Fig. 2. Frequency spectra of the normal surface velocity for the SAW pulse in aluminum at two propagation distances

2.3 Results

A nonlinear evolution equation [7] was used for evaluation of the experimental results. In order to describe the nonlinear propagation of a SAW in an isotropic solid three nonlinear acoustic parameters must be introduced: one parameter ϵ_1 of the local nonlinearity, and two parameters ϵ_2 and ϵ_3 of the

Table 1. Compilation of nonlinear acoustic and elastic parameters

	Acoustic constants (evaluated from our experiments)			Elastic constants (from [8] in GPa)		
	ϵ_1	ϵ_2	ϵ_3	C_{111}	C_{144}	C_{456}
Al	0.8(1)	−1.02(2)	3.2(8)	−1634	−125	− 85
Cu	1.0(1)	−0.98(2)	4.1(8)	−2745	172	−390
SiO_2	−0.8(1)	−0.25(1)	−4.0(5)	648	54	13.2

nonlocal nonlinearity. The evaluation yielded the nonlinear acoustic parameters compiled in Table 1. The dominant contribution to the time evolution of the SAW waveform is due to the main nonlinear acoustic parameter ϵ_1, which is positive for pulse compression and negative for stretching.

3 Removal and Adhesion of Fine Particles

The detachment of particles from a surface is the harder the smaller they are. The large accelerations (Fig. 1) of nonlinear SAW provide a unique opportunity to study adhesion for micron- and submicron-sized particles. The adhesion force created by the Van der Waals interaction is given by $F_{\mathrm{ad}} = A_{123}\, d/12z^2$ where A_{123} is the Hamaker constant with the indices (1) for particle, (2) for solid, and (3) for medium, d is the particle diameter, and z is the separation distance. Table 2 lists typical accelerations necessary to remove fine particles. The detachment of a particle from the surface occurs when the normal SAW acceleration is directed into the surface providing a detaching inertial force counteracting the adhesion force. We removed particles down to about 0.1 µm. Figure 3 shows the surface of a silicon wafer which after dusting with fine Al_2O_2 particles was exposed to about 50 laser pulses. The photograph demonstrates the removal of practically all dust articles in the area of SAW propagation of almost 1 cm in diameter. This shows the nonlocal character of the SAW cleaning techniques, considering that the

Table 2. Parameters of interest for the removal of fine particles

Size (µm)	Mass (g)	F_{ad} (dyne)	Acceleration (g=9.81m/s^2)
10	2×10^{-9}	9×10^{-2}	$4.5 \times 10^{4}\ g$
1	2×10^{-12}	9×10^{-3}	$4.5 \times 10^{6}\ g$
0.1	2×10^{-15}	9×10^{-4}	$4.5 \times 10^{8}\ g$

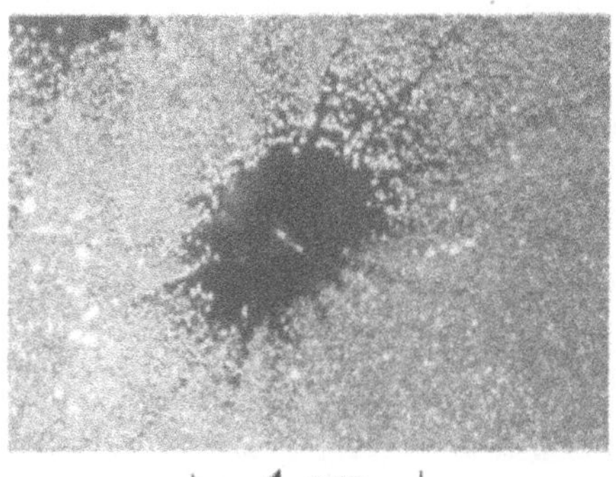

Fig. 3. Photograph of a dusted silicon surface with the dust particles removed in the central section by laser-generated SAW pulses exhibiting also phonon-focusing directions

focal spot of the laser beam was only about 10 μm. The dark rays are due to phonon focusing, since Si is an anisotropic material. In these directions, which are characteristic of a particular surface orientation, the SAW amplitude decreases much more slowly than in other directions and therefore particles are also removed along them.

4 Summary

In conclusion, both nonlinear pulse compression and stretching have been observed for giant SAWs propating in different materials. Such laser excited broadband giant SAWs produce in this nonlinear interaction process not only higher but also subharmonic acoustic frequencies. In addition, the high surface accelerations can be used to clean surfaces of particles down to about 0.1 μm in diameter and thereby study adhesion forces. On the applied side a novel nonlocal surface-cleaning technique for microcircuits resulted from our work, in which SAWs shake off the particles residing on the surface.

Acknowledgements

This work is supported by NSFgrants 9870143 and 9970241. The paper should be considered a late contribution to our interaction at the Mechanics Laboratory of the University of Heidelberg almost 40 years ago. Ever since, Theodor Hänsch's work has been a constant source of inspiration for me (H.S.).

References

1. A.A. Kolomenskii, A.A. Lomonosov, R. Kuschnereit, P. Hess, V.E. Gusev, Phys. Rev. Lett. **79**, 1325 (1997)
2. A.A. Kolomenskii, H.A. Schuessler, Phys. Rev. B **63**, 85413 (2001)
3. A.A. Maradudin, in *Nonequilibrium Phonon Dynamics*, ed. by W. E. Bron (Plenum, New York 1985) p. 395
4. A.A. Kolomenskii, H.A. Schuessler, V.G. Mikhalevich, A.A. Maznev, J. Appl. Phys. **84**, 2404 (1998)
5. A.A. Kolomenskii, A.A. Maznev, Phys. Rev. B **48**, 14502 (1993)
6. A.A.Kolomenskii, H.A. Schuessler, Phys. Lett. A **157**, 280 (2001)
7. V.E. Gusev, W. Lauriks, J. Thoen, Phys. Rev. B **55**, 9344 (1997)
8. *Landolt-Boernstein: Numerical Data and Functional Relationships in Science and Technology*, New Series, Group 3, Vol. **17** (Springer, Heidelberg 1984)

Part VI

Quantum Engineering

Radiative Control and Quantum Engineering: Single Atom Wants to Meet Single Photon

Dieter Meschede and Victor Gomer

1 Introduction

"An individual atomic particle kept at rest in free space for extended periods, is an ideal object for high resolution spectroscopy". This goal, the spectroscopist's dream as it was coined by Dehmelt [1], was first achieved with isolated barium ions in 1980 [2], establishing a striking contrast to the interpretations by Mach [3] and Schrödinger [4] that we will never experiment with single atoms. Today we are convinced that we will be able to fully control atoms as well as photons at their most fundamental level, i.e. at the level of single microscopic quantum objects, and useful devices may be constructed within the realm of emerging quantum technologies.

The most important prerequisite for this development is the exploration of new experimental techniques for more and more refined state preparations of fields and particles. During the 1970s, tunable laser radiation has experienced dramatic technological advances and has played a central role for experimental progress in studies on light–matter interactions: Coherent, narrowband excitation and fluorescence in combination with single photon detection provide an extremely sensitive system for the observation of microscopic particles with suitable resonance lines (see Fig. 1). Furthermore laser cooling [5, 7] which was suggested 25 years ago by Hänsch and Schawlow and independently by Wineland and Dehmelt [8] has become an indispensable tool to control all atomic degrees of freedom through light–matter interaction.

After more than 20 years of experimental work physicists have become used to routinely preparing numerous atomic particles in isolated and well controlled situations, not only with ions but also with neutral atoms. Initially the main incentive for experiments with individual particles was to remove systematic, in particular motional distortions, from precision measurements such as the first and second order Doppler effect prevailing in gaseous samples or limitations due to transit time. Motional atomic effects are fully suppressed in the so-called Lamb–Dicke regime where atoms are confined by external forces to a volume small compared to the optical wavelength. Here, at the optical Mößbauer, regime photons are absorbed and emitted without recoil.

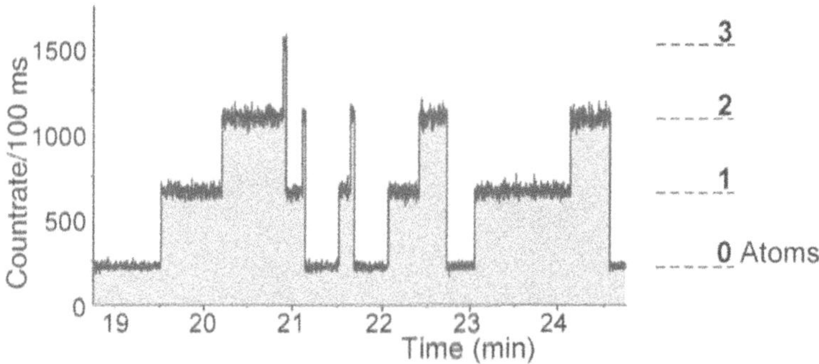

Fig. 1. Time chart of the fluorescence of cesiums atoms stored in a magneto-optical trap [5]. Each level corresponds to a small integer number of atoms which scatter up to 10^5 photons per second and atom into the photon counter. Atoms enter and leave the trap at random times. In this time chart five joint departures of two atoms are observed. They are caused by intratrap collisons which convert internal atomic energy into kinetic energy releasing the atoms from the trap [6]

2 Quantum Properties of Single Atom Radiation

Already classical radiation theory [9] gives hints that even a single, highly localized atom or ion interacting with a laser beam focused to the diffraction limit may act as an optically dense absorber due to the large resonant absorption cross-section $\sigma = 3\lambda^2/2\pi$, where λ is the resonant wavelength. Absorption has indeed been detected with a single trapped ion only [10]. It was found [11], however, that at this microscopic level interference of the atomic radiation field with the driving field needs to be taken into account. It effectively reduces the cross-section and inhibits straightforward application of a single atom in free space as an efficient quantum absorber, or switch for a weak travelling light field.

Numerous experiments have in recent times been carried out illustrating the quantum nature of light–matter interaction at the level of a single atom, most frequently with ions in traps but also with extremely dilute beams of neutral atoms [7].

Spontaneous emission is perhaps the simplest and most elementary process of light–matter interaction [12]. It involves one initially excited atom and a single photon, and it has played an important role in the early development of quantum electrodynamics [13].

The dynamics of spontaneous emission is manifestly dominated by quantum properties as was demonstrated by classic experiments exhibiting so-called "photon antibunching" [14, 15]: Fluctuations of the atomic radiation field are characterized through average intensity–intensity, or photon–photon correlation measurements $g^{(2)} = \langle I(t)I(t+\tau)\rangle/\langle I(t)\rangle^2$, where $I(t)$ is the instaneous intensity. While any classical fluctuating light source must have $g^{(2)}(0) \geq 1$, the radiation field from a single atom shows "antibunching" with

$g^{(2)}(0) = 0$ [16]. The "click" of the detector at photon arrival not only causes the photon to be detected, it also jointly projects the radiating atom into its ground state since the two quantum states involved are product states $|e\rangle|0\rangle$ and $|g\rangle|1\rangle$ where $|e\rangle, |g\rangle$ are the excited and ground atomic states and $|n\rangle$ is the photon number state of its radiation field. Following this projection, or measurement, atomic dynamics is governed by transient Rabi oscillations towards the equilibrium value of excitation.

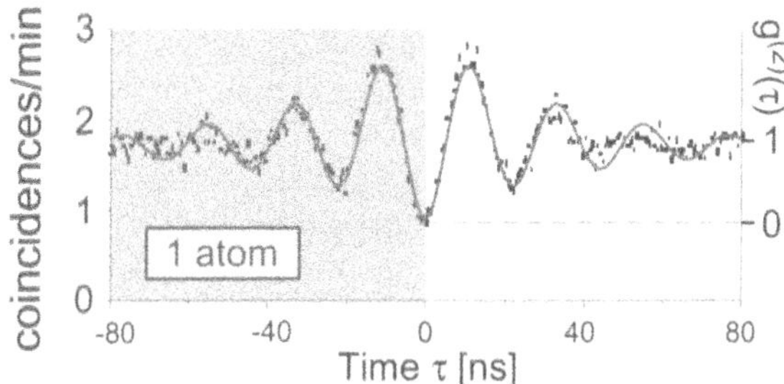

Fig. 2. Photon antibunching observed in the photon–photon correlation measurement ($g^{(2)}(\tau)$) of a single cesium atom trapped in a magneto-optical trap. A two-channel Hanbury-Brown and Twiss setup [16] was used for this measurement. The channels do not distinguish which photon was emitted first, hence data are symmetric with respect to $\tau = 0$. The nonzero coincidence rate at $\tau = 0$ is due to the straylight background. At $\tau = 0$ the atom is projected to the ground state. It relaxes through transient Rabi oscillations towards the equilibrium excitation value. All classcial light fields have $g^{(2)}(\tau) \geq 1$.
The graph shows raw data. The *solid line* represents a theory which is valid for a two-level system yielding surprisingly good agreement for this multilevel atom [15]

With a single atom three research groups successfully and simultaneously demonstrated in 1986 the appearance of the proverbial "quantum jumps" [17]: Quantum jumps are observed when in a V-shaped level fluorescence from a strong transition sharing the ground state with a weak transition is used to monitor whether the atom is in its ground state (fluorescence on) or in the long-lived excited state (fluorescence off) of the weak transition. In this method, which is called "electron shelving" [1], continuous observation ("measurement") of the strong transition projects the system onto the eigenstates of the measurement apparatus; the transition occurs instantaneously.

3 Manipulating Atomic Radiation

An atom radiating in free space is coupled to a continuum of field modes causing the dissipative nature of spontaneous emission. It was already recognized by Purcell and later by Kleppner [18], however, that spontaneous emission is not an invariable law of nature but can be controlled through a suitably shaped environment. Experimenters have succeeded in suppressing [19] and enhancing [20] spontaneous emission by confining the electromagnetic fields within reflecting walls and thus tailoring the spectrum of the electromagnetic field interacting with the atom.

A qualitatively new situation is obtained in cavity-QED when the photon emitted by an atom into the radiation field of a cavity is stored long enough to be reabsorbed by the same atom. In this case which is called the "strong coupling regime" the atom–field system can no longer be separated, its internal coupling exceeds any interaction with the environment. Such a concept is shown in Fig. 3, where on the left side the quantum states of light and matter are shown, and on the right side the quantum states of the combined atom–field system. Phenomena involving combined coherent states of light and matter are also known in other fields, including for instance polaritons in condensed matter or more recently the so-called "storage of light" in dilute atomic vapors [21].

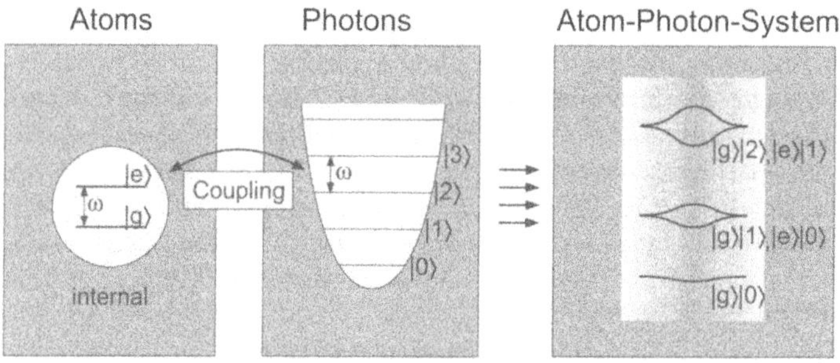

Fig. 3. Conceptual representation of atoms and photons interacting in an optical high finesse cavity. *Left side:* Atomic and field quantum states. *Right side:* Coupled quantum states in the strong coupling regime. The shaded area indicates the intensity distribution of the cavity field mode

In cavity-QED this situation is realized with single atoms and single photons only, i.e. at the most microscopic level. In a device now called a "micromaser" it was first realized with microwaves [22], then at optical frequencies [23].

Much progress has been obtained in controlling the photon state of such high-Q cavities and their interaction with atoms passing by. For instance, in the microwave region Fock states of the cavity field [24] and engineering of an entangled three-particle state involving atoms and photons were demonstrated [25]. Furthermore trapping of a single atom by the cavity field of a single optical photon was recently reported [26] and evidence for the controlled emission of single photons from atoms traversing the cavity has been achieved [27].

In all cavity–QED experiments carried out to date atoms were inserted into the electromagnetic field at random. Single-atom experiments were meant to reduce the flux of atoms drawn from atomic beams or atomic fountains until two-atom events became negligible because of Poissonian statistics. As a consequence experimental investigations of the quantum regime of light–matter interactions at optical frequencies in past decades concentrated on single–particle phenomena, and single trapped ions have dominated such experiments.

It appears almost natural that more recently researchers are focusing on many particle systems ruled by quantum interactions. Using constituents which are extremely well understood – atoms and photons – they have begun to construct several types of many particle systems. The generation of coherent matter waves [28] has become a celebrated example of many particle quantum engineering. Creation of multiparticle entanglement of microscopic particles poses yet another challenge for quantum engineers [29]. Controlled neutral atoms and photons meeting in high-Q cavities are good candidates for this objective.

4 An Optical Tweezer for Atom Delivery

In order to fully marry atoms and photons it has become necessary now to transport atoms in a controlled way into the field mode of an optical resonator. Alternatively it is of course possible to prepare atoms *in situ*, i.e. immediately within the cavity field mode. Separation of the preparation and application of atoms seems to be experimentally advantageous, however.

An array of optical dipole traps [30] can be constructed from counter-propagating laser beams (see Fig. 4). This can be used as a variant of an optical tweezer [31] for single atoms. If the trapping laser beam is detuned far from atomic resonance it provides a nearly conservative potential, and focusing causes very tight confinement of atoms not only along the direction of propagation but also in the tranverse direction [32–34]. It came indeed as a comforting surprise that atoms which had been prepared in exactly known numbers could be transferred from the MOT without loss to the dipole trap [32, 33].

A small detuning of one of the counterpropagating laser beams providing the dipole traps set the standing wave into motion and allows one to transport atoms over macroscopic distances. It was already shown that more than a centimeter can be achieved with again very high efficiency (Fig. 4b) [33]. This distance offers comfortable dimensions to implement a deterministic delivery system for cold atoms to a small volume cavity. We may soon see more steps of quantum engineering towards the creation of multiatom entanglement in cavity–QED systems [35] being taken.

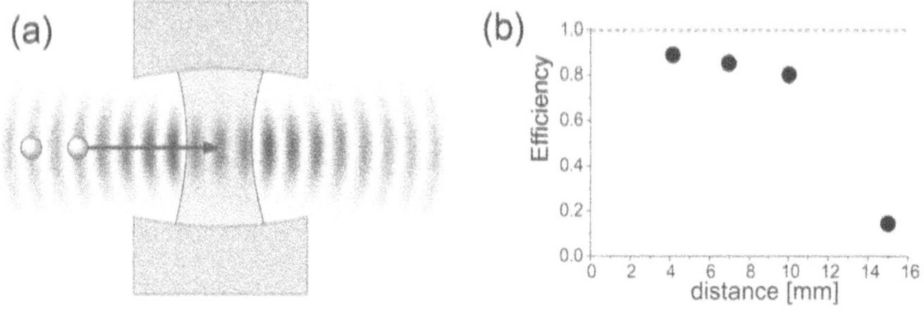

Fig. 4. (**a**) Cesium atoms are trapped at the antinodes of a standing wave neodym laser light field ($\lambda = 1.06\,\mu$m). For small detuning between the two counterpropagating beams this linear chain of dipole traps travels and acts as an "optical conveyor belt" for single atoms. A related concept to transport atoms was realized in the group of Hänsch [36] with a magnetic conveyor belt near a surface with microelectronic circuitry. The optical conveyor belt is a good candidate to deterministically insert a predetermined number of atoms into a high finesse resonator providing interaction with single photons. First experiments have shown (**b**) that the atoms can be transported with high efficiency over many millimeters. Due to laser beam divergence the trapping force declines along the beam. Near 15 mm the trapping forces are overcome by gravity [37]

Acknowledgements

Single-atom experiments derive their fascination from the thrill of manipulating microscopic, quantum objects. Experiments of this kind enjoy a very close relationship with precision measurements. Thus the playful and ingenious tools of laser spectroscopy invented by Theodor Hänsch have laid the groundwork for such experiments, too. From 1988 to 1990 one of us (DM) had the opportunity and lasting pleasure to experience and profit not only from those tools but also from Theodor Hänsch's deep appreciation of physical concepts and phenomena. He is not only a master of precision measurements but also of precision questioning. His creativity continues to benefit experimenters around the world much beyond his immediate scientific interests.

We cannot finish this article without further expressing our gratitude to the students and coworkers who have shared our enthusiasm for the quantum world over many years and helped advance the art of engineering neutral atom objects: Kai Dästner, Daniel Frese, Dietmar Haubrich, Stefan Kuhr, Svenja Knappe, Martin Müller, Arno Rauschenbeutel, Harald Schadwinkel, Dominik Schrader, Frank Strauch, Bernd Ueberholz, and Robert Wynands.

Also, we wish to acknowledge continued financial support by the Deutsche Forschungsgemeinschaft and by the state of Northrhine-Westfalia.

References

1. H. Dehmelt, Stored Ion Spectroscopy in *Advances in Laser Spectroscopy*, F.T. Arecchi, F. Strumia, H. Walther (eds.) (Plenum, New York 1983)
2. W. Neuhauser, M. Hohenstatt, P. Toschek, H. Dehmelt, Phys.Rev. A **22**, 1137 (1980)
3. E. Mach, *Mechanik in ihrer Entwicklung* (Wiss. Buchges., Leipzig 1912)
4. E. Schrödinger, Br. J. Philos. Sci. **3**, 109 (1952)
5. H. Metcalf, P. van der Straaten, *Laser cooling and Trapping*, (Springer, Berlin, Heidelberg 1999)
6. B. Ueberholz, S. Kuhr, D. Frese, D. Meschede, V. Gomer, J. Phys. B**33**, L135 (2000)
7. T.W. Hänsch, H. Walther, Rev. Mod. Phys. **71**, 242 (1999)
8. T.W. Hänsch, A. Schawlow, Opt. Commun. **13**, 68 (1975); D. Wineland, H. Dehmelt, Bull. Am. Phys. Soc. **20**, 637 (1975)
9. J.D. Jackson, *Classical Electrodynamics* (Wiley, New York 1975)
10. D. Wineland, W. Itano, J.C. Bergquist, Opt. Lett. **12**, 389 (1987)
11. S.J. van Enk, H.J. Kimble, Phys. Rev. A **61**, 051802-1 (2000)
12. D. Meschede, Phys. Rep. **211**, 210 (1992)
13. V. Weisskopf, E. Wigner, Z.Phys. **63**, 54 (1930)
14. H.J. Kimble, M. Dagenais, L. Mandel, Phys. Rev. Lett. **39**, 691 (1978); F.M. Rateike, G. Leuchs, H. Walther, In: *Dissipative Systems in Quantum Optics*, R. Bonifacio (ed.) (Springer, Berlin, Heidelberg 1982); F. Diedrich, H. Walther, Phys. Rev. Lett. **58**, 203 (1987)
15. V. Gomer, F. Strauch, B. Ueberholz, S. Knappe, D. Meschede, Phys. Rev. A **58**, R1657, (1998); V. Gomer, F. Strauch, B. Ueberholz, S. Knappe, D. Frese, D. Meschede, Appl. Phys. B **67**, 689 (1998)
16. R. Loudon, *The Quantum Theory of Light* (Clarendon, London 1983); D.F. Walls, *Quantum Optics* (Springer, Berlin, Heidelberg 1994)
17. W. Nagourney, J. Sandberg, H. Dehmelt, Phys. Rev. Lett. **56**, 2797 (1986); T. Sauter, W. Neuhauser, R. Blatt, P. Toschek, Phys. Rev. Lett. **57**, 1696 (1986); J.C. Bergquist, R.G. Hulet, W.M. Itano, D. Wineland, Phys. Rev. Lett. **57**, 1699 (1986)
18. E. Purcell, Phys. Rev. **69**, 681 (1946); D. Kleppner, Phys. Rev. Lett. **47**, 233 (1980)
19. R.G. Hulet, E. Hilfer, D. Kleppner, Phys. Rev. Lett. **55**, 2137 (1985); W. Jhe, A. Anderson, E. Hinds, D. Meschede, L. Moi, S. Haroche, Phys. Rev. Lett. **58**, 666 (1987)
20. P. Goy, J.M. Raimond, M. Gross, S. Haroche, Phys. Rev. Lett. **50**, 1903 (1983); G. Gabrielse, H. Dehmelt, Phys. Rev. Lett. **55**, 67 (1985)
21. M. Fleischhauer, M. Lukin, Phys. Rev. Lett. **84**, 5094 (2000); D.F. Phillips, A. Fleischhauer, A. Mair, R.L.L. Walsworth, Phys. Rev. Lett. **86**, 783 (2001); C. Liu, Z. Dutton, C.H. Behroozi, L.V. Hau, Nature **409**, 490 (2001)
22. D. Meschede, H. Walther, G. Müller, Phys. Rev. Lett. **54**, 551 (1985)
23. R.J. Thompson, G. Rempe, H.J. Kimble, Phys. Rev. Lett. **68**, 1132 (1992); H.J. Kimble, 'Structure and Dynamics in Cavity Quantum Electrodynamics.' In: *Cavity Quantum Electrodynamics*, P.R. Berman (ed.) (Academic, Boston 1994)
24. B.T.H. Varcoe, S. Brattke, M. Weidinger, H. Walther, Nature **403**, 743 (2000)

25. A. Rauschenbeutel, G. Nogues, G. Osnaghi, P. Bertet, M. Brune, J.M. Raimond, S. Haroche, Science **288**, 2024 (2000)
26. C.J. Hood, T.W. Lynn, A.C. Doherty, A.S. Parkins, H.J. Kimble, Science **287**, 1447 (2000); P.W.H. Pnkse, T. Fischer, P. Maunz, G. Rempe, Nature **404**, 365 (2000);
27. M. Hennrich, T. Legero, A. Kuhn, G. Rempe, Phys. Rev. Lett. **85**, 4872 (2000)
28. I. Bloch, T.W. Hänsch, T. Esslinger, Phys. Rev. Lett. **82**, 3008 (1999)
29. C.A. Sackett, D. Kielpinski, B.E. King, C. Langer, V. Meyer, C.J. Myatt, M. Rowe, Q.A. Turchette, W.M. Itano et al., Nature **404**, 256 (2000)
30. R. Grimm, M. Weidemüller, Y.B. Ovchinnikov, Adv. At. Mol. Opt. Phys. **42**, 95 (2000)
31. A. Ashkin, Phys. Rev. Lett. **24**, 156 (1970); S. Chu, J.E. Bjorkholm, A. Ashkin, A. Cable, Phys. Rev. Lett. **57**, 314 (1986)
32. D. Frese, S. Kuhr, W. Alt, D. Schrader, V. Gomer, D. Meschede, Phys. Rev. Lett. **84**, (2000)
33. S. Kuhr, W. Alt, D. Schrader, M. Müller, V. Gomer, D. Meschede, Science **293**, 278 (2001), published online June 14 2001; 10.1126/science.1062725
34. N. Schlosser, G. Reymond, I. Protsenko, P. Grangier, Nature **411**, 1024 (2001)
35. T. Pellizari, S.A. Gardiner, J.I. Cirac, P. Zoller, Phys. Rev. Lett. **75**, 3788 (1995); S.-B. Zheng, G.-C. Guo, Phys. Rev. Lett. **85**, 2392 (2001)
36. W. Hänsel, J. Reichel, P. Hommelhoff, T.W. Hänsch, Phys. Rev. Lett. **86**, 608 (2001)
37. D. Schrader, S. Kuhr, W. Alt, M. Müller, V. Gomer, D. Meschede, Appl. Phys. B, submitted (2001)

Optical Lattices as a Playground for Studying Multiparticle Entanglement

Hans J. Briegel, Robert Raussendorf, and Axel Schenzle

We discuss a class of multiparticle entangled states, the so-called cluster states, in the context of atom interferometers. We demonstrate how these states could be created efficiently in far-detuned optical lattices and what one could do with them experimentally. This includes the demonstration of novel effects of quantum mechanical nonlocality, such as teleportation and quantum computation, by simple measurements on trapped atoms.

1 Introduction

Quantum entanglement has been a central issue of long-lasting debates on the interpretation of quantum mechanics, initiated by the works of Einstein, Podolsky, Rosen [1] by Schrödinger [2], and later cast in more quantitative terms by Bell in his famous theorems on quantum nonlocality [3]. More recently, the discussion has been generalized to the situation of more than two particles by Greenberger, Horne and Zeilinger (GHZ) [4] and by Mermin [5].

Entanglement has been successfully created between particles (photons) many kilometres apart [6]. It can be very well controlled (the violation of Bell's inequalities has been demonstrated to exceed 200 standard deviations [7]) and forms the basis for new communicational protocols such as teleportation and entanglement-based quantum cryptography. Even though some aspects of bipartite entanglement – such as the phenomenon of bound entanglement [8, 9] – are not yet completely understood, for pure states at least there is both a unique measure for the degree of entanglement and a good understanding of its resource character for quantum communication.

For entanglement between three and more particles, much less is known. Also, since the creation of such states requires a highly controlled interaction between several particles, they are much harder to create in the laboratory. Recently, however, three- and four-particle entangled GHZ states have been created with photons [10] and with trapped ions [11]. With entangled states of many more particles, i.e. entangled states of *mesoscopic samples* of particles, two problems arise. The first problem is that it seems much harder to control a mesoscopic number of particles and their interactions. The relevant aspect of entanglement is the type of correlations one observes when one measures certain observables of the subsystems of a state. If the interaction is not controlled very well, or if there are additional degrees of freedom involved ("the environment"), our ignorance about the interaction and the state of the environmental degrees of freedom will not allow us to detect these quantum correlations let alone make use of them. The second problem is that it is not clear what types of entangled states are interesting in the first place, both from a conceptual perspective [12] and, in particular, for applications in quantum information theory.

In this paper, we will discuss a class of entangled states, the cluster states [14], with surprising and useful entanglement properties. It is shown that these states can be created efficiently in an optical lattice by a simple interferometric interaction. Cluster states exhibit highly nonclassical features which could be demonstrated experimentally, for example by a teleportation experiment involving several particles (or parties). In addition, these states form the resource for a new type of quantum computer [15].

The paper is organized as follows. In Sect. 2 we first review the proposal by Jaksch et al. [17] where cold controlled collisions have been introduced as a mechanism to create entangled states of atoms in an optical lattice. This proposal has been elaborated in [16], where it was shown how efficient quantum-error-correction techniques and, more generally, a quantum computer could be realized by using three-level atoms and sequences of interferometric lattice shifts to achieve selective quantum gate operations. In Sect. 3 we show that the simplest interferometric process that can be realized in an optical lattice, using only two-level atoms, formally corresponds to a *quantum Ising model.* We show how already in this scenario, without the possibility of selective quantum gate operations, an interesting type of multiparticle entangled state is created, which we call a cluster state. We discuss the entanglement properties of this state and illustrate the concepts of *maximal connectedness* and *persistency of entanglement*, first introduced in [14], in the context of the optical lattice. We then show how the cluster state can be used to store quantum information nonlocally and how to create other entangled quantum states by simple measurements on the cluster. This illustrates the capability of cluster states as a resource for quantum-state engineering. In Sect. 4, it is further shown that the cluster state can be used in a new model of a quantum computer. In this model [15], a quantum computation is realized by a mere sequence of 1-qubit measurements on an array of particles in the cluster state. In the concluding Sect. 5 a few remarks on experimental issues are made.

2 Optical Lattices as Multiatom Interferometers

In this section, the essential ingredients of the proposal of Jaksch et al. [17] are briefly summarized (see Fig. 1). The purpose of this summary is

(a) ^{87}Rb: F=2 $|0\rangle$
F=1 $|1\rangle$

(b)
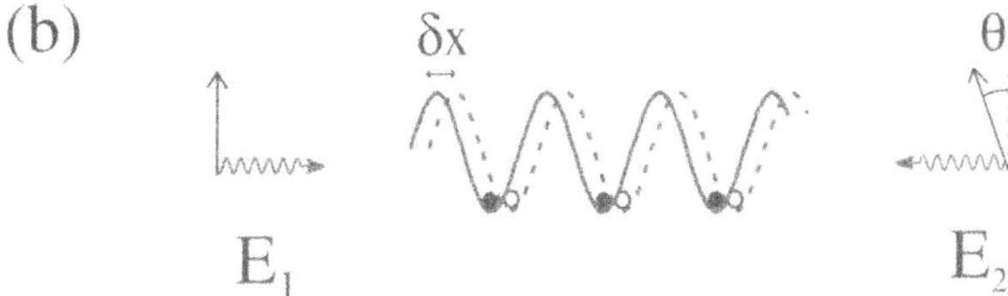

(c)
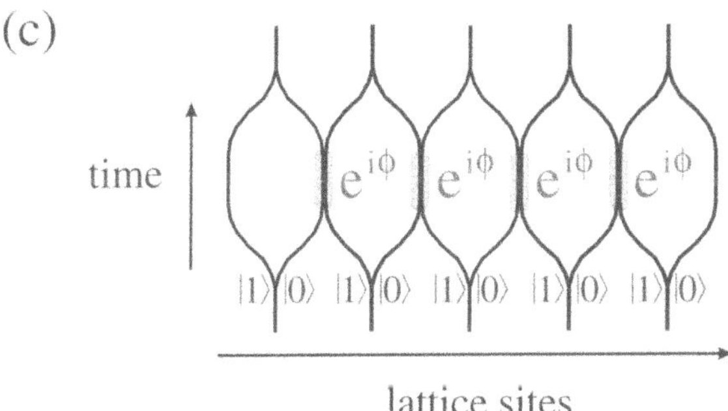

Fig. 1a–c. Optical-lattice based multiatom interferometer. See text

only to prepare the ground for the subsequent discussion of cluster states in Sect. 3. For a more detailed discussion we refer the reader to [17, 16]. The authors in [17] consider the example of alkali atoms with a nuclear spin equal to 3/2 (^{87}Rb, ^{23}Na) trapped by standing waves in three dimensions. The internal states of interest are hyperfine levels corresponding to the ground state $S_{1/2}$ as shown in Fig. 1(a). Along the z axis, the standing waves are in the lin$\angle$lin configuration (two linearly polarized counter-propagating waves with the electric fields $\boldsymbol{E}_1$ and $\boldsymbol{E}_2$ forming an angle θ [18]) as in Fig. 1(b). The total electric field is a superposition of right and left circularly polarized standing waves ($\sigma^{\pm}$) which can be shifted with respect to each other by changing θ, $\boldsymbol{E}^{+}(z,t) = E_0 \,\mathrm{e}^{-\mathrm{i}\,\nu t}\,[\boldsymbol{\varepsilon}_{+} \sin(kz+\theta/2) + \boldsymbol{\varepsilon}_{-} \sin(kz-\theta/2)]$, where $\boldsymbol{\varepsilon}_{\pm}$ denote unit right and left circular polarization vectors, $k = \nu/c$ is the laser wave vector and E_0 the amplitude. If the lasers are tuned appropriately between the $P_{1/2}$ and $P_{3/2}$ levels, atoms in a state with $m_s = +1/2$ couple only to the right-circularly polarized component σ^+ while atoms in a state with $m_s = -1/2$ couple only to the σ^--component [17]. The optical potentials for these two states are $V_{m_s=\pm 1/2}(z,\theta) = \alpha|E_0|^2 \sin^2(kz \pm \theta/2)$. If one chooses for the logical qubit states $|0\rangle$ and $|1\rangle$ the hyperfine structure states $|0\rangle \equiv |F=2, m_F=2\rangle$ and $|1\rangle \equiv |F=1, m_F=1\rangle$, these states are stable under collisions due to angular momentum conservation. The potentials "seen" by the atoms in these internal states are $V^0(z,\theta) = V_{m_s=1/2}(z,\theta)$ and $V^1(z,\theta) = \left[V_{m_s=1/2}(z,\theta) + 3V_{m_s=-1/2}(z,\theta)\right]/4$. If one stores atoms in these potentials and they are deep enough, there is no tunneling to neighboring wells and one can approximate them by harmonic potentials. By varying the angle θ from 0 to π, the potentials V^0 and V^1 move in opposite directions until their respective minima coincide. Then, going back to $\theta = 0$ the potentials return to their original positions where the minima coincide again. Even though the shape of the potential V^1 changes as it moves, by choosing particular motion profiles this process can be made sufficiently adiabatic such that the atoms stay in the ground state of their respective potential [17, 16, 19].

Near $\theta = \pi$, the spatial wavefunctions of a $|0\rangle$ atom in the jth well and a $|1\rangle$ atom in the $(j+1)$th well overlap. For very low temperatures, the repulsive interaction ("cold collision") between the atoms is coherent and gives rise to an energy shift. In first-order perturbation theory, it is given by

$$\Delta E(t) = \frac{4\pi a_s^{01}\hbar^2}{m} \int \mathrm{d}\mathbf{x} \left|\psi_j^0(\mathbf{x},t)\right|^2 \left|\psi_{j+1}^1(\mathbf{x},t)\right|^2, \tag{1}$$

where $\psi_j^\epsilon(\mathbf{x}) = \langle \mathbf{x}|\epsilon\rangle_j$ is the normalized one-particle wave function of the atom in well j and in the internal state $\epsilon \in \{0,1\}$, trapped in the time-dependent potential $V^\epsilon(\mathbf{x},t)$. Here a_s^{01} is the coherent s-wave scattering length for a pair of atoms in the (different) internal states 0 and 1. The phase shift acquired by this interaction is given by the time integral

$$\varphi = \frac{1}{\hbar} \int \mathrm{d}t \Delta E(t). \tag{2}$$

For two neighboring atoms in states $|\epsilon\rangle_j$ and $|\epsilon'\rangle_{j+1}$, respectively, a phase shift will be acquired only if $\epsilon = 0$ and $\epsilon' = 1$. This *conditional phase shift* is summarized by the transformation $|\epsilon\rangle_j|\epsilon'\rangle_{j+1} \longrightarrow \mathrm{e}^{-\mathrm{i}\,\phi\,\bar{\epsilon}\epsilon'}|\epsilon\rangle_j|\epsilon'\rangle_{j+1}$ with the notation $\bar{\epsilon} = \epsilon + 1 \bmod 2$. It has been shown that, for $\phi = \pi$ this conditional phase gate, together with the set of unitary SU(2) rotations in the internal state space of a single qubit, forms a universal set of quantum logic gates [20]. This means that, by performing sequences of such interferometric lattice shifts, together with the possibility of Raman transitions between the internal states of the atoms, one can in principle realize a quantum computer [16]. However, since the lattice shifts affect all atoms in the lattice in a similar way, an important element of control is still missing. To realize a quantum computer based on these elementary gate operations, it is necessary to achieve selective operations such that only particularly chosen atoms participate in an interaction, thereby realizing the desired two-qubit quantum gates. Selectivity can be achieved by using more than two internal atomic states. The reader who is interested in further details of this proposal for quantum computation with three-level atoms is referred to [16, 17].

3 Cluster States

In the following, we do not assume the possibility of selective interactions. This means that a lattice shift will aways affect all atoms simultaneously. The question we are going to investigate is this: "What quantum state is created at the output of the interferometer in Fig. 2 if all qubits enter it in the superposition $1/\sqrt{2}(|0\rangle + |1\rangle)$ of their internal states?" We will see that the systematic investigation of this simple question will shed new light on the issue of multiparticle entanglement and quantum mechanical nonlocality. It will also have far-reaching repercussions on the very *concept* of a quantum computation. For notational convenience, let us introduce the Pauli operators $\sigma_z^{(j)} = |0\rangle_j\langle 0| - |1\rangle_j\langle 1|$, $\sigma_x^{(j)} = |0\rangle_j\langle 1| + |1\rangle_j\langle 0|$, and $\sigma_y^{(j)} = \mathrm{i}\ \sigma_x^{(j)}\sigma_z^{(j)}$ associated with the qubit carried by atom j. The interaction Hamiltonian formally describing the interferometer in Fig. 1(c) is given by

$$H_{\text{int}} = \hbar g(t) \sum_j \frac{1 + \sigma_z^{(j)}}{2} \frac{1 - \sigma_z^{(j+1)}}{2}, \tag{3}$$

where j labels the lattice sites and the coupling strength $g(t)$ is, in the specific realization discussed earlier, controlled by the relative polarization angle θ of Fig. 1(b). In first-order perturbation theory we have, in fact, $g(t) = \Delta E(t)/\hbar$ with $\Delta E(t)$ given in (1). The operators $(1 \pm \sigma_z^{(j)})/2$ appearing in (3) are the projectors onto the logical states $|0\rangle_j$ and $|1\rangle_j$, respectively, and the reader who is more familiar with the "bra-ket" notation may prefer to write (3) in the form $H_{\text{int}} = \hbar g(t) \sum_j |0\rangle_j\langle 0| \otimes |1\rangle_{j+1}\langle 1|$. Here, as in (3), we assume that all sites (or "input ports") $1, 2, \ldots, N$ of the interferometer are occupied

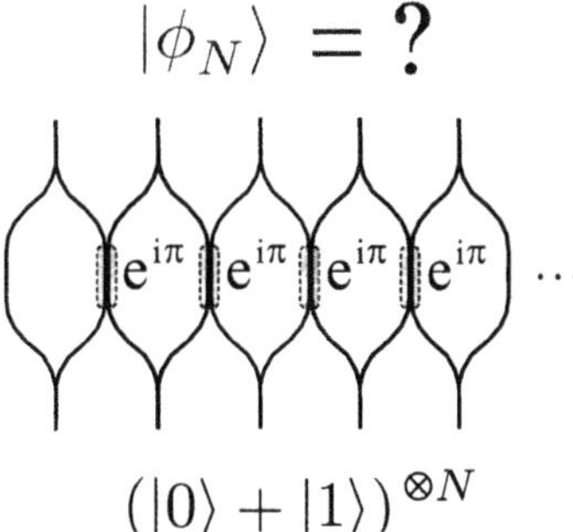

Fig. 2. Creation of a 1D cluster state

by an atom and the summation index j runs from 1 to $N-1$. The unitary transformation realized by the interferometer is thus given by

$$\begin{aligned} U(\varphi) &= \mathrm{e}^{-\frac{\mathrm{i}}{\hbar}\int \mathrm{d}t\, H_{\mathrm{int}}} \\ &= U_{\mathrm{local}}\, \mathrm{e}^{-\frac{\mathrm{i}\varphi}{4}\sum_{j=1}^{N-1}\sigma_z^{(j)}\sigma_z^{(j+1)}} \end{aligned} \tag{4}$$

with $\varphi = \int \mathrm{d}t\, g(t)$. The transformation $U_{\mathrm{local}} = \mathrm{e}^{-\frac{\mathrm{i}N}{4}\varphi}\, \mathrm{e}^{-\frac{\mathrm{i}\varphi}{4}\sigma_z^{(1)}}\, \mathrm{e}^{\frac{\mathrm{i}\varphi}{4}\sigma_z^{(N)}}$ contains a local rotation on atoms 1 and N and a global phase factor. The essential part of the interaction is contained in the second term in (4) which formally corresponds to the time evolution operator of the *quantum Ising model.* Since we are interested in the entanglement properties of the states generated under U, the extra local rotations are irrelevant and correspond only to a local change of basis. The statements to be made and conclusions to be drawn in the following are, in fact, not restricted to the optical lattice, but apply to any system of qubits with a quantum Ising interaction, whose strength can be controlled (and in particular switched on and off) by some external parameter.

In the following we set the phase φ of the interferometer equal to π and prepare all qubits in a superposition of $|0\rangle$ and $|1\rangle$ (i.e. in eigenstate of σ_x), as in Fig. 2. The state $|\phi_N\rangle$ at the output of the interferometer is then given by the expression

$$\begin{aligned} |\phi_N\rangle &= U(\pi)\frac{1}{2^{N/2}}\bigotimes_{j=1}^{N}(|0\rangle_j + |1\rangle_j) \\ &= \frac{1}{2^{N/2}}\bigotimes_{j=1}^{N}(|0\rangle_j\sigma_z^{(j+1)} + |1\rangle_j) \end{aligned} \tag{5}$$

with the convention that $\sigma_z^{(N+1)} \equiv 1$. We call $|\phi_N\rangle$ a *cluster state of N qubits.* The compact notation employed in (5) is easily understood by multiplying out the right-hand side. For $N = 2$, for example, one obtains $|\phi_2\rangle = \frac{1}{2}(|0\rangle_1\sigma_z^{(2)} + |1\rangle_1)(|0\rangle_2 + |1\rangle_2) = \frac{1}{2}\left[|0\rangle_1(|0\rangle_2 - |1\rangle_2) + |1\rangle_1(|0\rangle_2 + |1\rangle_2)\right]$ which is a maximally entangled Bell state. We may write it, up to a local unitary rotation on qubit 2, in the standard form

$$|\phi_2\rangle =_{\mathrm{l.u.}} \frac{1}{\sqrt{2}}(|0\rangle_1|0\rangle_2 + |1\rangle_1|1\rangle_2)\,, \tag{6}$$

where "l.u." indicates that equality holds up to a local unitary transformation on one or more of the qubits [22]. Similarly, one obtains for $N = 3, 4$,

$$\begin{aligned} |\phi_3\rangle &=_{\mathrm{l.u.}} \frac{1}{\sqrt{2}}\left(|0\rangle_1|0\rangle_2|0\rangle_3 + |1\rangle_1|1\rangle_2|1\rangle_3\right)\,, \\ |\phi_4\rangle &=_{\mathrm{l.u.}} \frac{1}{2}\left(|0\rangle_1|0\rangle_2|0\rangle_3|0\rangle_4 + |0\rangle_1|0\rangle_2|1\rangle_3|1\rangle_4\right. \\ &\quad \left. + |1\rangle_1|1\rangle_2|0\rangle_3|0\rangle_4 - |1\rangle_1|1\rangle_2|1\rangle_3|1\rangle_4\right)\,. \end{aligned} \tag{7}$$

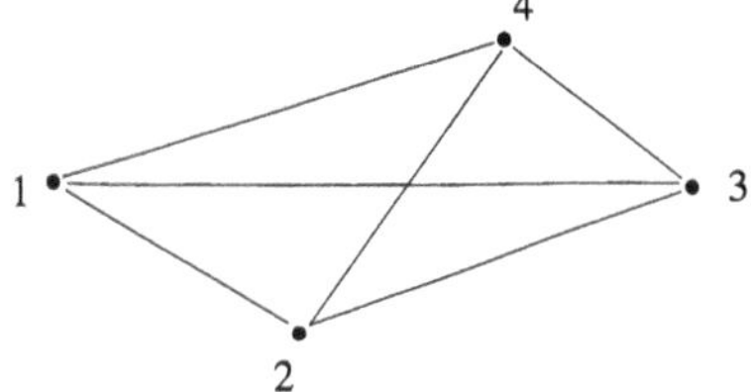

Fig. 3. Discussion of the entanglement properties of the 4-qubit states $|GHZ_4\rangle$ and $|\phi_4\rangle$ in the "LOCC" scenario, where the qubits are distributed between four remote parties, each holding one qubit. See text

While $|\phi_3\rangle$ corresponds to the well-known Greenberger–Horne–Zeilinger (GHZ) state of three qubits [4], $|\phi_4\rangle$ is *not equivalent* to a 4-qubit GHZ state [23]. More generally, the states $|\phi_N\rangle$ and the N-qubit GHZ state $|GHZ_N\rangle \equiv 2^{-1/2}(|0\rangle_1 \dots |0\rangle_N) + |1\rangle_1 \dots |1\rangle_N)$ are not equivalent for $N > 3$, i.e. cannot be transformed into each other if we allow only local rotations on individual qubits, as we shall see below.

How can we compare the entanglement of the qubits in state $|\phi_4\rangle$ with their entanglement in state $|GHZ_4\rangle$? To discuss entanglement properties, one usually considers a more general scenario, depicted in Fig. 3, where the qubits are distributed between remote parties, who may only perform *local operations* on their respective qubit and coordinate their actions by *classical communication*. This is the standard "LOCC" scenario of standard information theory. We can easily convince ourselves of the following two observations:

1. The states $|\phi_4\rangle$ and $|GHZ_4\rangle$ *share* the property that any two of the four qubits can be projected into a Bell state by measuring the other two qubits in an appropriate basis. As a consequence, the parties could use either of the states $|\phi_4\rangle$ or $|GHZ_4\rangle$ to *teleport* [24] a qubit between any of the four locations.
2. The states are *different* in that it is harder to *destroy* the entanglement of state $|\phi_4\rangle$ than that of $|GHZ_4\rangle$ by local operations. In fact, it is impossible to destroy all entanglement of $|\phi_4\rangle$ by a single local operation, such as a von Neumann measurement or complete depolarization of a qubit. For the state $|GHZ_4\rangle$, in contrast, a single local measurement suffices to bring it into a product state.

The observations about the states $|\phi_2\rangle$, $|\phi_3\rangle$ and, in particular, $|\phi_4\rangle$ motivate the subsequent discussion about the cluster state $|\phi_N\rangle$ for an arbitrary number N of qubits. Measurements on individual qubits and their effect on the entangled state play a central role in this discussion. The possibility to bring any two chosen qubits with certainty into a Bell state by local measurements (and rotations) on other qubits is an important concept. We have called this property *maximal connectedness*; it reveals some higher-order entanglement of the entire state which will turn out to be significant for the purpose of information processing. One can show that the cluster state $|\phi_N\rangle$, for arbitrary N, is indeed maximally connected. In the interferometer of Fig. 2

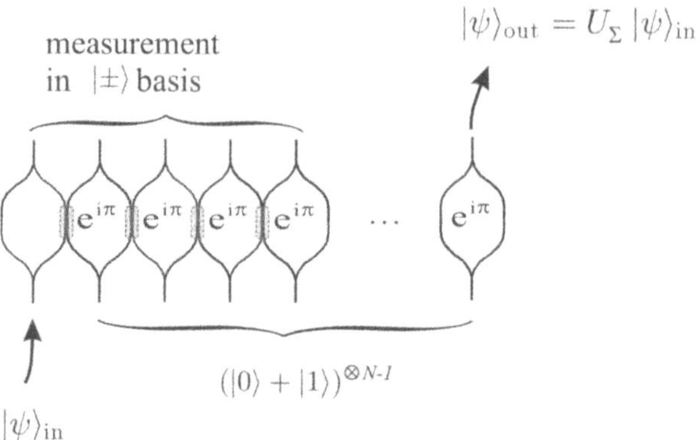

Fig. 4. Storage of quantum information in a state $|\phi_N\rangle$. The information is written onto the left qubit of a chain and extracted from the rightmost one. To accomplish this, the remaining qubits in the chain are measured in the σ_x-basis

this means that the atoms at any two output ports j and k can be projected into a Bell state by measuring σ_x at the ports between j and k and σ_z at the other ports except j and k.

Why is this property interesting for quantum information processing? As an immediate consequence, cluster states can be used to store and transmit quantum information in the interferometer. Suppose that a quantum state of a chain of N qubits, as displayed in Fig. 4, is prepared in the following way: qubit 1 is prepared in the state $|\psi\rangle_{\rm in}$ while the remaining qubits are prepared in $|+\rangle = 1/\sqrt{2}(|0\rangle + |1\rangle)$. Then, the chain of qubits is entangled by switching on the interferometric interaction (3) for a finite timespan such that conditional phase shifts $\varphi = \pi$ are acquired. An interesting question is this: From which atom at the output of the interferometer can the quantum information, written originally on particle 1, be extracted? On first sight one might think that the only possibilities were output ports 1 or 2, since the atom in well 1 has only interacted with its next neighbor in well 2. It turns out, however, that the state $|\psi\rangle_{\rm in}$ can be extracted from *any* output port. In Fig. 4, for example, the information is extracted at the rightmost port N! To accomplish this, atoms $1, \ldots, N-1$ are measured in the eigenbasis of σ_x. For atoms, this means that one would first apply a $\pi/2$ pulse and then perform a conditional resonance fluorescence measurement. After the measurements, qubit N is in state $|\psi\rangle_{\rm out}$ which is related to $|\psi\rangle_{\rm in}$, the state which was originally written onto qubit 1, via a known unitary transformation U_Σ: $|\psi\rangle_{\rm out} = U_\Sigma|\psi\rangle_{\rm in}$, where U_Σ depends on the results of the measurements made. It has the form: $U_\Sigma = {\sigma_x}^{s_x} H^{N-1}$ where $H = 1/\sqrt{2}(\sigma_x + \sigma_z)$ is the Hadamard transform and $s_x \in \{0,1\}$ is a function of the results obtained in the measurements. Alternatively, the state $|\psi\rangle_{\rm in}$ can be extracted at any other output port $k < N$. To do this, qubit $k+1$ is measured in the eigenbasis of σ_z (which, in effect, disentangles all qubits j with $j > k$ from the chain $1, \ldots, k$)

and the qubits $1, \ldots, k-1$ are measured in the basis of σ_x. The state of the qubit at port k is then related to the input state by a transformation similar to the one specified above. The described procedure resembles the process of teleportation, where the transmission of two classical bits is required, as well, to compensate for an extra unitary transformation on the teleported state, that arises due to the random result of a Bell measurement. In summary, by the interferometric process described in Fig. 4, the quantum information on qubit 1 becomes *stored* in the entire cluster. It can be retrieved from any qubit of the cluster by simple measurements on the other qubits and a subsequent unitary rotation on the readout qubit.

As we described in the context of Fig. 3, a distinguishing property of the cluster state compared to the GHZ state is the persistency of its entanglement against local measurements. We define the *persistency of entanglement* P_{e} of an entangled state of n qubits as the minimum number of local measurements such that, for all measurement outcomes, the state is completely *disentangled* [14]. In this terminology, $P_{\mathrm{e}}(|GHZ_4\rangle) = 1$ while $P_{\mathrm{e}}(|\phi_4\rangle) = 2$. For the cluster state $|\phi_N\rangle$ of N qubits, we find that $P_{\mathrm{e}}(|\phi_N\rangle) = \lfloor N/2 \rfloor$. In other words, at least about half of all atoms leaving the interferometer in Fig. 4 have to be subjected to a measurement in order to destroy all entanglement built up by the interferometric process. The persistency of entanglement quantifies, in very simple terms, the *operational effort* that is necessary *to disentangle* a given state. Its high persistency together with the property of being maximally connected, makes the cluster state a useful resource for a larger class of quantum state transformations, that go beyond the mere storage of an unknown quantum state. This will become clearer when we go to higher-dimensional lattices.

The cases of two- and three-dimensional lattices are different from the case $d = 1$ since there is no natural ordering of the qubits. Therefore, the concept of a "chain" of qubits docs not apply anymore. The natural generalization to higher dimensions is a "cluster" $\mathcal{C}$ of qubits as in Fig. 5. Two qubits are connected in a topological sense, if and only if one can get from one qubit site to the other via a sequence of occupied next neighbouring qubit sites. A d-dimensional *cluster* $\mathcal{C} \in \mathbb{Z}^d$ is a maximal set of qubits which are (topologically) connected. Irregular clusters are found in lattices with a finite site occupation probability $0 < p < 1$. For p above a certain critical value p_{c}, which depends on the dimension of the lattice, an infinitely extended cluster exists which is bounded in size only by the trap dimensions. For optical lattices in three dimensions, single-atom site occupation with a filling factor of 0.44 has been reported [25] which is significantly above the percolation threshold of 0.31 [26].

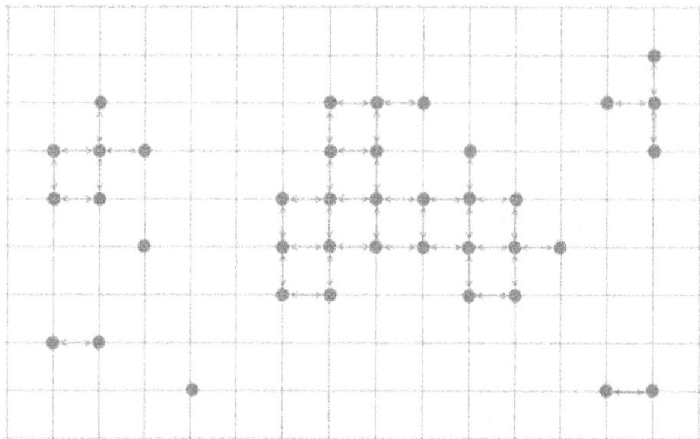

Fig. 5. Entangled clusters of two-state particles

The quantum mechanical state of a cluster C that is generated under the Hamiltonian (3) for $\varphi = \pi$ is, up to a normalization factor,

$$|\Phi\rangle_C = \bigotimes_{c \in C} \left(|0\rangle_c \bigotimes_{\gamma \in \Gamma} \sigma_z^{(c+\gamma)} + |1\rangle_c \right) \tag{8}$$

with the choice $\Gamma = \{1\}$ for $d = 1$, $\Gamma = \{(1,0),(0,1)\}$ for $d = 2$ and $\Gamma = \{(1,0,0),(0,1,0),(0,0,1)\}$ for $d = 3$. To realize the interaction required to create state (8) in an optical lattice, atoms are state-selectively moved along a certain path. This path takes the respective atoms in state $|0\rangle$ to a subset of neighboring atoms in the internal state $|1\rangle$. The set Γ specifies which neighboring atoms are "visited" and thus between which atoms conditional phase shifts are acquired. The sets Γ are nonordered since the order in which the phase shifts are acquired is of no concern. The conditional phase shifts mutually commute.

We find that all cluster states are maximally connected. It is noteworthy that the property of maximal connectedness of $|\Phi\rangle_C$ does not depend on the precise shape of the cluster, and not even on its topological characterization except for being a cluster.

Entanglement is often regarded as a resource and thus the question arises which states can be obtained from cluster states by local operations (LOCC). These are operations that are not capable of creating further entanglement between the particles (i.e. no interactions). A particularly simple subclass of local operations is the restriction to projective von Neumann measurements. We note without proof that from a square block C of qubits, one can obtain any state of the form $\alpha|00\ldots0\rangle_{C'} + \beta|11\ldots1\rangle_{C'}$ on a sublattice of qubits with double lattice spacing. For $\alpha = \beta = 1/\sqrt{2}$, this includes, in particular, the family of generalized (multiparticle) GHZ states. An illustration is given in Fig. 6. Even though the thereby obtained states are highly entangled, their

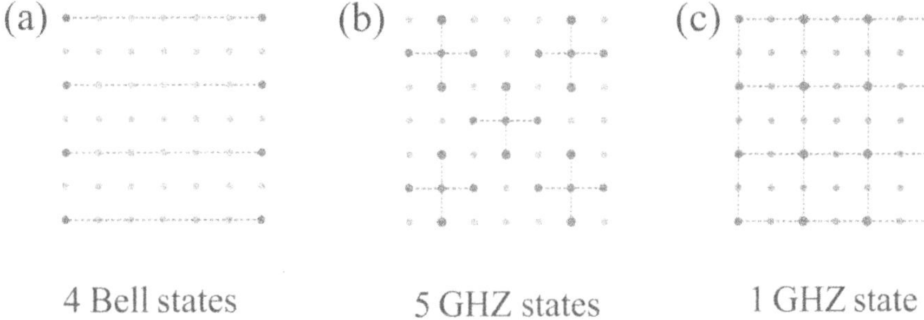

Fig. 6. States created by one-qubit measurements on a cluster state. (**a**) Four Bell states, (**b**) five generalized GHZ states of 5 qubits, (**c**) one generalized GHZ state of 16 qubits. The states are simply obtained by measuring the grey qubits in the appropriate basis [σ_x on the lines; σ_z between the lines], and by subsequent 1-qubit rotations on the remaining qubits

Schmidt entanglement measure [27] is always smaller than that of the original cluster state, and so the total amount of entanglement decreases, as it should.

4 The One-Way Quantum Computer

In the preceding section we have seen how a cluster state, by measuring some of its qubits, can be used to project the remaining (unmeasured) qubits into other entangled states. More generally, up to local unitaries U_Σ, *any* quantum state $|\psi_n\rangle$ of n qubits can be created via one-qubit measurements on a two- or three-dimensional cluster of sufficient size. The local unitaries U_Σ are of the form $U_\Sigma = \bigotimes_{i=1}^n {\sigma_x}^{x_i} {\sigma_z}^{z_i}$, with $x_i, z_i \in \{0,1\}$. In this sense, the cluster state can be regarded as a resource for quantum state engineering. This opens yet another application for cluster states. If any quantum state can be generated on a cluster state via local measurements, then of course also the state of the output register of a quantum computer, after running a quantum algorithm, can be obtained. This observation leads to a new model of quantum computation, the one-way quantum computer [15], see Fig. 7. In this model, the entire resource for the quantum computation is provided initially in the form of a specific entangled state, namely the cluster state [14], of a large number of qubits. Information is then written onto the cluster, processed, and read out from the cluster by just one-particle measurements. The entangled state of the cluster thereby serves as a universal "substrate" for any quantum computation. To process quantum information with a cluster state, it suffices to measure its particles in a certain order and in a certain basis. An example for a measurement pattern is given in Fig. 7. Quantum information is propagated horizontally through the cluster and processed by measuring the cluster qubits. The shaded structures form a circuit of

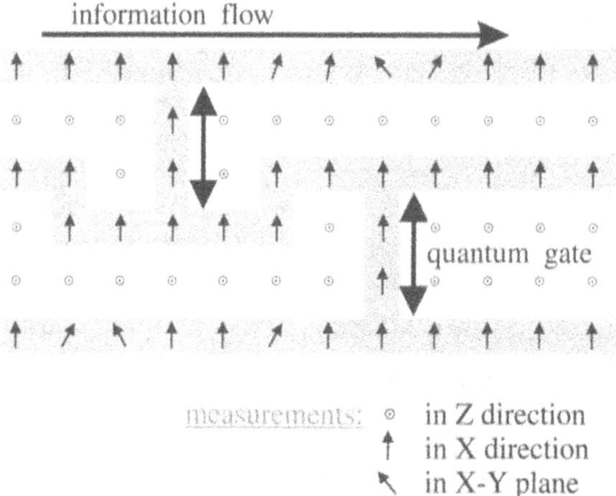

Fig. 7. Quantum computation by measuring two-state particles on a lattice. Before the measurements the qubits are in the cluster state $|\Phi\rangle_C$. *Circles* $\odot$ symbolize measurements of σ_z, *vertical arrows* are measurements of σ_x, while *tilted arrows* refer to measurements in the xy-plane

"wires" and one- and two-qubit gates. For example, the cluster qubits on the horizontal shaded structures which are measured in the eigenbasis of σ_x form wires and the cluster qubits on the vertical connections measured in the same basis realize CNOT-gates. For one-qubit rotations on the logical qubits, the basis in which a certain cluster qubit is measured depends in general on the results of preceeding measurements. The processing is finished once all qubits except the last one on each wire have been measured. At this point, the results of previous measurements determine in which basis these "output" qubits need to be measured for the final read-out. We note that, in the entire process, only 1-qubit measurements are required.

5 Experimental Issues

Having demonstrated the intriguing possibilities of entanglement generation and its transformations in an optical lattice, we are aware of the fact that most of the experimental requirements have yet to be realized. Given the impressive experimental advances made so far in the fields of neutral atom cooling and trapping [28, 29] and in atom interferometry [30], we are however confident that an optical-lattice based interferometer can be realized.

Essential for the realization of what has been said in this paper, using cold collisons, is the cooling of the atoms to the ground state (in all dimensions) of the optical lattice [31–33] and the achievement of a high site-occupation probability, with not more than a single atom per lattice site [25]. Numerical calculations as reported in [17, 19], using realistic parameters, give $k_\mathrm{B}T <$

$0.2\hbar\omega$ as a critical value, where ω is the trap frequency corresponding to each lattice well. Similarly, noise in the intensity, polarization, and phase of the trapping lasers has to be suppressed. Such noise translates into fluctuations of the position and the depth of the trapping wells and thus, in particular, to heating. The effect of these fluctuations has been analyzed in [34].

A second important ingredient is the possibility of addressing and, in particular, measuring every atom individually. This single-site addressability is obviously a problem. Recent progress has been reported in [35], where single-site resolution has been achieved for atoms trapped in a CO_2 far-detuned-infrared laser [36]. These lasers however cannot be used for state-dependent manipulation of the atoms, as the atomic polarizability does not, for these frequencies, depend on the internal atomic state [36]. A possible approach to overcome the addressability problem for atoms trapped in "near-resonant" lasers (i.e. with a detuning between the $P_{1/2}$ and $P_{3/2}$ levels, which is required for the interferometric lattice shift operation shown in Fig. 5) has been described in [35, 37]. After cooling the atoms in the CO_2-laser lattice, the atoms are transferred into a near-resonant standing wave, which leads to an occupation pattern with precisely every 13th trapping site occupied by an atom [35], such that selective addressing is no longer a problem. This lattice could then be used for state selective operations.

Some of the proposals described in this paper do not require a 2D or 3D lattice. In a 1D lattice, one could e.g. perform a teleportation experiment as shown schematically in Fig. 4. Another example would be an experiment where every second atom at the output of Fig. 3, using a chain with an odd number N of atoms, is measured in the eigenbasis of σ_x. As a result, the unmeasured atoms are projected into a "Schrödinger cat"-type state $|GHZ\rangle_M$ of $M = (N+1)/2$ atoms. All of the $N - M$ atoms can be measured simultaneously. The generation of the cat state can thus essentially be performed with a single interferometric lattice shift and a single step of measurements.

Acknowledgements

This paper is dedicated to Theodor W. Hänsch on the occasion of his 60th birthday. One of us (HJB) would like to thank Ignacio Cirac, Peter Zoller, and the Innsbruck group for many discussions and for fruitful collaboration at an earlier stage of this work. We thank T. Esslinger and M. Weitz for helpful comments. This work has been supported by the Deutsche Forschungsgemeinschaft (QIV) and the European Union (IST-1999-13021).

References

1. A. Einstein, B. Podolski, N. Rosen, Phys. Rev. **47**, 777 (1935)
2. E. Schrödinger, Naturwissenschaften **23**, 807–812, 823–828, 844-849 (1995)

3. J.S. Bell, Physics **1**, 195 (1964)
4. D.M. Greenberger, M.A. Horne, A. Zeilinger, Am. J. Phys. **58**, 1131 (1990)
5. N.D. Mermin, Phys. Rev. Lett. **65**, 3373 (1990)
6. W. Tittel, J. Brendel, H. Zbinden, N. Gisin, Phys. Rev. Lett. **81**, 3563 (1998)
7. Ch. Kurtsiefer, M. Oberparleitner, H. Weinfurter, Phys. Rev. A **64**, 023802 (2001)
8. M. Horodecki, P. Horodecki, R. Horodecki, Phys. Rev. Lett. **80**, 5239 (1998)
9. For a recent review of the theory of separability and distillability see e.g. the paper by M. Lewenstein *et al.*, J. Mod. Opt. **47**, 2841 (2000)
10. D. Bouwmeester, J.-W. Pan, M. Daniell, H. Weinfurter, A. Zeilinger, Phys. Rev. Lett. **82**, 1345 (1999)
11. C.A. Sackett, D. Kielpinski, B.E. King, C. Langer, V. Meyer, C.J. Myatt, M. Rowe, Q.A. Turchette, W.M. Itano, D.J. Winel, C. Monroe, Nature, **404**, 256 (2000)
12. Exceptions are, however, the celebrated GHZ state [4] and the so-called W state discussed in [13]
13. W. Dür, G. Vidal, J.I. Cirac, Phys. Rev. A **62**, 062314 (2000)
14. H.-J. Briegel, R. Raussendorf, Phys. Rev. Lett. **86**, 910 (2001)
15. R. Raussendorf, H.-J. Briegel, Phys. Rev. Lett. **86**, 5188 (2001)
16. H.-J. Briegel, T. Calarco, D. Jaksch, J.I. Cirac, P. Zoller, J. Mod. Opt., **47**, 415 (2000)
17. D. Jaksch, H.-J. Briegel, J.I. Cirac, C.W. Gardiner, P. Zoller, Phys. Rev. Lett., **82**, 1975 (1999)
18. V. Finkelstein, P.R. Berman, J. Guo, Phys. Rev. A, **45**, 1829 (1992)
19. D. Jaksch, *Bose–Einstein Condensation Applications*, Dissertation, University of Innsbruck, 1999
20. A. Barenco, C.H. Bennett, R. Cleve, D.P. DiVincenzo, N. Margolus, P. Shor, T. Sleator, J.A. Smolin, H. Weinfurter, Phys. Rev. A, **52**, 3457 (1995)
21. The influence of noise in the amplitude, phase, polarization of the trapping laser, which leads to fluctuations of the position and the depth of the potential wells, has been analyzed in [34]
22. N. Linden, S. Popescu, Fortschr. Phys. **46**, 567 (1998)
23. S. Wu, Y. Zhang, quant-ph/0004020 (2000)
24. C.H. Bennett, G. Brassard, C. Crepeau, R. Josza, A. Peres, W.K. Wootters, Phys. Rev. Lett., **70**, 1895 (1993)
25. M.T. DePue, C. McCormick, S.L. Winoto, S. Oliver, D.S. Weiss, Phys. Rev. Lett., **82**, 2262 (1999)
26. J.M. Ziman, *Models of Disorder* (Cambridge University Press, 1979)
27. J. Eisert, H.-J. Briegel, Phys. Rev. A **64**, 022306 (2001)
28. *Laser Manipulation of Atoms and Ions,* ed. by E. Arimondo, W.D. Phillips, F. Strumia (North Holland, Amsterdam 1992)
29. S. Chu, Rev. Mod. Phys. **70**, 686 (1998); C. Cohen-Tannoudji, *ibid.* **70**, 707 (1998); W.D. Phillips, *ibid.* **70**, 721 (1998)
30. See e.g. *Atom Interferometry*, ed. by P.R. Berman (Academic Press, 1997)
31. S.E. Hamann, D.L. Haycock, G. Klose, P.H. Pax, I.H. Deutsch, P.S. Jessen, Phys. Rev. Lett. **80**, 4149 (1998)
32. A.J. Kerman, V. Vuletic, C. Chin, S. Chu, Phys. Rev. Lett. **81**, 439 (2000)
33. M. Greiner, I. Bloch, O. Mandel, T.W. Hänsch, T. Esslinger, Preprint cond-mat/0105105 (2001)

34. D. Polivaev, *Einfluss von Rauschen in quantenmechanischen Verschränkungsoperationen mit ultrakalten Stößen* (in German), Master Thesis, LMU Munich 2000
35. R. Scheunemann, F.S. Cataliotti, T.W. Hänsch, M. Weitz, Phys. Rev. A, **62**, 051801(R) (2000)
36. S. Friebel, C. D'Andrea, J. Walz, M. Weitz, T.W. Hänsch, Phys. Rev. A, **57**, R20, (1998)
37. M. Weitz, IEEE J. Quantum Electron. **36**, 1346 (2000)

Efficient Generation of Polarization-Entangled Photon Pairs with a Laser Diode Source

Christian Kurtsiefer, Markus Oberparleiter, Jürgen Volz,
and Harald Weinfurter

Entanglement, as E. Schrödinger called it "the essence of quantum mechanics", is not only an essential ingredient of quantum effects but it became also the most important resource of quantum information processing and communication. Entangled photon pairs, created initially by electron–positron annihilation and later in atomic cascade decays, were first used in distinctive comparisons of various concepts of quantum mechanics [1, 2]. More recently, parametric fluorescence (spontaneous parametric down-conversion, SPDC) in nonlinear optical crystals as the source of entangled photon pairs [3–5] lead to a dramatic increase in count rates and experimental performance. This enabled a variety of experiments on the foundations of quantum mechanics [6–8] and the experimental realization of new concepts in quantum information [9–11]. In spite of that success, most of the experiments and potential applications still suffer from the low yield of the fluorescence process. When our new group joined the chair of Theodor Hänsch about two years ago, it was thus the motivation for our work to optimize collection efficiency and thereby the available rate of polarization-entangled photon pairs from parametric down-conversion and to make polarization-entangled photon pairs useful for real applications. Inspired by the many examples where Theodor Hänsch showed how to make experiments simpler, more reliable, and better, and comfortably backed by the huge experience of his groups with laser diode technology and nonlinear optics, we achieved, for the first time, the generation of polarization-entangled photon pairs from a compact laser diode source.

The performance of any application of entangled photon pairs in quantum communication and quantum metrology strongly depends on the single-photon detection efficiency. Silicon avalanche photodiodes (Si APDs) are widely used, because they show low noise and a detection efficiency of up to 70% for $\lambda = 600 - 900\,\mathrm{nm}$. To generate photon pairs in this regime, the pump wavelength for SPDC has to be shorter than 450 nm. Since currently no single-mode solid-state laser with an output power of more than a few mW is available for such wavelengths, large-frame ion lasers are usually used as the pump source. Yet, entangled photon pair sources based on ion lasers are large and expensive and not suited for practical applications.

In most of the experiments performed, the photons have been collected into spatial modes defined by apertures, and the spectrum of the collected light has been defined with optical filters to a given bandwidth $\Delta\lambda$ around the wavelengths λ_i and λ_s of that signal and idler photons. Yet, in many interferometric experiments requiring a well-defined spatial mode or for transport over larger distances it is essential to couple the light from the parametric down-conversion into single-mode optical fibers. In a recent experiment testing Bell inequalities with space-like separated observers [12], typical pair rates of $13\,000\,\mathrm{s}^{-1}$ for $\Delta\lambda_{\mathrm{FWHM}} = 3$ nm at a pump power of 400 mW from a large-frame argon ion laser have been achieved (crystal thickness 3 mm) [13]. The pair rate there was already an order of magnitude higher than in the initial experiments using type-II down-conversion sources [5], but for real applications of polarization-entangled photon pairs high rates from much simpler pump lasers are desirable. Different techniques have been implemented since then to increase the number of entangled photon pairs emitted in a single spatial mode, including the use of two type-I conversion crystals [14], focusing the pump beam [15, 16], or using resonant enhancement [17, 18]. Quite recently, several groups have succeeded in designing photon pair sources with confining waveguides in periodically poled crystal structures [19, 20], showing unprecedented efficiencies in the generation of correlated photon pairs. However, the achieved (time) entanglement is not satisfying yet.

Here, we first present a simple way to optimize the collection efficiency, using the fixed relation between emission direction and wavelength for a fixed pump frequency. In the second part we combine these improvements with resonant pump enhancement [18] to achieve an overall yield which is high enough to use the light from a frequency-doubled laser diode initially.

1 Efficient Collection of Photon Pairs

In type-II parametric fluorescence, a pump photon with energy $\hbar\omega_{\mathrm{p}}$ is converted in a nonlinear optical crystal into two orthogonally polarized photons, signal and idler, obeying energy and momentum conservation. For a fixed pump-photon momentum and a given idler frequency ω_{i}, there is (in the approximation of infinite crystals) a one-dimensional manifold of emission directions for the idler photon, and a corresponding one for the signal photon with a frequency ω_{s} with $\omega_{\mathrm{p}} = \omega_{\mathrm{i}} + \omega_{\mathrm{s}}$, forming two emission cones. To generate polarization-entangled photon pairs, the orientation of the nonlinear crystal is chosen such that the two cones intersect. This intersection defines two directions along which the polarization of each emitted photon is undefined, but perfectly anticorrelated with the polarization of the other one. Provided complete indistinguishability of which photon belongs to which emission cone, a polarization-entangled pair of photons is obtained [5].

For a first demonstration of this new scheme, we use an argon ion laser in our experiment at a wavelength of $\lambda_{\mathrm{p}} = 351.1$ nm to pump a BBO nonlinear optical crystal (thickness 2 mm), yielding down-converted photon pairs efficiently detectable with Si avalanche diodes. Figure 1 shows the emission of a particular wavelength spread of $\Delta\lambda = 5$ nm around $\lambda = 702$ nm observed for different orientations of the BBO crystal [22]. The intensity distribution $I(\theta, \phi)$ of photon pairs emitted at angles θ and ϕ was observed by mapping the spatial emission with a Fourier lens and a 2-dimensional translation of a silicon avalanche photodiode as single-photon detector.

To optimize collection efficiency, we match the angular distribution of the parametric fluorescence light for a given spectral bandwidth to the angular width of the spatial mode collected into a single-mode optical fiber. Consider the wavelength dependence of the opening angles of the emission cones: for a given spectral width $\Delta\lambda_{\mathrm{i}} = \Delta\lambda_{\mathrm{s}}$, the signal and idler light emitted along the intersection directions is dispersed over an angular width $\Delta\alpha_{\mathrm{i}}$ and $\Delta\alpha_{\mathrm{s}}$, respectively (Fig 2b). We use the approximate rotational symmetry of the emission cones and obtain

$$\Delta\alpha_{\mathrm{i}} = \Delta\alpha_{\mathrm{s}} \approx \Delta\theta_{\mathrm{s}} = \Delta\theta_{\mathrm{i}} = \frac{\mathrm{d}\theta_{\mathrm{i}}}{\mathrm{d}\lambda_{\mathrm{i}}}\Delta\lambda_{\mathrm{i}} \quad , \tag{1}$$

where θ_{s} and θ_{i} are the emission angles between the pump direction and signal and idler light, respectively, in the plane containing the optical axis of the (uniaxial) crystal (see the rings of intersection with a plane normal to the pump beam in Fig. 1). This expression can be obtained from energy and momentum conservation in a closed form, although the numerical solution is

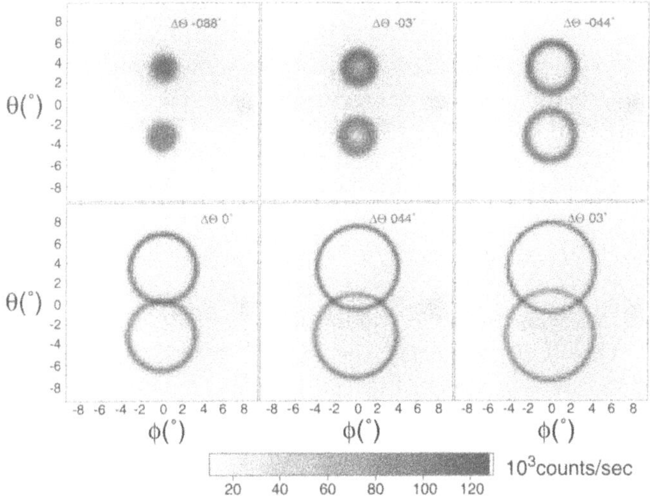

Fig. 1. Measured intensity distribution of down-converted light for different conversion crystal orientations; $\Delta\Theta = 0°$ corresponds to normal incidence on the crystal faces, corresponding to a configuration where the signal and idler components approximately co-propagate with the pump beam (corresponding to an angle $\phi = \theta = 0°$). In the first row, the cones the for signal and idler light never intersect, while for the last two distributions, there are two intersection directions for $\theta = 0°$, allowing polarization-entangled photon pairs to be collected

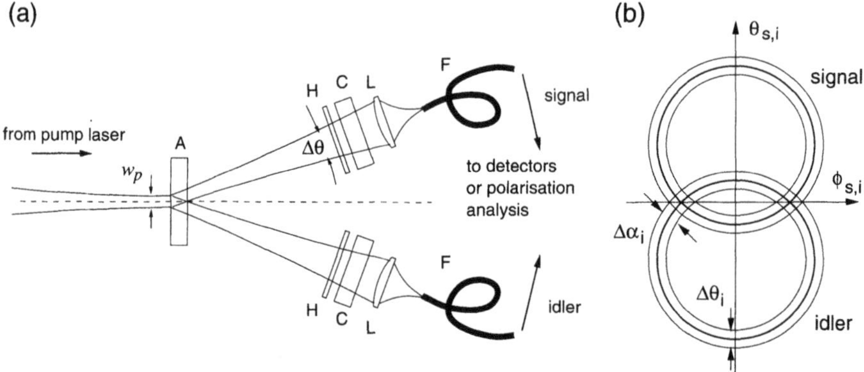

Fig. 2. Basic setup: (**a**) A UV pump laser beam is focused into a BBO crystal A to a waist w_{p}; light emitted by parametric fluorescence is mapped to the receiving modes of single-mode optical fibers F with lenses L. A combination of half-wave-plates H and additional BBO crystals C are used to compensate for walk-off in the first crystal. (**b**) Geometry for the emission of signal and idler photons for a finite bandwidth

usually faster. The pump light is considered as a plane wave propagating at an angle θ_{p} with respect to the optical axis of the nonlinear crystal. A more detailed discussion of the relation between spectral and angular distribution of the down-converted light can be found in [21].

For an appropriate choice of the crystal orientation, the two directional manifolds of signal and idler light intersect perpendicularly. Then, for a Gaussian spectral distribution of the light to be collected, the corresponding angular distributions are also Gaussian with characteristic widths $\Delta\alpha_{\mathrm{s}} = \Delta\alpha_{\mathrm{i}}$, and have rotational symmetry around the intersection directions. We now define Gaussian target modes aligned with the intersection directions of the emission cones with a divergence $\theta_{\mathrm{D}} = \Delta\alpha_{\mathrm{s,i}}$, which can be mapped optically to the collecting fibers. The beam waist of these Gaussian modes is given by $w_0 = \lambda/(\pi\theta_{\mathrm{D}})$. As the mode matching for the parametric down-conversion is described in a plane wave basis, we locate the waist of this Gaussian mode in the conversion crystal, as sketched in Fig. 2a. For our configurations, the crystal is always shorter than the Rayleigh length, $z_{\mathrm{r}} = \pi w_0^2/\lambda$, of the corresponding modes; thus we neglect possible effects due to wavefront curvature in the conversion crystal.

Having chosen the target modes, we only collect photons created in the region of overlap between the target modes and the pump field (Fig. 2a). Therefore, the pump field can be restricted to a region where the target modes have a significant field strength. To maximize overlap with the Gaussian pump field, we choose its waist w_{p} to be equal to the waist of the target modes in the crystal.

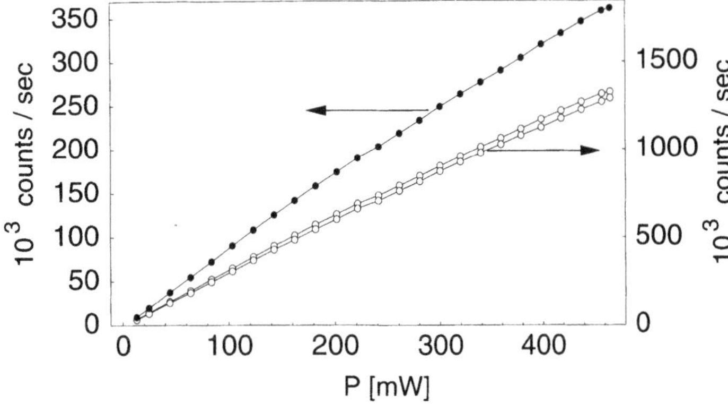

Fig. 3. Coincidence count rates (*left axis*) and individual count rates (*right axis*) of two photodetectors connected directly to the single mode fibers as a function of UV power.

For a pump wavelength of 351 nm obtained from an Ar ion laser, the two cones of signal and idler light at the degenerate wavelength $\lambda_s = \lambda_i =$ 702.2 nm intersect perpendicularly for a pump beam orientation of $\Theta_p =$ 49.7° with respect to the optical axis, resulting in an (external) angle of $\phi_{s,i} = \pm 3.1°$ with respect to the pump beam and an angular derivative of $|d\theta_i/d\lambda_i| = |d\theta_s/d\lambda_s| \approx 0.055°/\text{nm}$ (Fig. 2b).

Aiming for a spectral width of $\Delta\lambda_{\text{FWHM}} = 4$ nm we accordingly have chosen a divergence angle $\theta_D = 0.16°$ for the target modes, corresponding to a Gaussian beam waist of $w_0 = 82$ µm. These modes were geometrically mapped with aspheric lenses ($f = 11$ mm) to the receiving modes of single mode optical fibers with a Gaussian waist parameter determined to be $w_f =$ 2.3 µm.

To determine the collection efficiency, we connected the single mode optical fibers directly to two actively quenched silicon avalanche diodes. Figure 3 shows the coincidence count rates and the count rates for the individual detection events as a function of UV pump power. We determined a record coincidence/single count ratio (i.e. an overall efficiency) of 0.286 ± 0.001 for the whole range of pump power. The maximum coincidence count rate we observed was $360\,800\,\text{s}^{-1}$ at a pump power of $P = 465$ mW. The coincidence time window was measured to be $\tau_c = 6.8 \pm 0.1$ ns; thus accidental coincidence count rates were small over the whole range of pump power. For the low pump power regime, we obtain a slope of 900 coincidence counts per second and mW for our 2 mm long BBO crystal in the single-mode optical fibers.

The entanglement of the photon pairs is analyzed by polarization analysis at each of the fiber outputs. The measured correlation function exhibits a visibility of better than 97%. Using (raw) data from the high intensity experiment for the evaluation of a Clauser–Horne–Shimony–Holt-type (CHSH)

Bell inequality [23], we obtain $S = -2.6979 \pm 0.0034$, i.e. a violation of 204 standard deviations for a measurement time of only one second per angle setting.

2 The Laser Diode Source

This improved collection efficiency allowed us to incorporate one of the many developments of Theodor Hänsch, namely to replace bulky argon ion lasers with compact solid state based sources of blue light [24]. A frequency-doubled laser diode as the pump and an optical resonator for the pump to enhance the type-II SPDC process allowed us to set up an all-solid state source of polarization-entangled photon pairs.

We start with a single-mode cw laser diode (SDL-5431-G1, output power 175 mW) at a wavelength of 856 nm (Fig. 4). The collimated light passes through an anamorphic prism pair to compensate the elliptic beam profile and through optical isolators with a total isolation of more than 70 dB. To achieve high efficiency of frequency doubling, resonant enhancement of the IR light [25] and the high nonlinearity of a $KNbO_3$ crystal (a cut) in a semi-monolithic resonator configuration are used [26]. For this purpose, one end face of the crystal is flat and antireflection-coated (AR) for the fundamental and second harmonic, while the other end face is curved (radius $r = 5$ mm) and highly reflective for both wavelengths. This end face, together with the curved coupling mirror ($r = 25$ mm, AR at 428 nm, reflectivity $R = 95\%$ at 856 nm), forms the resonator with an optical length of 41.6 mm and a pump waist of 13 μm located inside the crystal.

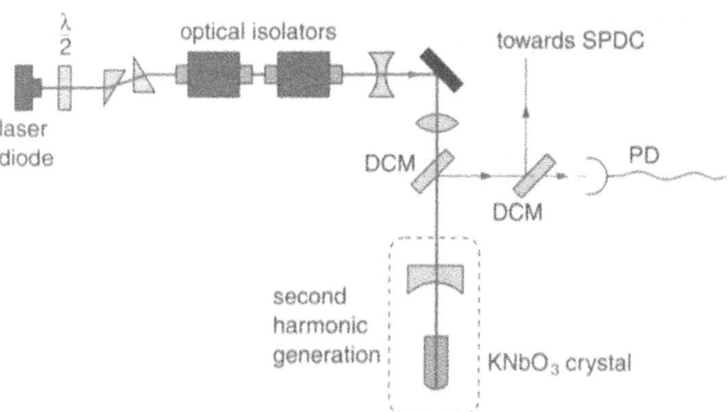

Fig. 4. Schematic setup of the frequency doubler. The light emitted from the laser diode passes through an anamorphic prism pair, two optical isolators and is mode-matched to a resonator, where second-harmonic generation takes place. (DCM: dichroic mirror; PD: photodiode; $\lambda/2$: half-wave plate at 45°; SPDC: spontaneous parametric down-conversion)

From the finesse $F = 80$ of the cavity we derive an enhancement of the pump intensity of 33. To maintain an optimal noncritical phase-matching condition, the temperatures of the crystal and laser diode are stabilized with Peltier elements to 18 and 22°C, respectively, to an accuracy of better than 0.1 K. A fraction of the back-reflected IR light is used to stabilize the cavity on the laser diode frequency with a Pound–Drever-like scheme with sidebands at 50 MHz obtained by current modulation of the laser diode [27].

From a power of 125 mW incident on the doubling cavity we obtained up to 12 mW for the second harmonic. Yet, thermal effects occurring along with the generation of the blue light made long-time stable operation difficult. Stable operation could be achieved by changing the external crystal temperature 1.5 K below the phase-matching temperature.

The generated second harmonic with $\lambda = 428$ nm and a power of about 6 mW is used as the pump light for SPDC, after separating it from the IR light by dichroic mirrors and a bandpass filter. To be competitive with conventional arrangements with typical pump powers of 100 – 500 mW, a significant increase of the photon pair yield is necessary for this source. Thus, the intensity of the second harmonic is enhanced in a second resonator, which has a similar semimonolithic design as above and which contains a BBO (β-BaB_2O_4) crystal for SPDC [18]. The frontside of the 2 mm long crystal (cut for type-II SPDC) is AR-coated for both 428 and 856 nm, whereas the backside is AR-coated for 856 nm and highly reflective for 428 nm. This backside, together with the curved coupling mirror (r=300 mm, R=97% at 428 nm and AR at 856 nm), forms the cavity for the blue light (Fig. 5). Optimization of the coupling efficiency of the down-converted photons into single-mode fibers determines the geometrical parameters of the cavity to a pump waist of 110 μm in the BBO crystal. The optical length of the resonator is, therefore, fixed to 31 mm. We measured a cavity finesse of $F = 67.5$, corresponding to an enhancement of the pump intensity of 13.7. The length of the resonator was stabilized to maximum transmission of the blue light using a dither-lock scheme. Impedance matching of the second cavity is important, because back-coupling of blue light into the frequency doubler leads to further instabilities of the second-harmonic power.

The angle between the extraordinarily polarized pump beam and the optical axis of the BBO crystal is 40.3°, so down-converted photons with twice the pump wavelength leave the crystal at a half-opening angle of $\alpha = 4°$. They pass through a half-wave plate and a second (1 mm long) BBO crystal to compensate walk-off [5], and are coupled into single-mode optical fibers, directing the light to Si APDs for detection. Figure 6 shows single-photon and photon-pair count rates measured as a function of SPDC cavity tuning. They exhibit Lorentzian dependency, consistent with the fact that SPDC fluorescence is proportional to the circulating power in the cavity. The achieved rates of about 10 000 coincidences per second in single-mode fibers on resonance (with 6.5 mW pump power) are comparable to or better than the count rates

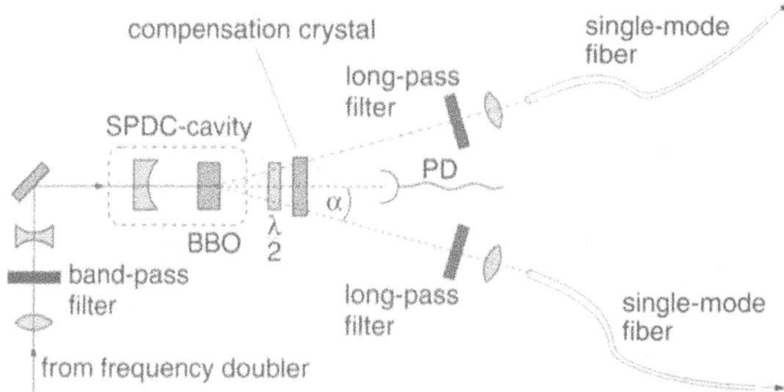

Fig. 5. Schematic view of the setup used to create entangled photon pairs. Blue light is incident on a cavity containing a nonlinear crystal (BBO). The photon pairs generated by SPDC are emitted at $\alpha = 4°$. After passing through a half-wave plate and the compensation crystal, they are coupled into single-mode optical fibers

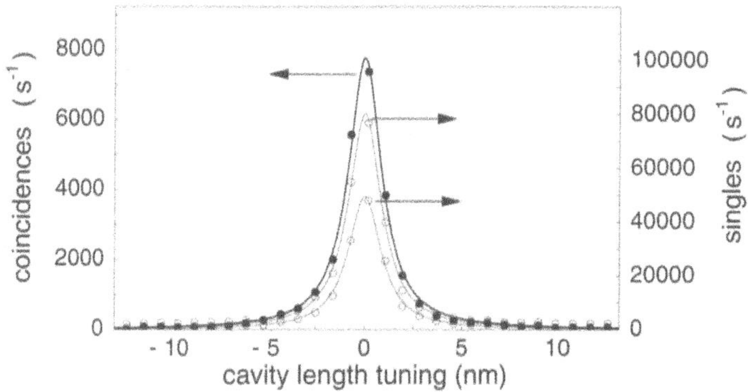

Fig. 6. Single (-○-) and coincidence (-•-) count rates as a function of length tuning of the SPDC resonator (with 5 mW pump power). The solid lines are least square fits of Lorentzians of the measured data.

of most recent down-conversion experiments [28], all operating with large-frame ion lasers.

Again the visibility of the entanglement was at least 95%, resulting in a value of $S = 2.629 \pm 0.0074$ for the standard CHSH-Bell correlation, which exceeds the maximal value of 2 allowed by local realistic theories by 85 standard deviations from an integration time of only 5 seconds per data point.

3 Conclusion

In this chapter we have presented a technique for optimizing the collection of entangled photon pairs created in a type-II parametric fluorescence process using mode-matching procedures. We thereby were able to increase the yield of photon pairs into single-mode optical fibers by two orders of magnitude [5], supplying a strong source for the quantum information experiments that have attracted attention recently. This high yield allowed the construction of an all-solid-state source of polarization-entangled photon pairs in the near-infrared region (NIR). The photon pair rates are comparable to those observed with ion laser systems, but at much lower costs and experimental effort. The advent of reliable blue single-mode laser diodes will allow a further simplification of the setup. This system, therefore, provides a compact, cheap, and easy-to-handle alternative for many applications in quantum information, photon pair enhanced microscopy, quantum metrology, or other experiments, where the continuous availability of polarization-entangled photon pairs in the NIR region is needed.

Acknowledgements

This work was supported by the European Union in the FET/ QuComm research project (IST-1999-10033) and the Deutsche Forschungsgemeinschaft.

References

1. J.F. Clauser, A. Shimony, Rep. Prog. Phys. **41**, 1881 (1978);
2. A. Aspect, P. Grangier, G. Roger, Phys. Rev. Lett. **47**, 460 (1981) **49**, 91 (1982); A. Aspect, J. Dalibard, R. Roger, Phys. Rev. Lett. **49**, 1804 (1982)
3. C.K. Hong, L. Mandel, Phys. Rev. A **31**, 2409 (1985)
4. J.G. Rarity, P.R. Tapster, Phys. Rev. Lett. **64**, 2495 (1990)
5. P.G. Kwiat, K. Mattle, H. Weinfurter, A. Zeilinger, A.V. Sergienko, Y. Shih, Phys. Rev. Lett. **75**, 4337 (1995)
6. L. Mandel, Rev. Mod. Phys. **71**, S274 (1999)
7. A. Zeilinger, Rev. Mod. Phys. **71**, S288 (1999)
8. A.E. Steinberg, P.G. Kwiat, R.Y. Chiao, in *Atomic, Molecular and Optical Physics Handbook* (ed. G. Drake) (AIP press, New York 1996) chap. 77, p. 901
9. D. Bouwmeester, J.-W. Pan, K. Mattle, M. Eibl, H. Weinfurter, A. Zeilinger, Nature **390**, 576 (1997)
10. D. Bouwmeester, J.-W. Pan, M. Daniell, H. Weinfurter, A. Zeilinger, Phys. Rev. Lett. **82**, 1345 (1999)
11. H. Weinfurter, Adv. At. Mol. Opt. Phys. **42**, 489 (2000)
12. G. Weihs, T. Jennewein, C. Simon, H. Weinfurter, A. Zeilinger, Phys. Rev. Lett. **81**, 5039 (1998)
13. G. Weihs, PhD thesis, Universty of Vienna (1999)
14. P.G. Kwiat, E. Waks, A.G. White, I. Appelbaum, P.H. Eberhard, Phys. Rev. A **60**, R773 (1999)

15. C.H. Monken, P.H. Suoto Ribeiro, S. Pádua, Phys. Rev. A **57**, R2267 (1998)
16. C.H. Monken, P.H. Suoto Riberio, S. Pádua, Phys. Rev. A **57**, 3123 (1998)
17. Z.Y. Ou, Y.J. Lu, Phys. Rev. Lett. **83**, 2556 (1999)
18. M. Oberparleiter, H. Weinfurter, Opt. Commun. **183**, 133(2000)
19. S. Tanzilli, H. De Riedmatten, W. Tittel, H. Zbinden, P. Baldi, M. De Micheli, D.B. Ostrowsky, N. Gisin, Electron. Lett. **37**, 26 (2001)
20. K. Sanaka, K. Kawahara, T. Kuga, Phys. Rev. Lett. **86**, 5620 (2001)
21. M.H. Rubin, Phys. Rev. A **54**, 5349 (1996)
22. Ch. Kurtsiefer, M. Oberparleiter, H. Weinfurter, J. Mod. Optics, in print
23. J.F. Clauser, M.H. Horne, A. Shimony, R.A. Holt, Phys. Rev. Lett. **23**, 880 (1969)
24. C. Zimmermann, V. Vuletic, A. Hemmerich, T.W. Hänsch, Appl. Phys. Lett. **66**, 2318 (1995).
25. A. Yarif, *Quantum Electronics*, 3rd ed. (Wiley, New York 1988) chap. 16
26. L. Goldberg, M.K. Chun, Appl. Phys. Lett. **55**, 218 (1989); K. Schneider, S. Schiller, J. Mlynek, M. Bode, I. Freitag, Opt. Lett. **21**, 1999 (1996)
27. W.J. Kozlovsky, W. Lenth, E.E. Latta, A. Moser, G.L. Bona, Appl. Phys. Lett. **56**, 2291 (1990)
28. P.G. Kwiat, E. Waks, A.G. White, I. Appelbaum, P.H. Eberhard, Phys. Rev. A **60**, R773 (1999); D.V. Strekalov, Y.-H. Kim, Y. Shih, Phys. Rev. A **60**, 2685 (1999); T. Jennewein, C. Simon, G. Weihs, H. Weinfurter, A. Zeilinger, Phys. Rev. Lett. **84**, 4729 (2000); D.S. Naik, C.G. Peterson, A.G. White, A.J. Berglund, P.G. Kwiat, Phys. Rev. Lett. **84**, 4733 (2000)

Small is Beautiful

Claus Zimmermann

Magnetic microtraps provide steep potentials for paramagnetic atoms. With microfabricated electromagnets trap oscillation frequencies of several 100 kHz can be obtained. With this technique it becomes feasible to guide atomic matter waves in the radial ground state of a magnetic guide, analogous to photons in a single mode fiber. "Quantum circuits" on the surface of an "atom chip" are conceivable. A breakthrough towards this goal has recently be achieved by loading a Bose–Einstein condensate into a surface microtrap.

1 Introduction

Physicists tend to go for the extreme. Questions like, "What is the largest object in the universe, what is the smallest, the hottest, the coldest?" very often lead to new frontiers of knowledge. Special and general relativity, quantum mechanics, particle physics, etc. have been some of the results so far. One can continue to ask more specific questions in the same spirit. What is the largest quantum object we can study? What is the highest laser power we can achieve? Which are the shortest light pulses we can make? What is the best frequency measurement we can do? If one proceeds along this line one is very soon led to the limits of experimental methods and instruments. Here, things become tricky. Theory can provide the relevant notions and interpretations but it does not tell you how to develop new methods and which technological approach will finally allow you to push the limits. What helps is intuition, experience, and luck. One of the lessons I learned from Theodor Hänsch is that experimental culture and progress is often based on a vivid curiosity in new tricks, devices and sometimes just entertaining gimmicks. Constructing Paul traps with a paper clip, or optically pumping jellies from the grocery store, looking for lasing operation (and eventually eating it) are only some famous examples. As a close coworker in his group for more than a decade I have been constantly fascinated by Theodor's wealth of ideas that inspired us every time we had the chance to discuss with him.

If you study something small your instruments should be small too. This sounds simple but it contradicts the common approach. Usually, one tends to start with dimensions that can be detected with the eye and manipulated with the hands. Lasers, for instance, dealing with light beams that are structured on the micrometer scale, were originally large instruments. Now, the trend goes to small semiconductor lasers and soon this laser type will probably dominate in almost all applications. Another example is the scanning tunneling microscope which has been scaled down from the initial size of a trunk to that of a cigarette box. For the subject I will discuss in this chapter, the reduction in size has been even more dramatic. Typical magnetic traps for atoms are still based on large, sometimes superconducting coils. However, in the rapidly growing field of microtraps we now obtain equivalent results with microfabricated setups on the micrometer scale and below.

For us in the Hänsch group the story began in 1990, with a discussion that we had at one of the group seminars routinely held in Schloss Ringberg, sited above one of the scenic Bavarian lakes near the Alps. We had just recently observed the 1s–2s transition of atomic hydrogen in an atomic beam with improved resolution and were now brainstorming about possible future ways to go. At this time hydrogen and sodium had been trapped at MIT and Amsterdam for spectroscopic experiments and the question was, what would be the influence of the Zeeman effect on the ultimate resolution. A theoretical study of Bergeman [1] on quantized states inside a magnetic quadrupole trap suggested the fantastic vision of cooling atoms into the vibrational ground state of motion inside a magnetic trap. After having discussed possible experimental scenarios, Theodor ask me what for years turned out to be the key question. "How can you ever hope to compress the atoms into such a tiny phase space volume". In fact, at that time, there was no way. Now, 11 years later, after optical cooling methods have been perfected, evaporative cooling is routinely used for various atomic species and Bose–Einstein condensation has been discovered, it is again miniaturization which provides the key technique.

to enter the regime of quantized motion for a noncondensed thermal cloud of atoms the energy separation between the harmonic oscillator states must be larger than any other energy of the system, in particular the thermal energy of the trapped atom and its chemical potential. In other words the trapping potential must be steep (not necessarily deep). Since the magnetic field strength of conventional traps can hardly be increased, the trick is to reduce the size of the trap and the elements that generate the magnetic fields. Four years ago we carried out a first experiment based on this approach by combining a permanent magnet with an electromagnet and called it a "tip trap" [2]. Soon we learned about a proposal for miniaturized magnetic traps based on current conductors [3]. In a second experiment we also demonstrated this concept [4], however the atomic motion inside these traps was still far away from being quantized. As often happens, ideas may rest for years only to become suddenly alive at different places at the same time. Now, there is a rapidly growing field of microtraps with experiments all over the world. Recently, a breakthrough has been achieved in Tübingen when we succeeded in loading a Bose–Einstein condensate into a linear microtrap [5]. Now one may pursue the dream of controlled dynamics of atomic matter waves above the surface of an atom chip where detection and manipulation is accomplished by microfabricated sources for electric and magnetic fields.

2 Magnetic Microtraps

Microtraps are instruments to generate magnetic fields with extreme gradients. Large gradients imply strong magnets and small dimensions. Permanent magnets are convenient sources for strong magnetic fields. Cobalt samarium permanent magnets, for instance, typically provide fields on the order of 0.5 tesla and, if combined with ferromagnetic pole pieces, the gradients may exceed several tens of tesla/cm. A possible setup for a spherical quadrupole trap uses a small permanent magnet with a tiny steel needle attached to it (Fig. 1). Steel can be sharpened to a radius below 0.01 mm whereas the brittle material of the permanent magnet does not allow for structuring below 0.05 mm. If placed in a homogeneous external field a spherical quadrupole field is generated in front of the needle with its minimum position and field gradient depending on the external field strength. We have tested this type of trap with a cloud of lithium atoms [2]. The atoms have been collected and cooled in a magneto-optical trap and then adiabatically compressed inside the trap. Temperature and density can be increased by orders of magnitude without reducing the phase space density. This would boost the elastic collision rate by more than a factor of 10 000, however, when collisions set in, the gas becomes unstable due to Majorana transitions: Near the center of the trap the orientation of the atomic magnetic moment cannot follow the magnetic field lines. The magnetic moment flips relative to the magnetic field and the atoms are expelled from the trap.

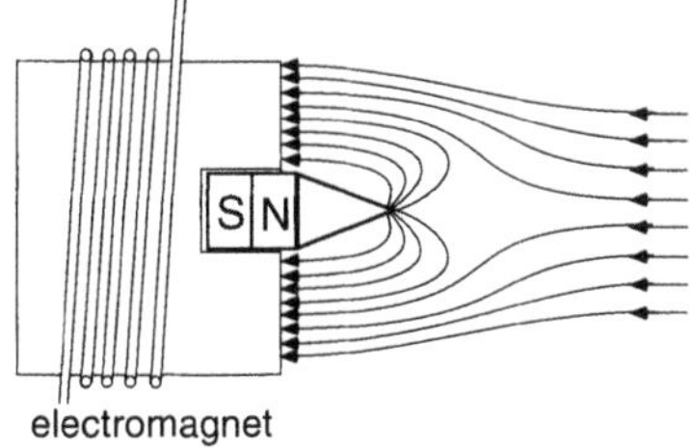

Fig. 1. A permanent magnet with an iron tip is placed in a homogeneous bias field that is generated by an electromagnet. The spherical quadrupole field in front of the tip features an ultra-large gradient and can be used for trapping atoms

An improved trap geometry makes use of a linear quadrupole field that confines the atoms only in two dimensions (x- and y-directions). An additional field oriented along the z-axis establishes the axial confinement and

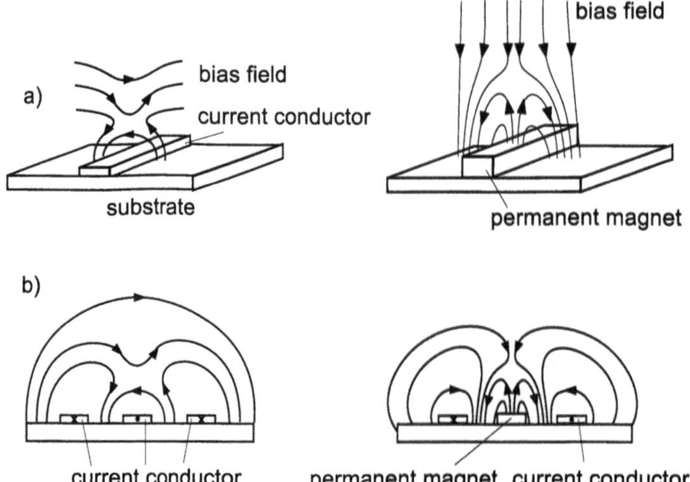

Fig. 2. (**a**) A current conductor in a homogeneous bias field provides a linear quadrupole field suitable for guiding atomic mater waves. A variation of this concept uses a permanent magnet made by a microfabricated strip of gadolinium doped iron. (**b**) The bias field may be generated "on chip" by additional conductors

simultaneously lifts the line of vanishing magnetic field to an offset value that guarantees suppression of Majorana flips. Such a trap can be realized in various ways and is particularly suitable for the design of linear microtraps (Fig 2). The basic setup consists of a thin wire and a more or less homogeneous external bias field. There will be a line of vanishing magnetic field parallel to the wire where the external field and that of the wire cancel. An expansion around this line reveals a linear radial increase and a circular symmetry of the magnetic field modulus (to first order). With a superimposed axial field the trapping potential obtains a parabolic shape and a stabilizing offset field (Fig. 3).

In contrast to a spherical microtrap the atoms in a linear trap are strongly confined only in two (radial) dimensions while the motion along the third (axial) dimension is quasi-free. If many atoms are to be trapped, a spherical microtrap is not very suitable since the density in the vibrational ground state rapidly reaches values such that three-body collisions act as a major loss

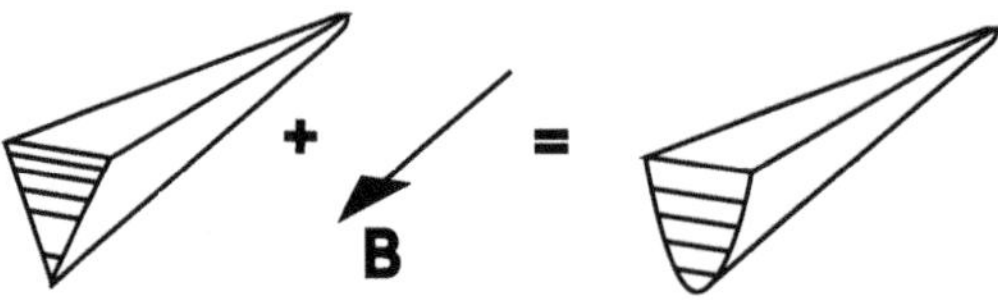

Fig. 3. A linear quadrupole field can be transformed into a harmonic guide by superimposing a homogeneous offset field

mechanism. The microtrap concept thus leads us very naturally to magnetic "fibers" which guide atomic matter waves very analogous to photons in an optical single mode fiber. If the conductors are microfabricated and combined on a substrate by standard etching and electroplating techniques one arrives at integrated "quantum circuits" for atom optics, sometimes dubbed as "atom chips" [6]. Linear quantum waveguides can now be loaded with atoms in the transverse ground state [5]. More complex structures remain to be realized.

3 Loading Microtraps

Quantized motion can only be observed if the atoms occupy a limited number of vibrational states. How many states are available for a single atom in a linear microtrap? If we restrict the atom to the radial ground state, only axial motion is possible. Assuming a parabolic potential in the axial direction, we obtain a nondegenerate equidistant energy spectrum. We can count the possible states by starting with the ground state until we reach the level of highest possible energy. It equals the energy separation between the ground state and the first excited state of the radial spectrum. If we could go to more energetic axial states the atom could use its axial energy to occupy the second radial state and the motion would no longer be one dimensional. Thus, the number of states N is identical to the frequency ratio between the radial and the axial motion, $N = \omega_r/\omega_a$ (Fig. 4).

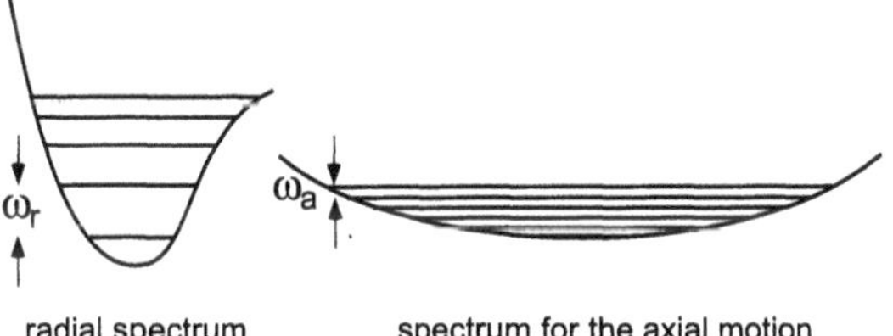

Fig. 4. For the atoms in the radial ground state the number of possible axial states is given by the energy ratio ω_r/ω_a

Technically this ratio is limited to 10^6 by a maximum radial frequency on the order of 1 MHz and a practical axial frequency of about 1 Hz. This means that filling at least one atom from a thermal gas into a microtrap requires a phase space density of better than 10^{-6}. This large value is barely achieved even in laser-cooled gases. Starting with optically cooled gases we cannot hope to get more than a few atoms into the microtrap. In order to load 10^6 atoms into the radial ground state of a linear microtrap one would need a thermal ensemble with a phase space density close to unity. The experimental task of loading a microtrap with a substantial number of atoms is equivalent to making a degenerate quantum gas. In 1990 this seemed to be out of reach for decades. Today, Bose–Einstein condensates and atomic Fermi gases are studied in more than 30 laboratories around the world and the stage is set for microtraps.

4 Adiabatic Transfer

Our approach to combining a Bose–Einstein condensate with a microtrap is conceptually simple. The trick is to construct a magnetic field that can be continuously changed from a rather shallow spherical quadrupole geometry into the tightly confining field of the microtrap. The atoms are initially collected in a MOT and subsequently loaded into the shallow quadrupole trap employing standard techniques of polarization gradient cooling and optical pumping. Now, the atoms are adiabatically compressed into the microtrap by gradual transformation of the magnetic field. The compression enhances the collision rate and thermalization is accelerated. When the collision rate reaches several 10/s the gas can be efficiently cooled by forced evaporation into the degenerate regime.

One example for a variable magnetic trapping field is the above described tip trap. By reducing the current in the electromagnet the position of the trap minimum moves away from the tip. At the same time the gradient flattens and eventually may be used to operate a MOT. Another less obvious example consists of a pair of anti-Helmholtz coils and a wire along the symmetry axis of the coils (z-axis). The total field is a superposition of the spherical quadrupole field generated by the coils and the circular field lines of the wire. The magnetic field lines of the wire have no component along the z-axis and the x-y-components of the quadrupole field are oriented perpendicular to those of the wire (Fig. 5). Thus, the modulus of the total field is simply the quadratic sum of the moduli of the two fields:

$$|B|^2 = |B_{\text{qp}}|^2 + |B_{\text{wire}}|^2 .$$

The potentials generated by each of the two fields alone simply add up and one obtains a rotational symmetric ring potential with nonzero field at the trap minimum.

Now, let us consider the case where the wire is displaced from the center. This destroys the rotational symmetry and the ring potential is lowered on one side and raised on the other side. The trap minimum reduces from a circular line to a single point and we obtain a Ioffe-type trap where the atoms can be stored without Majorana losses. What happens if we lower the current in the wire? It is clear that for vanishing current the spherical quadrupole field of the coils must be restored. Looking at it more closely, one finds that below a critical current the trap minimum splits up into two, each with a quadrupole field geometry [7]. One quadrupole moves to the wire where the minimum eventually hits the surface and the other one establishes the initial quadrupole between the coils. By running this process backwards,

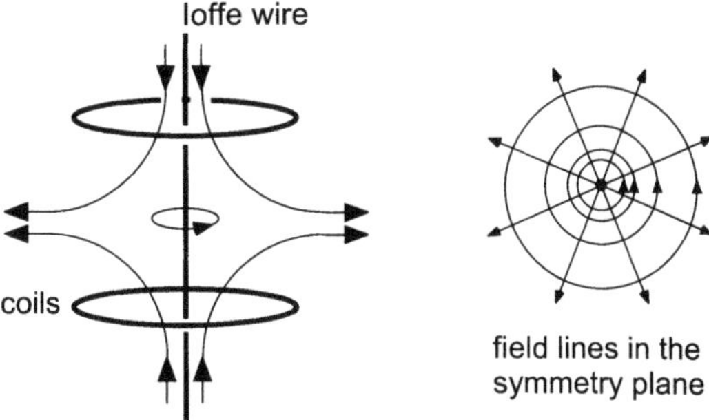

Fig. 5. The combination of two coils and a wire generates a circular trap potential with nonzero field minimum

it can be used to load atoms from a shallow quadrupole into a stable Ioffe-type trap.

We can go one step further. Let us consider the atoms being trapped in the Ioffe-type trap. Now, we add another wire in the plane of the upper coil and displace it by about the radius of the coil. If properly poled, the additional field of the wire will not change the geometric character of the total field (except for a slight distortion). Now, one can slowly turn off the upper coil. The trap minimum of the Ioffe-type trap moves closer to the wire and we arrive at a situation of a microtrap with the lower coil providing the bias field. In the experiment described below we use this transfer scheme to load a microtrap with a Bose–Einstein condensate.

To guarantee adiabaticity, the relative temporal change of the trap frequency during the transfer must be smaller than the frequency itself. Starting with trap frequencies of several 10 Hz the transfer can be completed in less than a second without detectable heating. A second criterion regards the force which originates from the accelerated motion of the trap center during the transfer. This force must be smaller than the restoring force of the trap which requires a smooth change of acceleration. In practice, both criteria are noncritical and easy to fulfill. It is more difficult to find a configuration with sufficient trap depth during the entire transfer. Since heating limits the currents in the electromagnets, the trap depth is often critical. It has to exceed the average kinetic energy $E = 3/2k_{\mathrm{B}}T$ at least by a factor 4 [8]. otherwise, atoms can be evaporated from the trap in an uncontrolled way. Microtraps only work with sufficiently precooled atoms.

5 Bose–Einstein Condensation in a Microtrap

The microstructure in our experiment consists of seven 2.5 cm long parallel copper conductors at a width of 3 μm, 11 μm and 30 μm which are electroplated on an Al_2O_3 substrate (Fig. 6). Test experiments with a set of

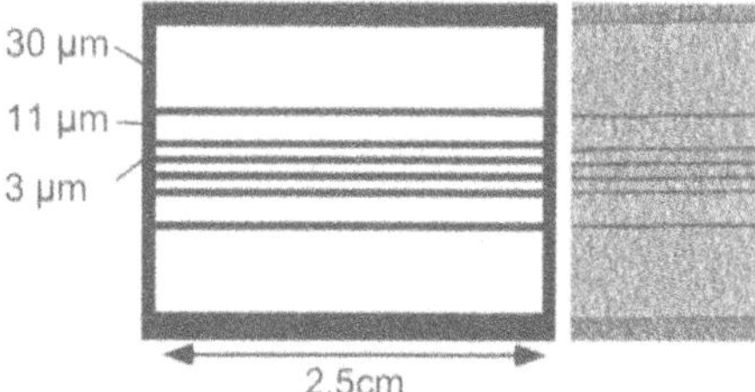

Fig. 6. The microstructure consists of seven electroplated conductors with different widths. A variety of trap geometries and on-chip interferometers can be realized by applying a suitable bias field and current distribution

identical microstructures revealed breakthrough current densities of typically 3×10^6 A/cm^2.

If placed in a bias field that is oriented perpendicular to the conductors and with the currents in all conductors driven in the same direction, the basic quadrupole trapping potential of a linear microtrap is formed. By subsequently turning off the current in the outer conductors the field gradient is increased while the trap depth is kept constant. Further compression can be achieved by inverting the currents in the outer conductors. Then, the bias field is formed by the outer conductors alone and no external field is required. In the final configuration the magnetic field gradient exceeds 50 T/cm. To avoid Majorana spin flips, this linear quadrupole field is superimposed with a magnetic offset field which is generated by two extra wire loops that are mounted directly onto the microstructure substrate, one at each end of the waveguide. With a longitudinal offset field of 1 G, the radial oscillation frequency of the trap amounts to $\omega_r = 2\pi \times 600\,000$ Hz. By choosing a small axial oscillation frequency (e.g. $\omega_a < 2\pi \times 5$ Hz), an aspect ratio of $\omega_r/\omega_a > 100\,000$ can be achieved. A set of parallel waveguides can be formed by a suitable choice of currents in the inner conductors of the microstructure [9]. The two guides may be merged and separated by varying the strength of the longitudinal offset field along the guides, thus allowing for the realization of on-chip interferometers. The microstructure is mounted horizontally on a 2×2 mm^2 copper rod ("compression wire") that is embedded in a heat sink at the bottom of the upper transfer coil (Fig. 7). The conductors on the microstructure are oriented parallel to the compression wire. The setup is completed by a vertical copper wire with 2 mm diameter ("Ioffe wire") that is oriented parallel to the symmetry axis of the transfer coils but displaced by 4 mm.

At a current of 3 A the transfer coils generate a spherical quadrupole field with a gradient of 58 G/cm along the symmetry axis. This field forms a relatively shallow magnetic trap where atoms can be conveniently stored with standard techniques. By increasing the current in the Ioffe wire, the center of the spherical quadrupole is shifted and transformed into a Ioffe-type trapping field, as described above [7]. At a current of 13 A in the Ioffe wire the resulting harmonic trap potential is characterized by an axial oscillation frequency of

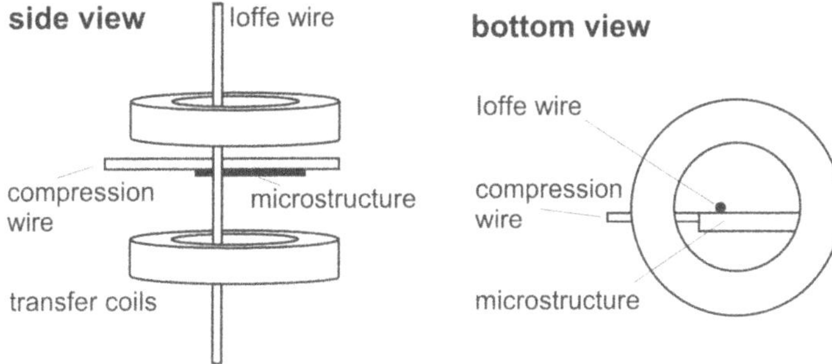

Fig. 7. Magnetic elements for loading the microtrap. The microstructure is mounted upside down on the compression wire. The microtrap is formed below the microstructure

$2\pi \times 14\,\mathrm{Hz}$, a radial oscillation frequency of $2\pi \times 110\,\mathrm{Hz}$ and an offset field of $0.7\,\mathrm{G}$. The transfer from the Ioffe-type trap into the microstructure is completed by changing the currents in the upper and lower coil. This shifts the magnetic field minimum onto the surface of the microstructure.

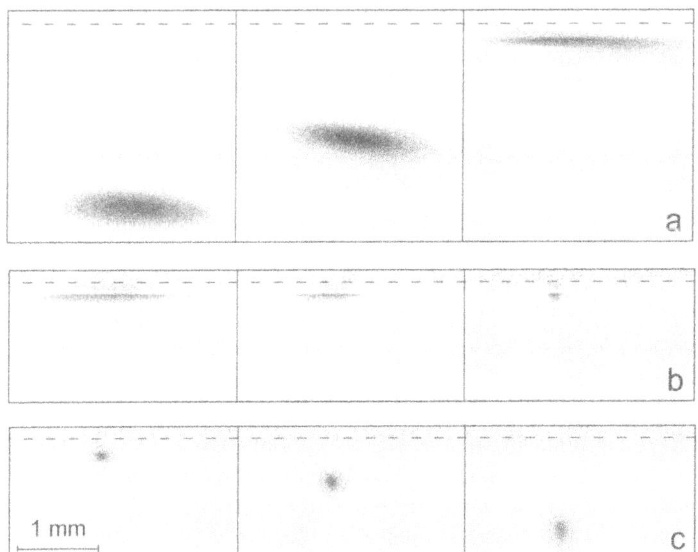

Fig. 8. Absorption images of the cloud during transfer and during compression. The dashed line indicates the surface of the micro structure. (**a**) Transfer and compression of the Ioffe-type trap into the microtrap. (**b**) RF cooling in the microtrap. The image on the right shows the condensate. (**c**) Release of the condensate. The images are taken after 5 ms, 10 ms, and 15 ms time of flight

In a current experiment we precool the atoms in the Ioffe-type trap by forced evaporation to a temperature of 5 μK within 20 s. The final cooling takes place after the atoms have been transferred into the microstructure (Fig. 8). Condensation is reached with 1×10^6 atoms at a critical temperature of 900 nK. For a pure condensate we obtain a chemical potential $\mu/k_B =$ 380 nK, a density $n_0 = 1 \times 10^{15}\,\mathrm{cm}^{-3}$ and number of atoms $N_0 = 400\,000$. The lifetime of the condensate is limited by three-body collisions to 100 ms. After relaxation of the trapping potential the lifetime is increased to 1 s. We don't observe any heating and applying a radio frequency shield has no effect. Further compression of the condensate into the microtrap leads to enhanced three-body collisions and loss of the condensate. To optimize the number of atoms which can be loaded into a magnetic waveguide it will be important to study the collisional properties of the condensate in strongly anisotropic traps and quasi-one-dimensional situations.

6 Final Remarks

How much smaller can one get? Even steeper traps are conceivable with ferromagnetic strips. Gadolinium doped iron can be microfabricated down to a width of 10 nm and magnetized with a coercivity field strength of 0.01 T. Field gradients on the order of 10 000 T/cm are not unrealistic. However, miniaturization is probably limited by the interaction between the atoms and the surface. The distance of the trap minimum from the surface is typically comparable to the width of the micromagnet. Below one micrometer, the atoms start to "see" the surface and heating is predicted due to induced dipole–dipole interactions [10]. This distance is comparable to the average separation between the atoms inside the trap and corresponds to only about 5000 atomic diameters. This marks a new regime. If we cannot clearly distinguish between the atoms in the trap and the atoms of the surface we enter a new field of research dealing with ultracold matter waves on surfaces. But this is another story.

Parallel to our experiments in Tübingen, also Theodor's group in Munich succeeded in making a condensate in a microtrap. Even with a relatively small initial number of atoms they managed to reach degeneracy in less than a second! This impressively demonstrates the enormous potential of microtraps for reducing the complexity of experiments with Bose–Einstein condensates. Such turbo condensation relaxes vacuum requirements substantially and it is not unrealistic to further simplify the setup until it can be made portable. This would bring sources for coherent matter waves in good neighborhood to the optical laser although it might be difficult to reach the same degree of miniaturization. It is hard to say what kind of applications are waiting for the on-chip atom laser. Some are quite obvious, as for instance interferometers for measuring rotations and forces. Also nanostructuring with diffraction limited matter waves is conceivable. However, keeping in mind that 40 years ago nobody would have expected laser bar code scanners in supermarkets, no application seems to be too fantastic to be excluded. I would not be too surprised if Theodor one day showed me during a conference break his new hand-held matter wave interferometer which can read the actual Earth rotation with several digits precision.

References

1. T.H. Bergeman, P. McNichols, J. Kycia, H. Metcalf, N.L. Balazs, J. Opt. Soc. Am. **B6**, 2249 (1989)
2. V. Vuletic, T. Fischer, M. Praeger, T. Hänsch, C. Zimmermann, Phys. Rev. Lett. **80**, 1634 (1998)
3. J.D. Weinstein, K.G. Libbrecht, Phys. Rev. A **52**, 4004 (1995)
4. J. Fortagh, A. Grossmann, T.W. Hänsch, C. Zimmermann, Phys. Rev. Lett. **81**, 5310 (1998)
5. H. Ott, J. Fortagh, G. Schlotterbeck, A. Grossmann, C. Zimmermann, Phys. Rev. Lett., in press
6. R. Folman *et al.*, Phys. Rev. Lett. **84**, 4749 (2000)
7. J. Fortagh, H. Ott, A. Grossmann, C. Zimmermann, Appl. Phys. B **70**, 701 (2000)
8. K.B. Davis, M.-O. Mewes, W. Ketterle, Appl. Phys. B **60**, 155 (1995)
9. E.A. Hinds, C.J. Vale, M.G. Boshier, Phys. Rev. Lett. **86**, 1462 (2001)
10. C. Henkel, M. Wilkens, Europhys. Lett. **47** 414 (1999)

Atomic Looping

Jakob Reichel and Wolfgang Hänsel

We describe an experiment in which cold netral atoms are transported along complex trajectories perpendicular to the surface of a microchip trap. The technique is general in nature, and we demonstrate two surprising particular cases.

1 Complex Magnetic Manipulation of Cold Neutral Atoms

The spectacular developments in laser spectroscopy – many of which are discussed in this volume – have led to the finest possible control over the internal degrees of freedom of neutral atomic ensembles. Methods for manipulating their external degrees of freedom have not quite reached the same level of sophistication yet. From the idea of laser cooling [1] to the achievement of Bose–Einstein condensation of atomic gases in magnetic traps [2–4], the history of this field has doubtlessly been an exciting one, too – but there is plenty of room for further development. For example, most of the magnetic potentials that have been used so far are of the most basic nature: linear or quadratic minima of the field modulus. While these are perfectly adequate for preparing ground-state wavefunctions, there is a huge and mostly unexplored area of cold-atom applications which involve more complex potentials. Theodor W. Hänsch pioneered the exploration of this area when he proposed magnetic devices such as the "atomic motor", also called "atomic conveyor belt". This device consists of a chain of potential wells in continuous movement, and is realized by running modulated currents through a pattern of lithographically created conductors on a chip (Fig. 1b). In its first experimental realization, a cloud of $\sim 10^5$ ^{87}Rb atoms was trapped at a distance of $\sim$ 300 µm from a microchip surface and transported over a distance of 4 mm parallel to that surface (Fig. 2). The wire pattern on the chip was only slightly more complicated than Theodor Hänsch's original sketch (Fig. 1a).

It is not a coincidence that such a device was realized with a planar arrangement of microscopic wires, rather than with macroscopic coils outside the vacuum system. Indeed, magnetostatics provides a strong argument why increasingly complex magnetic fields $\boldsymbol{B}(\boldsymbol{r})$ require current-carrying structures to be located increasingly close to the region of interest [5]. Consider a general

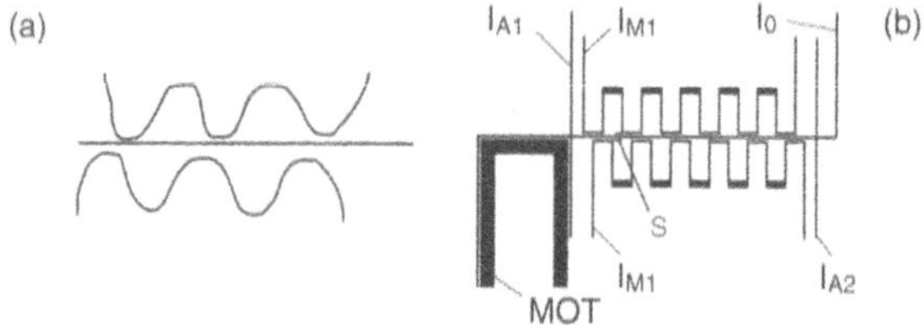

Fig. 1. **(a)** Sketch of the wire arrangement for the "atomic conveyor belt"; **(b)** Complete conductor pattern

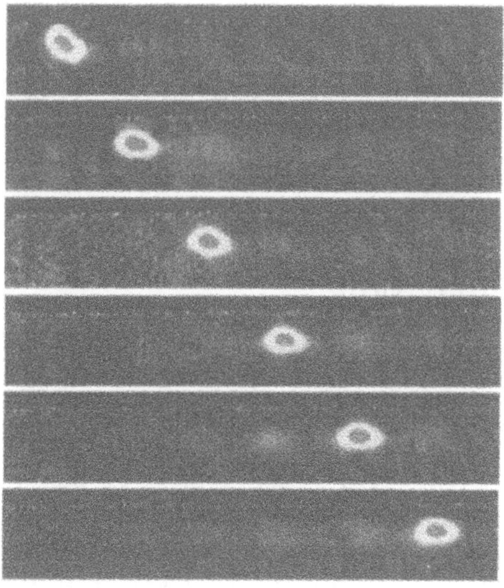

Fig. 2. The "atom motor" at work, transporting a cold atomic cloud over a distance of 4 mm

solution to the Laplace equation, $\triangle\Phi = 0$, with $\boldsymbol{B} = \boldsymbol{\nabla}\Phi$. We are interested in trapping potentials, which must possess a local minimum of B in a current-free region. Let us assume that such a minimum is located at $\boldsymbol{r} = 0$, and develop Φ around this point:

$$\Phi = \sum_{l=0}^{\infty} \sum_{m=-l}^{l} a_{lm} r^l Y_{lm}(\theta, \varphi) . \tag{1}$$

Each term $\Phi_{lm} = a_{lm} r^l Y_{lm}(\theta, \varphi)$ corresponds to a multipole component $\boldsymbol{B}_{lm}$ of the magnetic field. The complexity of the field translates into higher multipole orders l appearing in the sum. The multipole components can be expressed in cartesian coordinates:

$$\boldsymbol{r} \cdot \boldsymbol{B}_{lm} = r \frac{\partial \Phi_{lm}}{\partial r} = a_{lm} l r^l Y_{lm} = a_{lm} l \sum_{a+b+c=l} k_{a\ bc}\, x^a\ y^b z^c . \tag{2}$$

Thus, higher-order multipoles correspond to higher-order polynomials. Far away from the trap center $r = 0$, these will lead to high values of the magnetic field. Consequently, if the field is created by currents which are located far away from the center, then increasing complexity of the field requires increasingly high currents, eventually reaching practical limits. The widely used Ioffe–Pritchard traps employ quadrupole and hexapole fields and already need currents in the hundred ampere region – at the limit of what is practical in a laboratory of humane scale. Conversely, if the field-creating currents are at submillimeter distances from the field region of interest, then complex, high-order fields can be created using modest currents of only a few amperes.

2 A Novel Device

Let us apply the above considerations to a concrete example. The conductor pattern of Fig. 1b is a good starting point for many applications, as it contains a number of wires which are nonparallel and which can carry independently modulated currents. Indeed, on-chip Bose–Einstein condensation in record time [6] was achieved using this substrate. Here we modulate the currents $I_{M1}(t)$ and $I_{M2}(t)$, as well as the external bias field $\boldsymbol{B}(t) = B_x(t)\boldsymbol{e}_x + B_y(t)\boldsymbol{e}_y$ to realize an atomic trajectory which approximates the well-known function

$$\boldsymbol{r}(t) = \begin{pmatrix} x(t) \\ y(t) \\ z(t) \end{pmatrix} = \varrho \begin{pmatrix} \sin(\omega t) \\ \text{const} \\ 1 - \cos(\omega t) \end{pmatrix} , \tag{3}$$

where ϱ is a constant with the unit of length. The latter function is of immense importance in fields as diverse as gyrometry and roller-coaster design. We have chosen the longitudinal starting position ($x = 0$) at the point marked S in Fig. 1b. A numerical optimization of the currents and fields to approximate (3) with $\varrho = 100\,\mu\text{m}$ and $\omega = 2\pi/380\,\text{ms}$ yields the functions plotted in Fig. 3.

Experimentally, we first produce a Bose–Einstein condensate of ~ 3000 atoms as described in [6] and shift it to position S. Subsequent application

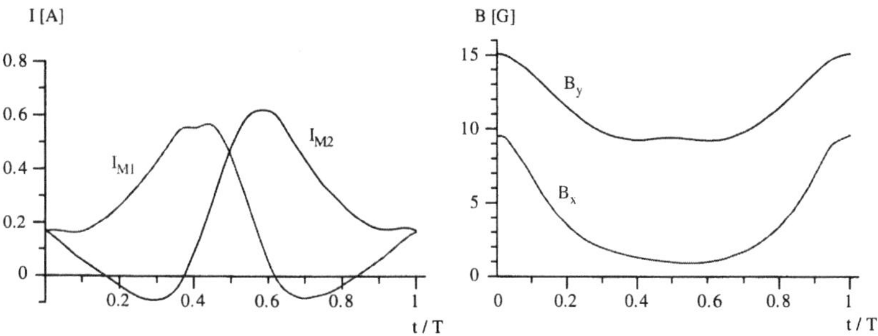

Fig. 3. The currents and external fields to approximate (3)

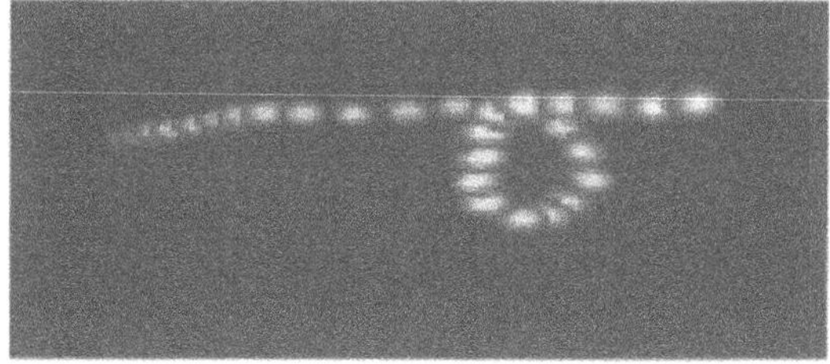

Fig. 4. Atomic trajectory. Superposition of absorption images taken at 30 ms intervals

of the currents and fields of Fig. 3 leads to the atomic trajectory shown in Fig. 4.

This result represents an important advance in the field of artistic cold-atom manipulation. Moreover, it arises some intriguing questions on a more general level. For example, consider an atomic ensemble which is split up coherently before reaching position S. (Suitable techniques for doing this are described in [7, 8].) With the help of additional wires, the left part of the wavepacket can be immobilized while the right part describes the closed trajectory as above. At the end of the cycle, the two parts are merged again. Clearly, the resulting vibrational state of the wavepacket will depend on the accumulated relative phase of the two partial wavepackets. The question thus arises whether this process can be used to realize an integrated atom interferometer with an enclosed area normal to the chip surface. The study of this process is beyond the scope of this article. A related and arguably more relevant system is treated in detail in [8].

We close by noting that even more complex parametrization can be realized if required by the circumstances. An example of a season-induced modulation is shown in Fig. 5, and it is obvious that this modulation causes a double loop structure in the atomic external degree of freedom. Successive absorption images, which are superposed into a single frame, clearly reveal the effect of composite atomic looping (Fig. 6).

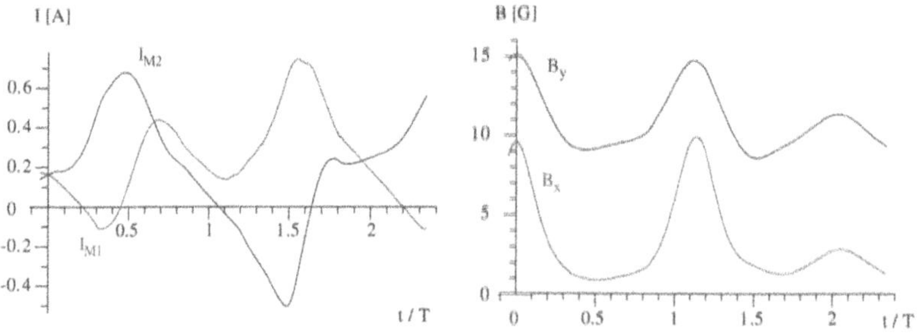

Fig. 5. The currents and external fields to approximate the path shown in Fig. 6

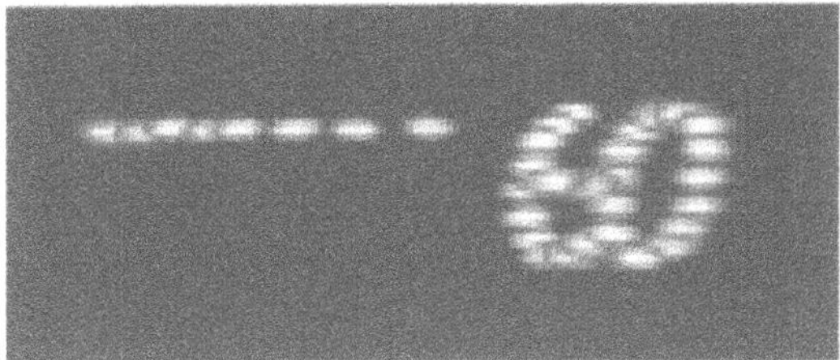

Fig. 6. Best wishes using composite atomic looping

References

1. T.W. Hänsch, A.L. Schawlow, Opt. Commun. **13**, 68 (1975)
2. M.H. Anderson et al. Science **269**, 198 (1995)
3. C.C. Bradley, C.A. Sackett, J.J. Tollett, R.G. Hulet, Phys. Rev. Lett. **75**, 1687 (1995)
4. K.B. Davis et al., Phys. Rev. Lett. **75**, 3969 (1995)
5. W. Hänsel, Ph. D. thesis, Ludwig-Maximilians-Universität München, 2000
6. W. Hänsel, P. Hommelhoff, T.W. Hänsch, J. Reichel, Nature, (2001) in press
7. J. Reichel, W. Hänsel, P. Hommelhoff, T.W. Hänsch, Appl. Phys. B **72**, 81 (2001)
8. W. Hänsel, J. Reichel, P. Hommelhoff, T.W. Hänsch, Phys. Rev. A, (2001) in press

A Toroidal Magnetic Guide for Neutral Atoms

Leonardo Ricci, Andrea Bertoldi, and Davide Bassi

An apparatus aimed at toroidal confinement and manipulation of cold neutral atoms is discussed. The set-up is based on a circular, quadrupolar guide. Atoms are directly loaded into the guide by means of an *in situ* MOT.

1 Introduction

The possibility of manipulating neutral atoms and eventually exploiting their wave properties opens extremely interesting perspectives in atomic physics and quantum optics. The spectacular advances over the last two decades in the fields of laser cooling and trapping of neutral atoms [1] have made possible the development of atom interferometers [2] which have been used for improved measurements of atomic properties, fundamental constants and inertial effects.

The rapidly growing field of atom optics relies on the development of basic devices such as beam-splitters and waveguides. Two main approaches use microfabricated mechanical configurations [2, 3] and photon–atom interactions [4–7]. More recently an approach based on the interaction between a magnetic field and the atomic dipole moment has received great interest. In this way, magnetic waveguides [8–11], mirrors [12], and lenses [13] have been demonstrated. Two main advantages of magnetic techniques are the variety of possible shapes for the field potential and the possibility of scaling down the size of the devices. This aspect is particularly important in order to yield modular components for integrated atom optics set-ups. By means of photolithographically fabricated current-carrying wires, a number of new experiments regarding waveguides [14–16], beam splitters [17, 18] and an atomic conveyor belt [19] have been reported.

Within this research context, we discuss here preliminary results and prospectives for a toroidal magnetic guide for neutral atoms. A planar magnet made of current-carrying wires produces a ring quadrupole field with the zero-field ring having a diameter of 52 mm. For a current of 1.3 A flowing within the wires, the resulting trapping potential has a depth of 1 mT and a gradient of 210 mT/m. The magnet has been designed by means of a Monte Carlo procedure in order to optimize the confinement performance. The trapping region lies 14 mm above the magnet upper surface and is enclosed within a box-shaped Pyrex cell. In this way the magnet lies outside the vacuum apparatus and excellent optical access is possible. A principal characteristic of this confinement geometry is the fact that in the neighborhood of every point of the zero-field ring the magnetic field is approximately linear-quadrupolar, with a nonzero gradient also in the tangential direction. This property has been exploited to realize a magneto-optical trap (MOT) directly within the guide. This provides an evident advantage with respect to similar atom guiding experiments, in which atoms have to be prepared outside and successively transferred into the guide. Once the MOT has been extinguished, the trap becomes a storage ring for the laser-cooled atoms.

By means of additional magnets, the azimuthal components of the field can be rendered non-zero without substantially affecting the radial and axial components. In this way atom losses via Majorana spin-flips are suppressed. Another consequence is the possibility of deforming the trapping potential in order to generate and move local minima along the ring. This suggests a scheme for manipulation of the trapped atoms similar to those used for charged particle [20–22].

Thanks to the field symmetry, several different MOTs can be operated simultaneously, thus increasing the atom occupation within the ring. Moreover, more sophisticated magneto-optical cooling and trapping schemes [23–25] could be used in order to achieve or improve the continuous loading of the ring. Compression of the atoms trapped within the whole ring via field deformation could then allow for achievement of very high atom densities.

The design of the magnet and the realization of the tangential MOT are described in the next two sections.

2 Development of the Toroidal Trap

A main goal in designing a magnetic trapping or guiding set-up is the optimization of the parameters defining the confinement efficiency. For translationally or circularly symmetric quadrupole fields, which provide the easiest magnetic potential with which to guide or trap low-field-seeking neutral atoms, the confinement efficiency is essentially expressed in terms of the field gradient of the points where the field vanishes and the depth of the magnetic potential well. If the field is generated by a current distribution, the gradient scales as I/S^2, I and S being respectively the current flowing in the wires and the effective distance between the magnet and the trapping region. Optimization of the trapping parameters requires the maximization of the first quantity and minimization of the second. However, the pursuit of this task has to comply with various experimental requirements.

The design of a magnet configuration may generally be divided into two stages. The first consists in determining the physical and geometric constraints of the problem. The most important physical constraint is the maximum temperature of the conductors. This parameter, together with the heat dissipation mechanism employed, sets the maximum operating power. For the subsequent determination of the maximum current (and thence of the characteristics of the power supply) one has also to take into account the conductor resistivity and the number of windings.[1] Geometric constraints then define the region in which the current distribution can be placed. The second step consists of finding the optimal current distribution within the boundaries of the preset region in order to maximize a given trapping parameter, for example the field gradient. This task can be carried out analytically only when the current is carried by a few conductors in a symmetrical arrangement. More generally, numerical optimization techniques are required.

In the present project, the aim was to generate a toroidal quadrupole field inside a box-shaped vacuum cell by means of a magnet situated outside the cell. The cell is placed horizontally and has a size of $80 \times 80 \times 40\,\mathrm{mm}^3$. The boundaries of the region hosting the current distribution have been assigned by taking into account the thickness and planarity of the Pyrex windows, requiring the radius of the zero-field circle to be about 25 mm and finally considering the necessity of providing suitable optical access for the laser beams to be used for laser cooling. A sectional view of the region, formed by two square tori, and the cell is shown in Fig. 1.

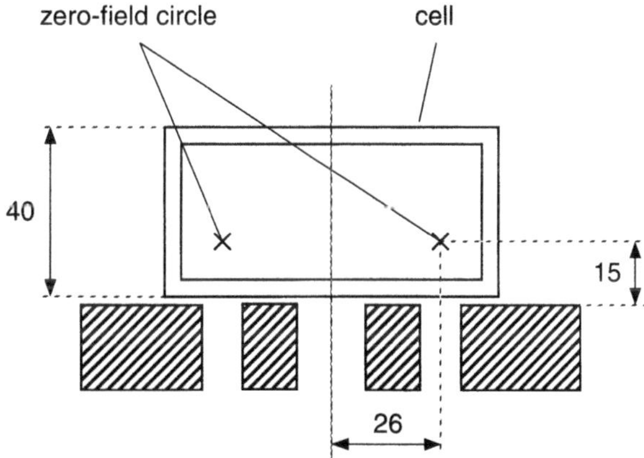

Fig. 1. Cross-section of the region available for the coils (*hatched area*) and the cell. All dimensions are in mm

[1] Given the maximum power and the conductor resistivity, the parameters to play with are the wire length and section. Thin wires provide better filling of the region available for the current distribution, thus providing superior performance if its boundaries have to be very close to the trapping region. However, this choice is limited by the difficulty of coiling a large number of windings and generating and controlling the high voltages required.

In the following, a description in terms of cylindrical coordinates is preferred, with the z-axis coinciding with the symmetry axis. In this way, as far as the cylindrical symmetry holds, the discussion can be restricted to the half-plane $\varphi = 0$, as the tangential components of the vectors concerned vanish.

The task consists of finding the optimal current distribution in such a way that both the magnetic field $\boldsymbol{B}$ and the radial gradient of the z-component of the field $\partial_\varrho B_z$ vanish at the assigned zero-field circle, while the axial gradient of the z-component of the field $\partial_z B_z$ is maximized.[2] The condition on the field ensures that the quadrupole term becomes the most significant in the multipole expansion of the local field around the zero-field circle.[3] The condition on the gradient forces the line-field asymptotes to be oriented along the $\widehat{\boldsymbol{\varrho}}$ and $\widehat{\boldsymbol{z}}$ directions. This aspect is important for the implementation of a MOT by exploting the magnetic field of the guide.

Let I_{max} identify the maximum current that can be tolerated by the wires. A suitable parametrization of the current distribution is obtained by approximating it by means of N ring wires. The k-th wire is then associated a current $I_k \cdot I_{\text{max}}$ and a two-dimensional vector $\boldsymbol{r}_k = [\varrho_k, z_k]$ describing the intersection of the wire with the half-plane $\varphi = 0$, and thus its radius ϱ_k and the distance z_k of its center from the origin. The zero-field circle is analogously identified by the vector $\boldsymbol{r}_0 = [\varrho_0, z_0]$. Each dimensionless parameter I_k must satisfy the requirement:

$$|I_k| \leq 1 , \forall k .$$

As the gradient scales linearly with current, it is convenient to carry out the calculation for a unitary I_{max} and then rescale the results for the actual value.

Both the magnetic field generated by each ring wire and the field gradient along any direction also have zero vanishing tangential components. Consequently, it is sufficient to have an arbitrary fourfold set of different wires and to weigh their currents properly in order to fulfil the requirements of vanishing $\boldsymbol{B}$ and $\partial_\varrho B_z$ at the assigned zero-field circle. As a result, three proportionality relations between the four currents are derived. These are therefore determined except for a scaling factor, which also determines the axial gradient $\partial_z B_z$ via a linear relation.

Optimization of the current distribution has accordingly been carried out by means of a Monte Carlo algorithm. The set of current values $\{I_k\}$, which is initially set to zero, undergoes a directed evolution as follows. At each step, a subset of four different wires is randomly chosen. The four respective currents are then evaluated by imposing the nulling conditions on field and gradient, scaling the current values in such a way that the highest current

[2] Due to static field divergence and rotation laws, this procedure provides complete knowledge of all gradient components.

[3] This occurs except for particular configurations in which also the quadrupole component vanishes.

modulus equals a preset value dI and choosing the current signs in order to have a positive contribution to $\partial_z B_z(\boldsymbol{r}_0)$. The resulting currents are added to the previous values thus generating a new set $\{I_k\}$. If one or more elements of the new set violate the condition on the current, the entire set is renormalized so that the largest current value has unitary modulus. Finally, the new set is accepted as a starting point for the subsequent step if the new gradient $\partial_z B_z(\boldsymbol{r}_0)$, evaluated by summing the contributions from all wires, exceeds the previous value. Because of the exclusive acceptance criterion, the algorithm can be classified as a zero-temperature Metropolis.

The choice of the parameter dI, i.e. the step length, is crucial in order to achieve fast convergence to the optimal solution. As a general rule, during the computation this parameter must be lowered from the starting value of 1 by several order of magnitude. The algorithm described rapidly provides the optimal solution, the resulting value for $\partial_z B_z(\boldsymbol{r}_0)$ being evaluated for a unitary $I_{\max}$.

An example of the current distribution evolution is shown in Fig. 2. Each toroidal region is split into two parts where the current flows in opposite directions. In each wire the intensity reaches the maximum value. The boundary profile between the parts makes the construction of such coils non-trivial.

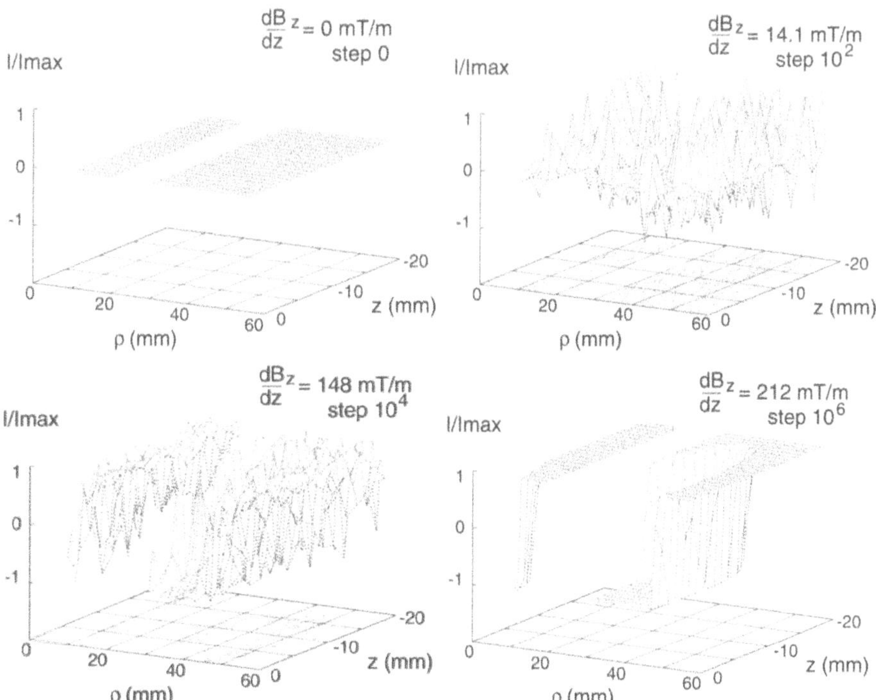

Fig. 2. Monte Carlo evolution of the current distribution over 10^6 steps. The gradient values refer to a maximum current $I_{\max}$ of 1.3 A

For this reason, a more practical design has been implemented. The magnet constructed is formed of three concentric coils with rectangular sections. The dimensions are chosen in order to best approximate the optimal distribution of Fig. 2. While the gradient remains basically unvaried with respect to the optimal value, the quadrupole asymptotes turn out to be tilted by about 6° with respect to the $\widehat{\boldsymbol{\varrho}}$ and $\widehat{\boldsymbol{z}}$ directions. However, as discussed later, this fact does not affect the functionality of a MOT on the ring guide.

The measured gradient $B' \equiv \partial_z B_z(\boldsymbol{r}_0)$ amounts to 210 mT/m for a maximum current $I_{\max}$ of 1.3 A, in very good agreement with the calculations. For the same current the trap depth is about 1 mT. The cross-section of the trapping region through the half-plane $\varphi = 0$ is egg-shaped, with radial and axial dimensions of about 10 mm and 16 mm, respectively.

The field in the neighborhood of the zero-magnetic field circle can be modelled as follows:

$$\boldsymbol{B} \cong B' \left[\frac{1}{2} \left(\varrho - \frac{\varrho_0}{\varrho^2} \right) \widehat{\boldsymbol{\varrho}} - z \widehat{\boldsymbol{z}} \right] .$$

3 Tangential MOT

Let P_0 be a point of the zero-magnetic field circle. It is convenient to take the point P_0 as the origin of a cartesian frame xyz, with directions $\widehat{\boldsymbol{x}}$ and $\widehat{\boldsymbol{y}}$ conciding respectively with the radial and the tangential directions at P_0, and $\widehat{\boldsymbol{z}}$ still indicating the direction parallel to the configuration axis.

The field in the neighborhood of P_0 can be expanded in multipole contributions, the n-th term being proportional to the n-th power of $1/\varrho_0$. The zeroth order term has the shape of a linear quadrupole,

$$\boldsymbol{B} \simeq B' \left(x \widehat{\boldsymbol{x}} - z \widehat{\boldsymbol{z}} \right) .$$

As a result, the field component B_y depends on higher-order contributions. In particular, on the y-axis the component B_y is given by:

$$B_y \simeq B' \frac{y^3}{2 \varrho_o^2} ,$$

thus showing an inflection point at P_0. For this reason, on the y-axis the field gradient $\partial_y B_y$ grows quadratically with the distance from the origin.

This field shape has been exploited to generate a tangential MOT. The glass cell is connected with a vacuum chamber (10^{-7} Pa) and a rubidium reservoir. Three retroreflected, mutually orthogonal laser beams with a diameter of 6 mm cross at the zero-magnetic field circle. The vertical beam is parallel to the configuration axis and enters the magnet through a suitable hole. The horizontal beams are placed radially and tangentially to the circle, respectively. All beams are slightly focused down in order to compensate the intensity imbalance due to uncoated cell windows. The light is derived from a grating-stabilized diode laser [26] providing a power of 8 mW and tuned close to the $5\,^2\mathrm{S}_{1/2}\,F=3 \rightarrow 5\,^2\mathrm{P}_{3/2}\,F'=4$ transition of ^{85}Rb. A second diode laser provides 2 mW of repumping light.

Approximately 60 % of the power is directed along $\widehat{\boldsymbol{y}}$, the remaining light being equally divided in both other directions. Figure 3 shows the MOT fluorescence. Atoms are collected from the background gas in about 2 s. The atom cloud is elongated in the tangential direction as the result of the quasi-translational symmetry of the local magnetic field. The effective MOT operation along the $\widehat{\boldsymbol{x}}$ and $\widehat{\boldsymbol{z}}$ directions has been verified by changing the polarization of the beams. However, along the tangential direction the atom cloud can be seen for any polarization of the correspondent retroreflected beam. The fluorescence intensity of the cloud reaches its maximum I_1 for the standard MOT σ^+–σ^- polarization pair, and its minimum I_2 for inverted polarizations ($I_2/I_1 \sim 0.2$). This behavior is consistent with a steadily working optical molasse, to which a MOT can be superposed in case of proper beam polarization. Due to the flatness of the tangential component of the field, the tangential magneto-optical confinement is effective only far away from the centre of the beam configuration.

Fig. 3. CCD image of the tangential MOT. The hole in the magnet, having a diameter of 8 mm, is visible in the *lower part* of the image

4 Conclusions

We have described the design of a novel neutral atom storage ring and preliminary experimental results concerning the preparation of cold atoms directly within the magnetic guide. The closed topology of the system presents several interesting aspects.[4] The next step in this work will consist in the demonstration of the ability of the apparatus to manipulate the external degrees of freedom of trapped atoms.

References

1. T.W. Hänsch, A.L. Schawlow, Opt. Commun. **13**, 68 (1975); D. Wineland, H. Dehmelt, Bull. Am. Phys. Soc. **20**, 637 (1975); H.J. Metcalf, P. van der Straten, *Laser Cooling and Trapping* (Springer, New York 1999); S. Chu, Rev. Mod. Phys. **70**, 685 (1998); C.N. Cohen-Tannoudji, Rev. Mod. Phys. **70**, 707 (1998); W.D. Phillips, Rev. Mod. Phys. **70**, 721 (1998).
2. P.R. Berman (ed.), *Atom Interferometry* (Academic Press, San Diego 1997) and references therein
3. O. Carnal, J. Mlynek, Phys. Rev. Lett. **66**, 2689 (1991); O. Carnal, M. Sigel, T. Sleator, H. Takuma, J. Mlynek, Phys. Rev. Lett. **67**, 3231 (1991)
4. M. Kasevich, S. Chu, Phys. Rev. Lett. **67**, 181 (1991)
5. G. Timp, R.E. Behringer, D.M. Tennant, J.E. Cunningham, M. Prentiss, K.K. Berggren, Phys. Rev. Lett. **69**, 1636 (1992)
6. W. Seifert, C.S. Adams, V.I. Balykin, C. Heine, Y. Ovchinnikov, J. Mlynek, Phys. Rev. A **49**, 3814 (1994)
7. M.J. Renn, D. Montgomery, O. Vdovin, D.Z. Anderson, C.E. Wieman, E.A. Cornell, Phys. Rev. Lett. **75**, 3253 (1995); M.J. Renn, E.A. Donley, E.A. Cornell, C.E. Wieman, D.Z. Anderson, Phys. Rev. A **53**, R648 (1996)
8. J. Schmiedmayer, Phys. Rev. A **52**, R13 (1995); J. Denschlag, D. Cassettari, J. Schmiedmayer, Phys. Rev. Lett. **82**, 2014 (1999); J. Denschlag, D. Cassettari, A. Chenet, S. Schneider, J. Schmiedmayer, Appl. Phys. B **69**, 291 (1999)
9. B. Ghaffari, J.M. Gerton, W.I. McAlexander, K.E. Strecker, D.M. Homan, R.G. Hulet, Phys. Rev. A **60**, 3878 (1999)
10. M. Key, I.G. Hughes, W. Rooijakkers, B.E. Sauer, E.A. Hinds, D.J. Richardson, P.G. Kazansky, Phys. Rev. Lett. **84**, 1371 (2000)
11. J. Fortagh, H. Ott, A. Grossmann, C. Zimmermann, Appl. Phys. B **70**, 701 (2000)
12. T.M. Roach, H. Abele, M.G. Boshier, H.L. Grossman, K.P. Zetie, E.A. Hinds, Phys. Rev. Lett. **75**, 629 (1995); C.V. Saba, P.A. Barton, M.G. Boshier, I.G. Hughes, P. Rosenbusch, B.E. Sauer, E.A. Hinds, Phys. Rev. Lett. **82**, 468 (1999)
13. W.G. Kaenders, F. Lison, I. Muller, A. Richter, R. Wynands, D. Meschede, Phys. Rev. A **54**, 5067 (1996)

[4] Recently, a toroidal confinement for Bose–Einstein condensate investigation has been proposed [27].

14. D. Müller, D.Z. Anderson, R.J. Grow, P.D.D. Schwindt, E.A. Cornell, Phys. Rev. Lett. **83**, 5194 (1999)
15. N.H. Dekker, C.S. Lee, V. Lorent, J.H. Thywissen, S.P. Smith, M. Drndic, R.M. Westervelt, M. Prentiss, Phys. Rev. Lett. **84**, 1124 (2000)
16. R. Folman, P. Krüger, D. Cassettari, B. Hessmo, T. Maier, J. Schmiedmayer, Phys. Rev. Lett. **84**, 4749 (2000)
17. D. Cassettari, B. Hessmo, R. Folman, T. Maier, J. Schmiedmayer, Phys. Rev. Lett. **85**, 5483 (2000)
18. D. Müller, E.A. Cornell, M. Prevedelli, P.D.D. Schwindt, A. Zozulya, D.Z. Anderson, Opt. Lett. **25**, 1382 (2000); D. Müller, E.A. Cornell, M. Prevedelli, P.D.D. Schwindt, Y.-J. Wang, D.Z. Anderson, Phys. Rev. A **63**, 041602 (2001)
19. J. Reichel, W. Hänsel, T.W. Hänsch, Phys. Rev. Lett. **83**, 3398 (1999); W. Hänsel, J. Reichel, P. Hommelhoff, T.W. Hänsch, LANL preprint server, quant-ph/0008111
20. V.I. Veksler, J. Phys. (Moscow) **9**, 153 (1945)
21. E.M. McMillan, Phys. Rev. **68**, 143 (1945)
22. H.L. Bethlem, G. Berden, A.J.A. van Roij, F.M.H. Crompvoets, G. Meijer, Phys. Rev. Lett. **84**, 5744 (2000)
23. W. Ketterle, K.B. Davis, M.A. Joffe, A. Martin, D.E. Pritchard, Phys. Rev. Lett. **70**, 2253 (1993)
24. Z.T. Lu, K.L. Corwin, M.J. Renn, M.H. Anderson, E.A. Cornell, C.E. Wieman, Phys. Rev. Lett. **77**, 3331 (1996)
25. L. Cacciapuoti, A. Castrillo, M. de Angelis, G.M. Tino, to be published in Eur. Phys. J. D
26. L. Ricci, M. Weidemüller, T. Esslinger, A. Hemmerich, C. Zimmermann, V. Vuletić, W. König, T.W. Hänsch, Opt. Commun. **117**, 541 (1995)
27. A.S. Arnold, E. Riis, in: *Proceedings of the XVII International Conference on Atomic Physics, Florence, Italy, June 4–9, 2000*

Si_{29} Nanoparticles: A New Form of Silicon

Munir H. Nayfeh

We dispersed bulk silicon into ultrasmall Si nanoparticles (Si_{29}, ~ 1 nm diameter) that have novel optical and electronic characteristics. The particles are ultrabright blue luminescent, two-fold brighter than fluorescein, and efficient under UV or near-infrared two-photon excitation. The blue emission from single particles is readily detectable. In addition to being ultrabright, reconstituted films exhibit stimulated and directed blue emission with high gain, and second harmonic generation. Using a density functional with exchange correlation, configuration interaction and Monte Carlo theory, we constructed a structural prototype for 29 atoms (magic number for Td symmetry). The particle is a filled fullerene, at the edge of the sp^3 diamond-like and the sp^2 graphite-like structures. Five atoms constitute the tetrahedral core. The remaining 24 atoms constitute a highly puckered (wrinkled) cage with a highly radiative quantum confinement-induced Si-Si reconstructed phase, found only in ultrasmall nanoparticles.

1 Synthesis of Si Nanoparticles

We produced silicon nanoparticles [1–8] by pulverizing bulk crystalline silicon using an electrochemical treatment. The process involves gradually immersing the wafer into a bath of HF and H_2O_2 while arranging for an electrical current to skim the top skin of the wafer [2]. H_2O_2 catalizes the etching producing ultrasmall stuctures [10] and cleans impurities and produces a higher electronic and chemical quality with ideal hydrogen termination and no oxygen [2]. This process erodes the surface layer, producing weakly interconnected nanostructures. The wafer is then immersed in an ultrasound bath [1, 2], causing the fragile nanostructure network to crumble into particles. The slightly larger, heavier particles precipitate out and can be turned into a yellow luminescent colloid, while the ultrasmall particles remain in suspension, where they can also be recovered to form a blue luminescent colloid. The process is efficient and the wafer can be recycled. We determined the size of the blue luminescent particles by direct imaging using high-resolution transmission electron microscopy (TEM). A thin graphite grid was coated with the colloid. Figure 1 shows the particles are 1 nm in diameter. Electron photospectroscopy shown in Fig. 2 shows that the particles are composed of silicon with less than 10 percent oxygen. In the procedure, the particles are produced with a hydrogen cap. After synthesis, the hydrogen cap can be removed and replaced by oxygen, nitrogen, or carbon [2, 8, 11]

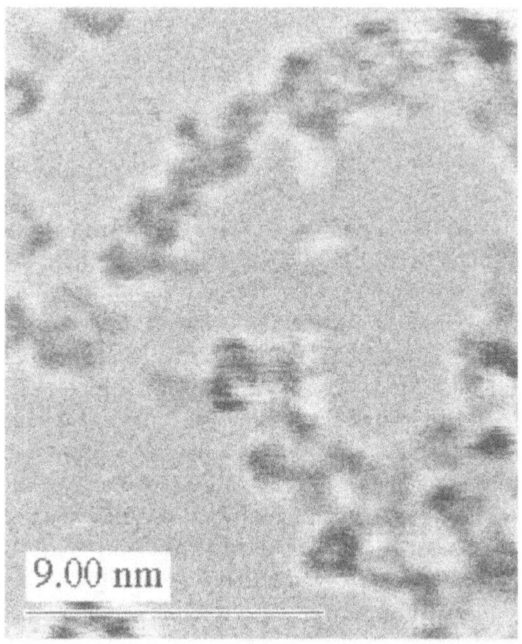

Fig. 1. Transmission electron microscopy image of the Si nanoparticles coating a thin graphite grid

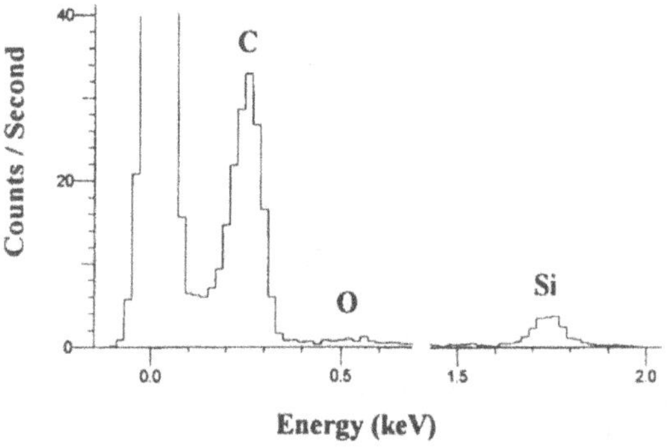

Fig. 2. Material analysis profile of the coated grid using electron photospectroscopy

2 Emission/Detection of Single Nanoparticles

We prepared a colloid of particles with a concentration of ~ 10 nM. The colloid was excited by 355 nm pulsed radiation. The blue emission is observable with the naked eye, in room light, as shown in Fig. 3. Figure 4 gives the spectrum for excitation wavelengths at 330, 350, 365, and 400 nm, showing a strong blue band that maximizes for 350 nm. Because UV is not friendly to biological molecules, we used near-infrared two-photon excitation (700 to 900 nm). Figures 5 and 6 give (at 780 nm) the autocorrelation function of

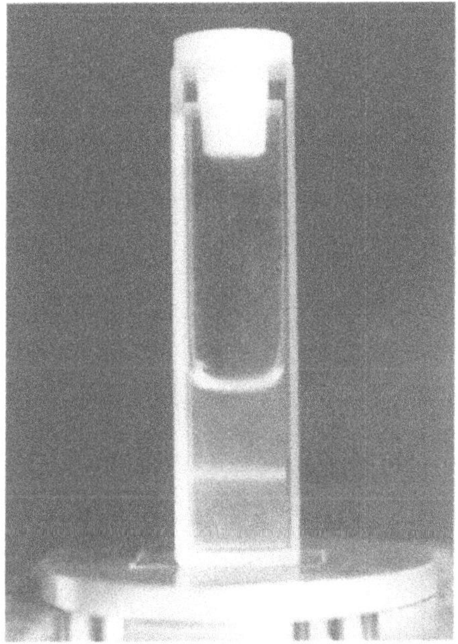

Fig. 3. Photo of a Si colloid excited by 355 nm radiation

the fluctuating time-series of the luminescence [1] with progressive dilution to demonstrate the sensitivity of the detection: Si with 5.4 particles; a fluorescein standard with 1.5 molecule; Si with 2.75 particles; and Si a single particle (0.75) in the focal volume. The measurements yield a particle size of ~ 1 nm, consistent with direct imaging by TEM. They also yield a brightness fourfold larger than that of fluorescein. The photostability was tested by targeting stationery particle frozen in a gel. Figure 7 gives examples of luminescent images, demonstrating the ability to observe/image single particles. "Parking" the excitation beam, focused to an average intensity as high as 10^6 W/cm^2, on stationary particles shows that the particles are photostable (see Fig. 8).

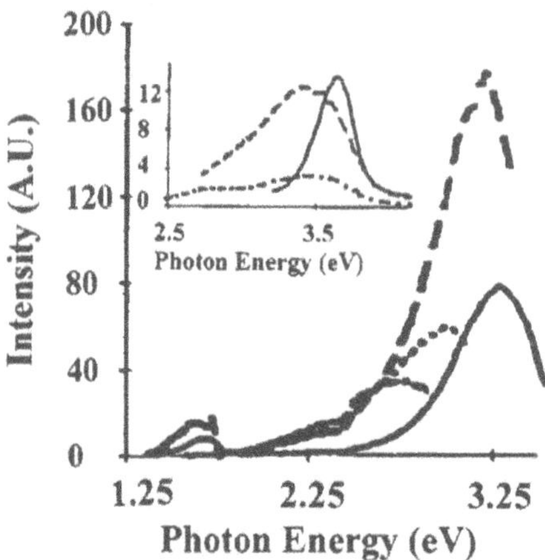

Fig. 4. Emission spectra of the particles at excitation photon energy of 330 nm (*solid*), 350 nm (*dash*), 365 nm (*dot*), and 400 nm (*thick dot*). Excitation spectra (*inset*) while monitoring the emission at 400 nm (*solid*), 500 nm (*dash*), and 600 nm (*dash-dot*)

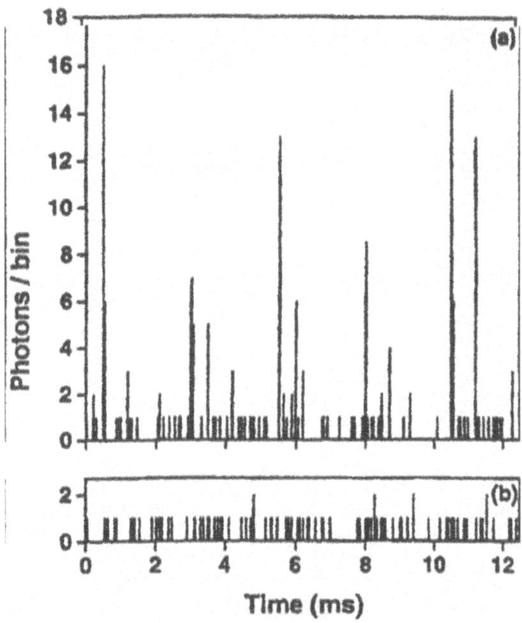

Fig. 5. Raw traces of (**a**) Si particles and (**b**) a fluorescein standard

3 Optical Properties of Superlattice Material

In addition to the novel properties that the individual particles provide, there is the potential to engineer additional properties by synthesizing two and

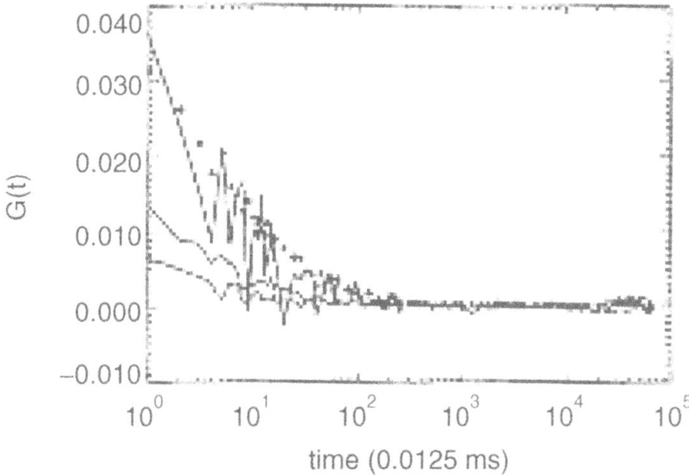

Fig. 6. Autocorrelation of 5.4, 2.75, and 0.75 Si particles (*bottom to top*), and 1.5 fluorescein molecules (*dot*)

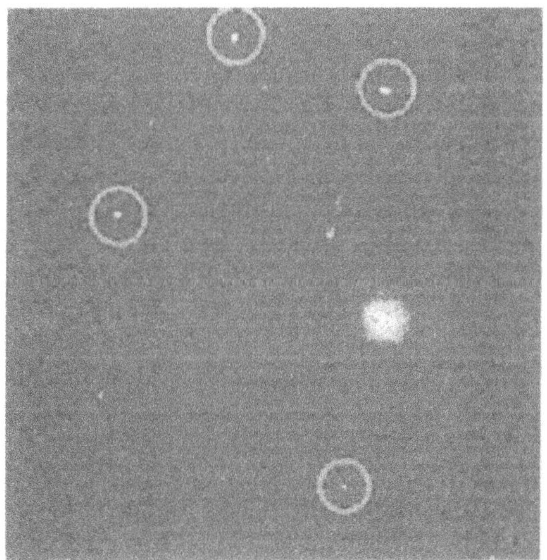

Fig. 7. Luminescent images of frozen Si particles in a gel, demonstrating the ability to observe/image single particles

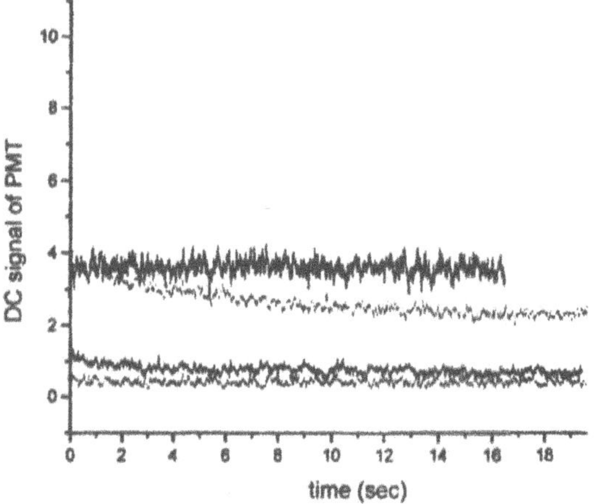

Fig. 8. Emission with time of frozen Si particles in a gel while "parking" the excitation beam on stationary particles shows that they are photostable

three dimensional arrays of the particles. The ability to produce monosize particles gives us the opportunity to manipulate interspacing with atomic precision to tailor new elemental silicon-based material with unique optical and electrical properties. The band gap of the new material is governed by the particle size. The conductivity is controlled by the interspacing (tunneling spacing). The transport is governed by the size, termination, and dielectric properties (charging) of the particles. We used precipitation from a volatile solvent to reconstitute the particles into thin films. By gentle evaporation, the existence of micocrystallites is demonstrated on high-quality Si or silicon oxide, or in free standing mode. The structures are optically clear. Optical imaging in initial experiments showed colloidal crystals of 5–50 μm across (Fig. 9) [4, 5]. We experimented with slower growth rates using slower evaporation rates at reduced temperatures to produce larger, flatter, clearer, and more uniform films. Careful regulation of the temperature allows precise adjustment of the destabilization.

We first examined dispersed particles in a colloid. We performed power dependence studies of the emission under a single-particle condition. Figure 10 shows quadratic dependence as in two-photon processes. Next we examined the dependence from aggregates of particles. Aggregates are found on the solid precursor before sonification and dispersion into individual particles. The intensity of the emission from such aggregates has a sharp threshold, with highly nonlinear emission, rising by several orders. Figure 11 shows the emission intensity under femtosecond pulsed excitation as a function of the average power of the incident radiation [3]. The inset is the log-log of the data. An average power of 15 mW corresponds to an average intensity of $5 \times 10^5\,\mathrm{W/cm^2}$. At the threshold there is a dramatic increase in the slope of the power pumping curve from ~ 1.6 to ~ 11–12, similar to stimulated

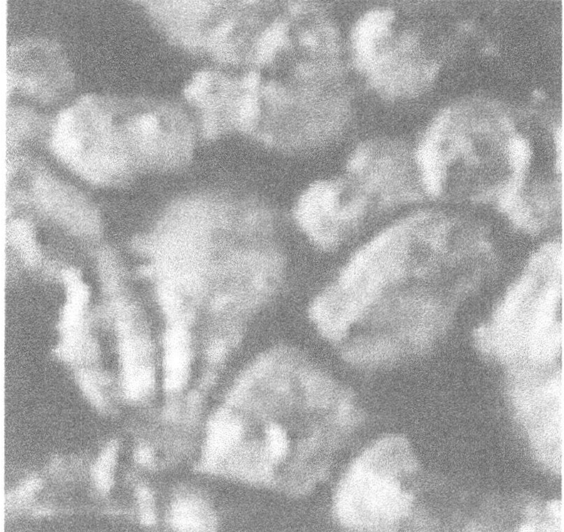

Fig. 9. Optical image of colloidal crystals of 5–20 μm across

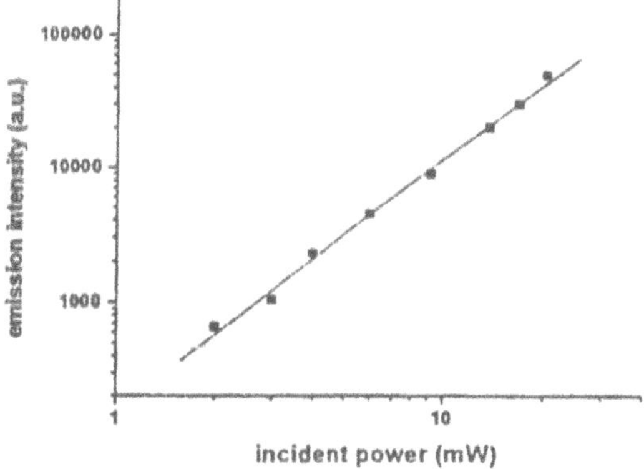

Fig. 10. Power dependence of emission of dispersed particles

emission in inverted systems. The contrast is more dramatic in the case of CW excitation (not shown). The fluorescence is extremely weak, basically unmeasurable for low powers. However, at an average comparable to the threshold intensity in the pulsed case, the CW exhibits somewhat similar threshold behavior.

The region that exhibits the threshold behavior on the solid precursor [3] is abundant, but it is spotty, shallow, and uncontrollable. We created large, thick, uniform and controled layers of the ultrasmall material by reconstituting the dispersed particles as described above. Figure 12 gives the emission intensity from such films as a function of the average incident intensity. It is typical of the response from any part of the film. For low intensity, the emission is finite, but at an average intensity of $\sim 10^6\,\mathrm{W/cm^2}$ ($\sim$ 20–25 mW) the

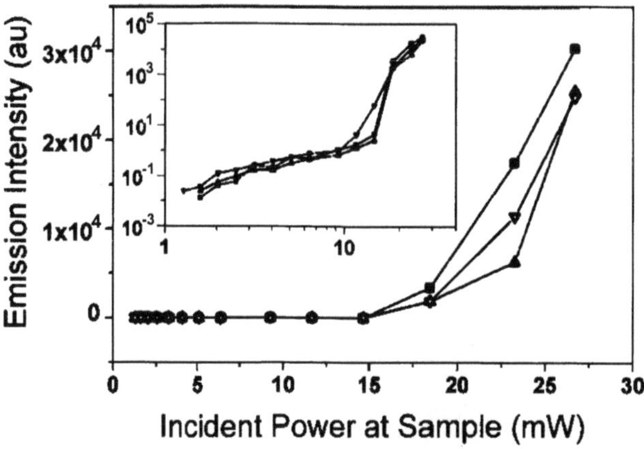

Fig. 11. The emission intensity from a solid precursor under femto-second pulsed excitation as a function of the average power of the incident radiation

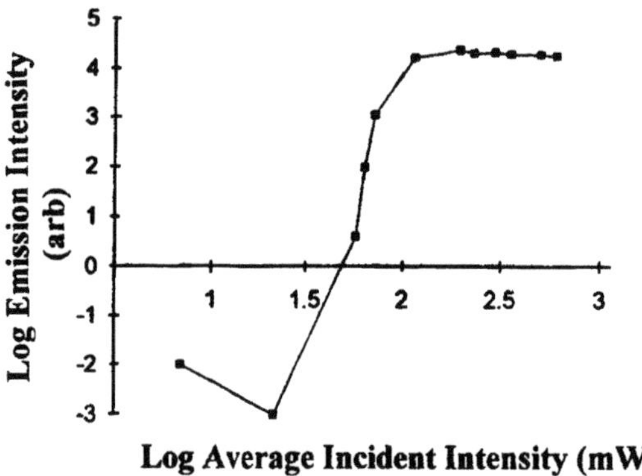

Fig. 12. Emission intensity as a function of the average incident intensity for a reconstituted thin film of Si nanoparticles close to one side provides a reflector of 2π solid angle, i.e., high feedback

emission exhibits a sharp threshold, rising by many orders of magnitude. Beyond the threshold, there sets in a low-order power dependence that saturates at the highest intensities. Figure 13 displays the interaction of the incident beam with the microcrystallites. It is normally incident (normal to the plane of the figure). The four examples in Fig. 13a show that directed blue beams emerge, propagating in the interior of the crystallite, in the plane of the sample, locally normal to the closest side. Due to re-absorption and scattering, it gets narrower as it fades away, propagating only 5–10 μm. Figure 13b shows two cases when the opposite faces of a crystallite are close. The beam in this case strikes the opposite face forming a weaker bright spot. The blue beams are characterized by a threshold. When the incident intensity is reduced, the beam fades away, and disappears, while the interaction spot remains bright

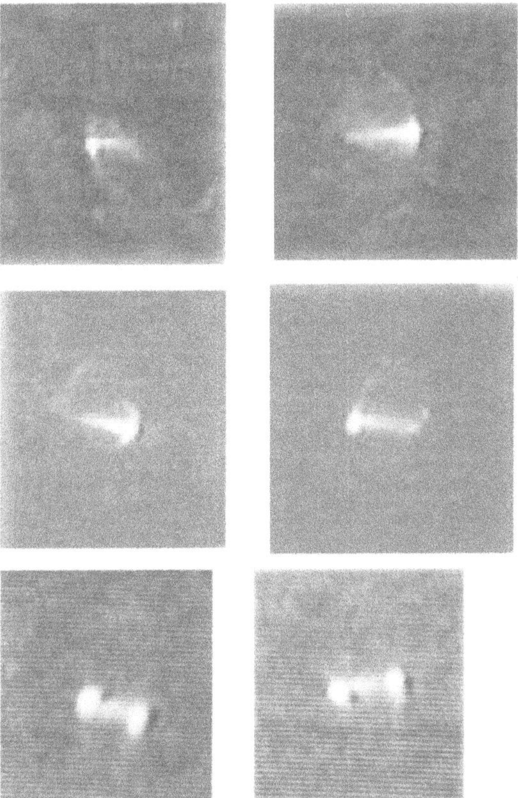

Fig. 13. Photoluminescence images from the microcrystallite region. The interaction spot appears as "white" spot. (**a**) Four examples of a blue beam (*top*). (**b**) Two examples of a blue beam between opposite faces (*bottom*)

(Fig. 14). Geometrically, the blue beam is favored when the incident beam strikes close to a face of a microcrystallite. If the beam requires feedback, then pumping bulk Si is known to have negligible nonlinearity, being zero at the second-order level (not allowed because of centrosymmetry), and very small at the third-order level. We recently reported the first observation of second harmonic generation in films of ultrasmall silicon nanoparticles. Figure 15 gives the emission spectra for the three excitation wavelengths 780, 800, and 832 nm (not of the same intensity). Each shows a peak with a shoulder on the red wing. The shoulders are at 390, 400, 416 nm, half the wavelengths of the incident beam. The peaks in the spectra are at 380, 390, and 406 nm, i.e., blue shifted by 10 nm from the shoulder in each of the spectra. These and other measurments show that the emitted wavelength tracks the incident wavelength. These results point to a mechanism in ultrasmall particles that break the centrosymmetry of bulk Si, the symmetry that inhibits second

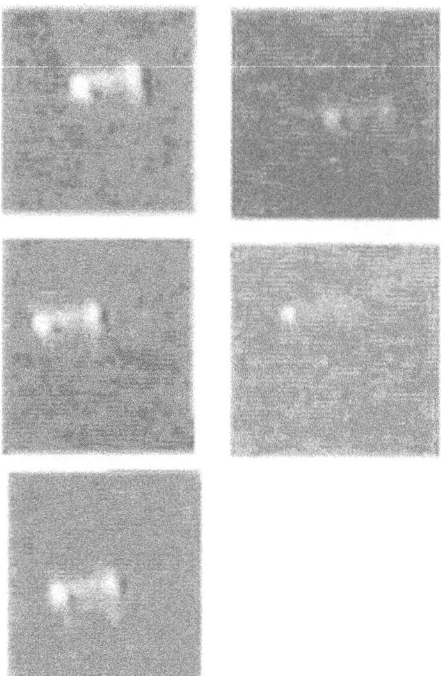

Fig. 14. Photoluminescence images from microcrystallites showing five frames of the blue beam taken under decreasing incident intensity

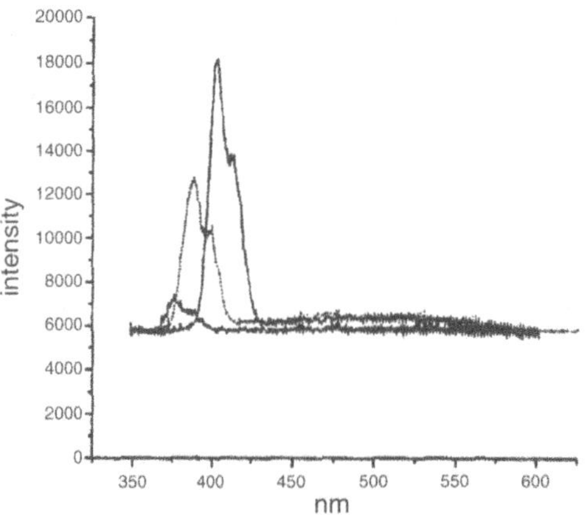

Fig. 15. The second harmonic emission spectra for the three excitation wavelengths (*right to left*) 780, 800, and 832 nm (at different incident intensities) respectively

harmonic generation. For reference, we show the absorption spectrum of the particles in Fig. 16.

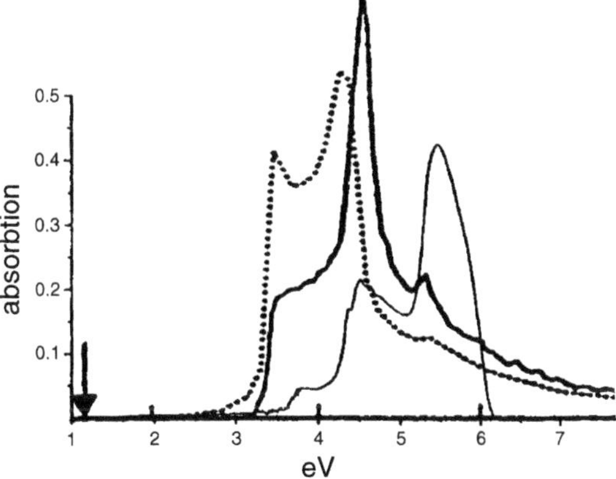

Fig. 16. The absorption coefficient spectrum of the particle colloid (*light solid*), along with the experimental (*dot*) and theoretical (*solid*) results of bulk silicon

4 Structural Prototype of the Particles

The search [7] for a realistic structural prototype started from a spherical piece of crystalline Si which, for the experimentally observed size of ~ 1 nm, contains 29 atoms (magic number for Td symmetry and the spherical shape). All dangling bonds were terminated by hydrogen. However, the corresponding electronic energy gap was larger than 6 eV, suggesting that the observed clusters possess smaller number of terminating hydrogen/oxygen, with part of the dangling bonds saturated by nanocrystal surface reconstruction. By eliminating 12 H atoms we arrived at a structure of $Si_{29}H_{24}$ with six reconstructed surface Si-Si dimers similar to Si(001) surface 2×1 reconstruction. The resulting $Si_{29}H_{24}$ system was then relaxed using the density functional theory (DFT) with the PW91 exchange-correlation functional (Fig. 17), giving, after correcting for the well-known DFT gap underestimation, a band gap of 3.5 eV, close to that observed. We next calculated the optical absorption spectrum as shown in Fig. 18. We employed the configuration interaction singles (CIS) method which constructs the wavefunctions with correct spatial and spin symmetries and which accommodates, to first order, the excitonic effects. The correct overall shift of the spectrum due to the incomplete treatment of the electron correlation effects was estimated from a preliminary quantum Monte Carlo calculations of the first two transitions (T1→ T2) which correspond to the edge of the absorption spectrum. The resulting spectrum with a Gaussian broadening of 0.14 eV, which qualitatively represents the temperature and size averaging, given in Fig. 18, shows agreement with our measurement.

We evaluated the cluster polarizability, a uniquely defined property, using a GGA/PW91 and 6-31G* basis set, giving ~ 793 (a.u.). A general definition of a dielectric constant in semiconductor nanoparticles is not possible since the energy levels are discrete. To arrive at an effective "dielectric constant"

Fig. 17. Configuration structure of a particle that contains 29 silicon atoms, with five core atoms and 24 on the surface. Hydrogen terminating atoms are in white

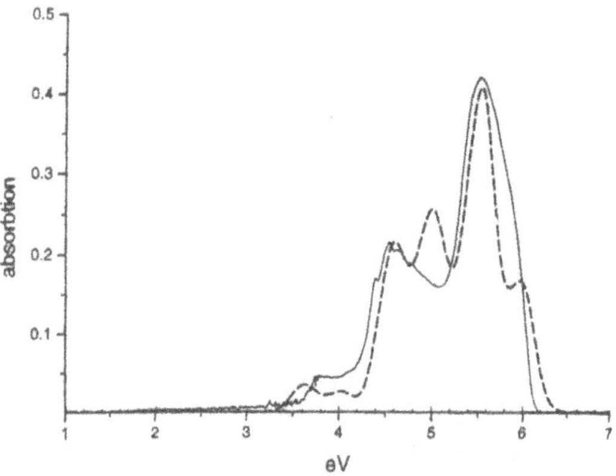

Fig. 18. The calculated (*dot*) and experimental (*solid*) absorption (normalized) of the 29-atom particle

one has to define an appropriate "cluster interior volume". For an effective interior radius (~ 4.2 Å) we get a dielectric constant of ~ 5.7. This value is close to estimations done before by Allan et al. [12] and by Zunger et al. [13]. The Penn model-based analysis [14] gives, for a diameter of 0.8 nm, a more reduced static dielectric constant of 2.

Figure 19 gives a schematic of the interatomic potential of the Si-Si reconstructed bond [12] calculated by Allan et al. using LDF in a 1.03 nm diameter particle, and the various pathways for absorption and emission calculated by Nayfeh et al. [15]. The excited state is a double well with a potential barrier. The inner well (at 2.35 Å) is associated with the tetrahedral configuration and radiates on a long time-scale of milliseconds. The outer well, a new state found only in ultrasmall nanoparticles (< 1.75 nm across), is a trap well (at 3.85 Å) that radiates with lifetimes of 5 ns to 100 µs.

The calculations were performed with the silicon atoms terminated by hydrogen. Only minor changes such as a shift of the bottom of the outer well inward are expected for oxygen terminated dimers since they are less amenable to expansion. The state to which the outer well radiates vertically down is the ground electronic state. But at extended bond lengths the ground state is high-lying and unpopulated, hence this system constitutes a stimulated emission channel, and possible gain. The blue emission proceeds at an interatomic distance at the top of the barrier (~ 3 Å), where the lifetime is in the nanosecond regime and mixing between the two states is most significant. Emission from the bottom of the outer well is of longer time characteristic and of longer wavelength (in the red or near infrared). According to the Frank–Condon principle, absorption proceeds vertically up into the inner well at a bond of 2.35 Å, followed by transfer into the outer by bond expansion via tunneling (double well vibrations) or thermal activation (i.e., self-trapping) as shown in Fig. 20. Also, above-barrier absorption followed by relaxation populates the trapping well. For sizes less than a critical size of ~ 1.4 nm, the trapping edge is lower than the absorption edge, allowing strong transfer to the outer well [15]. Evidence for a potential barrier was demonstrated by manipulation of the material by metal and oxide coatings [16]. We now examine the prospect of gain. The initial gain coefficient [17] is $\gamma = \Delta N \lambda^2 \Delta\nu / (8\pi n^2 \tau)$ where $\Delta N = N_2 - N_1$ is the population inversion, n is the refractive index, $\Delta\nu$ is the emission width, λ is the emission wavelength, and τ is the spontaneous lifetime. Measurements in the precursor colloid show several emission channels with lifetimes of 1 ns, 5 ns, 10–15 ns, and ~ 100 microseconds, with most of the blue emission being in the 10–15 ns time scale. With near saturated absorption, followed by strong transfer to the outer well, we expect the density of the excited emitters to be nearly 25 percent of the atomic solid density ($\sim 1.5 \times 10^{22}/\mathrm{cm}^3$). We use $\tau = 0.01$ to 1 µs, $\lambda = 400$ nm, $n = 2$, and $\Delta\nu = 100$ nm ($\sim 10^{14}$ Hz). With these parameters, $\gamma \sim 1.5 \times 10^3$ to $1.5 \times 1m^5\,\mathrm{cm}^{-1}$. This gain allows considerable growth over microscopic distances. The number of spontaneous emission modes is given

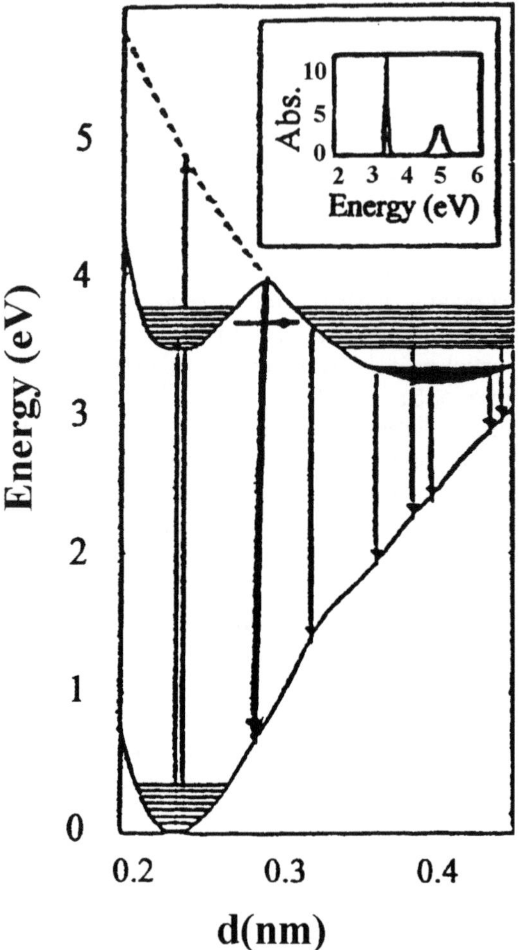

Fig. 19. Interatomic potential of the dimers in 1.03 nm crystallites showing the ground and first excited electronic states, along with the pathways for excitation and emisison. The inset is the transition probability in absorption

by $p = 8\pi\Delta\nu n^3 V/\lambda^2 c$ where V is the active volume. Using an active cross-section of $(2\,\mu\text{m})^2$ and an active sample thickness of $0.2\,\mu\text{m}$, we get $p \sim 10^3$, sizable even for microscopic volumes. The restructured Si-Si bonds provide anharmonicity and optical nonlinearity. To make an estimate, we expand the interatomic potential as a function of the bond length (Fig. 19) about its minimum: $V(r) = ar^2 + Dr^3$ where r is from the potential minimum. The fit gives $mD \sim 5.1 \times 10^{13}\,\text{V/m}^3$, where m is the mass of the electron. From D we can determine the frequency independent nonlinear optical coefficient $\delta = mD\,/\,2e^3N^2$ where N is the density of electrons that contribute to the polarization [17]. The polarization at the second harmonic is proportional to

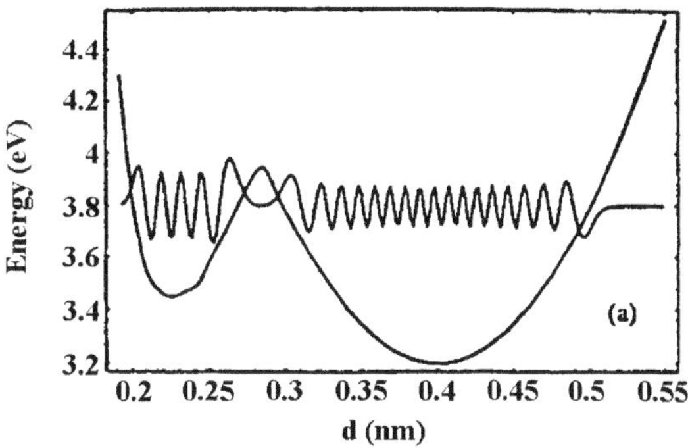

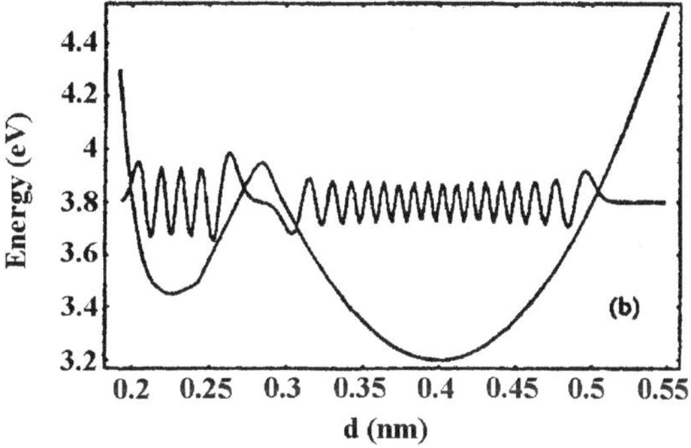

Fig. 20. Double well vibrational wave function (8.33) in a 1.03 nm crystallite: (**a**) bonding and (**b**) antibonding

δ and to the susceptibilities at ω and 2ω: $P^{(2\omega)} = 1/2d^{(2\omega)}\ E_0^2 \cos 2\omega$, where $d^{(2\omega)} = \delta(\chi^{(\omega)})^2\chi^{(2\omega)}$, and $\chi^{(\omega)}$ is the linear susceptibility at frequency ω. This gives $\delta \sim 7 \times 10^{13}$ for $N = 6 \times 10^{28}/\mathrm{m}^3$. Note that the mean value of the parameter δ for 25 noncentrosymmetric crystals known for second harmonic generation is 2×10^9. The blue peak (shifted by 10 nm from the second harmonic) suggests a higher-order nonlinear coherent process such as stimulated anti-Stokes scattering [18]. We previously calculated 42 meV for the vibrational spacing using a numerical matrix method [15]. For an incident wavelength of ~780 nm, this would produce an anti-Stokes line at 759.4 nm, with a shift of 20.6 nm. Harmonic generation of the Raman shifted radiation gives 380 nm, with a shift of 10.3 nm.

5 Conclusion

We converted bulk silicon into ultrasmall nanoparticles. The particles, which are about one billionth of a meter in diameter, contain 29 silicon atoms. We have observed some most unexpected and totally surprising phenomena.

Unlike bulk Si, an optically inactive indirect gap material, ultrasmall Si nanoparticles are extremely active optically, exceeding the level of fluorescein, produce stimulated emission with blue collimated beams, and exhibit harmonic generation. The phenomenon occurs when the crystallite dimensions are brought down to one nanometer or less. In these particles, the atoms, especially the surface ones, are subjected to unbalanced forces, strong enough to reshape or re-engineer the molecular structure. Under these conditions, a new configuration or phase of Si, which otherwise is nonexistent, forms. In a very recent development, we have prepared green, yellow, and red bright particles, and observed red laser action from reconstituted crystallites.

Acknowledgements

The authors acknowledge the State of Illinois Grant IDCCA No. 00-49106, US NSF Grant BES-0118053, the US DOE Grant DEFG02-ER9645439, NIH Grant RR03155, Motorola, and the University of Illinois at Urbana-Champaign.

References

1. O. Akcakir, J. Therrien, G. Belomoin, N. Barry, E. Gratton, M. Nayfeh, Appl. Phys. Lett. **76**, 1857 (2000)
2. G. Belomoin, J. Therrien, M. Nayfeh, Appl. Phys. Lett. **77**, 779 (2000)
3. M. Nayfeh, O. Akcakir, J. Therrien, Z. Yamani , N. Barry, W. Yu, E. Gratton, Appl. Phys. Lett. **75**, 4112 (1999)
4. M. Nayfeh, N. Barry, J. Therrien, O. Akcakir, E. Gratton, G. Belomoin, Appl. Phys. Lett. **78**, 1131 (2001)
5. M. Nayfeh, O. Akcakir, G. Belomoin, N. Barry, J. Therrien, E. Gratton, Appl. Phys. Lett. **77**, 4086 (2000)
6. J. Therrien, G. Belomoin, M. Nayfeh, Appl. Phys. Lett. **77**, 1668 (2000)
7. L. Mitas, J. Therrien, R. Twesten, G. Belomoin, O. Akcakir, M. Nayfeh, Appl. Phys. Lett. **78**, 1918 (2001)
8. E. Rogozhina, G. Belomoin, A. Smith, L. Abuhassan, N. Barry, O. Akcakir, P.V. Braun, M.H. Nayfeh, Appl. Phys. Lett. **78**, 3711 (2001)
9. Z. Yamani, H. Thompson, L. Abuhassan, M.H. Nayfeh, Appl. Phys. Lett. **70**, 3404 (1997); D. Andsager, J. Hilliard , J.M. Hetrick, L.H. Abuhassan, M. Plisch, M.H. Nayfeh, J. Appl. Phys. **74**, 4783 (1993)
10. Z. Yamani, S. Ashhab, A. Nayfeh, M. Nayfeh, J. Appl. Phys. **83**, 3929 (1998); Z. Yamani, A. Alaql, J. Therrien, O. Nayfeh, M. Nayfeh, Appl. Phys. Lett. **74**, 3483 (1999)
11. W.H. Thompson, Z. Yamani, L. Abuhassan, O. Gurdal, M. Nayfeh, Appl. Phys. Lett. **73**, 841 (1998); W.H. Thompson, Z. Yamani, L. Abuhassan, J. Green, M. Alhassan, M. Nayfeh, J. Appl. Phys. **80**, 5415 (1996)
12. G. Allan, C. Delerue, M. Lannoo, Phys. Rev. Lett. **76**, 2961 (1996)
13. L.-W. Wang, A. Zunger, Phys. Rev. Lett. **73**, 1039 (1994)

14. D.R. Penn, Phys. Rev. **128**, 2093 (1962); R. Tsu, L. Ioriatti, J. Harvey, H. Shen, R. Lux, Mater. Res. Soc. Symp. Proc. **283**, 437 (1993)
15. M. Nayfeh, N. Rigakis, Z. Yamani, Phys. Rev. B **56**, 2079 (1997); MRS Proceedings, **486**, 243 (1998)
16. Z. Yamani, A. Alaql, J. Therrien, O. Nayfeh and M Nayfeh, Appl. Phys. Lett. **74**, 3483 (1999)
17. A. Yariv, *Quantum Electronics* (John Wiley & Sons, 1975)
18. Y. Shen, N. Bloembergen, Phys. Rev. **137**, A1787 (1965); J.A. Armstrong, N. Bloembergen, J. Dunning, P.S. Pershan, Phys. Rev. **127**, 1918 (1962)

Molecular Self-Assembly

Wolfgang M. Heckl

The formation of highly ordered monolayers of organic molecules through physisorption-mediated molecular self-assembly at the solid–liquid interface is an example of the spontaneous creation of order. The resulting two-dimensional molecular crystal structures can be determined using the combination of scanning tunneling real-space analysis and diffraction methods together with molecular modeling. Here we present examples ranging from liquid crystals to heterocyclic organic molecules like trimesic acid, representative of a molecular host–guest system. In the case of DNA molecules we have proposed a functional role for this process of spontaneous self-organization for the emergence of terrestrial life, which may also lead towards the construction of genetically based supramolecular architectures for modern technical applications, such as heterogeneous catalysis.

Foreword

It has been exactly 10 years since we published the first real space images of DNA molecules obtained by scanning tunneling microscopy (STM) [1]. After postdoctoral study with Gerd Binnig at the IBM research laboratory, Munich, which was located at the institute of Theodor Hänsch, I was happy to be able to organize a small group under the supervision of Theodor, who had gained much faith in the new technology of scanning probe microscopy. Not only was the laser not at the very heart of our research, it was almost a paradigm shift that Theodor, who became later in 1993 my habilitation father, would enter the field of biomolecules. I am grateful for his critical support within this exciting new field of high-resolution microscopy, where often the artifacts still had to excluded in intense discussions with Christoph Gerber, Gerd Binnig, Doug Smith, Franz Giessibl, Martin Specht and Johannes Pedarnig. What I have enjoyed most in working with Theodor were the afternoons in his private laboratory, where he imparted to me his joy in playing with physics. STM-imaging of self-assembled organic molecules on a crystalline solid surface still has some features of black magic, but since the early times quite a remarkable standard has been achieved [2]. From the early images of liquid crystals [3, 4] to hydrocarbons [5] to bucky balls [6] and all sorts of heterocyclic flat organic molecules [7] progress has been made not only in

STM-imaging and structure determination of the two-dimensional molecular adlayers, but the STM tip has been used for single-molecule manipulation [8, 9]. Here I will present some examples ranging from nucleic acids [1, 10–12] to liquid crystals [13] and to recently achieved progress in substituted benzene molecules [14]. These later systems may act as structuring templates for nanoparticles in their function as two-dimensional host-guest systems.

1 Introduction

Nanoarchitectonics is a new interdisciplinary field within the nanosciences, which investigates the principles responsible for the formation of higher-ordered functional structures starting from their nanoscopic building blocks like atoms and molecules. Such a bottom-up approach is new within the field of contemporary technology, which has used very successfully the top-down strategy for the miniaturization of fabrication processes during the last hundred years. However, nature has always worked bottom-up, where the principles of self-assembly lead to crystal growth in the inorganic world, and, via molecular self-assembly, to functional structures in biology. For instance, the three-dimensional architecture of a nanomachine called the ribosome comprises the natural molecular assembler, which organizes the transition from the DNA informational blueprint into polypeptides and other functional units. To understand and make technological use of the underlying mechanism of this process is one of the major goals of modern proteomics, where the relation between the DNA base sequence and the respective protein must be mastered. One approach to this question is to simplify the process by transferring it into a two-dimensional scenario, thus reducing the complexity of the three-dimensional architecture to an in-plane problem. Such a reduced coordination space may also be adequate for a primordial-soup scenario, where the spontaneous self-assembly of abiotically produced organic compounds may be facilitated. The formation of highly ordered monolayers of the purine and pyrimidine DNA bases through physisorption-mediated molecular self-assembly at a solid–liquid mineral interface and the subsequent stereospecific adsorption of amino acids is an example of the spontaneous creation of nanoscale order. We have proposed a functional role for this process for the emergence of life that may also lead towards the construction of genetically based supramolecular architectures for modern technical applications [15–18].

2 Directed Molecular Self-Assembly

Molecular self-assembly has been defined as the spontaneous emergence of highly organized functional supramolecular architectures from single components of a system under certain external conditions [19]. Such conditions may include, for example, a suitable template, where molecules can adsorb to, and the appropriate environmental conditions, such as the right mixture and concentration of molecules in a solvent like water, temperature, etc. In contrast to molecular chemistry, which has established its power over the covalent bond, non-covalent intermolecular forces prevail in the field of supramolecular chemistry. The energetic and stereochemical properties of non-covalent intermolecular forces such as electrostatic interactions, van der Waals forces and, most prominently, hydrogen bridge bonding act like the joints of Lego game pieces, because they direct the monomeric building blocks to spontaneously assemble into supramolecular structures with inherently higher order. The stereospecific hydrogen bridge bonds determine the structure that comprises the blueprint for this type of transition from monomers to two-dimensional polymeric flat crystalline organic layers.

3 STM-Imaging of Self-Assembled Organic Molecules

3.1 Nucleic Acid Bases as Genetically-Based Supramolecular Architectures Coding for Amino Acids

With the advent of scanning tunneling microscopy [20] it was possible to observe directly DNA base molecules with possible hydrogen bridge donors and acceptors self-organized into such periodic organic molecular layers on mineral template surfaces [1, 21]. Since then a large number of studies of these self-assembled DNA-base systems have been published (for a comprehensive overview refer to [16]). In nature, the main example for coded self-organization is the DNA double helix. Molecules capable of making hydrogen bridge bonds, such as the DNA bases, may self-organize into two-dimensional crystals as shown in Fig. 1.

Here, the steric arrangement of functional groups such as the possible H-bridge donors N–H, O–H and the possible H-bridge acceptors such as N, O moieties are responsible for the specific periodic DNA-base surface. The electrostatic potential of adenine, where N–H and H interactions are used to build a monolayer, is shown here as an example. It has been calculated in an ab initio molecular orbital calculation to exactly display H-bond acceptors and H-bond donors [22]. Such a potential map can be considered as a representation of the chemical activity of the molecule. The arrangement of the molecules shown here is only one possible pattern, which fulfills the periodic space-filling requirement of an energetically minimized two-dimensional crystal structure. Subsequent experimental observations must verify the proposed models. Purine and pyrimidine base layers comprise a unique model system for STM studies for several reasons:

- Nucleic acid bases belong to one of the most important classes of organic molecules with outstanding properties.
- They allow three- and two- dimensional molecular self-assembly through molecular recognition due to the specific H-bonding. Specific H-bond donor or acceptor sites exist within one molecule.

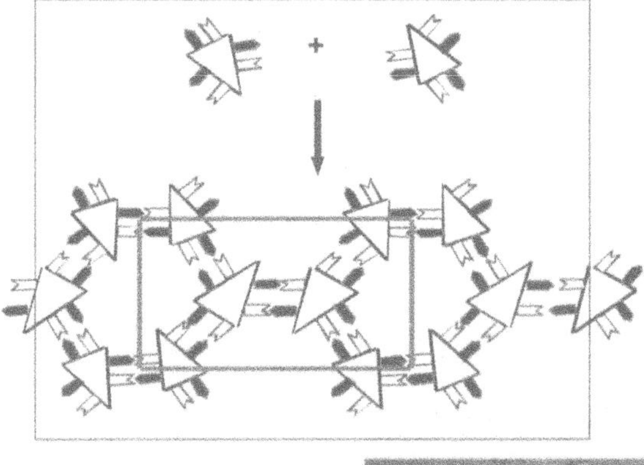

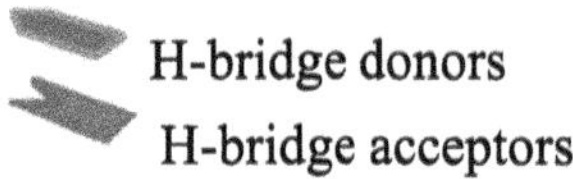

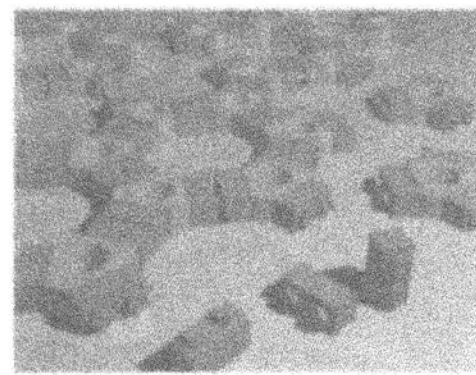

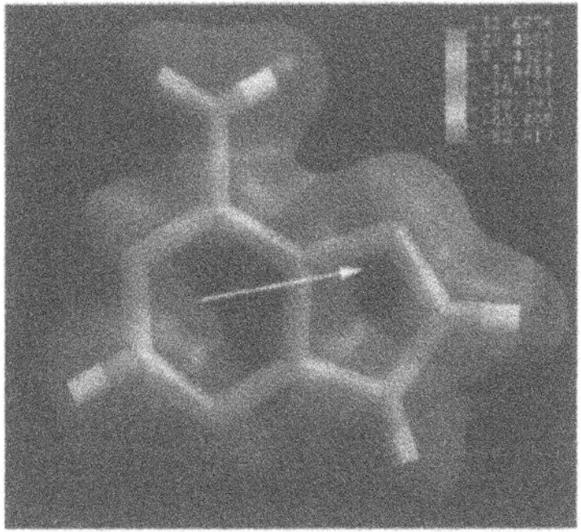

Fig. 1. Schematic of recognition-directed molecular self-assembly of monomeric DNA-based molecules into a two-dimensional organic crystalline layer

- There are no tautomers observed assuring the proper H-bond geometry over a large pH range. The substituents -NH-2 and = O are in the amino and keto forms, never in the imino or enol configuration [23].
- Planar π-electronic ring systems build delocalized electronic systems with potentially anisotropic electronic properties within and perpendicular to the molecular monolayers, giving rise to particular molecular electronic effects.
- Nucleic acid bases crystallize spontaneously via self-assembly (SA) from liquid solution or via epitaxial growth through organic molecular beam epitaxy (OMBE) in ultra-high vacuum (UHV) first found by Freund et al. [24].
- The result of structure determination of two-dimensional nucleic acid base crystals may be compared to the three-dimensional crystal structures.

The transition from two-dimensional to three-dimensional growth (i.e. the build up of multilayers) can even be observed in real-space STM experiments and in thermal desorption spectroscopy.

- Modifications of specific sites within the base molecules allow for the contrast variation for STM as well as X-ray diffraction experiments. Additionally, the character of the H-bond type can be modified on purpose and its influence on the crystal structure is studied directly.
- Molecular mechanics simulations can be applied to the DNA-base layer system to predict and verify different crystal structures and to compare it to STM and LEED structure determination [24], as well as to measurements of the adsorption energy through thermal desorption spectroscopy (TDS) and to find the energetically most-favorable crystal structure to be observed in nature [22, 25, 26].
- These well-known and stable two-dimensional molecular crystals are subject to molecular structuring experiments via nanoscablation and molecular writing as well as nano-contact printing.

Stable two-dimensional base crystals comprise biocompatible surface films for the template-induced construction of biopolymers and may have applications in separation science.

Experimentally, the sizzling technique [1, 27] has been applied to initiate the self-organization of the molecules. Here the molecules are dissolved in water and applied to a slightly heated (up to 130 °C) inorganic template surface, providing the energy to facilitate the two-dimensional crystal growth via molecular surface diffusion. In order to be error tolerant, with respect to the right-docking position of a new molecule adsorbing to a growing two-dimensional seed crystal, it is important that the molecular interactions are of a weak second-order chemical-bond nature. Simultaneously occurring growth and dissolution processes are shifted towards the formation of the molecular crystal because the mineral template forces a seed crystal to grow due to the fixation of molecules upon physisorption. The key to the spontaneous emergence of these supramolecular structures is the appropriate energetic interplay between intermolecular interactions in two dimensions. The in-plane molecular H-bonding is responsible for creating the repetitive pattern of the base layer. This is comparable to the pattern which arises through complementary H-bonding in the three-dimensional DNA-polymer. The physisorption energy of the mineral surface determines the exact surface atomic positions at which adsorption occurs. From an experimental point of view the inorganic template should have a surface energy low enough to facilitate physisorption, in contrast to chemisorption, where the growth of two-dimensional organic crystals is hindered by the restricted surface mobility of the molecules. Therefore, and in order to be able to do STM, we have mainly used the conductive natural crystalline mineral surfaces graphite (001) and molybdenite (001).

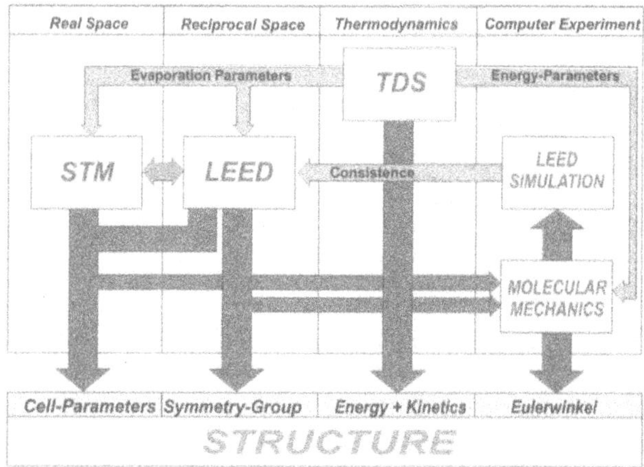

Fig. 2. Concept for the structure determination of self-assembled molecular layers by the combination of different complementary techniques

In the following example the concept of organic monolayer structure determination as shown in Fig. 2 has been applied onto adenine physisorbed onto graphite [24].

Here the real-space STM technique provides high-resolution microscopic images of the molecules on a local scale and is combined with the low-energy electron diffraction (LEED) technique or the surface X-ray diffraction technique [12], providing a reciprocal-space image of the average two-dimensional molecular crystal structure. Together with molecular modeling force field calculations [22] and the determination of the preparation parameters and thermodynamic variables, such as adsorption energy, measured by thermal desorption spectroscopy (TDS), clear models can be derived as a result. Figure 3 shows the adenine on graphite layer in a high-resolution STM image together with the LEED pattern. Single molecules can be identified to form a periodic molecular arrangement with a rectangular unit cell with vectors $a = 22.1\,\text{Å}$ and $b = 8.5\,\text{Å}$. The flat lying centrosymmetric adenine dimers with p2gg-symmetry grow heteroepitaxially on the mineral surface with 4 molecules per coincident unit cell with

$$\begin{pmatrix} a \\ b \end{pmatrix} = \begin{pmatrix} 9 & 0 \\ 2 & 4 \end{pmatrix} \cdot \begin{pmatrix} g_1 \\ g_2 \end{pmatrix}$$

metric, where g_1 and g_2 are the hexagonally arranged graphite unit-cell vectors of 2.46 Å in length. The coordination number for a molecular dimer is 6, representative of the densely packed energetically minimized stable structure where each dimer is surrounded by 6 next-neighbor dimer molecules. Each molecule is surrounded by 3 neighboring molecules with the maximum number of 6 hydrogen bonds per molecule. This type of arrangement also leads to cyclic hydrogen bonding, which stabilizes the monolayer via π-electron coop-

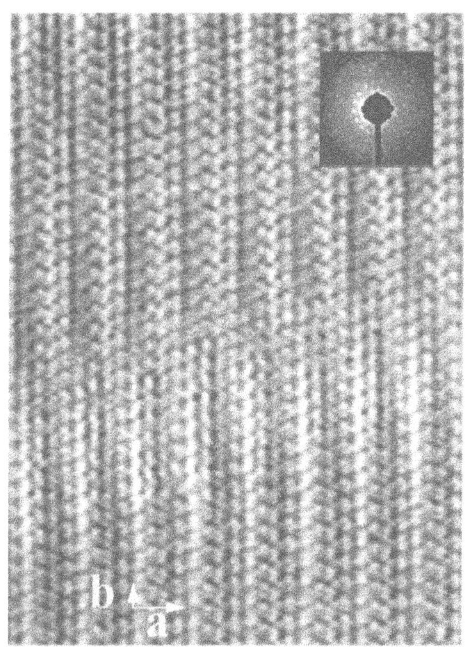

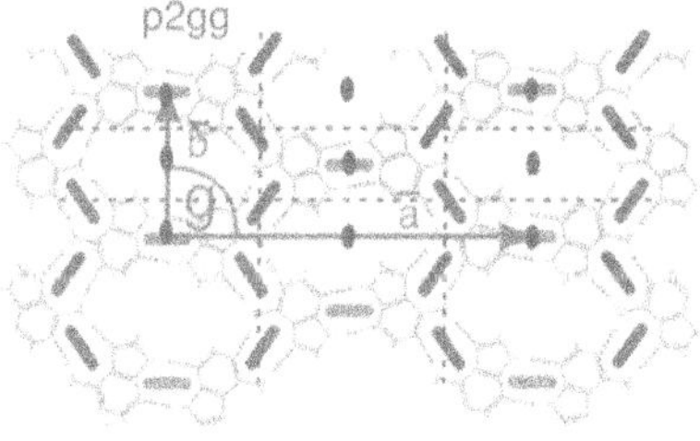

Fig. 3. STM image of a spontaneously self-organized adenine monolayer physisorbed onto the mineral surface of graphite, together with the structure (*below*) determined from reciprocal-space LEED analysis (diffraction pattern shown *above right*) and the direct-space STM image with rectangular unit cell with vectors $a = 22.1$ Å and $b = 8.5$ Å

erativity. In some cases the spontaneous molecular self-assembly from liquid solution leads to a localized chiral-symmetry break, which may have some role in the origin of biomolecular optical asymmetry. Sowerby et al. [15] have observed such a spontaneous symmetry break, for example upon adsorption of adenine on molybdenite. Although the adenine molecule is achiral by the usual definition, whether a stereochemical center is present or not, there are two mirror-symmetric possible ways in which the molecules can adsorb to the (symmetric) mineral template surface. Whenever the unit cell is oblique, it exists in possible left-handed and right-handed molecular crystals within the definition of two-dimensional chirality of not being superimposable by any translations or rotations in two dimensions.

Whether such a symmetry break upon adsorption may obtain a bias towards one type of handedness over the other due to a force from an external source such as, for example, polarized light, is not known to date. STM, however, is the only technique for observing such differences on the local scale, and the possibility that purine–pyrimidine arrays assembled in such a chiral fashion on naturally occurring mineral surfaces might act as chiral organic templates for subsequent biomolecular assembly of higher-ordered compounds is intriguing. We have presented an organic template model [16] where the spontaneous molecular self-assembly of DNA-base molecules from liquid solution on a mineral surface may lead to a (localized chiral?) organic template surface, capable of stereospecific interaction with subsequently adsorbed amino acids which may self-organize on top of the templating nucleic acid layer. Thus, the construction of supramolecular complexity through the control of the intermolecular bond, preferentially the H-bond, is achieved. The proposed architecture may facilitate the polycondensation of the amino acid monomers and lead to polypeptides, leaving the surface after loss of the ammonium proton upon formation of the peptide bond, which can then no longer form a stabilizing H-bond to the base template. As a consequence, this process is capable of catalyzing the formation of polypeptides in liquid. Such a scenario can be regarded as primitive self-programmable, self-assembling two-dimensional genetic matter. An aperiodic mixture of different nucleic acids has been observed and may be used to construct a variable peptide library [15, 16], thus reducing the complexity of the ribosomal RNA-mediated process of polypeptide synthesis in nature. Whether the two-dimensional DNA-base layer can act as a primitive coding mechanism depends on the exact adsorption process of amino acids on top of the nucleic acid layer. This has recently been tested by predicting the adsorption of lysine on a flat adenine template surface via molecular mechanics simulation [28]. Here the lysine molecules order, as shown in Fig. 4, with a calculated adsorption energy of 13 kcal/mol. The energetic landscape for this process exhibits a very quick and steep descent to a stable energetic minimum in total energy, which demonstrates the robustness of the proposed process. It has recently been shown experimentally that the adsorption of different nucleic acids onto graphite surfaces is specific, demonstrating the influence of the energy of adsorption, which is the prerequisite for a rational coding mechanism [29]. A coding-like discrimination of amino acids by purine bases adsorbed on an inorganic surface has just been shown [30].

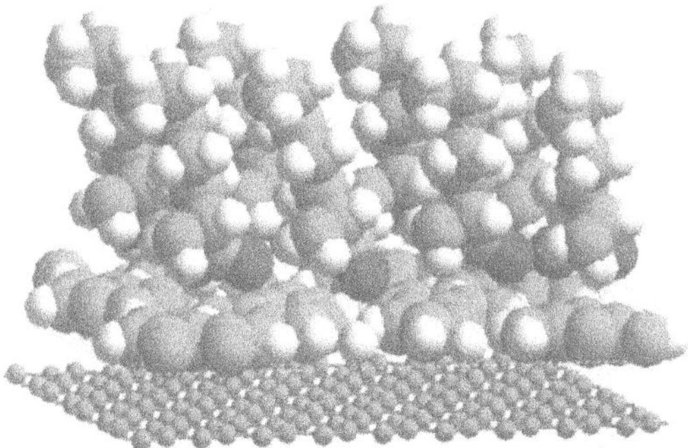

Fig. 4. Molecular mechanics-calculated model of the observed self-assembly of nucleic acid adenine (first layer) and subsequent self-assembly of amino acid lysine on top of the templating nucleic acid layer

3.2 Coadsorption of 8-Cyanobyphenyle (8 CB) and Perylen-Tetracarboxyl-Dianhydrid (PTCDA) on HOPG (Highly Oriented Pyrolythic Graphite)

Figure 5 shows an STM image of 8 CB used as a solvent for PTCDA. Clearly, two phases can be distinguished. Phase separation allows for the two molecules to crystallize in their respective two-dimensional crystal structures as shown. PTCDA mainly interact via weak van der Waals and electrostatic interactions and are therefore ideal candidates for nanoablation using the

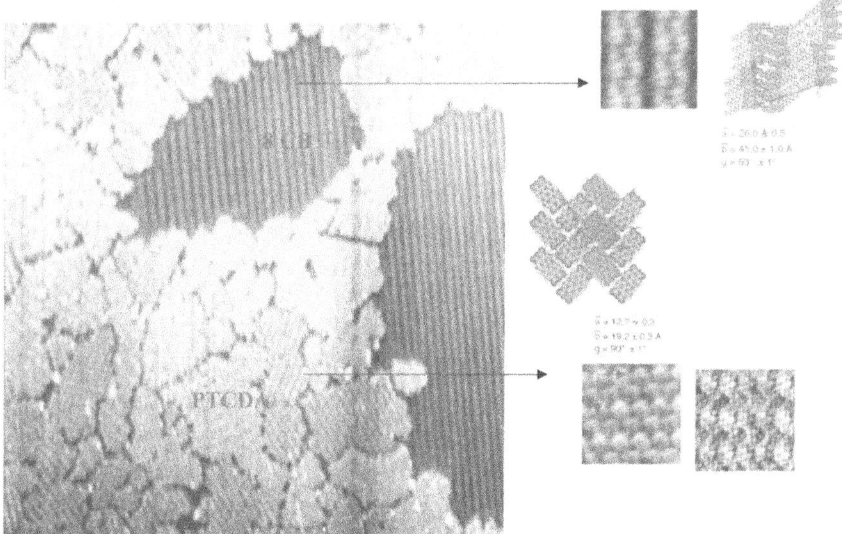

Fig. 5. STM image of 8 CB coadsorbed with PTCDA together with lattice models.

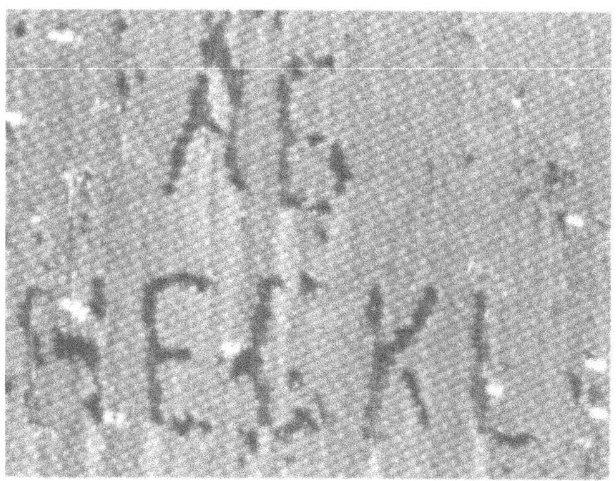

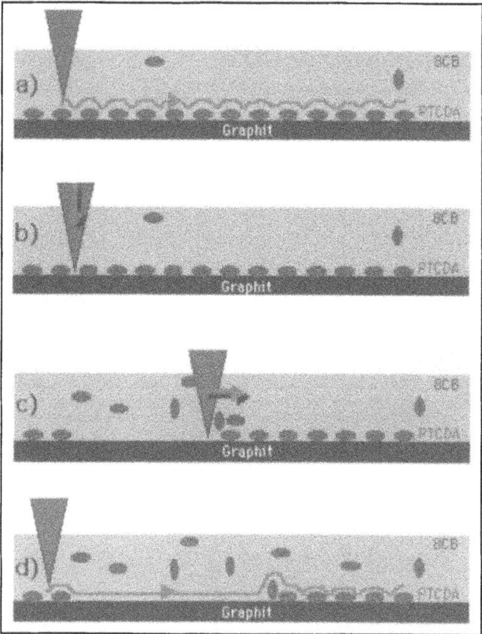

Fig. 6. STM image of a monolayer of PTCDA on HOPG. Individual molecules have been removed to form a pattern

STM tip. Lowering the tip bias can lead to the erosion of the film shown in Fig. 6. Here the tunneling gap resistance was reduced from 10^{10} ohm (imaging mode) to 1.4×10^9 ohm until the contact of the tip with the molecules was able to extract single molecules in a predefined manner (nanostructuring mode). It was possible to control the extraction in order to write a pattern into the layer by computer-programmed movement of the tip motion via a home-built nanomanipulator. In order to subsequently image the pattern consisting of around 100 removed molecules the tunneling bias was raised again to its initial value. This procedure may ultimately lead to molecular storage devices with terabyte capacity. Read, write and temperature-induced erase procedures have just been demonstrated in our laboratory.

3.3 Trimesic Acid (TMA) as an Example for a Molecular Host-Guest System

TMA has been grown on the surface of natural graphite by organic molecular beam epitaxy [14] in two modifications. The one shown here demonstrates the ability of STM to show submolecular resolution by imaging the benzene ring as an extended donut-shaped electron density map. Figure 7 shows that the resolution chicken-wire structure is an ideal construction for a two-dimensional periodic array of guest molecules in an extended host structure. Here the 6-membered molecular caves are only filled in two positions with TMA molecules. Coadsorption of metal atoms may lead to a periodic array of nanodots with nanometer periodicity. Changing the side groups and the atom type then can lead to tailored nanoscopic arrays for quantum-dot applications.

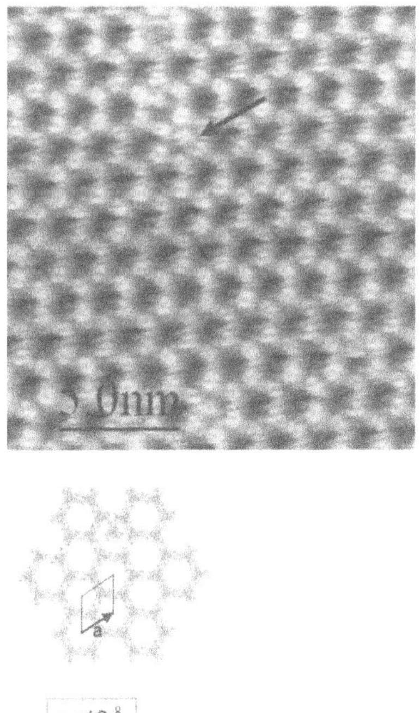

Fig. 7. STM image of TMA self-assembled on natural graphite in UHV. Arrow marks host molecule in TMA-guest cave

4 Conclusion

The application of near-field microscopy techniques, namely scanning tunneling microscopy (STM), to self-assembled two-dimensional organic molecular crystals has allowed for the first time real-space analysis of these systems with molecular-scale resolution. Together with the ability to manipulate on a single-molecule scale using STM, this has stimulated the development of new concepts regarding the possible role of molecular self-assembly in the development of nanoscale manufacturing of devices. Additionally, the de novo emergence of higher-ordered supramolecular architectures, comprised of today's DNA and protein building blocks, may eventually guide us to a route to life under prebiotic conditions. We have suggested that purine and pyrimidine monolayers could be candidates for a stationary phase in organic molecule separation systems, and as templates for the assembly of higher-ordered polymers at the prebiotic solid–liquid interface.

Acknowledgements

The base coding theory described in this paper has been developed as the result of collaboration with Stephen Sowerby and George Petersen of the Department of Biochemistry, University of Otago, New Zealand. My students, M. Reiter, M. Edelwirth, F. Jamitzky, F. Trixler and F. Griessl, have contributed with their work and various images.

References

1. W.M. Heckl, D.P.E. Smith, G. Binnig, H. Klagges, T.W. Hänsch, J. Maddocks, Proc. Natl. Acad. Sci. USA **88**, 8003 (1991)
2. *Procedures in Scanning Probe Microscopies*, ed. by Colton et al. (John Wiley, 1998)
3. J.S. Foster, J.E. Frommer, Nature, **333**, 542 (1988)
4. D.P.E. Smith, H. Hörber, Ch. Gerber, G. Binnig, Science **245**, 43 (1989)
5. J.P. Rabe, S. Buchholz, Science **253**, 424 (1991)
6. M.T. Cuberes, R.R. Schlittler, J.K. Gimzewski, Appl. Phys. A **66**, 745 (1998)
7. K. Gemzewski, C. Joachim, Science **283**, 1683 (1999)
8. W.M. Heckl, in *The Human Genome*, ed. by E.P. Fischer, S. Klose (Piper, München 1995)
9. T.A. Jung, R.R. Schlittler, J.K. Gimzewski, H. Tang, C. Joachim, Science **271**, 181 (1996)
10. W.M. Heckl, A. Engel, 'Imaging Nucleic Acids with Scanning Probe Microscopes', in *Visualization of Nucleic Acids*, ed. by G. Morel (CRC Press, Boca Raton, CA 1995)
11. M. Reiter, M. Edelwirth, W.M. Heckl, S.J. Sowerby, Probe Microscopy, **1**, 1291 (1999)
12. H.L. Meyerheim, F. Trixler, W. Stracke, W.M. Heckl, Z. Kristallogr. **214**, 771 (1999)

13. F. Trixler, W.M. Heckl, to be published
14. S. Griessl, M. Edelwirth, R. Schloderer, M. Hietschold, W.M. Heckl, to be published
15. S.J. Sowerby, W.M. Heckl, G.B. Petersen, J. Mol. Evol. **43**, 419 (1996)
16. S.J. Sowerby, W.M. Heckl, Origin of Life and Evolution of the Biosphere, **28**, 283 (1998)
17. S.J. Sowerby, M. Edelwirth, W.M. Heckl, J. Phys. Chem. **102**, 5914 (1998)
18. S.J. Sowerby, P.A. Stockwell, W.M. Heckl, G.B. Petersen, Origin of Life and Evolution of the Biosphere, **30**, 81 (2000)
19. J.-M. Lehn, in *Supramolecular Chemistry, Concepts and Perspectives* (Wiley VCH, Weinheim 1995)
20. G. Binnig, H. Rohrer, C. Gerber, E. Weibel: Phys. Rev. Lett. **49**, 57 (1982)
21. M.J. Allen, M. Balooch, S. Subbiah, R.J. Tench, W. Siekhaus, R. Balhorn, Scanning Microscopy **5**, 625 (1991)
22. M. Edelwirth, J. Freund, S.J. Sowerby, W.M. Heckl, Surf. Sci. **417**, 201 (1998)
23. W. Saenger in *Principles of Nucleic Acid Structure* (Springer, Berlin, Heidelberg 1984)
24. J. Freund, M. Edelwirth, W.M. Heckl, Phys. Rev. B **55**, 5394 (1997)
25. S.J. Sowerby, M. Edelwirth, W.M. Heckl, Appl. Phys. Lett. A **66**, 649 (1998)
26. M. Dove, W.M. Heckl, submitted
27. W.M. Heckl, 'Molecular Self-Assembly and the Origin of Life', in *Lecture Notes in Physics* (Springer, Berlin, Heidelberg 2001)
28. F. Jamitzky, W.M. Heckl, to be published
29. S.J. Sowerby, C.A. Cohn, W.M. Heckl, N.G. Holm, Proc. Natl. Acad. Sci. USA, **98.3** 820 (2001)
30. S.J. Sowerby, G.B. Petersen, N.G. Holm, private communication

Subject Index

GPSR Compliance
The European Union's (EU) General Product Safety Regulation (GPSR) is a set of rules that requires consumer products to be safe and our obligations to ensure this.

If you have any concerns about our products, you can contact us on

ProductSafety@springernature.com

In case Publisher is established outside the EU, the EU authorized representative is:

Springer Nature Customer Service Center GmbH
Europaplatz 3
69115 Heidelberg, Germany

www.ingramcontent.com/pod-product-compliance
Ingram Content Group UK Ltd.
Pitfield, Milton Keynes, MK11 3LW, UK
UKHW021901190726
13853UKWH00003B/1370

* 9 7 8 3 6 6 2 0 4 8 9 8 6 *